Boundary Value Problems, Integral Equations and Related Problems

Proceedings of the Third International Conference

Boundary Value Problems, Integral Equations and Related Problems

Proceedings of the Third International Conference

Beijing and Baoding, China 20 – 25 August 2010

Chief Editor

Guo Chun Wen *(Peking University, P R China)*

World Scientific

NEW JERSEY • LONDON • SINGAPORE • BEIJING • SHANGHAI • HONG KONG • TAIPEI • CHENNAI

Published by

World Scientific Publishing Co. Pte. Ltd.
5 Toh Tuck Link, Singapore 596224
USA office: 27 Warren Street, Suite 401-402, Hackensack, NJ 07601
UK office: 57 Shelton Street, Covent Garden, London WC2H 9HE

British Library Cataloguing-in-Publication Data
A catalogue record for this book is available from the British Library.

BOUNDARY VALUE PROBLEMS, INTEGRAL EQUATIONS AND RELATED PROBLEMS

ISBN-13 978-981-4327-85-5
ISBN-10 981-4327-85-9

Printed in Singapore by B & Jo Enterprise Pte Ltd

Preface

The first international conference on boundary value problems, integral equations and related problems was convened from September 2 to 7, 1990 in Beijing, China. The second international conference on the same subjects was convened from August 8 to 14, 1999 in Beijing and Chengde, Hebei, China. From August 20 to 25, 2010, we held the third international conference on boundary value problems, integral equations and related problems in Beijing and Baoding, Hebei, China.

In this third conference, the following topics were discussed:

1. Various boundary value problems for partial differential equations and functional equations.

2. The theory and methods of integral equations and integral operators including singular integral equations.

3. Applications of boundary value problems and integral equations to mechanics and physics.

4. Numerical methods of integral equations and boundary value problems.

5. Theory and methods for inverse problems in mathematical physics.

6. Clifford analysis, and some related problems.

The conference was organized by Peking University, Fudan University, Wuhan University, Beijing Normal University, Sun Yat-Sen University, Ningxia University, Hebei University, Hebei Normal University and Renmin University of China. The conference was supported by the Education Ministry, the Chinese Mathematical Society, the Mathematical Center of Ministry of Education and the National Natural Science Foundation of China, and also supported by many colleagues in China and abroad. There were about 110 attendants at the conference and 16 of them were foreign experts from USA, Russia, Greece, Poland, Japan, France, Iran, Canada, etc. The attendants proposed about 70 talks.

Now our international conference has just come to a successful close. Through the one-week interaction, we exchanged the views of mathematical problems through talks and academical materials, we can gather to learn the recent achievements, to explore the future, to share our knowledge, we promoted mutual understanding and friendship among the colleagues and made progress on further development of research on boundary value problems, integral equations and related problems. We often organize symposia, but this was the third international conference. This proceedings volume contains 44 papers. The editors apologize for being unable to include all the talks given during the conference.

Finally, as editors we express our gratitude to the contributors of the volume for sending us their manuscripts. We also thank the editorial staff of World Scientific Publishing Company for making the publication of this volume possible.

Cheif Editor:

Guo Chun Wen (Peking University)

Editorial Board:

Jin Cheng (FudanUniversity)

Dai Qing Dai (Sun Yat-Sen University)

Jin Yuan Du (Wuhan University)

Hong Ya Gao (Hebei University)

Xing Li (Ningxia University)

Yu Ying Qiao (Hebei Normal University)

Zuo Liang Xu (Remin University of China)

Zhen Zhao (Beijing Normal University)

Contents

Part 2 Inverse Problems in Mathematical Physics, Engineering and Finance

Part 3 Geometry Function Theory and Nonlinear Analysis

Part 5 Theory and Applications of Integral Equations

Part 6 Numerical Analysis of Differential and Integral Equations

GALERKIN METHOD FOR A QUASILINEAR PARABOLIC EQUATION WITH NONLOCAL BOUNDARY CONDITIONS

BAI-LI CHEN

Gustavus Adolphus College, MN 56082, USA

E-mail: bchen@gustavus.edu

Existence of a generalized solution is proved for a quasilinear parabolic equation with nonlocal boundary conditions using Faedo-Galerkin approximation.

Keywords: Faedo-Galerkin method, nonlocal boundary conditions, a priori estimates, quasilinear parabolic equations, generalized solution.

AMS No: 35K20, 35K59, 35B45, 35D30.

1. Introduction

In this paper, we apply Faedo-Galerkin method to the following quasilinear parabolic equation with nonlocal boundary conditions, and establish the existence of a generalized solution for the problem

$$\frac{\partial u}{\partial t}-\sum_{i=1}^{n}\frac{\partial}{\partial x_i}(|u|^{p-2}\frac{\partial u}{\partial x_i})+|u|^{p-2}u=f(x,t),\ x\in\Omega,\ t\in[0,T], \tag{1.1}$$

$$u(x,t)=\int_{\Omega}k(x,y)u(y,t)dy,\ \ x\in\Gamma, \tag{1.2}$$

$$u(x,0)=u_0(x). \tag{1.3}$$

It is well known that Faedo-Galerkin method is used to prove the existence of solutions for linear parabolic equations in [6]. In [5], Faedo-Galerkin method is coupled with contraction mapping theorems to prove the existence of weak solutions of semilinear wave equations with dynamic boundary conditions. Bouziani et al. use Faedo-Galerkin method to show the existence of a unique weak solution for a linear parabolic equation with nonlocal boundary conditions (see [2]). J. L. Lion's book (see Ch I, [7]), collects the work of Dubinskii and Raviart, in which they use Faedo-Galerkin method to prove the existence and uniqueness of weak solution for a quasilinear parabolic equation with homogeneous boundary condition.

However, to the author's best knowledge, the literatures studying quasilinear parabolic equations with nonlocal boundary conditions are very limited. In [4], T. D. Dzhuraev et al. study a quasilinear parabolic equation with nonlocal boundary conditions of a different type.

Problem (1.1)–(1.3) is the extension of the problem in (p. 140, [7]), in which the boundary conditions are homogeneous.

The paper is organized as follows: in Section 2, we give the definition of the generalized solution of problem (1.1)–(1.3) and introduce the function spaces related to the generalized solution. In Section 3, we demonstrate the construction of approximation solutions by Faedo-Galerkin method, and derive a priori estimates for the approximation solution. Section 4 is devoted to the proof of existence of the generalized solution by compactness arguments.

2. Preliminaries

Throughout this paper, we use the following notations:

Ω	: regular and bounded domain of $\mathbb{R}^n$.
Γ	: boundary of Ω
$(\, , \,)$	: usual inner product in $L^2(\Omega)$.
$W^{k,p}(\Omega)$	: Sobelev space on Ω.
$H^r(\Omega)$	: Sobelev space $W^{r,2}(\Omega)$.
$L^p(\Omega)$	: L^p space defined on Ω.
$L^p(\Gamma)$	: L^p space defined on Γ.
$\| \, \|_p$	: norm in $L^p(\Omega)$.
$\| \, \|_{p,\Gamma}$	: norm in $L^p(\Gamma)$.
$H^{-r}(\Omega)$	: dual space of $H^r(\Omega)$.
$\| \, \|_{H^{-r}(\Omega)}$	: norm in $H^{-r}(\Omega)$.
c	: nonzero constant which may take different values on each occurrence.
C	: nonnegative constant which may take different values on each occurrence.
$\hookrightarrow$	: continuous embedding
$K(x)$	: norm of $k(x,y)$ in $L^q(\Omega)$ with respect to y, i.e. $K(x) = (\int_\Omega \|k(x,y)\|^q dy)^{\frac{1}{q}}$.
$K_i(x)$	: norm of $D_i k(x,y)$ in $L^q(\Omega)$ with respect to y, i.e. $K_i(x) = \int_\Omega \|\frac{\partial k(x,y)}{\partial x_i}\|^q dy)^{\frac{1}{q}}$.

Throughout this paper, we make the following assumptions:

(A1) $n \geq 2, \; p > n, \; r > \dfrac{n}{2} + 2.$

(A2) $\dfrac{1}{p} + \dfrac{1}{q} = 1.$

(A3) $f \in L^q(0,T; L^q(\Omega))$ and $u_0 \in L^\infty(\Omega)$.

(A4) For any $x \in \Gamma$, $K(x) < \infty, \; K_i(x) < \infty$.

(A5) $\displaystyle\sum_{i=1}^n \int_\Gamma K(x)^{p-1} K_i(x) d\Gamma < 1 - \frac{1}{p}.$

With assumption (A1), using Sobelev embedding theorems (see [1]), we have

$$H^r(\Omega) \hookrightarrow W^{2,p}(\Omega) \hookrightarrow W^{1,p}(\Omega) \hookrightarrow L^p(\Omega) \hookrightarrow L^2(\Omega).$$

Define a space V:

$$V = \{v \in H^r(\Omega) : \ v(x) = \int_\Omega k(x,y)v(y)dy, \ for \ x \in \Gamma\}. \tag{2.1}$$

It is easy to see that V is a subspace of $H^r(\Omega)$.

Definition 2.1. Define a generalized solution of problem (1.1)–(1.3) as a function u, such that

(1) $u \in L^\infty(0,T; L^2(\Omega)) \cap C([0,T], H^{-r}(\Omega))$.
(2) $\frac{du}{dt} \in L^q(0,T; H^{-r}(\Omega))$.
(3) $u(x,0) = u_0(x)$.
(4) The identity:

$$(\frac{du}{dt}, v) - (\sum_{i=1}^n \frac{\partial}{\partial x_i}(|u|^{p-2}\frac{\partial u}{\partial x_i}), v) + (|u|^{p-2}u, v) = (f, v) \tag{2.2}$$

holds for all $v \in V$ and a.e. $t \in [0,T]$.

Remark 2.1. From the proof of existence theorem in Section 4, we will see that each inner product in the identity (2.2) is a function of t in $L^q(0,T)$, hence the identity holds for a.e. $t \in [0,T]$. On the other hand, since $u(t) \in V$, the boundary condition (1.2) is satisfied.

3. Construction of an Approximate Solution and a Priori Estimates

Since V is a subspace of $H^r(\Omega)$, which is separable. We can choose a countable set of distinct basis elements w_j, $j = 1, 2, \cdots$, which generate V and are orthonormal in $L^2(\Omega)$. Let V_m be the subspace of V generated by the first m elements: $w_1, w_2, \cdots, w_m$. We construct the approximate solution of the form:

$$u_m(x,t) = \sum_{j=1}^m g_{jm}(t)w_j(x), \ \ (x,t) \in \Omega \times [0,T], \tag{3.1}$$

where $(g_{jm}(t))_{j=1}^m$ remains to be determined.

Denote the orthogonal projection of u_0 on V_m as $u_m^0 = P_{V_m}u_0$, then $u_m^0 \to u_0$ in V, as $m \to \infty$. Let $(g_{jm}^0)_{j=1}^m$ be the coordinate of u_m^0 in the basis $(w_j)_{j=1}^m$ of V_m, i.e. $u_m^0 = \sum_{j=1}^m g_{jm}^0 w_j$, let $g_{jm}(0) = g_{jm}^0$. We need to determine $(g_{jm}(t))_{j=1}^m$ to satisfy

$$
\begin{aligned}
(u_m', w_j) - (\sum_{i=1}^{n} \frac{\partial}{\partial x_i}(|u_m|^{p-2}\frac{\partial u_m}{\partial x_i}), w_j) + (|u_m|^{p-2}u_m, w_j) \\
= (f, w_j), \ \ 1 \le j \le m.
\end{aligned} \tag{3.2}
$$

Do integration by parts on the second term of LHS, we have

$$
\begin{aligned}
(u_m', w_j) + \sum_{i=1}^{n} \int_\Omega (|u_m|^{p-2}D_i u_m)(D_i w_j)dx \\
-\sum_{i=1}^{n}\int_\Gamma (|u_m|^{p-2}D_i u_m)w_j d\Gamma + (|u_m|^{p-2}u_m, w_j) = (f, w_j), 1 \le j \le m.
\end{aligned} \tag{3.3}
$$

The above system is a system of ordinary differential equations in $(g_{jm}(t))_{j=1}^{m}$. By Caratheodory theorem (see [3]), there exists solution $(g_{jm}(t))_{j=1}^{m}$, $t \in [0, t_m)$. We need a priori estimates that permit to extend the solution to the whole domain $[0, T]$.

We derive a priori estimates for the approximate solution as follows: Multiply (3.3) by $g_{jm}(t)$, then sum over j from 1 to m, we have

$$
\begin{aligned}
\frac{1}{2}\frac{d}{dt}|u_m(t)|_2^2 + \frac{4}{p^2}\sum_{i=1}^{n}\int_\Omega (D_i(|u_m|^{\frac{p-2}{2}}u_m))^2 dx + |u_m(t)|_p^p \\
= (f, u_m) + \sum_{i=1}^{n}\int_\Gamma (|u_m|^{p-2}D_i u_m)u_m d\Gamma.
\end{aligned} \tag{3.4}
$$

Integrate with respect to t from 0 to T on both sides, we obtain

$$
\begin{aligned}
\frac{1}{2}|u_m(T)|_2^2 + \int_0^T \frac{4}{p^2}\sum_{i=1}^{n}\int_\Omega (D_i(|u_m|^{\frac{p-2}{2}}u_m))^2 dx dt + \int_0^T |u_m(t)|_p^p dt \\
= \int_0^T (f, u_m)dt + \int_0^T \sum_{i=1}^{n}\int_\Gamma (|u_m|^{p-2}D_i u_m)u_m d\Gamma dt + \frac{1}{2}|u_m(0)|_2^2.
\end{aligned} \tag{3.5}
$$

This gives

$$
\begin{aligned}
\frac{1}{2}|u_m(T)|_2^2 + \int_0^T \frac{4}{p^2}\sum_{i=1}^{n}\int_\Omega (D_i(|u_m|^{\frac{p-2}{2}}u_m))^2 dx dt + \int_0^T |u_m(t)|_p^p dt \\
\le \int_0^T |(f, u_m)|dt + \int_0^T \sum_{i=1}^{n}\int_\Gamma |(|u_m|^{p-2}D_i u_m)u_m| d\Gamma dt + \frac{1}{2}|u_m(0)|_2^2.
\end{aligned} \tag{3.6}
$$

The first term in the RHS of (3.6) can be estimated as follows:

$$\begin{aligned}\int_0^T |(f,u_m)|dt &= \int_0^T \int_\Omega |fu_m|dxdt \\ &\le \int_0^T |f|_q |u_m|_p dt \text{ (Hölder's inequality)} \\ &\le \int_0^T \left(\frac{1}{p}|u_m|_p^p + \frac{p-1}{p}|f|_q^{\frac{p}{p-1}}\right)dt.\end{aligned} \tag{3.7}$$

The Young's inequality can be seen as in [6].

Next, we estimate second term in the RHS of (3.6): For $x \in \Gamma$, we have

$$|u_m(x,t)| = \int_\Omega |k(x,y)u_m(y,t)|dy \le |k(x,y)|_q |u_m|_p. \tag{3.8}$$

Then we have $|u_m(x,t)| \le K(x)|u_m|_p$ for $x \in \Gamma$. Similarly, we have $|D_i u_m(x,t)| \le K_i(x)|u_m|_p$ for $x \in \Gamma$. Moreover we have

$$\begin{aligned}&\int_0^T \left|\sum_{i=1}^n \int_\Gamma (|u_m|^{p-2} D_i u_m) u_m d\Gamma\right| dt \\ &\le \int_0^T \sum_{i=1}^n \int_\Gamma K(x)^{p-1} |u_m|_p^{p-1} K_i(x)|u_m|_p d\Gamma dt \\ &\le \int_0^T \left(\sum_{i=1}^n \int_\Gamma K(x)^{p-1} K_i(x) d\Gamma\right)|u_m|_p^p dt \\ &= \left(\sum_{i=1}^n \int_\Gamma K(x)^{p-1}K_i(x)d\Gamma\right)\left(\int_0^T |u_m|_p^p dt\right) \\ &= C\left(\int_0^T |u_m|_p^p dt\right).\end{aligned} \tag{3.9}$$

With the above estimates, together with (3.6), we have

$$\begin{aligned}&\frac{1}{2}|u_m(T)|_2^2 + \int_0^T \frac{4}{p^2}\sum_{i=1}^n \int_\Omega (D_i(|u_m|^{\frac{p-2}{2}}u_m))^2 dxdt \\ &+ \int_0^T \left(1 - \frac{1}{p} - C\right)|u_m(t)|_p^p dt \le \int_0^T \left(\frac{p-1}{p}|f|_q^{\frac{p}{p-1}}\right)dt + \frac{1}{2}|u_m(0)|_2^2\end{aligned}$$

holds for any finite $T > 0$.

Under the assumption (A1)–(A5), we have the following a priori estimates:

(B) u_m is bounded in $L^\infty(0,T;\ L^2(\Omega))$.
(C) $|u_m|^{\frac{p-2}{2}}|u_m|$ is bounded in $L^2(0,T;H^1(\Omega))$.
(D) u_m is bounded in $L^p(0,T;L^p(\Omega))$.
Since T is an arbitrary positive number, we have $|u_m|_p^p < \infty$ *a.e.* t.

4. Existence of a Generalized Solution

To prove the existence of a generalized solution, we first prove there exists subsequence of u_m, still denoted as u_m, such that

(E) $u_m \to u$ in $L^p(0,T;\ L^p(\Omega))$ strongly and almost everywhere.

To prove (E), we need the following lemma:

Lemma 4.1. *Let u_m, constructed as in* (3.1), *be the approximate solution of* (1.1)–(1.3) *in the sense of definition* 2.1. *Then u'_m is bounded in $L^q(0,T;\ H^{-r}(\Omega))$.*

Proof. For $v \in V \subset H^r$, from (3.3), we have

$$\begin{aligned}(u_m', v) + (\sum_{i=1}^{n}\int_\Omega (|u_m|^{p-2}D_i u_m)(D_i v)dx \\ -\sum_{i=1}^{n}\int_\Gamma (|u_m|^{p-2}D_i u_m)v d\Gamma + (|u_m|^{p-2}u_m, v) = (f,v).\end{aligned} \tag{4.1}$$

The last term in the LHS can be estimated as in [7]:

$$\begin{aligned}|(|u_m|^{p-2}u_m, v)| &\le |\ |u_m|^{p-1}|_q |v|_p \\ &\le (|u_m|_p^p)^{\frac{1}{q}}|v|_p \\ &\le (|u_m|_p^p)^{\frac{1}{q}} C|v|_{H^r} \text{ (since } H^r \hookrightarrow L^p).\end{aligned}$$

Hence $||u_m|^{p-2}u_m|_{H^{-r}(\Omega)} \le C(|u_m|_p^p)^{\frac{1}{q}} < \infty$, and the norm of $|u_m|^{p-2}u_m$ in $L^q(0,T;\ H^{-r}(\Omega))$ is bounded by

$$(\int_0^T (C(|u_m|_p^p)^{\frac{1}{q}})^q dt)^{\frac{1}{q}} = (\int_0^T C^q |u_m|_p^p dt)^{\frac{1}{q}} < \infty.$$

Therefore, $|u_m|^{p-2}u_m$ is bounded in $L^q(0,T;\ H^{-r}(\Omega))$.

Next, we consider the term $\sum_{i=1}^n \int_\Gamma (|u_m|^{p-2}D_i u_m)v d\Gamma$ in the LHS of (4.1):

$$v \longrightarrow \sum_{i=1}^{n}\int_\Gamma (|u_m|^{p-2}D_i u_m)v d\Gamma = (a(u_m), v).$$

We have

$$\begin{aligned}
&\sum_{i=1}^{n}\int_{\Gamma}(|u_m|^{p-2}D_iu_m)vd\Gamma \\
&\leq \sum_{i=1}^{n} |\ (|u_m|^{p-2}D_iu_m)|_{q,\Gamma}|v|_{p,\Gamma} \\
&\leq \sum_{i=1}^{n} \left|(K(x)^{p-2}K_i(x)|u_m|_p^{p-1})\right|_{q,\Gamma} \left|(K(x)|v|_p)\right|_{p,\Gamma} \\
&\leq \sum_{i=1}^{n} \left|K(x)^{p-2}K_i(x)\right|_{q,\Gamma} |K(x)|_{p,\Gamma}|u_m|_p^{p-1}|v|_p \\
&\leq \sum_{i=1}^{n} |K(x)^{p-2}K_i(x)|_{q,\Gamma}|K(x)|_{p,\Gamma}|u_m|_p^{p-1}C|v|_{H^r}.
\end{aligned}$$

Then

$$|a(u_m)|_{H^{-r}(\Omega)} \leq \sum_{i=1}^{n} |K(x)^{p-2}K_i(x)|_{q,\Gamma}|K(x)|_{p,\Gamma}|u_m|_p^{p-1}C < \infty.$$

Moreover the norm of $a(u_m)$ in $L^q(0,T;H^{-r}(\Omega))$ is bounded by

$$(\int_0^T \sum_{i=1}^{n}(|K(x)^{p-2}K_i(x)|_{q,\Gamma}|K(x)|_{p,\Gamma}C)^q|u_m|_p^p dt)^{\frac{1}{q}} < \infty.$$

Hence, $a(u_m)$ is bounded in $L^q(0,T;H^{-r}(\Omega))$.

Next, we consider the second term in the LHS of (4.1). Integrating by parts gives

$$\begin{aligned}
&\sum_{i=1}^{n}\int_{\Omega}(|u_m|^{p-2}D_iu_m)(D_iv)dx \\
&= \frac{1}{c}(\sum_{i=1}^{n}\int_{\Gamma}|u_m|^{p-2}u_mD_ivd\Gamma - \int_{\Omega}|u_m|^{p-2}u_m\Delta vdx).
\end{aligned} \tag{4.2}$$

Consider $v \longrightarrow \sum_{i=1}^{n}\int_{\Gamma}|u|^{p-2}uD_ivd\Gamma = (I_1(u),v)$, we have

$$|(I_1(u),v)| \leq \sum_{i=1}^{n} |\ |u|^{p-2}u|_{q,\Gamma}|D_iv|_{p,\Gamma}$$

$$
\begin{aligned}
&= \sum_{i=1}^{n} \left| \left(\int_\Omega k(x,y)u(y,t)dy \right)^{p-1} \right|_{q,\Gamma} \left| \int_\Omega D_i k(x,y)v(y,t)dy \right|_{p,\Gamma} \\
&\le \sum_{i=1}^{n} \left| (K(x)^{p-1}|u|_p^{p-1}) \right|_{q,\Gamma} \left| (K_i(x)|v|_p) \right|_{p,\Gamma} \\
&= \sum_{i=1}^{n} |K(x)^{p-1}|_{q,\Gamma} |K_i(x)|_{p,\Gamma} |u|_p^{p-1} |v|_p \\
&\le \sum_{i=1}^{n} |K(x)^{p-1}|_{q,\Gamma} |K_i(x)|_{p,\Gamma} |u|_p^{p-1} C |v|_{H^r}.
\end{aligned}
$$

So we have

$$
|I_1(u_m)|_{H^{-r}(\Omega)} \le \sum_{i=1}^{n} |K(x)^{p-1}|_{q,\Gamma} |K_i(x)|_{p,\Gamma} |u_m|_p^{p-1} C < \infty.
$$

With this, it is easy to see that norm of $I_1(u_m)$ in $L^q(0,T;H^{-r}(\Omega))$ is bounded.

Next, consider $v \longrightarrow \int_\Omega |u|^{p-2} u \Delta v dx = (I_2(u), v)$. From the proof of Theorem 12.2 in [7], we know $I_2(u_m)$ is bounded in $L^q(0,T;H^{-r}(\Omega))$. Since $f \in L^q(0,T;L^q(\Omega)) \subset L^q(0,T;H^{-r}(\Omega))$, from (4.1) and the above discussion, we have u_m' is bounded in $L^q(0,T;H^{-r}(\Omega))$. This is the end of proof. With Lemma 4.1, we can make use of Theorem 12.1 in [7]. We quote the theorem here:

Theorem 4.2. *Let B, B_1 be Banach space, S is a set, define $M(v) = (\sum_{i=1}^{n} \int_\Omega |v|^{p-2} (\frac{\partial v}{\partial x_i})^2 dx)^{\frac{1}{p}}$ on S with*

(a) *$S \subset B \subset B_1$, and $M(v) \ge 0$ on S, $M(\lambda v) = |\lambda| M(v)$.*

(b) *The set $\{v | v \in S, M(v) \le 1\}$ is relatively compact in B.*

Define the set F as follows:

$$
\begin{aligned}
F = \{ v :\ & v \text{ locally summable on } [0,T] \text{ with value in } B_1, \\
& \int_0^T (M(v(t)))^{p_0} dt \le C,\ v' \text{ bounded in } L^{p_1}(0,T;B_1) \},
\end{aligned}
$$

where $1 < p_i < \infty$, $i = 0, 1$. Then $F \subset L^{p_0}(0,T;B)$ and F is relatively compact in $L^{p_0}(0,T;B)$.

To apply Theorem 4.2, we let $S = \{v :\ |v|^{\frac{p-2}{2}} v \in H^1(\Omega)\}$. Since $H^1(\Omega)$ is also compactly embedded in $L^2(\Omega)$, the proof of Proposition 12.1 in (p. 143, [7]) also works for $|v|^{\frac{p-2}{2}} v \in H^1(\Omega)$, then (b) holds.

Let $B=L^p(\Omega)$, $B_1=H^{-r}(\Omega)$, $p_0=p$, $p_1=q$, and u_m be the approximate solution of equations (1.1)–(1.3), constructed as in Section 3, we have

$$\begin{aligned}&\int_0^T (M(u_m))^{p_0}dt\\&=\int_0^T (\sum_{i=1}^n \int_\Omega |u_m|^{p-2}(\frac{\partial u_m}{\partial x_i})^2 dx)dt\\&=C\int_0^T \sum_{i=1}^n \int_\Omega (D_i(|u_m|^{\frac{p-2}{2}}u_m))^2 dxdt\\&<\infty.\end{aligned}$$

Hence together with Lemma 4.1 and a priori estimates, we finish the proof of (E).

Next, we prove that we can pass the limit in (4.1), to prove this, we consider each term in the LHS of (4.1):

(F) $(|u_m|^{p-2}u_m, v)\longrightarrow(|u|^{p-2}u, v)$.

To prove (F), we need to show that $|u_m|^{p-2}u_m \rightharpoonup |u|^{p-2}u$ in $L^q(\Omega)$ weakly, this is a consequence of Lemma 1.3 in [7].

(G) $\displaystyle\int_\Gamma (|u_m|^{p-2}D_i u_m)vd\Gamma \longrightarrow \int_\Gamma (|u|^{p-2}D_i u)vd\Gamma.$

Proof of (G). By a priori estimates, u_m is bounded in $L^p(\Omega)$ for almost every t, then there exists subsequence of u_m, still denoted as u_m, converges to u weak star in $L^p(\Omega)$ (Alaoglu's Theorem) for almost every $t\in[0,T]$.

Under the assumption that for fixed x, $|k(x,y)|_q=(\int|k(x,y)|^q dy)^{\frac{1}{q}}<\infty$, i.e. $k(x,y)\in L^q(\Omega)$ for fixed $x\in\Gamma$, we have

$$\int_\Omega k(x,y)u_m(y,t)dy \longrightarrow \int_\Omega k(x,y)u(y,t)dy \text{ as } m\longrightarrow\infty.$$

Similarly,

$$\int_\Omega D_i k(x,y)u_m(y,t)dy \longrightarrow \int_\Omega D_i k(x,y)u(y,t)dy \text{ as } m\longrightarrow\infty.$$

Therefore, for $x\in\Gamma$, we have

$$|u_m(x,t)|^{p-2}D_i u_m(x,t)\longrightarrow |u(x,t)|^{p-2}D_i u(x,t) \;\; a.e.$$

Next, we prove that $\left|\,(|u_m(x,t)|^{p-2}D_i u_m(x,t))\right|_{q,\Gamma}<\infty$: For $x\in\Gamma$, we have

$$\begin{aligned}u_m(x,t)&=\int_\Omega k(x,y)u_m(y,t)dy,\\|u_m(x,t)|&<|k(x,y)|_q|u_m|_p\le K(x)C.\end{aligned}$$

Since $K(x) \in L^p(\Gamma)$, we have $|u_m|_{p,\Gamma} < \infty$. Similarly, we have $|D_i u_m|_{p,\Gamma} < \infty$ and $|v|_{p,\Gamma} < \infty$. Then

$$\begin{aligned} &\big|\, |u_m|^{p-2} D_i u_m \big|_{q,\Gamma} \\ &\leq \big|\, |u_m|^{p-2} \big|_{\frac{p}{p-2},\Gamma} |D_i u_m|_{p,\Gamma} \ (\text{since } \tfrac{1}{q} = \tfrac{p-2}{p} + \tfrac{1}{p}, \text{ see [p. 25, 1]}) \\ &= (|u_m|_{p,\Gamma})^{p-2} |D_i u_m|_{p,\Gamma} < \infty. \end{aligned}$$

By Lemma 1.3 in [7], we get $|u_m|^{p-2} D_i u_m \rightharpoonup |u|^{p-2} D_i u$ weakly in $L^q(\Gamma)$ for a.e. $t \in [0,T]$. Since $|v|_{p,\Gamma} < \infty$, (G) is proved.

(H) $\displaystyle\int_\Omega (|u_m|^{p-2} D_i u_m)(D_i v)dx \longrightarrow \int_\Omega (|u|^{p-2} D_i u)(D_i v)dx.$

From (4.2) we know, we need to prove:

(i) $\displaystyle\int_\Gamma |u_m|^{p-2} u_m D_i v d\Gamma \longrightarrow \int_\Gamma |u|^{p-2} u D_i v d\Gamma,$

and

(ii) $\displaystyle\int_\Omega |u_m|^{p-2} u_m \Delta v dx \longrightarrow \int_\Omega |u|^{p-2} u \Delta v dx.$

Proof of (i). From the proof of (G), we have, for $x \in \Gamma$,

$$|u_m(x,t)|^{p-2} u_m(x,t) \longrightarrow |u(x,t)|^{p-2} u(x,t) \text{ almost everywhere,}$$
$$\big|\, |u_m|^{p-2} u_m \big|_{q,\Gamma} = |u_m|_{p,\Gamma}^{p-1} < \infty.$$

Therefore, we can apply Lemma 1.3 in [7] to conclude that $|u_m(x,t)|^{p-2} \times u_m(x,t) \rightharpoonup |u(x,t)|^{p-2} u(x,t)$ weakly in $L^q(\Gamma)$. Since $D_i v \in L^p(\Gamma)$, (i) is proved.

Proof of (ii). From (E), we have for $x \in \Omega$, $|u_m(x,t)|^{p-2} u_m(x,t) \longrightarrow |u(x,t)|^{p-2} u(x,t)$ almost everywhere. Since $\big|\, |u_m|^{p-2} u_m \big|_q = |u_m|_p^{p-1} < \infty$, by Lemma 1.3 in [7], we have:

$$|u_m|^{p-2} u_m \rightharpoonup |u|^{p-2} u \text{ weakly in } L^q(\Omega).$$

Since $\Delta v \in L^p(\Omega)$, we finish the proof of (ii).

(I) $(u_m', v) \longrightarrow (u', v)$ and $u(t)$ is continuous on $[0,T]$.

Proof of (I). Since u_m' is bounded in $L^q(0,T;H^{-r}(\Omega))$, by Alaoglu's theorem, there exists a subsequence of u_m', still denoted as u_m', converges to χ weak star in $L^q(0,T;H^{-r}(\Omega))$. By slightly modifying the proof of Theorem 1 in [2] (consider the space $L^q(0,T;H^{-r}(\Omega))$, instead of $L^2(0,T;B_2^{\frac{1}{2}}(0,1))$.), we have $\chi = u'$ and $u(t)$ is continuous on $[0,T]$.

Based on the above discussion, we summerize the existence theorem as follows:

Theorem 4.3. *Under the assumptions* (A1)−(A5), *there exists a generalized solution* u, *such that*

(1) $u \in L^\infty(0,T;L^2(\Omega)) \cap C([0,T],H^{-r}(\Omega))$.

(2) $|u|^{\frac{p-2}{2}}u$ *is bounded in* $L^2(0,T;H^1(\Omega))$.

(3) $\frac{du}{dt} \in L^q(0,T;H^{-r}(\Omega))$.

(4) $u(x,0) = u_0(x)$.

(5) *The identity*

$$(\frac{du}{dt},v) - (\sum_{i=1}^{n}\frac{\partial}{\partial x_i}(|u|^{p-2}\frac{\partial u}{\partial x_i}),v) + (|u|^{p-2}u,v) = (f,v)$$

holds for all $v \in V$ *and a.e.* $t \in [0,T]$.

References

[1] R. A. Adams and J. F. Fournier, *Sobolev spaces,* Elsevier/Academic Press, Amsterdam, 2003.

[2] A. Bouziani, N. Merazga and S. Benamira, Galerkin method applied to a parabolic evolution problem with nonlocal boundary conditions, *Nonlinear Anal.,* 69 (2008), 1515–1524.

[3] E. A. Coddington and N. Levinson, *Theory of ordinary differential equations,* McGraw-Hill Book Company, New York, 1955.

[4] T. D. Dzhuraev and J. O. Takhirov, A problem with nonlocal boundary conditions for a quasilinear parabolic equation, *Georgian Math. J.,* 6 (1999), 421–428.

[5] S. Gerbi and B. Said-Houari, Local existence and exponential growth for a semilinear damped wave equation with dynamic boundary conditions, *Adv. Differential Equations,* 13 (2008), 1051–1074.

[6] O. A. Ladyzenskaja, V. A. Solonnikov and N. N. Uralceva, *Linear and quasi-linear equations of parabolic type*, Nauka, Moscow 1967.

[7] J. L. Lions, *Quelques méthodes de résolution des problémes aux limites non linéaires,* (French) Dunod; Gauthier-Villars, Paris, 1969.

[8] P. A. Raviart, Sur la résolution et l'approximation de certaines équations paraboliques non linéaires dégénérées, *Arch. Rational Mech. Anal.,* 25 (1967), 64–80.

SOME BOUNDARY VALUE PROBLEMS FOR NONLINEAR ELLIPTIC SYSTEMS OF SECOND ORDER IN HIGH DIMENSIONAL DOMAINS[1]

DE-CHANG CHEN[a] and GUO-CHUN WEN[b]

[a] *Uniformed Services University of the Health Sciences, MD 20814, USA*
E-mail: dchen@usuhs.mil

[b] *School of Mathematical Sciences, Peking University, Beijing 100871, China*
E-mail: wengc@math.pku.edu.cn

This paper mainly concerns oblique derivative problems for nonlinear non-divergent elliptic systems of several second order equations with measurable coefficients in a multiply connected domain. Firstly, we give a priori estimates of solutions for the above boundary value problems with some conditions, and then by using the above estimates of solutions and the Leray-Schauder theorem, the existence of solutions for the above problems is proved.

Keywords: Oblique derivative problems, nonlinear elliptic systems, high dimensional domain.
AMS No: 35J65, 35J55, 35J45.

1. Formulation of Oblique Derivative Problems

Let Q be a bounded domain in $\mathbb{R}^N$ and the boundary $\partial Q \in C^2_\alpha(0 < \alpha < 1)$. We consider the nonlinear elliptic equation of second order

$$F^{(k)}(x, u, D_x u, D_x^2 u) = 0 \ \text{ in } \ Q, \ \ k = 1, ..., m.$$

Under certain conditions, the system (1.1) can be reduced to the form

$$\sum_{i,j=1}^{N} a_{ij}^{(k)} u_{kx_ix_j} + \sum_{h=1}^{m}[\sum_{i=1}^{N} b_{hi}^{(k)} u_{hx_i} + c_h^{(k)} u_h] = f^{(k)} \ \text{ in } Q, \ k=1, ..., m, \quad (1.1)$$

where $u = (u_1, ..., u_m)^T$ is the transposition of $(u_1, ..., u_m)$, $D_x u = (u_{x_i})$, $D_x^2 u = (u_{x_ix_j})$, and

$$a_{ij}^{(k)} = \int_0^1 F^{(k)}_{\tau r_{kij}}(x, u, p, \tau r) d\tau, \ b_{hi}^{(k)} = \int_0^1 F^{(k)}_{\tau p_{hi}}(x, u, \tau p, 0) d\tau,$$

$$c_h^{(k)} = \int_0^1 F^{(k)}_{\tau u_h}(x, \tau u, 0, 0) d\tau, \ f^{(k)} = -F^{(k)}(x, 0, 0, 0), \ r = D_x^2 u,$$

$$p = D_x u, \ r_{hij} = u_{hx_ix_j}, \ p_{hi} = u_{hx_i}, \ h, k = 1, ..., m.$$

[1]This research is supported by NSFC (No.10971224)

In this paper we consider

$$\sum_{i,j=1}^{N} a_{ij}^{(k)} u_{kx_ix_j} + \sum_{h=1}^{m}[\sum_{i=1}^{N} b_{hi}^{(k)} u_{hx_i} + \hat{c}_h^{(k)} u_h] = f^{(k)} \text{ in } Q, k=1,...,m, \quad (1.2)$$

in which $\hat{c}_h^{(k)} = c_h^{(k)} - |u_h|^{n_h}, h, k = 1, ..., m$, $n_h\,(h = 1, ..., m)$ are positive numbers. Suppose that (1.2) satisfies Condition C, i.e. for arbitrary functions $u_k^1(x)$, $u_k^2(x) \in C_\beta^1(\overline{Q}) \cap W_2^2(Q)$, $F^{(k)}(x, u, D_x u, D_x^2 u)\,(k = 1, ..., m)$ satisfy the conditions

$$F^{(k)}(x, u^1, D_x u^1, D_x^2 u^1) - F^{(k)}(x, u^2, D_x u^2, D_x^2 u^2)$$
$$= \sum_{i,j=1}^{N} \tilde{a}_{ij}^{(k)} u_{kx_ix_j} + \sum_{h=1}^{m}[\sum_{i=1}^{N} \tilde{b}_{hi}^{(k)} u_{hx_i} + \tilde{c}_h^{(k)} u_h],\ k = 1, ..., m,$$

where $\beta\,(0 < \beta < 1)$ is a constant, $u = u^1 - u^2$ and

$$\tilde{a}_{ij}^{(k)} = \int_0^1 F_{u_{kx_ix_j}}^{(k)}(x, \tilde{u}, \tilde{p}, \tilde{r})d\tau,\ \tilde{b}_{hi}^{(k)} = \int_0^1 F_{u_{hx_i}}^{(k)}(x, \tilde{u}, \tilde{p}, \tilde{r})d\tau,$$
$$\tilde{c}_h^{(k)} = \int_0^1 F_{u_h}^{(k)}(x, \tilde{u}, \tilde{p}, \tilde{r})d\tau,\ \tilde{r} = D_x^2[u^2 + \tau(u^1 - u^2)],$$
$$\tilde{p} = D_x[u^2 + \tau(u^1 - u^2)], \tilde{u} = u^2 + \tau(u^1 - u^2), h, k = 1, ..., m,$$

and $\tilde{a}_{ij}^{(k)}, \tilde{b}_{hi}^{(k)}, \tilde{c}_h^{(k)}, f^{(k)}$ satisfy the conditions

$$q_0 \sum_{j=1}^{n} |\xi_j|^2 \le \sum_{i,j=1}^{N} \tilde{a}_{ij}^{(k)} \xi_i \xi_j \le q_0^{-1} \sum_{j=1}^{n} |\xi_j|^2,\ 0 < q_0 < 1, \quad (1.3)$$

$$\sup_Q [\sum_{i,j=1}^{N} (\tilde{a}_{ij}^{(k)})^2] / \inf_Q [\sum_{i=1}^{N} \tilde{a}_{ii}^{(k)}]^2 \le q_1 < \frac{2N-1}{2N^2 - 2N - 1}, \quad (1.4)$$

$$|\tilde{a}_{ij}^{(k)}| \le k_0,\ |\tilde{b}_{hi}^{(k)}| \le q_h^{(k)} \le k_0,\ L_p(f^{(k)}, Q) \le k_1,$$
$$|c_h^{(k)}| \le q_h^{(k)} \le k_0\,(k \ne h),\ |c_h^h| \le k_0, \quad (1.5)$$
$$\sup_Q c_k^{(k)} < 0,\ i, j = 1, ..., N,\ h, k = 1, ..., m,$$

in which $q_0, q_1, q_h^{(k)}\,(k, h = 1, ..., m)$, $k_0, k_1, p\,(> N + 2)$ are non-negative constants. Moreover, for almost every point $x \in Q$ and $D_x^2 u$, $\tilde{a}_{ij}^{(k)}(x, u, D_x u, D_x^2 u)$, $\tilde{b}_{hi}^{(k)}(x, u, D_x u)$, $\tilde{c}_h^{(k)}(x, u)$ are continuous in $u \in \mathbb{R}$, $D_x u \in \mathbb{R}^N$.

Now we explain the condition (1.4). It is enough to consider the linear elliptic equation of (1.1) with $m = 1$, namely

$$\sum_{i,j=1}^{N} a_{ij}(x)u_{x_ix_j} + \sum_{i=1}^{N} b_i(x)u_{x_i} + c(x)u = f(x) \text{ in } Q. \tag{1.6}$$

Let (1.6) be divided by $\Lambda = \tau \inf_Q \sum_{i=1}^{N} a_{ii}$, here τ is an undetermined positive constant, and denote $\hat{a}_{ij} = a_{ij}/\Lambda$, $\hat{b}_i = b_i/\Lambda\,(i,j = 1,...,N)$, $\hat{c} = c/\Lambda$, $\hat{f} = f/\Lambda$, equation (1.6) is reduced to the form

$$\sum_{i,j=1}^{N} \hat{a}_{ij}(x)u_{x_ix_j} + \sum_{i=1}^{N} \hat{b}_i(x)u_{x_i} + \hat{c}(x)u = \hat{f}(x), \text{ i.e.}$$

$$\Delta u = -\sum_{i,j=1}^{N}[\hat{a}_{ij}(x) - \delta_{ij}]u_{x_ix_j} - \sum_{i=1}^{N} \hat{b}_i(x)u_{x_i} - \hat{c}(x)u + \hat{f}(x) \text{ in } Q.$$

We require that the above coefficients satisfy

$$\begin{aligned} &\sup_Q[2\sum_{i,j=1,i<j}^{N} \hat{a}_{ij}^2 + \sum_{i=1}^{N}(\hat{a}_{ii}-1)^2] \\ &= \sup_Q[\sum_{i,j=1}^{N} \hat{a}_{ij}^2 + N - 2\sum_{i=1}^{N} \hat{a}_{ii}] < \frac{N+1}{2N-1}, \text{ i.e.} \\ &\sup_Q[\sum_{i,j=1}^{N} \hat{a}_{ij}^2 - 2\sum_{i=1}^{N} \hat{a}_{ii}] < \frac{N+1}{2N-1} - N, \end{aligned} \tag{1.7}$$

with the constant $\tau = (2N-1)/(2N^2-2N-1)$, it can be derived from the condition in (1.4). In fact, we consider

$$\sup_Q \sum_{i,j=1}^{N} \hat{a}_{ij}^2 - 2\inf_Q \sum_{i=1}^{N} \hat{a}_{ii} < \frac{N+1}{2N-1} - N, \text{ i.e.}$$

$$\frac{\sup_Q \sum_{i,j=1}^{N} a_{ij}^2}{\tau^2 \inf_Q[\sum_{i=1}^{N} a_{ii}]^2} < \frac{2}{\tau} + \frac{N+1}{2N-1} - N, \text{ i.e. } \frac{\sup_Q \sum_{i,j=1}^{N} a_{ij}^2}{\inf_Q[\sum_{i,j=1}^{N} a_{ii}]^2} < f(\tau),$$

and can find the maximum of $f(\tau) = 2\tau + (1+2N-2N^2)\tau^2/(2N-1)$ on $[0,\infty)$ at the point $\tau = (2N-1)/(2N^2-2N-1)$, and the maximum of $f(\tau)$ equals $f((2N-1)/(2N^2-2N-1)) = (2N-1)/(2N^2-2N-1)$, the above inequality with $\tau = (2N-1)/(2N^2-2N-1)$ is just the inequality in (1.4). From the inequality (1.4), it follows that (1.7) with $\tau = (2N-1)/(2N^2-2N-1)$ holds.

The so-called oblique derivative boundary value problem (Problem O) is to find a continuously differentiable solution $u = u(x) = (u_1, ..., u_m)^T \in B^* = C^1_\beta(\overline{Q}) \cap W^2_2(Q)$ satisfying the boundary conditions

$$\begin{gathered} lu = d\frac{\partial u}{\partial \nu} + \sigma u = \tau(x), \quad x \in \partial Q, \quad \text{i.e.} \\ lu = \sum_{j=1}^{N} d_j \frac{\partial u}{\partial x_j} + \sigma u = \tau(x), \quad x \in \partial Q, \end{gathered} \tag{1.8}$$

in which $d(x), d_j(x)(j = 1, ..., N), \sigma(x), \tau(x)$ satisfy the conditions

$$\begin{gathered} C^1_\alpha[\sigma(x), \partial Q] \le k_0, \ C_\alpha[d_j(x), \partial Q] \le k_0, \ C^1_\alpha[\tau(x), \partial Q] \le k_2, \\ \cos(\nu, \mu) \ge q_0 > 0, \ \ d > 0, \ \ \sigma \ge 0, \ \ x \in \partial Q, \end{gathered} \tag{1.9}$$

where μ is the unit outward normal on $\partial Q, \alpha, \beta\,(0 < \beta \le \alpha < 1), k_0, k_2, q_0\,(0 < q_0 < 1)$ are non-negative constants. In particular, Problem O with the condition $\nu = \mu$, $\sigma = 0$ on ∂Q is the initial-Neumenn problem (Problem N).

Theorem 1.1. *If system* (1.2) *satisfies Condition* C *and* $f^{(k)} = 0$, $q_h^{(k)}$ $(k \ne k, k, h = 1, ..., m,\ i, j = 1, ..., N)$ *are small enough, then the solutions of Problem* O *for the above equation only has the trivial solution.*

Proof. Let $u(x) = [u_1(x), ..., u_m(x)]^T$ be two solutions of Problem O. It is easy to see that $u(x)$ satisfies the following boundary value problem

$$\sum_{i,j=1}^{N} a_{ij}^{(k)} u_{kx_ix_j} + \sum_{h=1}^{m} [\sum_{i=1}^{N} b_{hi}^{(k)} u_{hx_i} + \hat{c}_h^{(k)} u_h] = 0, \ x \in Q, k = 1, ..., m, \tag{1.10}$$

$$d\frac{\partial u}{\partial \nu} + \sigma(x) u = 0, \quad x \in \partial Q, \tag{1.11}$$

where $a_{ij}^{(k)}, b_{hi}^{(k)}, \hat{c}_h^{(k)}$ are as stated in (1.2). Let us multiply u_k to each equation of system (1.10). Thus a system for $u_k^2 = (u_k)^2\ (k = 1, ..., m)$ can be obtained, i.e.

$$\begin{gathered} \frac{1}{2} \sum_{i,j=1}^{N} a_{ij}^{(k)} (u_k^2)_{x_ix_j} = \sum_{i,j=1}^{N} a_{ij}^{(k)} u_{kx_i} u_{kx_j} \\ - \sum_{h=1}^{m} [\sum_{i=1}^{N} b_{hi}^{(k)} u_k u_{hx_i} + \hat{c}_h^{(k)} u_k u_h] = 0 \text{ in } Q, \quad k = 1, ..., m. \end{gathered} \tag{1.12}$$

Noting that Condition C and $q_h^{(k)}$ $(h \neq k, h, k = 1, ..., m)$ are small enough, there exists a small positive number ε_0 such that $|b_{hi}^{(k)}| \leq \varepsilon_0$, $|\hat{c}_h^{(k)}| \leq \varepsilon_0$ $(h \neq k,\ h, k = 1, ..., m, j = 1, ..., N)$, if the maximum of $u_k^2 = (u_k)^2$ attains at an inner point $P_0 = x_0 \in Q$, and

$$M_0 = \max[|u_k(P_0)|^{n_k+1}, |u_k(P_0)|^2] > 0,$$

we choose a sufficiently small neighborhood G_0 of P_0 such that $2|u_k u_{kx_i}| \leq \varepsilon_0^2/NM_0$ in G_0 and $\min_{1\leq h\leq m}(-\sup_Q \hat{c}_h^{(h)}) - [m^2N/4q_0 + k_0 + 1]\varepsilon_0^2 > 0$, thus we have

$$\sum_{i,j=1}^{N} a_{ij}^{(k)} u_{kx_i} u_{kx_j} \geq q_0 \sum_{i=1}^{N} |u_{kx_i}|^2,$$

$$-\sum_{h=1}^{m}[\sum_{i=1}^{N} b_{hi}^{(k)} u_k u_{hx_i}] \geq -\sum_{h=1}^{m}\sum_{i=1}^{N}[\frac{q_0}{m}|u_{kx_i}|^2 - [\frac{m}{4q_0}u_k^2 + \frac{k_0}{NM_0}]\varepsilon_0^2$$

$$\geq -q_0 \sum_{h=1}^{m}\sum_{i=1}^{N}|u_{hx_i}|^2 - [\frac{m^2N}{4q_0} + k_0]\varepsilon_0^2 u_k^2,$$

$$-\sum_{h=1}^{m} \hat{c}_h^{(k)} u_k u_h \geq \sum_{h=1}^{m}[\min_{1\leq h\leq m}(-\sup_Q \hat{c}_h^{(h)}) - \varepsilon_0^2]u_k^2, k = 1, ..., m,$$

and then

$$\sum_{i,j=1}^{N} a_{ij}^{(k)} u_{kx_i} u_{kx_j} \geq \{\min_{1\leq h\leq m}(-\sup_Q \hat{c}_h^{(h)}) - [\frac{m^2N}{4q_0} + k_0 + 1]\varepsilon_0^2\}u^2 > 0$$

$$\text{in } G_0,\ 1 \leq k \leq m.$$

On the basis of the maximum principle of the solution u_k^2 $(1 \leq k \leq m)$ for equation (1.12), we see that u_k^2 $(1 \leq k \leq m)$ cannot take a positive maximum in Q. Hence $u^2(x) = \sum_{k=1}^{m} u_k^2 = 0$, namely $u(x) = 0$ in Q.

2. Estimates of Solutions of Oblique Derivative Problems

Theorem 2.1. *Suppose that (1.2) satisfies Condition C and q_{kh} $(k \neq h, k, h = 1, ..., N)$ are small enough. Then any solution $u(x)$ of Problem O satisfies the estimates*

$$\hat{C}_\beta^1(u, \overline{Q}) = \sum_{k=1}^{m} C_\beta^1(|u_k|^{n+1}, \overline{Q}) \leq M_1 = M_1(q, p, \alpha, k, Q), \tag{2.1}$$

$$||u||_{W_2^2(Q)} = \sum_{k=1}^{m} ||u||_{W_2^2(Q)} \le M_2 = M_2(q,p,\alpha,k,Q), \tag{2.2}$$

where $\beta\,(0 < \beta \le \alpha)$, $q = (q_0, q_1)$, $k = (k_0, k_1, k_2)$, M_1, M_2 *are non-negative constants.*

Proof. First of all, under Condition C, we substitute the solution $u(x)$ of Problem O into system (1.2), thus we can only discuss the linear elliptic system in the form

$$\sum_{i,j=1}^{N} a_{ij}^{(k)} u_{kx_ix_j} + \sum_{h=1}^{m}[\sum_{i=1}^{N} b_{hi}^{(k)} u_{hx_i} + \hat{c}^{(k)} u_h] = f^{(k)} \text{ in } Q, k = 1, ..., m, \tag{2.3}$$

and verify the boundedness estimate of the solution $u(x)$

$$\hat{C}^1[u,\overline{Q}] = \sum_{k=1}^{m}[C^1[|u_k|^{n+1}, \overline{Q}] \le M_3, \tag{2.4}$$

where $M_3 = M_3(q,p,\alpha,k,Q)$. Suppose that (2.4) is not true, then there exist sequences of functions $\{a_{ij}^{(kl)}\}$, $\{b_{hi}^{(kl)}\}$, $\{\hat{c}_h^{(kl)}\}$, $\{f^{(kl)}\}$ and $\{\sigma^{(l)}\}$, $\{\tau^{(l)}\}$, which satisfy Condition C and (1.5), (1.9), and $\{a_{ij}^{(k)}\}$, $\{b_{hi}^{(kl)}\}$, $\{\hat{c}_h^{(kl)}\}$, $\{f^{(kl)}\}$ weakly converge to $a_{ij}^{(0)}$, $b_{hi}^{(k0)}$, $\hat{c}_h^{(k0)}$, $f^{(k0)}$, and $\{\sigma^{(l)}\}$, $\{\tau^{(l)}\}$ uniformly converge to σ^0, $\tau^{(0)}$ in Ω or ∂Q respectively, and the boundary value problem

$$\sum_{i,j=1}^{N} a_{ij}^{(kl)} u_{kx_ix_j} + \sum_{h=1}^{m}[\sum_{i=1}^{N} b_{hi}^{(kl)} u_{hx_i} + \hat{c}_h^{(kl)} u_h] = f^{(kl)} \text{ in } Q, k=1, ..., m, \tag{2.5}$$

$$d\frac{\partial u^{(l)}}{\partial \nu} + \sigma^{(l)} u^{(l)} = \tau^{(l)}(x) \text{ on } \partial Q, \tag{2.6}$$

have solutions $u^{(l)}(x)\,(l = 1, 2, ...)$ such that $||u^{(l)}||_{\hat{C}^1(\overline{Q})} = h^{(l)}$ is unbounded. There is no harm assuming that $h^{(l)} \ge 1$, and $\lim_{l\to\infty} h^{(l)} = +\infty$. It is easy to see that $U^{(l)} = u^{(l)}/h^{(l)}$ is a solution of the following boundary value problem

$$\sum_{i,j=1}^{N} a_{ij}^{(kl)} U_{kx_ix_j}^{(l)} + \sum_{h=1}^{m}[\sum_{i=1}^{N} b_{hi}^{(kl)} U_{hx_i}^{(l)} + \hat{c}_h^{(kl)} U_h^{(l)}] = \frac{f^{(kl)}}{h^{(l)}} \text{ in } Q, k=1, ..., m, \tag{2.7}$$

$$d\frac{\partial U^{(l)}}{\partial \nu} + \sigma^{(l)} U^{(l)} = \frac{\tau^{(l)}}{h^{(l)}} \text{ on } \partial Q. \tag{2.8}$$

Noting that $\sum_{h=1}^m[\sum_{i=1}^N b_{hi}^{(kl)}U_{hx_i}^{(l)}+\hat{c}_h^{(kl)}U_h^{(l)}]$ in (2.8) are bounded, by the method in Theorem 3.1, Chapter IV, [5], we can obtain the estimate

$$\hat{C}_\beta^1(U^{(l)},\overline{Q})\leq M_4, ||U^{(l)}||_{W_2^2(Q)}\leq M_5, \tag{2.9}$$

where $\beta\,(0<\beta\leq\alpha), M_j=M_j(q,p,\alpha,k,Q,M_3)\,(j=4,5)$ are non-negative constants. Hence from $\{U^{(l)}\}$, we can choose a subsequence $\{U^{(l_k)}\}$ such that $\{U^{(l_k)}\}$, $\{U_{x_i}^{(l_k)}\}$ uniformly converge to $U^{(0)}, U_{x_i}^{(0)}$ in $\overline{Q}$ respectively, and $\{U_{x_ix_j}^{(l_k)}\}$ weakly converge to $U_{x_ix_j}^{(0)}$ in Q respectively, and $U^{(0)}$ is a solution of the following boundary value problem

$$\sum_{i,j=1}^N a_{ij}^{(k0)}U_{kx_ix_j}^{(0)}+\sum_{h=1}^m[\sum_{i=1}^N b_{hi}^{(k0)}U_{hx_i}^{(0)}+\hat{c}^{(k0)}U_h^{(0)}]=0 \text{ on } Q, k=1,...,m, \tag{2.10}$$

$$d\frac{\partial U^{(0)}}{\partial\nu}+\sigma^{(0)}U^{(0)}=0 \text{ on } \partial Q. \tag{2.11}$$

According to Theorem 1.1, we can get $U^{(0)}(x)=0$, $x\in\overline{Q}$. However, from $\hat{C}^1[U^{(l)},\overline{Q})]=1$, there exists a point $x^*\in\overline{Q}$, such that $\sum_{k=1}^m[|U_k^{(0)}(x^*)|+\sum_{i=1}^N|U_{kx_i}^{(0)}(x^*)|]>0$. This contradiction proves that (2.4) is true. By the same way as stated before, we can derive the estimates (2.1) and (2.2).

Moreover by using the method in Chapter IV, [5], we can prove the following theorem.

Theorem 2.2. *Suppose that equation* (1.2) *satisfies the same as in Theorem* 2.1. *Then any solution $u(x)$ of Problem O satisfies the estimates*

$$\hat{C}_\beta^1[u,\bar{Q}]\leq M_6(k_1+k_2),\ ||u||_{W_2^2(Q)}\leq M_7(k_1+k_2), \tag{2.12}$$

where $\beta\,(0<\beta\leq\alpha)$, $M_j=M_j(q,p,\alpha,k_0,Q)\,(j=6,7)$ are non-negative constants.

3. Solvability of Oblique Derivative Problems

We first consider the other form of system (1.2), namely

$$\begin{aligned}\Delta u_k=g^{kl}(x,u,Du,D^2u),\ g^{kl}=\Delta u_k&+\sum_{i,j=1}^N a_{ij}^{kl}u_{kx_ix_j}\\&+\sum_{h=1}^m[\sum_{i=1}^N b_{hi}^{kl}u_{hx_i}+\hat{c}_h^{kl}u_h]=f^{kl} \text{ in } Q,\end{aligned} \tag{3.1}$$

where $k=1,...,m$, $\Delta u_k=\sum_{i=1}^N\partial^2u_k/\partial x_i^2$ and the coefficients

$$a_{ij}^{kl}=\begin{cases}a_{ij}^{(k)},\\ \delta_{ij},\end{cases} b_{hi}^{kl}=\begin{cases}b_{hi}^{(k)},\\ 0,\end{cases} \hat{c}_h^{kl}=\begin{cases}\hat{c}_h^{(k)},\\ 0,\end{cases} f^{kl}=\begin{cases}f^{(k)} \text{ in } Q_l,\\ 0 \text{ in } \mathbb{R}^N\backslash Q_l,\end{cases} \tag{3.2}$$

where $Q_l = \{x \in Q \,|\, \text{dist}(x, \partial Q) \geq 1/l\}$, l is a positive integer.

Theorem 3.1. *Under the same conditions in Theorem* 2.1, *if* $u(x) = [u^1(x), ..., u_m(x)]$ *is any solution of Problem O for system* (3.1), *then* $u(x)$ *can be expressed in the form*

$$u(x)=U(x)+V(x)=U(x)+v_0(x)+v(x),\ v(x)=H\rho$$
$$=\int_{Q_0} G(x-\zeta)\rho(\zeta)d\zeta, G=\begin{cases} |x-\zeta|^{2-N}/N(2-N)\omega_N, N>2, \\ \log|x-\zeta|/2\pi,\ N=2, \end{cases} \tag{3.3}$$

where $\omega_N = 2\pi^{N/2}/(N\Gamma(N/2))$ *is the volume of unit ball in* $\mathbb{R}^N$, $\rho(x) = \Delta u$ *and* $V(x)$ *is a solution of Problem* D_0 *for* (3.1), *namely the equation* (3.1) *in* $Q_0 = \{|x| < R\}$ *with the boundary condition* $V(x) = 0$ *on* ∂Q, *here* R *is a large number, such that* $Q_0 \supset \overline{Q}$, *and* $U(x)$ *is a solution of Problem O for* $\Delta U = 0$ *in* Q *with the boundary condition* (3.11) *below, which satisfy the estimates*

$$\hat{C}^1_\beta[U,\overline{Q}]+\parallel U \parallel_{W^2_2(Q)}\leq M_8,\ C^1_\beta[V,\overline{Q_0}]+\parallel V \parallel_{W^2_2(Q_0)}\leq M_9, \tag{3.4}$$

where $\beta\,(0 < \beta \leq \alpha)$, $M_j = M_j(q,p,\alpha,k,Q_l)\,(j = 8,9)$ *are non-negative constants,* $q = (q_0, q_1)$, $k = (k_0, k_1, k_2)$.

Proof. It is easy to see that the solution $u(x)$ of Problem O for system (3.1) can be expressed by the form (3.3). Noting that $a^{kl}_{ij} = 0\,(i \neq j)$, $b^{kl}_{hi} = 0$, $c^{kl}_h = 0$, $f^{kl}(x) = 0$ in $\mathbb{R}^N\backslash Q_l$ and $V(x)$ is a solution of Problem D_0 for (3.1) in Q_0, we can obtain that $V(x)$ in $\hat{Q}_{2l} = \overline{Q}\backslash Q_{2l}$ satisfies the estimate

$$C^2[V(x), \hat{Q}_{2l}] \leq M_{10} = M_{10}(q,p,\alpha,k,Q_l).$$

On the basis of Theorem 2.1, we can see that $U(x)$ satisfies the first estimate in (3.4), and then $V(x)$ satisfies the second estimate in (3.4).

Theorem 3.2. *If system* (1.2) *satisfies the same conditions as in Theorem* 2.1, *then Problem O for* (3.1) *has a solution* $u(x)$.

Proof. In order to prove the existence of solutions of Problem O for the nonlinear system (3.1) by using the Larey-Schauder theorem, we introduce the system with the parameter $h \in [0,1]$:

$$\Delta u_k = hg^{kl}(x,u,Du,D^2u) \text{ in } Q,\ k = 1,...,m. \tag{3.5}$$

Denote by B_M a bounded open set in the Banach space $B = \hat{W}^2_2(Q_0) = \hat{C}^1_\beta(\overline{Q}_0) \cap W^2_2(Q_0)(0 < \beta \leq \alpha)$, the elements of which are real functions $V(x)$ satisfying the inequalities

$$\parallel V \parallel_{\hat{W}^2_2(Q_0)}= \hat{C}^1_\beta[V,\overline{Q_0}] + ||V||_{W^2_2(Q_0)} < M_{11} = M_9 + 1, \tag{3.6}$$

in which M_9 is a non-negative constant as stated in (3.4). We choose any function $\tilde{V}(x) \in \overline{B_M}$ and substitute it into the appropriate positions in the right hand side of (3.5), and then we make an integral $\tilde{v}(x) = H\rho$ as follows:

$$\tilde{v}(x) = H\tilde{\rho},\ \tilde{\rho}(x) = \Delta\tilde{V}. \tag{3.7}$$

Next we find a solution $\tilde{v}_0(x)$ of the boundary value problem in Q_0:

$$\Delta\tilde{v}_0 = 0 \text{ on } Q_0, \tag{3.8}$$

$$\tilde{v}_0(x) = -\tilde{v}(x) \text{ on } \partial Q_0, \tag{3.9}$$

and denote the solution $\hat{V}(x) = \tilde{v}(x) + \tilde{v}_0(x)$ of the corresponding Problem D_0. Moreover on the basis of the result in [5], we can find a solution $\tilde{U}(x)$ of the corresponding Problem $\tilde{O}$ in Q:

$$\Delta\tilde{U} = 0 \text{ on } Q, \tag{3.10}$$

$$d\frac{\partial\tilde{U}}{\partial\vec{\nu}} + \sigma(x)\tilde{U} = \tau(x) - d\frac{\partial\hat{V}}{\partial\vec{\nu}} + \sigma(x)\hat{V} \text{ on } \partial Q. \tag{3.11}$$

Now we discuss the system

$$\Delta V = hg^{kl}(x, \tilde{u}, D\tilde{u}, D^2\tilde{U} + D^2V),\ k = 1, ..., m,\ 0 \le h \le 1, \tag{3.12}$$

where $\tilde{u} = \tilde{U} + \hat{V}$. By Condition C, applying the principle of contracting mapping, we can find a unique solution $V(x)$ of Problem D_0 for system (3.12) in Q_0 satisfying the boundary condition

$$V(x) = 0 \text{ on } \partial Q_0. \tag{3.13}$$

Denote $u(x) = U(x) + V(x)$, where the relation between U and V is same that between $\tilde{U}$ and $\tilde{V}$, and by $V = S(\tilde{V}, h)$, $u = S_1(\tilde{V}, h)$ $(0 \le h \le 1)$ the mappings from $\tilde{V}$ onto V and u respectively. Furthermore, if $V(x)$ is a solution of Problem O in Q_0 for the system

$$\Delta V = hg^{kl}(x, u, Du, D^2(U + V)),\ 0 \le h \le 1, \tag{3.14}$$

where $u = S_1(V, h)$, then from Theorem 3.1, the solution $V(x)$ of Problem D_0 for (3.14) satisfies the second estimate (3.4), consequently $V(x) \in B_M$. Set $B_0 = B_M \times [0, 1]$, we can verify that the mapping $V = S(\tilde{V}, h)$ satisfies the three conditions of Leray-Schauder theorem:

1) For every $h \in [0, 1]$, $V = S(\tilde{V}, h)$ continuously maps the Banach space B into itself, and is completely continuous on B_M. Besides, for every function $\tilde{V}(x) \in \overline{B_M}$, $S(\tilde{V}, h)$ is uniformly continuous with respect to $h \in [0, 1]$.

2) For $h = 0$, from (3.6) and (3.12), it is clear that $V = S(\tilde{V}, 0) \in B_M$.

3) From Theorems 3.1, we know that the system of functions $V = S(\tilde{V}, h)\,(0 \le h \le 1)$ satisfies the second estimate in (3.4). Moreover by the inequality (3.6), it is not difficult to see that the functional equation $V = S(\tilde{V}, h)\,(0 \le h \le 1)$ has not a solution $V(x)$ on the boundary $\partial B_M = \overline{B_M} \backslash B_M$.

Hence by the Leray-Schauder theorem, we know that Problem D_0 for system (3.12) with $h = 1$ has a solution $V(x) \in B_M$, and then Problem O of system (3.5) with $h = 1$, i.e. (3.1) has a solution $u(x) = S_1(V, h) = U(x) + V(x) = U(x) + v_0(x) + v(x) \in B$.

Theorem 3.3. *Under the same conditions in Theorem* 2.1, *Problem O for system* (1.2) *has a solution.*

Proof. By Theorems 2.1 and 3.2, Problem O for system (3.1) possesses a solution $u^l(x)$, and the solution $u^l(x)$ of Problem O for (3.1) satisfies the estimates (2.1) and (2.2), where $l = 1, 2, \ldots$. Thus, we can choose a subsequence $\{u^{l_k}(x)\}$, such that $\{u^{l_k}(x)\}$, $\{u^{l_k}_{x_i}(x)\}\,(i = 1, \ldots, N)$ in $\overline{Q}$ uniformly converge to $u^0(x)$, $u^0_{x_i}(x)\,(i = 1, \ldots, N)$ respectively. Obviously, $u^0(x)$ satisfies the boundary condition of Problem O. On the basis of principle of compactness of solutions for system (3.1) (see Theorem 5.5, Chapter I, [5]), it is easy to see that $u^0(x)$ is a solution of Problem O for (1.2).

References

[1] H. O. Codres, Über die Randwertaufgabe bei quasilinearen Differentialgleichungen zweiter Ordnung in mehr als zwei Variablen, *Math. Ann.* 131 (1956), 278–312.

[2] Yu. A. Alkhutov and I. T. Mamedov, The first boundary value problem for nondivergence second order parabolic equations with discontinuous coefficients, *Math. USSR Sbornik* 59 (1988), 471–495.

[3] O. A. Ladyshenskaja and N. N. Uraltseva, *Linear and quasilinear elliptic equations,* Academic Press, New York, 1968.

[4] G. C. Wen and H. Begehr, *Boundary value problems for elliptic equations and systems,* Longman Scientific and Technical, Harlow, 1990.

[5] G. C. Wen , Z. L. Xu and H. Y. Gao, *Boundary Value problems for nonlinear elliptic equations in high dimensional domains,* Nonlinear Sciences and Numerical Simulation, Research Information LTD UK, 2004.

[6] G. C. Wen and D. C. Chen, Oblique derivative problems for nonlinear elliptic equations of second order with measurable coefficients in high dimensional domains, *Nonlinear Sciences and Numerical Simulation* 10 (2005), 559–570.

WHEN DOES A SCHRÖDINGER HEAT EQUATION PERMIT POSITIVE SOLUTIONS

Qi S. ZHANG

Department of Mathematics, University of California, Riverside, CA 92521, USA

E-mail: qizhang@math.ucr.edu

We introduce some new classes of time dependent functions whose defining properties take into account of oscillations around singularities. We study properties of solutions to the heat equation with coefficients in these classes which are much more singular than those allowed under the current theory. In the case of L^2 potentials and L^2 solutions, we give a characterization of potentials which allow the Schrödinger heat equation to have a positive solution. This provides a new result on the long running problem of identifying potentials permitting a positive solution to the Schrödinger equation. We also establish a nearly necessary and sufficient condition on certain sign changing potentials such that the corresponding heat kernel has Gaussian upper and lower bound. An application to Navier Stokes equation is also given.

Keywords: Schrödinger equation, positive solution.

AMS No: 35K05.

1. Introduction

In the first part of the paper we would like to study the heat equation with a singular L^2 potential $V = V(x,t)$, i.e.

$$
\begin{aligned}
&\Delta u + Vu - u_t = 0 \text{ in } \mathbb{R}^n \times (0,\infty),\ V \in L^2(\mathbb{R}^n \times (0,\infty)),\ n \geq 3,\\
&u(x,0) = u_0(x),\ x \in \mathbb{R}^n,\ u_0 \in L^2(\mathbb{R}^n).
\end{aligned} \tag{1.1}
$$

Since we are only concerned with local regularity issue in this paper, we will always assume that V is zero outside of a cylinder in space time: $B(0,R_0) \times [0,T_0]$, unless stated otherwise. Here R_0 and T_0 are fixed positive number. The L^2 condition on the potential V is modeled after the three dimensional vorticity equation derived from the Navier-Stokes equation. There the potential V is in fact the gradient of the velocity which is known to be in L^2. The unknown function u in (1.1) corresponds to the vorticity which is also known to be a L^2 function.

We will use the following definition of weak solutions.

Definition 1.1. Let $T > 0$. We say that $u \in L^1_{loc}(\mathbb{R}^n \times (0,T))$ is a solution to (1.1), if $V(.), u(.,t) \in L^1_{loc}(\mathbb{R}^n \times (0,T))$ and

$$
\int_{\mathbb{R}^n} u_0(x)\phi(x)dx + \int_0^T \int_{\mathbb{R}^n} u\phi_t dxdt + \int_0^T \int_D u\Delta\phi dxdt + \int_0^T \int_{\mathbb{R}^n} Vu\phi dxdt = 0
$$

for all smooth, compactly supported ϕ vanishing on $\mathbb{R}^n \times \{T\}$.

It is well known that L^2 potentials in general are too singular to allow weak solutions of (1.1) to be bounded or unique. Therefore further assumptions must be imposed in order to establish a regularity theory. The classical condition on the potential V for Hölder continuity and uniqueness of weak solutions is that $V \in L^{p,q}_{loc}$ with $\frac{n}{p} + \frac{2}{q} < 2$. This condition is sharp in general since one can easily construct a counter example. For instance for $V = a/|x|^2$ with $a > 0$, then there is no bounded positive solution to (1.1) (see [2]). In fact in that paper, it is shown that if a is sufficiently large, then even weak positive solutions can not exist. There is a long history of finding larger class of potentials such that some regularity of the weak solutions is possible. Among them is the Kato class, time independent or otherwise. Roughly speaking a function is in a Kato type class if the convolution of the absolute value of the function and the fundamental solution of Laplace or the heat equation is bounded. This class of functions are moderately more general than the standard $L^{p,q}$ class. However, it is still far from enough for applications in such places as the vorticity equation mentioned above. We refer the reader to the papers [1,11,16,18] and reference therein for results in this direction. The main results there is the continuity of weak solutions with potentials in the Kato class. In addition, equation (1.1) with V in Morrey or Besov classes are also studied. However, these classes are essentially logarithmic improvements over the standard $L^{p,q}$ class. In the paper [17], K. Sturm proved Gaussian upper and lower bound for the fundamental solution when the potential belongs to a class of time independent, singular oscillating functions. His condition is on the L^1 bound of the fundamental solution of a slightly "larger" potential.

In this paper we introduce a new class of *time dependent* potentials which can be written as a nonlinear combination of derivatives of a function. The general idea of studying elliptic and parabolic equations with potentials as the spatial derivative of some functions is not new. This has been used in the classical books [7,12,13]. Here we also allow the appearance of time derivative which can not be dominated by the Laplace operator. Another innovation is the use of a suitable combination of derivatives. The class we are going to define in section 2 essentially characterize all L^2 potentials which allow (1.1) to have positive L^2 solutions.

The question of whether the Laplace or the heat equation with a potential possesses a positive solution has been a long standing one. For the Laplace equation, when the potential has only mild singularity, i.e. in the Kato class, a satisfactory answer can be found in the Allegretto-Piepenbrink theory. See Theorem C.8.1 in the survey paper [16]. Brezis and J. L. Lions (see [2], p.122) asked when (1.1) with more singular potential has a positive solutions. This problem was solved in [2] when $V = a/|x|^2$ with $a > 0$. In the case of general time independent potentials $V \geq 0$, it was solved in [3] and later generalized in [8]. However the case of time dependent or sign changing potentials is completely open. One of the main results of the paper (Theorem 2.1) gives a solution of the problem with L^2 potentials. The main advantage of the new class of potentials is that it correctly captures the cancelation effect of sign changing functions. Moreover, we show in Theorems 2.2–2.3 below that, if we just narrow the class a little, then the weak fundamental solutions not only exist but also have Gaussian upper bound. A Gaussian lower bound is also established under further but necessary restrictions.

Some of the results of the paper can be generalized beyond L^2 potentials. However we will not seek full generalization this time.

Before proceeding further let us fix some notations and symbols, to which will refer the reader going over the rest of the paper.

Notations. We will use $\mathbb{R}^+$ to denote $(0, \infty)$. The letter C, c with or without index will denote generic positive constants whose value may change from line to line, unless specified otherwise. When we say a time dependent function is in L^2 we mean its square is integrable in $\mathbb{R}^n \times \mathbb{R}^+$. We use G_V to denote the fundamental solution of (1.1) if it exists. Please see the next section for its existence and uniqueness. The symbol G_0 will denote the fundamental solution of the heat equation free of potentials. Give $b > 0$, we will use g_b to denote a Gaussian with b as the exponential parameter, i.e.

$$g_b = g_b(x, t; y, s) = \frac{1}{(t-s)^{n/2}} e^{-b|x-y|^2/(t-s)}.$$

Given a L^1_{loc} function f in space time, we will use $g_b \star f(x, t)$ to denote

$$\int_0^t \int_{\mathbb{R}^n} g_b(x, t; y, s) f(y, s) dy ds.$$

When we say that G_V has Gaussian upper bound, we mean that exists $b > 0$ and $c > 0$ such that $G_V(x, t; y, s) \leq c g_b(x, t; y, s)$. The same goes for the Gaussian lower bound.

When we say a function is a positive solution to (1.1) we mean it is a nonnegative weak solution which is not identically zero.

Here is the plan of the paper. In the next section we provide the definitions, statements and proofs of the main results. In Section 3, we define another class of singular potentials, called heat bounded class. Some applications to the Navier-Stokes equation is given in Section 4.

2. Singular Potentials as Combinations of Derivatives

2.1. Definitions, statements of theorems

Definition 2.1. Given two functions $V \in L^2(\mathbb{R}^n \times \mathbb{R}^+)$, $f \in L^1_{loc}(\mathbb{R}^n \times \mathbb{R}^+)$ and $\alpha > 0$, we say that

$$V = \Delta f - \alpha|\nabla f|^2 - f_t,$$

if there exists sequences of functions $\{V_i\}$ and $\{f_i\}$ such that the following conditions hold for all $i = 1, 2...$:

(i) $V_i \in L^2(\mathbb{R}^n \times \mathbb{R}^+)$, $\Delta f_i \in L^2(\mathbb{R}^n \times \mathbb{R}^+)$, $\partial_t f_i \in L^2(\mathbb{R}^n \times \mathbb{R}^+)$, $f_i \in C(\mathbb{R}^n \times \mathbb{R}^+)$.

(ii) $V_i \to V$ strongly in $L^2(\mathbb{R}^n \times \mathbb{R}^+)$, $|V_i| \le |V_{i+1}|$, $f_i \to f$ a.e. and $f_i(x, 0) = f_{i+1}(x, 0)$.

(iii) $V_i = \Delta f_i - \alpha|\nabla f_i|^2 - \partial_t f_i$.

Here we remark that we do not assume Δf, $|\nabla f|^2$ or $\partial_t f$ are in L^2 individually. This explains the lengthy appearance of the definition.

The main results of Section 2 are the next three theorems. The first one states a necessary and sufficient condition such that (1.1) possesses a positive solution. The second theorem establishes Gaussian upper and lower bound for the fundamental solutions of (1.1). The third theorem is an application of the second one in the more traditional setting of $L^{p,q}$ conditions on the potential. It will show that our conditions are genuinely much broader than the traditional ones.

It should be made clear that there is no claim on uniqueness in any of the theorems. In the absence of uniqueness how does one define the fundamental solution? This is possible due to the uniqueness of problem (1.1) when the potential V is truncated from above. This fact is proved in Proposition 2.1 below. Consequently we can state

Definition 2.2. The fundamental solution G_V is defined as the pointwise limit of the (increasing) sequence of the fundamental solution G_{V_i} where $V_i = \min\{V, i\}$ with $i = 1, 2, ...$

We remark that G_V thus defined may be infinity somewhere or everywhere. However we will show that they have better behavior or even Gaussian bounds under further conditions.

Theorem 2.1. (i) *Suppose* (1.1) *with some* $u_0 \ge 0$ *has a positive solution. Then*

$$V = \Delta f - |\nabla f|^2 - \partial_t f,$$

with $e^{-f} \in L^2(\mathbb{R}^n \times \mathbb{R}^+)$. *Moreover* $f \in L^1_{loc}(\mathbb{R}^n \times \mathbb{R}^+)$ *if* $\ln u_0 \in L^1_{loc}(\mathbb{R}^n)$.

(ii) *Suppose* $V = \Delta f - |\nabla f|^2 - \partial_t f$ *for some* f *such that* $e^{-f} \in L^2(\mathbb{R}^n \times \mathbb{R}^+)$. *Then the equation in* (1.1) *has a positive* L^2 *solution for some* $u_0 \in L^2(\mathbb{R}^n)$.

Theorem 2.2. (i) *Suppose* $V = \Delta f - \alpha|\nabla f|^2 - \partial_t f$ *for one given* $\alpha > 1$ *and* $f \in L^\infty(\mathbb{R}^n \times \mathbb{R}^+)$. *Then* G_V *has Gaussian upper bound in all space time.*

(ii) *Under the same assumption as in* (i), *if* $g_{1/4}\star|\nabla f|^2 \in L^\infty(\mathbb{R}^n \times \mathbb{R}^+)$, *then* G_V *has Gaussian lower bound in all space time.*

(iii) *Under the same assumption as in* (i), *suppose* G_V *has Gaussian lower bound in all space time. Then there exists* $b > 0$ *such that* $g_b\star|\nabla f|^2 \in L^\infty(\mathbb{R}^n \times \mathbb{R}^+)$.

Remark 2.1. At the first glance, Theorem 2.1 may seem like a restatement of existence of positive solutions without much work. However Theorem 2.2 shows that if one just puts a little more restriction on the potential V, then the fundamental solution actually has a Gaussian upper bound. Under an additional but necessary assumption, a Gaussian lower bound also holds. Even the widely studied potential $a/|x|^2$ in $\mathbb{R}^n$ can be recast in the form of Theorem 1.1, as indicated in the following

Example 2.1. For a real number b, we write $f = b \ln r$ with $r = |x|$. Then direct calculation shows, for $r \neq 0$,

$$\frac{b(n-2-b)}{r^2} = \Delta f - |\nabla f|^2.$$

Let $a = b(n-2-b)$. Then it is clear that the range of a is $(-\infty, (n-2)^2/4]$. In this interval (1.1) with $V = a/|x|^2$ permits positive solutions. This recovers the existence part in the classical result [2]. Highly singular, time dependent examples can be constructed by taking $f = \sin(\frac{1}{|x|-\sqrt{t}})$ e.g.

Moreover the corollaries below relate our class of potentials with the traditional "form bounded" or domination class (2.1) below (see also [16]). In the difficult time dependent case, Corollary 2.1 shows that potentials permitting positive solutions, can be written as the sum of one form bounded potentials and the time derivative of a function almost bounded from above by a constant.

Remark 2.2. From the proof, it will be clear that under the assumption of part (i) of Theorem 2.2, one has

$$\int G_{\alpha V}(x, t; y, s) dy \leq C < \infty.$$

This is one of the main assumptions used by Sturm [17] in the time independent case (Theorem 4.12). If $\alpha = 1$, then the conclusion of Theorem 2.2 may not hold even for time independent potentials (see [17]). Also note that this theorem provides a nearly necessary and sufficient condition on certain sign changing potential such that the corresponding heat kernel has

Gaussian upper and lower bound. The only "gap" in the condition is the difference in the parameters of the kernels $g_{1/4}$ and g_b. It is well know and easy to check if $|\nabla f| \in L^{p,q}$ with $\frac{n}{p} + \frac{2}{q} < 1$ and $f = 0$ outside a compact set, then $g_b \star |\nabla f|^2$ is a bounded function for all $b > 0$.

Corollary 2.1. *Let* $V \in L^2(\mathbb{R}^n \times (0,\infty))$. (a) *Suppose*

$$\int_0^T \int V\phi^2 \leq \int_0^T \int |\nabla\phi|^2 + b\int_0^T \int \phi^2 \tag{2.1}$$

for all smooth, compactly supported function ϕ *in* $\mathbb{R}^n \times (0,T)$ *and some* $b > 0$ *and* $T > 0$. *Then* (1.1) *has a positive solution when* $u_0 \geq 0$ *and moreover*

$$V = \Delta f - |\nabla f|^2 - \partial_t f.$$

(b) *Suppose* $V = \Delta f - |\nabla f|^2$ *then* V *is form bounded, i.e. it satisfies* (2.1).

(c) *Let* V *be a* L^2 *potential permitting positive* L^2 *solutions for* (1.1). *Then* V *can be written as the sum of one form bounded potentials and the time derivative of a function almost bounded from above by a constant.*

In the next corollary, we consider only time-independent, nonnegative potentials. Here the definition of $V = \Delta f - |\nabla f|^2$ is slightly different from that of Definition (1.1) since we do not have to worry about time derivatives. One interesting consequence is that these class of potentials is *exactly* the usual form boundedness potentials.

Corollary 2.2. *Suppose* $0 \leq V \in L^1(\mathbb{R}^n)$. *Then the following statements are equivalent.* (1) *For some* $f \in L^1_{loc}(\mathbb{R}^n)$ *and a constant* $b > 0$,

$$V = \Delta f - |\nabla f|^2 + b.$$

This mean there exist $V_j \in L^\infty$ *such that* $V_j \to V$ *in* $L^2(\mathbb{R}^1)$ *and* $V_j = \Delta f_j - |\nabla f_j|^2 + b$ *for some* $f_j \in W^{2,2}(\mathbb{R}^n)$, $j = 1, 2, \ldots$.

(2)

$$\int V\phi^2 dx \leq \int |\nabla\phi|^2 dx + b\int \phi^2 dx.$$

for all smooth, compactly supported function ϕ *in* $\mathbb{R}^n$ *and some* $b \geq 0$.

Remark 2.3. Condition (2) in the above corollary just means that the bottom of the spectrum for the operator $-\Delta - V$ is finite. This condition is the same as those given in [3,8].

It is a fact that most people feel more familiar with the case when the potential V is written as $L^{p,q}$ functions. Also there may be some inconvenience about the presence of the nonlinear term $|\nabla f|^2$ in the potential in Theorem 2.2. Therefore in our next theorem, we will use only $L^{p,q}$ conditions on f without nonlinear terms.

Theorem 2.3. *Suppose*

(a) $f \in L^\infty(\mathbb{R}^n \times \mathbb{R}^+)$;

(b) $f = 0$ *outside a cylinder* $B(0, R_0) \times [0, T_0]$, $R_0, T_0 > 0$;

(c) $|\nabla f| \in L^{p,q}(\mathbb{R}^n \times \mathbb{R}^+)$ *with* $\frac{n}{p} + \frac{2}{q} < 1$;

(d) $V = \Delta f - \partial_t f \in L^2(\mathbb{R}^n \times \mathbb{R}^+)$.

Then there exists a constant A_0 *depending only on* n, p, q *such that the following statements hold, provided that*

$$\| \, |\nabla f| \, \|_{L^{p,q}(\mathbb{R}^n \times \mathbb{R}^+)} < A_0.$$

(i) *The kernel* G_V *has Gaussian upper and lower bound in all space time.*

(ii) *The kernel* $G_{\partial_t f}$ *has Gaussian upper and lower bound in all space time.*

Remark 2.4. If V is independent of time, then Theorem 2.3. reduces to the known classical fact:

If a potential V is the derivative a of a small $L^{n+\epsilon}$ function, then G_V has Gaussian upper and lower bound (see [11] e.g.).

In the time dependent case our result is genuinely new due to the presence of the term $\partial_t f$. Let us mention that some smallness condition on the potential is needed for the existence of Gaussian bounds for G_V. This is the case even for time independent, smooth potentials due to the possible presence of ground state.

Preliminaries. In order to prove the theorems we need to prove a proposition concerning the existence, uniqueness and maximum principle for solutions of (1.1) under the assumptions that V is bounded from above by a constant. The result is standard if one assumes that the gradient of solutions are L^2. However we only assume that solutions are L^2. Therefore a little extra work is needed.

Proposition 2.1. *Suppose that* $V \in L^2(\mathbb{R}^n \times \mathbb{R}^+)$ *and that* $V \le b$ *for a positive constant* b. *Then the following conclusions hold.*

(i) *The only* L^2 *solution to the problem*

$$\Delta u + Vu - \partial_t u = 0, \ \mathbb{R}^n \times (0, T), \ T > 0, \ u(x, 0) = 0$$

is zero.

(ii) *Let* u *be a solution to the problem in* (i) *such that* $u(\cdot, t) \in L^1(\mathbb{R}^n)$, $u \in L^1(\mathbb{R}^n \times (0, T))$ *and* $Vu \in L^1(\mathbb{R}^n \times (0, T))$. *Then* u *is identically zero.*

(iii) *Under the same assumptions as in* (i), *the problem*

$$\Delta u + Vu - \partial_t u = 0, \ \mathbb{R}^n \times (0, T), \ T > 0, \ u(\cdot, 0) = u_0(\cdot) \ge 0, \ u_0 \in L^2(\mathbb{R}^n)$$

has a unique L^2 nonnegative solution.

(iv) *Under the same assumptions as in* (i), *let u be a L^2 solution to the following problem*

$$\Delta u + Vu - \partial_t u = f, \quad \mathbb{R}^n \times (0,T),\ T>0,\ u(\cdot,0)=0.$$

Here $f \le 0$ and $f \in L^1(\mathbb{R}^n \times (0,T))$. Then $u \ge 0$ in $\mathbb{R}^n \times (0,T)$.

Proof of (i). Let u be a L^2 solution to the problem in (i). Choose a standard mollifier ρ and define, for $j = 1, 2, ...,$

$$u_j(x,t) = j^n \int \rho(j(x-y))u(y,t)dy \equiv \int \rho_j(x-y)u(y,t)dy.$$

Then ∇u_j and Δu_j exist in the classical sense. From the equation on u, it holds

$$\Delta u_j + \int \rho_j(x-y)V(y,t)u(y,t)dy - \partial_t u_j = 0,$$

where $\partial_t u_j$ is understood in the weak sense.

Given $\epsilon > 0$, we define

$$h_j = \sqrt{u_j^2 + \epsilon}.$$

Let $0 \le \phi \in C_0^\infty(\mathbb{R}^n)$. Then direct calculation shows

$$\int \phi h_j(x,s)|_0^t dx = \int_0^t \int \frac{u_j \Delta u_j}{\sqrt{u_j^2 + +\epsilon}}(x,t)\phi(x)dxds$$

$$+ \int_0^t \int \phi(x) \frac{u_j(x,s)}{\sqrt{u_j^2+\epsilon}} \int \rho_j(x-y)V(y,t)u(y,t)dydxdt \equiv T_1 + T_2.$$

Using integration by parts, we deduce

$$T_1 = -\int_0^t \int \frac{|\nabla u_j|^2}{\sqrt{u_j^2 + \phi(x,s)}} dxds$$

$$+ \int_0^t \int \frac{u_j^2}{u_j^2+\epsilon} \frac{|\nabla u_j|^2}{\sqrt{u_j^2+\epsilon}} \phi(x,s)dxds - \int_0^t \int u_j \frac{\nabla u_j \nabla \phi}{\sqrt{u_j^2+\epsilon}} dxds.$$

Since the sum of the first two terms on the right-hand side of the above inequality is non-positive, we have

$$T_1 \le \int_0^t \int u_j \frac{|\nabla u_j|\ |\nabla \phi|}{\sqrt{u_j^2+\epsilon}} dxds.$$

Taking ϵ to zero, we obtain

$$\int \phi |u_j(x,t)| dx \leq \int_0^t \int |\nabla u_j| |\nabla \phi| dx ds + \int_0^t \int \phi(x) | \int \rho_j V^+ u(y,s) dy | dx ds$$
$$- \int_0^t \int \phi(x) \frac{u_j}{|u_j|}(x,s) \int \rho_j (x-y) V^- u(y,s) dy dx ds.$$

Here and later we set $\frac{u_j}{|u_j|}(x,s) = 0$, if $u_j(x,s) = 0$.

Next, since $u, V \in L^2(\mathbb{R}^n \times (0,T))$, one has $uV \in L^1(\mathbb{R}^n \times (0,T))$. From the equation

$$\Delta u + Vu - \partial_t u = 0,$$

one deduces

$$u(x,t) = \int_0^t \int G_0(x,t;y,s)(Vu)(y,s) dy ds.$$

Here, as always, G_0 is the fundamental solution of the free heat equation. Hence $u(\cdot,t) \in L^1(\mathbb{R}^n)$ and $u \in L^1(\mathbb{R}^n \times (0,T))$. Therefore, for any fixed j, there holds

$$|\nabla u_j| \in L^1(\mathbb{R}^n \times (0,T)).$$

Now, for each $R > 0$, we choose ϕ so that $\phi = 1$ in $B(0,R)$, $\phi = 0$ in $B(0,R+1)^c$ and $|\nabla \phi| \leq 2$. Observing

$$\int_0^t \int |\nabla u_j| \, |\nabla \phi| dx ds \to 0, \quad R \to \infty,$$

we deduce, by letting $R \to \infty$,

$$\begin{aligned} \int |u_j(x,t)| dx \leq & \int_0^t \int | \int \rho_j V^+ u(y,s) dy | dx ds \\ & - \int_0^t \int \frac{u_j}{|u_j|}(x,s) \int \rho_j (x-y) V^- u(y,s) dy dx ds. \end{aligned} \tag{2.2}$$

By the fact that $V^- u, V^+ u \in L^1(\mathbb{R}^n \times (0,T))$, we know that

$$\int \rho_j(\cdot - y) V^- u(y,\cdot) dy \to V^- u(\cdot,\cdot), \quad \int \rho_j(\cdot - y) V^+ u(y,\cdot) dy \to V^+ u(\cdot,\cdot),$$

in $L^1(\mathbb{R}^n \times (0,T))$. Since $\frac{u_j}{|u_j|}(x,s)$ is bounded and converges to $\frac{u}{|u|}(x,s)$ a.e. in the support of u, we have

$$\left| \int_0^t \int [\frac{u_j}{|u_j|}(x,s) \int \rho_j (x-y) V^- u(y,s) dy - \frac{V^- u^2}{|u|}] dx ds \right|$$
$$\leq \left| \int_0^t \int (\frac{u_j}{|u_j|} - \frac{u}{|u|}) V^- u dx ds \right| + \left| \int_0^t \int \frac{u_j}{|u_j|} (\rho_j \star V^- u - V^- u) dx ds \right| \to 0.$$

Substituting this to (2.2), we deduce, by taking $j \to \infty$,

$$\int |u(x,t)|dx \le \int_0^t \int V^+ |u(x,s)| dxds - \int_0^t \int \frac{V^- u^2}{|u|}(x,s)dxds.$$

Therefore

$$\int |u(x,t)|dx \le \int_0^t \int |u(x,s)|dxds \; \|V^+\|_\infty.$$

By Gronwall's inequality $u(x,t) = 0$ a.e. This proves part (i).

Proof of (ii). Notice that the only place we have used the L^2 boundedness of u is to ensure that $Vu \in L^1(\mathbb{R}^n \times (0,T))$. But this a part of the assumptions in (ii). Therefore (ii) is also proven.

Proof of (iii). The uniqueness is an immediate consequence of part (i). So we only need to prove existence. This follows from a standard limiting process. For completeness we sketch the proof.

Given $k = 1, 2, ...$ let V_k be the truncated potential

$$V_k = \sup\{V(x,t), -k\}.$$

Since V_k is a bounded function there exists a unique, nonnegative solution u_k to the following problem.

$$\begin{cases} \Delta u_k + V_k u_k - \partial_t u_k = 0, & \text{in } \mathbb{R}^n \times (0,\infty), \\ u(x,0) = u_0(x) \ge 0, \; x \in \mathbb{R}^n, \; u_0 \in L^2(\mathbb{R}^n). \end{cases}$$

By the standard maximum principle, $\{u_k\}$ is a nonincreasing sequence and

$$\frac{1}{2}\int u_k^2|_0^T dx + \int_0^T \int |\nabla u_k|^2 dxdt = \int_0^T \int v_k u_k^2 dxdt$$
$$\le \int_0^T \int V_k^+ u_k^2 dxdt \le \|V^+\|_\infty \int_0^T \int u_k^2 dxdt.$$

By Gronwall's lemma, we have

$$\int_0^T \int |\nabla u_k|^2 dxdt + \int u_k^2(x,T)dx \le \int u_0^2(x)dx + \|v^+\|_\infty \int u_0^2(x)dx e^{2\|v^+\|_\infty T}.$$

It follows that u_k converges pointwise to a function u which also satisfies the above inequality. Let ϕ be a test function with compact support. Then

$$\int (u_k\phi)|_0^T dx - \int_0^T \int u_k \phi_t dxdt - \int_0^T \int u_k \Delta\phi dxdt - \int_0^T \int V_k u_k \phi dxdt = 0.$$

Since $\{u_k\}$ is a monotone sequence and also since $|V_k u_k| \le |V| u_1 \in L^1(\mathbb{R}^n \times (0,T))$, the dominated convergence theorem implies that

$$\int (u\phi)|_0^T dx - \int_0^T \int u\phi_t dxdt - \int_0^T \int u\Delta\phi dxdt - \int_0^T \int Vu\phi dxdt = 0.$$

This shows that u is a nonnegative solution. It is clear that u is not identically zero since u_0 is not. This proves part (iii) of the proposition.

Proof of (iv).

Let V_k be a truncated potential as in part (iii). Since V_k is bounded, the standard maximum principle shows that there exists a unique, nonnegative solution to the following problem

$$\Delta u_k + V_k u_k - \partial_t u_k = f \le 0, \ \mathbb{R}^n \times (0,T), \ T > 0, \ \ u_k(\cdot, 0) = 0.$$

Moreover $\{u_k\}$ forms a decreasing sequence. Since V_k is a bounded function, the standard parabolic theory shows that

$$\int u_k(x,t)dx = \int_0^t \int V_k^+ u_k dxds - \int_0^t \int V_k^- u_k dxds + \int_0^t \int f dxds.$$

Therefore

$$\int u_k(x,t)dx \le \|V^+\|_\infty \int_0^t \int u_k dxds + \int_0^t \int f dxds.$$

This implies

$$\int u_k(x,t)dx + \int_0^t \int u_k dxds \le C(t, \|V^+\|_\infty, \|f\|_1).$$

It follows that

$$\int u_k(x,t)dx + \int_0^t \int u_k dxds + \int_0^t \int V_k^- u_k dxds \le C(t, \|V^+\|_\infty, \|f\|_1).$$

Let w be the pointwise limit of the decreasing sequence $\{u_k\}$. Then we have

$$\int w(x,t)dx + \int_0^t \int wdxds \le C(t, \|V^+\|_\infty, \|f\|_1).$$

It is straight forward to check that w is a nonnegative solution to the problem

$$\Delta w + Vw - \partial_t w = f \le 0, \ \mathbb{R}^n \times (0,T), \ T > 0, \ w(\cdot, 0) = 0.$$

Hence

$$\Delta(w-u)+V(w-u)-\partial_t(w-u)=0,\ \mathbb{R}^n\times(0,T),\ T>0,\ (w-u)(\cdot,0)=0.$$

Recall that u is assumed to be a L^2 solution and that $V\in L^2$. We have that $Vu\in L^1$ and consequently $u(\cdot,t)\in L^1(\mathbb{R}^n)$ and $u\in L^1(\mathbb{R}^n\times(0,T))$. Now by Part (ii) of the proposition, we deduce $w=u$ since w is also L^1. Hence $u\geq 0$. This finishes the proof of the proposition.

2.3. Proofs of theorems

Proof of Theorem 2.1 (i).

For $j=1,2...$, let $V_j=\min\{V(x,t),j\}$. Since V_j is bounded from above, Proposition 2.1 shows that there exists a unique solution u_j to the following problem.

$$\Delta u_j+V_ju_j-\partial_tu_j=0,\ (x,t)\in\mathbb{R}^n\times(0,\infty),\ u_j(x,0)=u_0(x),\ x\in\mathbb{R}^n.$$

Notice that u_j-u_{j-1} is a solution to the problem

$$\begin{gathered}\Delta(u_j-u_{j-1})+V_j(u_j-u_{j-1})-\partial_t(u_j-u_{j-1})=(V_{j-1}-V_j)u_{j-1},\\ (x,t)\in\mathbb{R}^n\times(0,\infty),\\ (u_j-u_{j-1})(x,0)=0,\ x\in\mathbb{R}^n.\end{gathered}$$

Notice that

$$(V_{j-1}-V_j)u_{j-1}\leq 0,\ (V_{j-1}-V_j)u_{j-1}\in L^1(\mathbb{R}^n\times(0,T)),\ T>0.$$

We can then apply Proposition 2.1 (iv) to conclude that

$$u_j\geq u_{j-1}.$$

Moreover

$$\begin{aligned}&\Delta(u-u_j)+V_j(u-u_j)-\partial_t(u-u_j)=(V_j-V)u,\ , &&(x,t)\in\mathbb{R}^n\times(0,\infty)\\ &(u-u_j)(x,0)=0, &&x\in\mathbb{R}^n,\end{aligned}$$

with

$$(V_j-V)u\leq 0,\ (V_j-V)u\in L^1(\mathbb{R}^n\times(0,T)),\ T>0.$$

By Proposition 2.1 (iv) again we know that $u\geq u_j$. Therefore $\{u_j\}$ is a non-decreasing sequence of nonnegative functions bounded from above by a L^2 function. Let w be the pointwise limit of u_j. The w is L^2 and $|V_ju_j|\leq|Vu|\in L^1(\mathbb{R}^n\times(0,T))$, $T>0$. By the dominated convergence theorem, it is straight forward to check that w is a nonnegative L^2 solution to the equation

$$\Delta w + Vw - \partial_t w = 0, \quad (x,t) \in \mathbb{R}^n \times (0,\infty),$$
$$w(x,0) = u_0(x), \quad x \in \mathbb{R}^n.$$

Fixing j, for any $k = 1, 2, ...$ Let $V_{jk} = \max\{V_j(x,t), -k\}$. Since V_{jk} is bounded, the following problem has a unique L^2 solution.

$$\Delta u_{jk} + V_{jk} u_{jk} - \partial_t u_{jk} = 0, \quad (x,t) \in \mathbb{R}^n \times (0,\infty),$$
$$u_j(x,0) = u_0(x), \quad x \in \mathbb{R}^n.$$

Due to the fact that $\{V_{jk}\}$ is a decreasing sequence of k, the maximum principle shows that $\{u_{jk}\}$ is also a decreasing sequence of k. Since $V_{jk}u_{jk} \in L^2(\mathbb{R}^n \times (0,T))$, $T > 0$, the parabolic version of the Calderon-Zygmond theory shows

$$\Delta u_{jk}, \quad \partial_t u_{jk} \in L^2(\mathbb{R}^n \times (0,T)), \ T > 0.$$

Since

$$0 \le u_{jk} - u_j \le u_{j1} - u_j, \ 0 \le w - u_j \le w - u_1, \ k = 1, 2, 3, ...,$$

we can apply the dominated convergence theorem to conclude that

$$\lim_{k\to\infty} \int_0^T \int (u_{jk} - u_j)^2 dx dt = 0, \ \lim_{j\to\infty} \int_0^T \int (w - u_j)^2 dx dt = 0.$$

Therefore we can extract a subsequence $\{u_{jk_j}\}$ such that

$$\lim_{j\to\infty} \int_0^T \int (w - u_{jk_j})^2 dx dt = 0.$$

Hence there exists a subsequence, still called $\{u_{jk_j}\}$ such that $u_{jk_j} \to w$ *a.e.*

Recall that $u_0 \ge 0$, $u_0 \ne 0$ and V_{jk_j} is bounded. It is clear that $u_{jk_j} > 0$ when $t > 0$. Now we define

$$f_j = -\ln u_{jk_j}, \quad f = -\log w.$$

Then

$$V_{jk_j} = \Delta f_j - |\nabla f_j|^2 - \partial_t f_j.$$

Clearly $f_j \to f$ a.e. and $V_{jk_j} \to V$ in L^2 as $j \to \infty$. By Definition 2.1, this means

$$V = \Delta f - |\nabla f|^2 - \partial_t f.$$

It is clear that $e^{-f}b = w$ is L^2 by construction.

Proof of Theorem 2.1 (ii).

By assumption, there exist sequences of functions $\{V_j\}$ and $\{f_j\}$ such that

$$\|V_j\|_{L^2} \le C, \ |V_j| \le |V_{j+1}|, \ f_j \in L^\infty,$$

$$\|V_j - V\|_{L^2} \to \infty, \ V_j = \Delta f_j - |\nabla f_j|^2 - \partial_t f_j.$$

Then for $u_j \ e^{-f_j}$, we have

$$\Delta u_j + V_j u_j - \partial_t u_j = 0.$$

We will show that $\|u_j\|_{L^2}$ is uniformly bounded. To this end, we observe that

$$\Delta(u_j - u_{j+1}) + V_j(u_j - u_{j+1}) - \partial_t(u_j - u_{j+1}) = -(V_j - V_{j+1})u_{j+1} \ge 0.$$

Recall from Definition 2.1 that $f_j(x,0)$ is independent of j. Hence $u_j(x,0) = u_{j+1}(x,0)$. Therefore $0 \le u_j \le u_{j+1}$. By the assumption that $f_j \to f$ a.e., we know that $u_j = e^{-f_j} \to e^{-f}$ a.e. Note that $e^{-f} \in L^2(\mathbb{R}^n \times (0,\infty))$. Hence $\|u_j\|_{L^2}$ is uniformly bounded.

By weak compactness in L^2, there exists a subsequence, still called $\{u_j\}$ such that u_j converges weakly to a L^2 function which we will call u. Observe that, for any compactly supported test function ϕ, there holds

$$\begin{aligned}\|u_j V_j \phi - uV\phi\|_{L^1} &\le \|u_j (V_j - V)\phi\|_{L^1} + \|(u_j - u)V\phi\|_{L^1}\\ &\le \|u_j\|_{L^2} \|V_j - V\|_{L^2} \ \|\phi\|_{L^\infty} + \|(u_j - u)V\phi\|_{L^1}.\end{aligned}$$

Hence

$$\|u_j V_j \phi - uV\phi\|_{L^1} \to 0,$$

when $j \to \infty$. From here it is easy to check that u is a nonnegative solution to (1.1) with $u_0 = e^{-f_j(x,0)}$ as the initial value. Note the u_0 is independent of j. If $u_0 \in L^2(\mathbb{R}^n)$, then we are done. Otherwise, we can selection a L^2 function dominated by u_0 to serve as the initial value.

Next we will provide a

Proof of Theorem 2.2 (i).

We will use an idea based on an argument in [17] where the heat equation with some singular, time independent potentials are studied.

By virtue of Proposition 2.1, the fundamental solution G_V is defined as the limit of fundamental solutions of the equation in (1.1), where V is replaced by nonsingular potentials. Therefore we can and will assume that

V is smooth in this subsection. The constants involved will be independent of the smoothness.

Since, by assumption

$$V = \Delta f - \alpha|\nabla f|^2 - \partial_t f, \tag{2.3}$$

one has

$$\alpha V = \Delta(\alpha f) - |\nabla(\alpha f)|^2 - \partial_t(\alpha f).$$

Writing $F = e^{-\alpha f}$, it is easy to show that

$$\Delta F + \alpha V F - \partial_t F = 0. \tag{2.4}$$

Let us denote the fundamental solution of the equation in (2.4) by $G_{\alpha V}$. Since f is bounded, we know that F is bounded between two positive constants. Therefore it is clear that

$$0 < \frac{\inf F}{\sup F} \le \int G_{\alpha V}(x, t; y, s) dy \le \frac{\sup F}{\inf F}, \tag{2.5}$$

for all $x \in \mathbb{R}^n$ and $t > s$. Here $\inf F$ and $\sup F$ are taken over the whole domain of F.

By Feynman-Kac formula and Hölder's inequality, for a given $\phi \in C_0^\infty(\mathbb{R}^n)$, there holds

$$\begin{aligned} &|\int G_V(x, t; y, s)\phi(y)dy| \\ \le &\Big[\int G_{\alpha V}(x, t; y, s)dy\Big]^{1/\alpha}\Big[\int G_0(x, t; y, s)|\phi(y)|^{\alpha/(\alpha-1)}dy\Big]^{(\alpha-1)/\alpha}. \end{aligned}$$

By (2.5), we deduce

$$|\int G_V(x, t; y, s)\phi(y)dy| \le \frac{c s_0^{1/\alpha}}{(t-s)^{(\alpha-1)n/(2\alpha)}} \|\phi\|_{\alpha/(\alpha-1)},$$

where $s_0 = \frac{\sup F}{inf F}$. The norm on ϕ means the $L^{\alpha/(\alpha-1)}(\mathbb{R}^n)$ norm. Hence

$$\|G_V(\cdot, t; \cdot, s)\|_{\alpha/(\alpha-1),\infty} \le \frac{c s_0^{1/\alpha}}{(t-s)^{(\alpha-1)n/(2\alpha)}}. \tag{2.6}$$

Here and later the norm $\|\cdot\|_{p,q}$ stands for the operator norm from $L^p(\mathbb{R}^n)$ to $L^q(\mathbb{R}^n)$ for p, q between 1 and ∞.

Without loss of generality we assume that $\alpha/(\alpha-1)$ is an integer. This is so because otherwise we can choose one $\alpha_1 \in (1, \alpha)$ such that $\alpha_1/(\alpha_1 - 1)$

is an integer. Then interpolating between $G_{\alpha V}$ and G_0 by Feynman-Kac formula again, we know that

$$\int G_{\alpha_1 V}(x,t;y,s)dy \le C(F,\alpha,\alpha_1).$$

Then we can just work with $G_{\alpha_1 V}$ instead of $G_{\alpha V}$ in the above.

Using the reproducing property of G_V we deduce

$$\|G_V(\cdot,t;\cdot,s)\|_{1,\infty} \le \prod_{j=1}^{m} \|G_V(\cdot,s+t_j;\ \cdot,t_{j-1})\|_{p_j,q_j}, \tag{2.7}$$

where

$$m = \frac{\alpha}{\alpha-1},\ p_j = \frac{m}{m-l+1},\ q_j = \frac{m}{m-l},\ t_j = s + \frac{(t-s)j}{m}.$$

For each j between 1 and m, we apply the Riesz-Thorin interpolation theorem to deduce

$$\begin{aligned} &\|G_V(\cdot,t_j;\ \cdot,t_{j-1})\|_{p_j,q_j} \\ \le\ &\|G_V(\cdot,s+t_j;\ \cdot,t_{j-1})\|_{1,m/(m-1)}^{1-\lambda_j}\ \|G_V(\cdot,t_j;\ \cdot,t_{j-1})\|_{m,\infty}^{\lambda_j}. \end{aligned}$$

Here the parameters are determined by the following relations

$$\frac{1}{p_j} = \frac{1-\lambda_j}{1} + \frac{\lambda_j}{m},\ \frac{1}{q_j} = \frac{1-\lambda_j}{m/(m-1)} + \frac{\lambda_j}{\infty},\ \lambda_j = \frac{j-1}{m-1}.$$

It follows that

$$\|G_V(\cdot,t_j;\ \cdot,t_{j-1})\|_{p_j,q_j} \le \|G_V(\cdot,t_j;\ \cdot,t_{j-1})\|_{m,\infty}. \tag{2.8}$$

Substituting (2.6) to (2.8), we deduce, after noticing that $t_j - t_{j-1} = (t-s)/m$,

$$\|G_V(\cdot,t_j;\ \cdot,t_{j-1})\|_{p_j,q_j} \le \frac{c s_0^{1/\alpha} m^{(\alpha-1)n/(2\alpha)}}{(t-s)^{(\alpha-1)n/(2\alpha)}}.$$

This and (2.7) imply that

$$\|G_V(\cdot,t;\cdot,s)\|_{1,\infty} \le \frac{c s_0^{m/\alpha} m^{n/2}}{(t-s)^{n/2}}.$$

Here we just used the relation $m = \alpha/(\alpha-1)$. This yields the on-diagonal upper bound

$$G_V(x,t;y,s) \le \frac{c s_0^{1/(\alpha-1)} (\alpha/(\alpha-1))^{n/2}}{(t-s)^{n/2}}. \tag{2.9}$$

In order to obtain the full Gaussian bound, we observe that, for any $p > 1$, the Feynman-Kac formula implies

$$G_V(x,t;y,s) \le \left[G_{pV}(x,t;y,s)\right]^{1/p} \left[G_0(x,t;y,s)\right]^{(p-1)/p}. \tag{2.10}$$

Notice also

$$pV = p(\Delta f - \alpha|\nabla f|^2 - \partial_t f) = \Delta(pf) - \frac{\alpha}{p}|\nabla(pf)|^2 - \partial_t(pf).$$

Taking $p = (1+\alpha)/2$, then $\frac{\alpha}{p} > 1$. Therefore pV also satisfies the condition of Theorem 2.2 (i). Hence, the on-diagonal bound (2.9) holds for G_{pV}, i.e., there exists a constant $C(\alpha, e^{\sup f - \inf f})$ such that

$$G_{pV}(x,t;y,s) \le \frac{C(\alpha, e^{\sup f - \inf f})}{(t-s)^{n/2}}.$$

Substituting this to the inequality (2.10), we obtain the desired Gaussian upper bound for G_V.

Proof of Theorem 2.2 (ii).

In this part we prove the Gaussian lower bound. We will follow Nash's original idea. The novelty is a way of handling the potential term even, if it is very singular. The main idea is to exploit the structure of the potential when it is written as a combination of derivatives.

Since the setting of our problem is invariant under the scaling, for $r > 0$,

$$V_r(x,t) = r^2 V(rx, r^2 t), \ f_r(x,t) = f(rx, r^2 t), \ u_r(x,t) = r^2 u(rx, r^2 t),$$

we can just prove the lower bound for $t = 1$ and $s = 0$. We divide the proof into three steps.

Step 1. Fixing $x \in \mathbb{R}^n$, let us set

$$u(y,s) = G_V(y,s;x,0), \ H(s) = \int e^{-\pi|y|^2} \ln u(y,s) dy.$$

Differentiating $H(s)$, one obtains

$$H'(s) = \int e^{-\pi|y|^2} \frac{\partial_s u(y,s)}{u(y,s)} dy = -\int \nabla\left(\frac{e^{-\pi|y|^2}}{u}\right) \nabla u dy + \int e^{-\pi|y|^2} V(y,s) dy.$$

Estimating the first term on the righthand side of the above inequality as in [6], Section 2, one arrives at

$$H'(s) \ge -C + \frac{1}{2} \int e^{-\pi|y|^2} |\nabla \ln u(y,s)|^2 dy + \int e^{-\pi|y|^2} V(y,s) dy. \tag{2.11}$$

Here C is a positive constant. Since,

$$V = \Delta f - \alpha|\nabla f|^2 - \partial_t f,$$

we know that

$$f(x,t) = \int G_0(x,t;y,0)f(y,0)$$
$$-\int_0^t\int G_0(x,t;y,s)V(y,s)dyds - \alpha\int_0^t\int G_0(x,t;y,s)|\nabla f|^2 dyds.$$

By our assumption

$$\int_0^t\int G_0(x,t;y,s)|\nabla f|^2 dyds < \infty.$$

Hence the boundedness of f implies that

$$[m(x,t) \equiv -\int_0^t\int G_0(x,t;y,s)V(y,s)dyds \in L^\infty.$$

Moreover

$$\Delta m - \partial_t m = V. \tag{2.12}$$

Therefore

$$\int e^{-\pi|y|^2}V(y,s)dy = \int e^{-\pi|y|^2}[\Delta m - \partial_s m](y,s)dy$$
$$= \int[\Delta e^{-\pi|y|^2}]mdy - \partial_s\int e^{-\pi|y|^2}m(y,s)dy \geq -C - \partial_s\int e^{-\pi|y|^2}m(y,s)dy.$$

Here we have used the boundedness of m. Substituting the above to the right-hand side of (2.11), we obtain

$$H'(s) \geq -C + \frac{1}{2}\int e^{-\pi|y|^2}|\nabla \ln u(y,s)|^2 dy - M'(s), \tag{2.13}$$

where

$$M(s) = \int e^{-\pi|y|^2}m(y,s)dy. \tag{2.14}$$

Step 2. By Poincaré's inequality with $e^{-\pi|y|^2}$ as weight, we deduce, for some $B > 0$,

$$H'(s) \geq -C + B\int e^{-\pi|y|^2}[\ln u(y,s) - H(s)]^2 dy - M'(s).$$

Next, observe that $(\ln u - H(s))^2/u$ is non-increasing as a function of u when u is between $e^{2+H(s)}$ and ∞. Also from the Gaussian upper bound,

$$\sup_{1/2 \le s \le 1} u(y,s) \le K < \infty.$$

Therefore

$$H'(s) \ge -C + CBK^{-1}(\ln K - H(s))^2 \int_{u(y,s)\ge \exp(2+H(s))} e^{-\pi|y|^2} u(y,s)dy - M'(s). \tag{2.15}$$

Using the Gaussian upper bound again, we know that $H(s) \le C$ for some $C > 0$ and that

$$\begin{aligned}\int_{u(y,s)\ge exp(2+H(s))} e^{-\pi|y|^2} u(y,s)dy &\ge \int e^{-\pi|y|^2} u(y,s)dy - ce^{2+H(s)} \\ &\ge e^{-\pi r^2} \int_{|y|<r} u(y,s)dy - ce^{2+H(s)} \\ &= e^{-\pi r^2} \Big[\int u(y,s)dy - \int_{|y|<r} u(y,s)dy \Big] - ce^{2+H(s)}.\end{aligned} \tag{2.16}$$

We aim to find a lower bound for the right-hand side of (2.16). By (2.12),

$$V = \Delta m - \partial_t m \ge \Delta m - |\nabla m|^2 - \partial_t m \equiv V_1.$$

Write $h = e^{-m}$. Then

$$\Delta h + V_1 h - \partial_t u = 0.$$

Since $m \in L^\infty$, we know that h is bounded between two positive constants. Observe that

$$h(x,t) = \int G_{V_1}(x,t;y,s)h(y,s)dy.$$

Hence

$$0 < c_1 < \int G_{V_1}(x,t;y,s)dy \le c_2.$$

By the maximum principle, we have

$$\int G_V(x,t;y,s)dy \ge \int G_{V_1}(x,t;y,s)dy \ge c_1 > 0. \tag{2.17}$$

Recall that $u(y,s) = G_V(y,s;x,0)$. Substituting (2.17) to (2.16) and applying the Gaussian upper bound on $u(y,s)$, we deduce

$$\int_{u(y,s)\ge exp(2+H(s))} e^{-\pi|y|^2} u(y,s)dy \ge ce^{-\pi r^2} c_1 - ce^{2+H(s)}, \tag{2.18}$$

when r is sufficiently large. Substituting (2.18) to (2.15), we arrive at

$$H'(s) \geq -C + CBK^{-1}(\ln K - H(s))^2 e^{-\pi r^2} c_1 - ce^{2+H(s)} - M'(s). \quad (2.19)$$

We claim that $H(1) \geq -c_0$ for some sufficiently large $c_0 > 0$. Suppose otherwise, i.e. $H(1) < -c_0$. From (2.19), for some $C > 0$,

$$H'(s) \geq -C - M'(s).$$

Hence

$$H(1) - H(s) \geq -C(1-s) - (M(1) - M(s)).$$

Therefore

$$H(s) \leq H(1) + C(1-s) + (M(1) - M(s)) \leq -c_0/2,$$

when c_0 is chosen sufficiently large. It follows from (2.19) that

$$H'(s) \geq -c_1 + c_2 H^2(s) - M'(s).$$

This shows

$$(H(s) + M(s))' \geq -c_3 + c_4 (H(s) + M(s))^2.$$

From here, one immediately deduces

$$H(1) \geq -A, \quad A > 0.$$

The claim is proven. Thus

$$\int e^{-\pi|y|^2} \ln G_V(y, s; x, 0) dy \geq -c_0,$$

where $|x - y| \leq 1$.

Using the reproducing property of G_V and Jensen's inequality, we have, when $|x - y| \leq 1$,

$$\ln G_V(x, 2; y, 0) = \ln \int G_V(x, 2; z, 1) G_V(z, 1; y, 0) dz$$

$$\geq \int e^{-\pi|y|^2} \ln G_V(x, 2; z, 1) dz + \int e^{-\pi|y|^2} \ln G_V(z, 1; y, 0) dz \geq -C.$$

This proves the on-diagonal lower bound. The full Gaussian lower bound now follows from the standard argument in [6].

Proof of Theorem 2.2 (iii).

Since,

$$V = \Delta f - \alpha|\nabla f|^2 - \partial_t f,$$

we have

$$V + (\alpha - 1)|\nabla f|^2 = \Delta f - |\nabla f|^2 - \partial_t f.$$

Let $u = e^{-f}$, by direct calculation,

$$\Delta u + Vu - \partial_t u + (\alpha - 1)|\nabla f|^2 u = 0.$$

Hence

$$u(x,t) = \int G_V(x,t;y,0)u(x,0)dy$$

$$+(\alpha - 1)\int_0^t \int G_V(x,t;y,s)|\nabla f|^2(y,s)u(y,s)dyds.$$

Since $f \in L^\infty$, we know that u is bounded between two positive constants. If, by assumption, G_V has a Gaussian lower bound, then, for some $b > 0$, we have

$$\int_0^t \int g_b(x,t;y,s)|\nabla f|^2(y,s)dyds \leq C\frac{\sup u}{\inf u}.$$

This completes the proof of part (iii) of Theorem 2.2.

Proof of Theorem 2.3.

(i) We write

$$V_1 = 2(\Delta f - \partial_t f - 3|\nabla f|^2), \quad V_2 = 6|\nabla f|^2.$$

Let G_{V_i} $(i = 1, 2)$, be the fundamental solution of $\Delta u + V_i u - \partial_t u = 0$. Since

$$V = \Delta f - \partial_t f = (V_1/2) + (V_2/2),$$

the Feynman-Kac formula implies

$$G_V(x,t;y,s) \leq \left[G_{V_1}(x,t;y,s)\right]^{1/2} \left[G_{V_2}(x,t;y,s)\right]^{1/2}.$$

Observe that

$$V_1 = \Delta(2f) - \partial_t(2f) - \frac{3}{2}|\nabla(2f)|^2.$$

Hence, by Theorem 2.2, we know that G_{V_1} has Gaussian upper bound. Under the smallness assumption on the $L^{p,q}$ norm of $|\nabla f|^2$ in the theorem, it is well known that G_{V_2} also has a Gaussian upper bound. Therefore G_V has Gaussian upper bound.

In order to prove the Gaussian lower bound, we observe that

$$V = \Delta f - \partial_t f \geq V_3 \equiv \Delta f - \partial_t f - 2|\nabla f|^2.$$

Under our assumption on the $L^{p,q}$ norm of $|\nabla f|$, it is straight forward to check that

$$g_{1/4} \star |\nabla f|^2 \in L^\infty.$$

Hence Theorem 2.2 (ii) shows G_{V_3} has Gaussian lower bound. Clearly this Gaussian lower bound of G_{V_3} is also a Gaussian lower bound of G_V by the maximum principle. This proves part (a).

(ii) Clearly we can choose A_0 sufficiently small so that all the following kernels have global Gaussian upper and lower bound:

$$G_{2V}, \quad G_{V/2}, \quad G_{2\Delta f}, \quad G_{\Delta f/2}. \tag{2.20}$$

The bounds on the first two kernels follow from part (i). The bounds on the last two kernels follow from standard theory since $\Delta f = \mathrm{div}(\nabla f)$ with ∇f has a small norm in the suitable $L^{p,q}$ class (see [11] e.g.).

Now observe that

$$-\partial_t f = \Delta f - \partial_t f - \Delta f = V - \Delta f,$$

$$\frac{V}{2} = \frac{\Delta f}{2} - \frac{\partial_t f}{2}.$$

By Feynman-Kac formula

$$G_{(-\partial_t f)} \leq \big(G_{2V}\big)^{1/2}\big(G_{2\Delta f}\big)^{1/2};$$

$$G_{V/2} \leq \big(G_{2\Delta f}\big)^{1/2}\big(G_{-\partial_t f}\big)^{1/2}.$$

Hence (2.20) show that $G_{(-\partial_t f)}$ also has global Gaussian upper and lower bound. Since the setting of the Theorem is invariant under the reflection $f \to -f$ the result follows.

We close this section by giving proofs of the corollaries.

Proof of Corollary 2.1.

(a) Let $V_k = \min\{V(x,t), k\}$, $k = 1, 2, \ldots$. Then (1.1) with V replaced by V_k has a unique solution.

Let $J(t) \equiv \int_{\mathbb{R}^n} u_k^2(x,t)dx$. Then

$$J'(t) = 2\int_{\mathbb{R}^n} [-\nabla u_k \nabla u_k + V_k u_k^2]dx.$$

By our assumption on V, $J'(t) \leq 2bJ(t)$ which implies

$$\int_{\mathbb{R}^n} u_k^2(x,t)dx \leq \int_{\mathbb{R}^n} u_0^2(x)dx \; e^{2bt}.$$

Therefore if $u_0 \in L^2(D)$, we conclude that $u_k(x,t)$ increases to a finite positive limit $u(x,t)$ as $k \to \infty$, for all t and for a.e. x. Moreover $u(\cdot,t) \in L^2(\mathbb{R}^n)$. We show that the above u is a positive L^2 solution to (1.1).

Since u_k is a solution to (1.1) with V replaced by V_k, for any $\psi \in C_0^\infty(\mathbb{R}^n \times (0,T))$, we have

$$\int (u_k\psi)|_{t_1}^{t_2} dx - \int_{t_1}^{t_2}\int u_k\psi_t dxdt - \int_{t_1}^{t_2}\int u_k \Delta\psi dxdt - \int_{t_1}^{t_2}\int V_k u_k \psi dxdt = 0,$$

for all $t_1, t_2 \in (\delta, t_0)$.

By our assumption $|V_k u_k| \le |Vu| \in L^1(\mathbb{R}^n \times (0,T))$. Taking $k \to \infty$ and using the dominated convergence theorem, we obtain

$$\int (u\psi)|_{t_1}^{t_2} dx - \int_{t_1}^{t_2}\int u\psi_t dxdt - \int_{t_1}^{t_2}\int u \Delta\psi dxdt - \int_{t_1}^{t_2}\int V u \psi dxdt = 0.$$

This shows that u is a positive solution to (1.1). By Theorem 2.1

$$V = \Delta f - |\nabla f|^2 - \partial_t f.$$

(b) Suppose

$$V = \Delta f - |\nabla f|^2.$$

Due to the L^2 convergence, it suffices to prove that V_j in Definition 2.1 satisfies (2.). Let ϕ be a test function, then

$$\begin{aligned}
&\int_0^\infty \int V_j \phi^2 dxdt = \int_0^\infty \int [\Delta f_j - |\nabla f_j|^2]\phi^2 dxdt \\
&= -2\int_0^\infty \int \nabla f_j \nabla\phi\phi dxdt - \int_0^\infty \int |\nabla f_j|^2 \phi^2 dxdt \\
&\le \int_0^\infty \int |\nabla \phi|^2 dxdt\phi.
\end{aligned}$$

(c) The statement is self-evident by part (b) and Theorem 2.1.

Proof of Corollary 2.2.

Suppose $V = \Delta f - |\nabla f|^2 + b$. Then, by the same limiting argument as above, we have

$$\begin{aligned}
&\int V\phi^2 dx = \lim_{j\to\infty} \int [\Delta f_j - |\nabla f_j|^2]\phi^2 dx + b\int \phi^2 dx \\
&= \lim_{j\to\infty}\Big(-2\textstyle\int \nabla f_j \nabla \phi \ \phi dx - \int |\nabla f_j|^2 \phi^2 dx\Big) + b \int \phi^2 dx.
\end{aligned}$$

Therefore

$$\int V\phi^2 dx \le \int |\nabla \phi|^2 dx + b\int \phi^2 dx.$$

Also by part (a) of Corollary 1, (1.1) has a positive solution when $u_0 \geq 0$.

On the other hand, suppose V satisfies

$$\int V\phi^2 dx \leq \int |\nabla \phi|^2 dx + b \int \phi^2 dx.$$

Write $V_j = \min\{V, j\}$ with $j = 1, 2, \ldots.$ Then

$$\int (V_j - b)\phi^2 dx \leq \int |\nabla \phi|^2 dx.$$

Notice that $V_j - b$ is a bounded function. Hence we can apply Theorem C.8.1 in [16] to conclude that there exists $u_j > 0$ such that

$$\Delta u_j + (V_j - b)u_j = 0.$$

Writing $f_j = -\ln u_j$, we have

$$V_j = \Delta f_j - |\nabla f_j|^2 + b.$$

By definition, this means

$$V = \Delta f - |\nabla f|^2 + b.$$

3. Heat Bounded Functions and the Heat Equation

Here we introduce another class of singular functions that has its origin in the Kato type class. As mentioned in the introduction, a function is in a Kato type class if the convolution of the absolute value of the function and the fundamental solution of Laplace or the heat equation is bounded. Here we generalize this notion by a simple but key stroke, i.e., we delete the absolute value sign on the function in the definition of the Kato class. More precisely, we have

Definition 3.1. Let $f = f(x,t)$ be a local L^1 function in space time and G_0 be the standard Gaussian in $\mathbb{R}^n$. We say that f is *heat bounded* in a domain $\Omega \subset \mathbb{R}^n \times \mathbb{R}^1$, if

$$G_0 \star f(x,t) \equiv \int_0^t \int_{\mathbb{R}^n} G_0(x,t;y,s) f(y,s) dy ds$$

is a bounded function in Ω.

We say that f is *almost heat bounded* in a domain $\Omega \subset \mathbb{R}^n \times \mathbb{R}^1$ if

$$G_0 \star f(x,t) \in L^p(\Omega)$$

for all $p > 1$.

Example 3.1. The function $V(x) = a\frac{\chi_{B(0,1)}}{|x|^2}$ is not heat bounded but is almost heat bounded in $\mathbb{R}^n$. Here a is a nonzero constant.

In the next two propositions, we provide a comparison between the heat bounded class and more familiar classes of functions.

Proposition 3.1. *Suppose, in the distribution sense, $V = \partial^2_{ij} f$ with $f \in \cap_{p>1} L^p(\mathbb{R}^n \times (0,T))$. Then V is almost heat bounded in $\mathbb{R}^n \times (0,T)$.*

Proof. Let G_0 be the free heat kernel in $\mathbb{R}^n \times (0,\infty)$. By the assumption on f, the function $u = u(x,t)$, defined by

$$u(x,t) = \int_0^t \int_{\mathbb{R}^n} G_0(x,t;y) f(y,s) dyds$$

is a solution to the equation

$$\begin{cases} \Delta u(x,t) - u_t(x,t) = -f(x,t), \ x \in \mathbb{R}^n, \ t > 0, \\ u(x,0) = 0. \end{cases}$$

By the parabolic version of the Calderon-Zygmond inequality (see [13] e.g.), we know that

$$u \in W^{2,p}(\mathbb{R}^n), \quad \forall p > 1.$$

Hence

$$\partial^2_{ij} \int_0^t \int_{\mathbb{R}^n} G_0(x,t;y) f(y,s) dyds \in L^p(\mathbb{R}^n), \quad \forall p > 1.$$

Proposition 3.2. *Suppose, $0 \le V \in L^1_{loc}$ is form bounded in $D \times [0,T]$. i.e.*

$$\int_0^T \int_D V\phi^2 \le b_1 \int_0^T \int_D |\nabla \phi|^2 + b_2 \int_0^T \int_D \phi^2$$

for all smooth, compactly supported function $\phi \in D \times [0,T] \subset \mathbb{R}^n \times [0,T]$. Then V is almost heat bounded in $D \times [0,T]$.

Proof. We will only consider the case when $D = \mathbb{R}^n$. The other cases follow from the full space case by a standard comparison method.

Since one can consider cV with c sufficiently small otherwise, we can choose the constant b_1 in the definition of form boundedness to be $1/2$, i.e. we assume that

$$\int_0^T \int_D V\phi^2 \le \frac{1}{2} \int_0^T \int_D |\nabla \phi|^2 + b_2 \int_0^T \int_D \phi^2,$$

for all smooth, compactly supported function ϕ.

Let u_k be the solution of

$$\begin{aligned} &\Delta u_k + V_k u_k - (u_k)_t = 0, \text{ in } \mathbb{R}^n \times (0,\infty),\ V \in L^2_{loc}(\mathbb{R}^n \times (0,\infty)) \\ &\qquad u_k(x,0) = u_0(x) > 0,\ x \in \mathbb{R}^n,\ u_0 \in L^2(\mathbb{R}^n). \end{aligned} \tag{3.1}$$

Here V_k is the truncated potential $V_k = \min\{V, k\}$ with k being positive integers. Clearly $V_k \le V_{k+1}$.

We show that u_k converge pointwise to a locally integrable function.

Let $J(t) \equiv \int_D u_k^2(x,t)dx$. Then

$$J'(t) = 2\int_D [-\nabla u_k \nabla u_k + V_k u_k^2]dx.$$

By our assumption on V, $J'(t) \le 2bJ(t)$ which implies

$$\int_D u_k^2(x,t)dx \le \int_D u_0^2(x)dx\ e^{2bt}.$$

Therefore if $u_0 \in L^2(D)$, we conclude that $u_k(x,t)$ increases to a finite positive limit $u(x,t)$ as $k \to \infty$, for all t and for a.e. x. Moreover $u(\cdot,t) \in L^2(\mathbb{R}^n)$.

Write $w_k = \log u_k$. From (3.1), one deduces

$$\Delta w_k + |\nabla w_k|^2 + V_k - (w_k)_t = 0.$$

Therefore

$$w_k(x,t) = \int_D G_0(x,t;y,0)w_k(x,0)dy$$

$$+\int_0^t\int_D G_0(x,t;y,s)|\nabla w_k(x,s)|^2 dyds + \int_0^t\int_D G_0(x,t;y,s)V_k(y,s)dyds.$$

Therefore

$$\int_0^t\int_D G_0(x,t;y,s)V_k(y,s)dyds \le w_k(x,t) - \int_D G_0(x,t;y,0)w_k(x,0)dy.$$

By the monotone convergence theorem

$$\begin{aligned} \int_0^t\int_D G_0(x,t;y,s)V(y,s)dyds &\le w(x,t) - \int_D G_0(x,t;y,0)w(y,0)dy \\ &\le \log(1+u(x,t)) - \int_D G_0(x,t;y,0)\log u_0(y)dy. \end{aligned}$$

Now we take

$$u_0(x) = \frac{1}{1+|x|^n}.$$

Then, since $u_0 \in L^2(\mathbb{R}^n)$, we have

$$u(\cdot,t) \in L^2(\mathbb{R}^n).$$

By Jensen's inequality,

$$\log(1+u(\cdot,t)) \in L^p(\mathbb{R}^n), \ \forall p > 1.$$

It is also clear that

$$\int_D G_0(\cdot,t;y,0)\log(1+|y|^n)dy \in L^p_{loc}(\mathbb{R}^n), \ \forall p > 1.$$

The result follows.

4. Applications to the Navier-Stokes Equation

In this section, we establish a new a priori estimate for a certain quantity involving the velocity and vorticity of the 3 dimensional Navier-Stokes equation.

$$\begin{aligned} &u_t - \Delta u(x,t) + u\cdot\nabla u(x,t) + \nabla p = 0, \\ &\nabla\cdot u = 0, \ u(x,0) = u_0(x), \end{aligned} \tag{4.1}$$

for $(x,t) \in \mathbb{R}^3 \times (0,\infty)$, where Δ is the standard Laplacian, a vector field u represents the velocity of the fluid, and a scalar field p the pressure. (The viscosity is normalized, $\nu = 1$.)

There has been an extensive and rapidly growing literature on the equation, which is impossible to quote extensively here. Let us just mention that weak solutions are known to exist due to the seminal work of Leray (see [10]). However it is not known if the weak solution is smooth everywhere. Several sufficient conditions implying smoothness of weak solutions have been made. See for example [14,15]. In these two papers, it was shown that if the velocity u is in $L^{p,q}$ class with $\frac{3}{p}+\frac{2}{q} < 1$, then u is actually smooth. For more sufficiency results in various other spaces we refer the reader to the more recent survey paper [4]. However it is only known that $u \in L^{10/3,10/3}$. Therefore there is a gap in between the a priori estimate and the sufficiency condition.

What we will prove here is a different sufficiency condition and a priori estimate using the heat bounded and almost heat bounded potentials defined in the previous section. There is still a gap between the two conditions. However the gap seems logarithmic. More precisely, we have

Theorem 4.1. *Let u be a Leray-Hopf solution of the Navier-Stokes equation, which is classical in $\mathbb{R}^3 \times (0,T)$. Let w be the vorticity $\nabla \times u$. Define*

the quantity

$$\mathbf{Q} = \mathbf{Q}(x,t) \equiv \frac{curl(u \times w) \cdot w + 2|\nabla \sqrt{|w|^2+1}|^2 - |\nabla w|^2}{|w|^2+1}(x,t).$$

Then the following statements hold for any $\delta \in (0,T)$.

(1) *The quantity $\mathbf{Q}$ is almost heat bounded in $(\mathbb{R}^3 \times (\delta, T]) \cap \{|w| \geq 1\}$.*

(2) *u is a classical solution of the Navier-Stokes equation in $\mathbb{R}^3 \times (\delta, T]$ if and only if $\mathbf{Q}$ is heat bounded in $(\mathbb{R}^3 \times (\delta, T]) \cap \{|w| \geq 1\}$.*

Remark 4.1. The quantity Q is well defined since we assume that u is smooth for $t \in (0,T)$. The first term in Q is essentially the vortex stretching factor which is the hardest to control. The point of the theorem is that if there is blow up at time T, then the blow up just happens barely.

Proof of Theorem 4.1. We will just prove (1) since (2) is self-evident afterward. We divide the proof into three steps.

Step 1. Rewriting the vortex equation in the log form. Let $w = w(x,t)$ be the vortex. It is well known that $|w|^2$ satisfies the following scalar heat equation with lower order terms

$$\Delta |w|^2 - u \cdot \nabla |w|^2 + 2\alpha |w|^2 - 2|\nabla w|^2 - (|w|^2)_t = 0. \tag{4.2}$$

Here α is the vortex stretching potential given by (c.f. [5])

$$\alpha(x,t) = \frac{3}{4\pi} P.V. \int_{\mathbb{R}^3} D[\tilde{y}, \tilde{\omega}(x+y), \tilde{\omega}(x)]\ |\omega(x+y,t)| \frac{dy}{|y|^3} = \frac{w \nabla u \cdot w}{|w|^2}. \tag{4.3}$$

A straightforward computation from (4.2) shows

$$\Delta \ln(|w|^2+1) - u \cdot \nabla \ln(|w|^2+1) + 2\frac{\alpha |w|^2}{|w|^2+1}$$
$$-2\frac{|\nabla w|^2}{|w|^2+1} + \frac{|(\nabla |w|^2)|^2}{(|w|^2+1)^2} - \partial_t(\ln(|w|^2+1)) = 0.$$

Write $f = \frac{1}{2}\ln(|w|^2+1)$. We deduce

$$\Delta f - u \cdot \nabla f + \frac{\alpha |w|^2}{|w|^2+1} + 2|\nabla f|^2 - \frac{|\nabla w|^2}{|w|^2+1} - f_t = 0. \tag{4.4}$$

Step 2. A representation formula. By our assumption, for $t \in (0,T)$, u and w are classical functions and f vanishes near infinity. This shows

$$f(x,t) = \int G_0(x,t;y,s) f_0(y) dy$$
$$+ \int_0^t \int G_0(x,t;y,s) \left[\frac{\alpha |w|^2}{|w|^2+1} - u \cdot \nabla f + 2|\nabla f|^2 - \frac{|\nabla w|^2}{|w|^2+1} \right](y,s) dy ds. \tag{4.5}$$

Step 3. Apply Jensen's inequality. For convenience, we write

$$Q \equiv \frac{\alpha|w|^2}{|w|^2+1} - u \cdot \nabla f + 2|\nabla f|^2 - \frac{|\nabla w|^2}{|w|^2+1}. \tag{4.6}$$

It is clear that

$$\begin{aligned} Q &= \frac{w\nabla u \cdot w}{|w|^2+1} - \frac{1}{2}u_j\partial_j \ln(|w|^2+1) + 2|\nabla f|^2 - \frac{|\nabla w|^2}{|w|^2+1} \\ &= \frac{u\nabla w \cdot w}{|w|^2} - \frac{u_i\partial_j w_i w_i}{|w|^2+1} + 2|\nabla f|^2 - \frac{|\nabla w|^2}{|w|^2+1} \\ &= \frac{w\nabla u \cdot w - u\nabla w \cdot w + 2|\nabla\sqrt{|w|^2+1}|^2 - |\nabla w|^2}{|w|^2+1}. \end{aligned}$$

Following the well known vector identity, we have

$$Q = \frac{\text{curl}(u \times w) \cdot w + 2|\nabla\sqrt{|w|^2+1}|^2 - |\nabla w|^2}{|w|^2+1}. \tag{4.7}$$

It is well known that $w \in L^2(\mathbb{R}^3 \times \mathbb{R}^+)$. Using Jensen's inequality, it is easy to show that $f(\cdot, t) \in L^p$ for any $p > 1$, in the region where $|w| \geq 1$. Hence the quantity Q is almost heat bounded.

References

[1] M. Aizenman and B. Simon, Brownian motion and Harnack inequality for Schrödinger operators, *Comm. Pure Appl. Math.* 35 (1982), 2:209–273.

[2] Pierre Baras and Jerome A. Goldstein, The heat equation with a singular potential, *Trans. Amer. Math. Soc.* 284 (1984), 1:121–139.

[3] Cabré Xavier and Yvan Martel, Existence versus instantaneous blowup for linear heat equations with singular potentials, *C. R. Acad. Sci., I Math.* 329 (1999), 1:973–978 (French).

[4] Marco Cannone, Harmonic analysis tools for solving the incompressible Navier-Stokes equations, *Handbook of mathematical fluid dynamics,* Vol. III, 161–244.

[5] Peter Constantin, Geometric statistics in turbulence, *SIAM Rev.* 36 (1994), 73–98.

[6] E. B. Fabes and D. W. Stroock, A new proof of Moser's parabolic Harnack inequality using the old ideas of Nash, *Arch. Rational Mech. Anal.* 96 (1986), 4:327–338.

[7] David Gilbarg and Neil S. Trudinger, *Elliptic partial differential equations of second order*, Second edition. Grundlehren der Mathematischen Wissenschaften [Fundamental Principles of Mathematical Sciences], 224, Springer-Verlag, Berlin, 1983.

[8] Jerome A. Goldstein and Qi S. Zhang, Linear parabolic equations with strong singular potentials, *Trans. Amer. Math. Soc.* 355 (2003), 1:197–211.

[9] Ismail Kombe, The linear heat equation with highly oscillating potential, *Proc. Amer. Math. Soc.* 132 (2004), 9:2683–2691.

[10] J. Leray, Essai sur le mouvements d' un liquide visqueux emplissant l' espace, *Acta Math.* 63 (1934), 193-248.

[11] Vitali Liskevich and Semenov Yuli, Estimates for fundamental solutions of second-order parabolic equations, *J. London Math. Soc.* 62 (2000), 2:521–543.

[12] O. A. Ladyzenskaja, V. A. Solonnikov and N. N. Uralceva, *Linear and quasilinear equations of parabolic type,* Translations of Mathematical Monographs, Vol. 23, American Mathematical Society, Providence, RI, 1967.

[13] Gary M. Lieberman, *Second order parabolic differential equations,* World Scientific Publishing Co., Inc., Singapore, 1996.

[14] G. Prodi, Un teorema de unicita per le equazioni di Navier-Stokes, *Annali di Mat.* 48 (1959), 173–182.

[15] James Serrin, On the interior regularity of weak solutions of the Navier-Stokes equations, *Arch. Rational Mech. Anal.* 9 (1962), 187–195.

[16] B. Simon, Schrödinger semigroups, *Bull. A.M.S.* 7 (1982), 447–526.

[17] Karl-Theodor Sturm, Schrödinger semigroups on manifolds, *J. Funct. Anal.* 118 (1993), 2:309–350.

[18] Qi S. Zhang, On a parabolic equation with a singular lower order term II: The Gaussian bounds, *Indiana Univ. Math. J.* 46 (1997), 3:989–1020.

CHARACTERIZATION OF THE SPECTRUM OF AN IRREGULAR BOUNDARY VALUE PROBLEM FOR THE STURM-LIOUVILLE OPERATOR[1]

ALEXANDER MAKIN

Moscow State University of Instrument-Making and Informatics, Moscow 107996, Russia

E-mail: alexmakin@yandex.ru

Keywords: Eigenvalue problem, Sturm-Liouville operator, irregular boundary conditions.

AMS No: 34A55, 34B09.

In the present paper, we consider the eigenvalue problem for the Sturm-Liouvulle equation

$$u'' - q(x)u + \lambda u = 0 \tag{1}$$

on the interval $(0, \pi)$ with the boundary conditions

$$u'(0) + (-1)^{\theta} u'(\pi) + bu(\pi) = 0, \;\; u(0) + (-1)^{\theta+1} u(\pi) = 0, \tag{2}$$

where b is a complex number, $\theta = 0, 1$, and the function $q(x)$ is an arbitrary complex-valued function of the class $L_2(0, \pi)$.

Denote by $c(x, \mu), s(x, \mu)$ $(\lambda = \mu^2)$ the fundamental system of solutions to (1) with the initial conditions $c(0, \mu) = s'(0, \mu) = 1$, $c'(0, \mu) = s(0, \mu) = 0$. The following identity is well known

$$c(x, \mu)s'(x, \mu) - c'(x, \mu)s(x, \mu) = 1. \tag{3}$$

Simple calculations show that the characteristic equation of (1), (2) can be reduced to the form $\Delta(\mu) = 0$, where

$$\Delta(\mu) = c(\pi, \mu) - s'(\pi, \mu) + (-1)^{\theta+1} bs(\pi, \mu). \tag{4}$$

The characteristic determinant $\Delta(\mu)$ of problem (1), (2), given by (4), is referred to as the characteristic determinant corresponding to the triple $(b, \theta, q(x))$. We denote $< q >= \frac{1}{\pi} \int_0^{\pi} q(x)dx$.

By PW_σ we denote the class of entire functions $f(z)$ of exponential type $\leq \sigma$ such that $||f(z)||_{L_2(\mathbb{R})} < \infty$, and by PW_σ^- we denote the set of odd functions in PW_σ.

The following two assertions provide necessary and sufficient conditions to be satisfied by the characteristic determinant $\Delta(\mu)$.

[1]This research is supported by RFBR (No. 10-01-00411)

Theorem 1. *If a function $\Delta(\mu)$ is the characteristic determinant corresponding to the triple $(b, \theta, q(x))$, then*

$$\Delta(\mu) = (-1)^{\theta+1} b \frac{\sin \pi\mu}{\mu} + \frac{f(\mu)}{\mu},$$

where $f(\mu) \in PW_{\pi}^{-}$.

Proof. Let $e(x, \mu)$ be a solution to (1) satisfying the initial conditions $e(0, \mu) = 1$, $e'(0, \mu) = i\mu$, and let $K(x, t)$, $K^{+}(x, t) = K(x, t) + K(x, -t)$, and $K^{-}(x, t) = K(x, t) - K(x, -t)$ be the transformation kernels [1] that realize the representations

$$\begin{aligned} e(x, \mu) &= e^{i\mu x} + \int_{-x}^{x} K(x, t) e^{i\mu t} dt, \\ c(x, \mu) &= \cos \mu x + \int_{0}^{x} K^{+}(x, t) \cos \mu t dt, \qquad (5) \\ s(x, \mu) &= \frac{\sin \mu x}{\mu} + \int_{0}^{x} K^{-}(x, t) \frac{\sin \mu t}{\mu} dt. \end{aligned}$$

It was shown in [2] that

$$c(\pi, \mu) = \cos \pi\mu + \frac{\pi}{2} < q > \frac{\sin \pi\mu}{\mu} - \int_{0}^{\pi} \frac{\partial K^{+}(\pi, t)}{\partial t} \frac{\sin \mu t}{\mu} dt, \qquad (6)$$

$$s'(\pi, \mu) = \cos \pi\mu + \frac{\pi}{2} < q > \frac{\sin \pi\mu}{\mu} + \int_{0}^{\pi} \frac{\partial K^{-}(\pi, t)}{\partial x} \frac{\sin \mu t}{\mu} dt. \qquad (7)$$

Substituting the right-hand sides of expressions (5), (6), (7) into (4), we obtain

$$\Delta(\mu) = (-1)^{\theta+1} b \frac{\sin \pi\mu}{\mu}$$

$$+ \frac{1}{\mu} \int_{0}^{\pi} [-\frac{\partial K^{+}(\pi, t)}{\partial t} - \frac{\partial K^{-}(\pi, t)}{\partial x} + (-1)^{\theta+1} b K^{-}(\pi, t)] \sin \mu t dt.$$

This relation, together with the Paley-Wiener theorem implies the assertion of Theorem 1.

Theorem 2. *Let a function $u(\mu)$ have the form*

$$u(\mu) = (-1)^{\theta+1} b \frac{\sin \pi\mu}{\mu} + \frac{f(\mu)}{\mu},$$

where $f(\mu) \in PW_{\pi}^{-}$, b is a complex number. Then, there exists a function $q(x) \in L_2(0, \pi)$ such that the characteristic determinant corresponding to the triple $(b, \theta, q(x))$ satisfies $\Delta(\mu) = u(\mu)$.

Thus, Theorems 1 and 2 give necessary and sufficient conditions for a function $u(\mu)$ to be the characteristic determinant of problem (1), (2).

Further we consider problem (1), (2) under the supplementary condition $b \neq 0$.

Theorem 3. *For a set Λ of complex numbers to be the spectrum of problem (1), (2) it is necessary and sufficient that it has the form $\Lambda = \{\lambda_n\}$, where $\lambda_n = \mu_n^2$,*

$$\mu_n = n + r_n,$$

where $\{r_n\} \in l_2$, $n = 1, 2, \ldots$.

Proof. Necessity. It follows from Theorem 1 that the characteristic equation of problem (1), (2) can be reduced to the form

$$(-1)^\theta b \frac{\sin \pi\mu}{\mu} = \frac{f(\mu)}{\mu}, \tag{8}$$

where $f(\mu) \in PW_\pi^-$. It was shown in [1] that equation (8) has the roots $\mu_n = n + r_n$, where $r_n = o(1)$, $n = 1, 2, \ldots$. Hence it follows that

$$\sin \pi r_n = (-1)^{\theta+n} \frac{f(n + r_n)}{b}.$$

Since $\{f(n + r_n)\} \in l_2$, by [3], it follows that $\{r_n\} \in l_2$.

Sufficiency. Let the set Λ admit the representation of the above-mentioned form. We denote

$$u(\mu) = (-1)^{\theta+1} b\pi\mu \prod_{n=1}^{\infty} \left(\frac{\lambda_n - \mu^2}{n^2} \right).$$

It follows from [4] and the conditions of the theorem that the infinite product in the right-hand side of the last equality converges uniformly in any bounded domain. We denote $\phi(\mu) = (-1)^{\theta+1} b \sin \pi\mu - u(\mu)$. One can prove that $\phi(\mu) \in PW_\pi^-$. It follows from Theorem 2 that there exists a function $q(x) \in L_2(0, \pi)$ such that the characteristic determinant corresponding to the triple $(b, \theta, q(x))$ satisfies $\Delta(\mu) = u(\mu)$.

Consider problem (1), (2) if $b = 0$. Substituting the functions $c(x, \mu)$, $s(x, \mu)$ into boundary conditions and taking into account (3), we find that each root subspace contains one eigenfunction and, possibly, associated functions. The characteristic equation has the form

$$\frac{f(\mu)}{\mu} = 0,$$

where $f(\mu) \in PW_\pi^-$.

Consider several examples.

1) Let

$$f(\mu) = \frac{\sin^3 \frac{\pi\mu}{3}}{\mu^2}.$$

It follows from well-known identity

$$\sin \pi\mu = \pi\mu \prod_{n=1}^{\infty} \frac{n^2 - \mu^2}{n^2}, \tag{9}$$

that

$$f(\mu) = \frac{\pi^3}{27}\mu \prod_{n=1}^{\infty} \left(\frac{n^2 - (\frac{\mu}{3})^2}{n^2} \right)^3.$$

We set

$$f(\mu) = \frac{\pi^3}{27}\mu \prod_{n=1}^{\infty} \left(\frac{n^2 - (\frac{\mu}{3})^2}{n^2} \right)^2 \left(\frac{n^2 - (-1)^{\theta_n} (\frac{\mu}{3})^2}{n^2} \right),$$

where $\theta_n = 0$ if $n \neq 2^p$ and $n = 1$ if $n = 2^p$, $p = 1, 2 \ldots$. Let us prove that $u(\mu) \in PW_\pi^-$.

By Theorem 2, there exists a potential $q(x) \in L_2(0, \pi)$ such that the corresponding characteristic determinant is $\Delta(\mu) = u(\mu)/\mu$. Boundary value problem (1), (2) with this potential has a subsequence of real eigenvalues $\lambda_{n_k} \to -\infty$ as $k \to \infty$. One can prove that the system of eigen- and associated functions of problem (1), (2) is complete in $L_2(0, \pi)$.

2) Set

$$f(\mu) = \frac{\sin^k(\alpha\pi\mu/k) \sin^k((1 - \alpha)\pi\mu/k)}{\mu^{2k-1}},$$

where k is an arbitrary natural number, and α is an irrational number, $0 < \alpha < 1$. Obviously, $f(\mu) \in PW_\pi^-$. Then, by Theorem 2, there exists a potential $q(x) \in L_2(0, \pi)$, such that the corresponding characteristic determinant $\Delta(\mu) = f(\mu)/\mu$. Since the equations $\sin(\alpha\pi\mu/k) = 0$ and $\sin((1 - \alpha)\pi\mu/k) = 0$ have no common roots, except zero, we see that each root subspace of problem (1), (2) with potential $q_1(x)$ contains one eigenfunction and associated functions up to order $k - 1$. One can readily see that $|\Delta(\mu)| \geq ce^{|\mathrm{Im}\mu|\pi}|\mu|^{1-2k}$ $(c > 0)$, if μ belongs to a sequence of infinitely expanding contours. Then, by [5], the system of eigen-and associated functions of problem (1), (2) is complete in $L_2(0, \pi)$.

3) Set $f(\mu) = \sin^2(\pi\mu/2)/\mu$. It follows from (9) that

$$f(\mu) = \frac{\pi^2}{4}\mu \prod_{n=1}^{\infty} \left(\frac{(2n)^2 - \mu^2}{(2n)^2} \right)^2.$$

We denote

$$u(\mu) = \frac{\pi^2}{4}\mu \prod_{n=1}^{\infty} \left(\frac{\mu_n^2 - \mu^2}{(2n)^2} \right)^2, \tag{10}$$

where $\mu_n = 2n$, if $n \neq 2^p + k$, $k = 1, \ldots, [\ln p]$, $p = p_0, p_0 + 1, \ldots$. $\mu_n = 2^{p+1}$, if $n = 2^p + k$, $k = 1, \ldots, [\ln p]$, $p = p_0, p_0 + 1, \ldots$ $(p_0 \geq 10)$. It can easily be checked that

$$\lim_{n\to\infty} \frac{\mu_n}{n} = 2, \quad \prod_{n=1}^{\infty} \frac{\mu_n}{2n} = c_1 \neq 0.$$

This, together with [6] implies that the infinite product in right-hand side of (10) uniformly converges in any bounded domain of the complex plane, therefore, $u(\mu)$ is an entire analytical function.

One can prove that $u(\mu) \in PW_\pi^-$. Then by theorem 2 there exists a potential $q(x) \in L_2(0, \pi)$, such that the corresponding characteristic determinant $\Delta(\mu) = u(\mu)/\mu$. This yields that the dimensions of root subspaces of problem (1), (2) with potential $q(x)$ increase infinitely, and the system of root functions contains associated functions of arbitrarily high order. One can prove that the system of eigen-and associated functions of problem (1), (2) is complete in $L_2(0, \pi)$.

Further we consider the following eigenvalue problem for operator (1) with boundary conditions

$$u'(0) + du'(\pi) = 0, \quad u(0) - du(\pi) = 0, \tag{11}$$

where $d \neq 0$. By [1], the conditions in (11) are degenerate boundary conditions. Simple calculations show that the characteristic equation of (1), (11) can be reduced to the form $\Delta(\mu) = 0$, where

$$\Delta(\mu) = \frac{d^2 - 1}{d} + c(\pi, \mu) - s'(\pi, \mu). \tag{12}$$

Theorem 4. *If a function $u(\mu)$ can be represented in the form*

$$u(\mu) = \gamma + \frac{f(\mu)}{\mu},$$

where $f(\mu) \in PW_\pi^-$, γ is a complex number, then, there exists a function $q(x) \in L_2(0, \pi)$ such that the characteristic determinant $\Delta(\mu)$ of the problem (1), (11) with the potential $q(x)$, where $d = (\gamma + \sqrt{\gamma^2 + 4})/2$ or $d = (\gamma - \sqrt{\gamma^2 + 4})/2$, is identically equal to the function $u(\mu)$.

Proof. By Theorem 2, there exists a function $q(x) \in L_2(0, \pi)$ such that $c(\pi, \mu) - s'(\pi, \mu) = f(\mu)/\mu$. Evidently, $(d^2 - 1)/d = \gamma$. This relation, together with (12), implies the assertion of Theorem 4.

References

[1] V. A. Marchenko, *Sturm-Liouville operators and their applications,* Izdat. "Naukova Dumka", Kiev, 1977.

[2] L. A. Pastur and V. A. Tkachenko, Spectral theory of a class of one-dimensional Schrödinger operators with limit-peroodic potentials, *Trans. Moscow Math. Soc.* 1989, 115-166.

[3] V. A. Tkachenko, Discriminants and generic spectra of nonselfadjoint Hill's operators, *Adv. Sov. Math.* 19 (1994), 41-71.

[4] M. A. Evgrafov, *Asymptotic estimates and entire functions,* Gosudarstv. Izdat. Fiz-Mat. Lit., Moscow, 1962.

[5] M. M. Malamud, On completeness on the system of root vector Sturm-Liouville operator subject to general boundary conditions, *Dokl. Math.* 419 (2008), 19-22.

[6] M. A. Lavrentiev and B. V. Shabat, *Methods of the theory of a complex variable,* GITTL, Moscow, 1973.

UNIQUENESS OF SOLUTIONS FOR THE EXTERIOR QUATERELLIPTIC-QUATERHYPERBOLIC TRICOMI PROBLEM WITH EIGHT PARABOLIC LINES

JOHN MICHAEL RASSIAS

National and Capodistrian University of Athens Pedagogical Department
4, Agamemnonos Str., Aghia Paraskevi Athens, Attikis 15342, GREECE
E-mail: jrassias@primedu.uoa.gr

The famous Tricomi equation was established in 1923 by F. G. Tricomi who is the pioneer of parabolic elliptic and hyperbolic boundary value problems and related problems of variable type. In 1945 F. I. Frankl established a generalization of these problems for the well-known Chaplygin equation subject to a certain Frankl condition. In 1953 and 1955 M. H. Protter generalized these problems even further by improving the Frankl condition. In 1977 we generalized these results in several n-dimensional simply connected domains. In 1990 we proposed the exterior Tricomi problem in a doubly connected domain. In 2002 we considered uniqueness of quasi-regular solutions for a bi-parabolic elliptic bi-hyperbolic Tricomi problem. In 2006 G. C. Wen investigated the exterior Tricomi problem for general mixed type equations. In this paper we establish uniqueness of quasi-regular solutions for the exterior Tricomi and Frankl problems for quaterelliptic-quaterhyperbolic mixed type partial differential equations of second order with eight parabolic degenerate lines and propose certain open problems. These mixed type boundary value problems are very important in fluid mechanics.

Keywords: Quasi-regular solution, Tricomi equation, Chaplygin equation, quaterelliptic equation, quaterhyperbolic equation, Tricomi problem.
AMS No: 35MO5.

1. Introduction

In 1904 S. A. Chaplygin [11] pointed out that the nonlinear equation of an adiabatic potential perfect gas is closely connected with the study of a linear mixed type equation named Chaplygin equation. In 1923 F. G Tricomi [19] initiated the work on boundary value problems for linear partial differential mixed type equations of second order and related equations of variable type. The well-known mixed type partial differential equation was called *Tricomi equation* after F. G. Tricomi, who introduced this equation. In 1945 F. I. Frankl [3] drew attention to the fact that the Tricomi problem was closely connected to the study of gas flow with nearly sonic speeds. In 1953 and 1955 M. H. Protter [7] generalized and improved the afore-mentioned results in the euclidean plane. In 1977 we [8] generalized these results in multi-dimensional domains. In 1982 we [9] established a maximum principle of the Cauchy problem for hyperbolic equations in multi-dimensional domains. In 1983 we [10] solved the Tricomi problem with two parabolic lines of degeneracy and, in 1992, we [12] established the well-posedness of the Tricomi problem in euclidean regions. Interesting results for the Tricomi problem were achieved by G. Baranchev [1] in 1986, and M. Kracht and E. Kreyszig [4] in 1986, as well. Related information was reported by G. Fichera [2] in 1985, and E. Kreyszig ([5–6]) in 1989 and 1994. Our ([11, 14–15]) work, in 1990 and 1999, was in analogous areas of mixed type equations. In 1990-2009, G. C. Wen et al. ([17, 20–28]) have applied the complex analytic method and achieved fundamental uniqueness and existence results for solutions of the Tricomi and Frankl problems for classical mixed type partial differential equations with boundary conditions. In 1993 R. I. Semerdjieva [18] introduced the hyperbolic equation $K_1(y)u_{xx} + (K_2(y)u_y)_y + ru = f$ in the lower half-plane. In 1997 we [13] considered the more general case of the above hyperbolic equation, so that it was elliptic in the upper half-plane and parabolic on the line $y = 0$. In 2002, we [16] considered the more general Tricomi problem with partial differential equation the new *bi-parabolic elliptic bi-hyperbolic equation*

$$Lu \equiv K_1(y)(M_2(x)u_x)_x + M_1(x)(K_2(y)u_y)_y + r(x,y)u = f(x,y), \quad (1.1)$$

which is parabolic on both segments $x = 0,\ 0 < y \leq 1;\ y = 0,\ 0 < x \leq 1$, elliptic in the euclidean region $G_e = \{(x,y) \in G(\subset \mathbb{R}^2) : x > 0,\ y > 0\}$ and hyperbolic in both regions $G_{h_1} = \{(x,y) \in G(\subset \mathbb{R}^2) : x > 0,\ y < 0\}$; $G_{h_2} = \{(x,y) \in G(\subset \mathbb{R}^2) : x < 0,\ y > 0\}$, with G the mixed domain of (1.1). In 1999 we [15] proved existence of weak solutions for a particular Tricomi problem. Then we established uniqueness of quasi-regular solutions ([8, 10–13, 16]) for the Tricomi problem. However, the question about the uniqueness of quasi-regular solutions and the existence of weak solutions for the Tricomi problem associated to the said mixed type equation (1.1) for even more general doubly connected mixed domain is still *open*.

In particular via this paper we propose and investigate the exterior Tricomi problem for quaterelliptic and quaterhyperbolic equations with eight parabolic lines of degeneracy and establish uniqueness of quasi-regular solutions (see Figure 1). Also we propose new open problems.

2. The Exterior Tricomi Problem

Consider the quaterelliptic-quaterhyperbolic equation (1.1) with eight parabolic lines of degeneracy in a bounded doubly connected mixed domain D with a piecewise smooth boundary ∂D, where $f = f(x,y)$ is continuous in D, $r = r(x,y)$ is once-continuously differentiable in D, $K_i = K_i(y)$ $(i = 1,2)$ are once-continuously differentiable for $y \in [-k_1, k_2]$ with

$-k_1 = \inf\{y : (x,y) \in D\}$ and $k_2 = \sup\{y : (x,y) \in D\}$, and $M_i = M_i(x)$ $(i = 1,2)$ are once-continuously differentiable for $x \in [-m_1, m_2]$ with $-m_1 = \inf\{x : (x,y) \in D\}$ and $m_2 = \sup\{x : (x,y) \in D\}$.

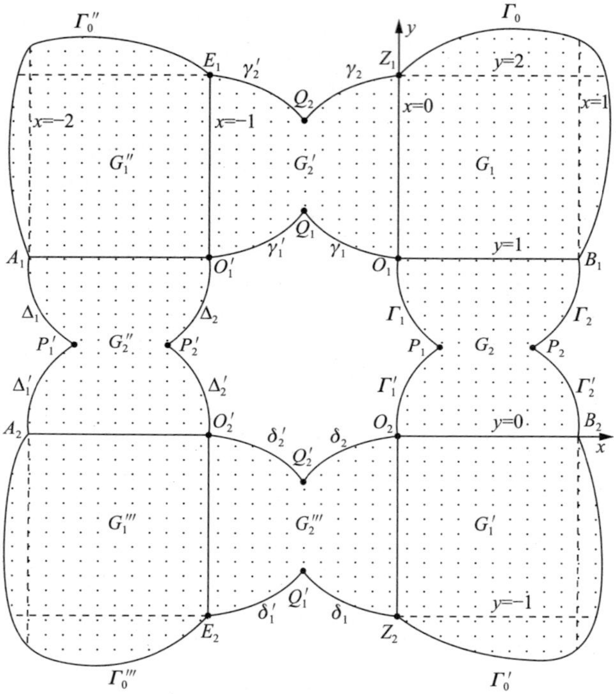

Figure 1: The exterior Tricomi problem with eight parabolic lines

Besides,

$$K_1(y) \begin{cases} > 0 & \text{for} \quad \{y < 0\} \cup \{y > 1\}, \\ = 0 & \text{for} \quad \{y = 0\} \cup \{y = 1\}, \\ < 0 & \text{for} \quad \{0 < y < 1\}, \end{cases}$$

$$M_1(x) \begin{cases} > 0 & \text{for} \quad \{x < -1\} \cup \{x > 0\}, \\ = 0 & \text{for} \quad \{x = 0\} \cup \{x = -1\}, \\ < 0 & \text{for} \quad \{-1 < x < 0\}, \end{cases}$$

as well as $K_2 = K_2(y) > 0$, $M_2 = M_2(x) > 0$, everywhere in D, so that

$$K = \frac{K_1(y)}{K_2(y)} \begin{cases} > 0 & \text{for} \quad \{y < 0\} \cup \{y > 1\}, \\ = 0 & \text{for} \quad \{y = 0\} \cup \{y = 1\}, \\ < 0 & \text{for} \quad \{0 < y < 1\}, \end{cases}$$

$$M = \frac{M_1(x)}{M_2(x)} \begin{cases} > 0 & \text{for} \quad \{x < -1\} \cup \{x > 0\}, \\ = 0 & \text{for} \quad \{x = 0\} \cup \{x = -1\}, \\ < 0 & \text{for} \quad \{-1 < x < 0\}. \end{cases}$$

The boundary $\partial D = Ext(D) \bigcup Int(D)$ of the doubly connected domain D is formed by the following two exterior and interior boundaries

$$Ext(D) = (\Gamma_0 \cup \Gamma_0{}' \cup \Gamma_0{}'' \cup \Gamma_0{}''') \cup (\Gamma_2 \cup \Gamma_2{}') \cup (\gamma_2 \cup \gamma_2{}') \cup (\Delta_1 \cup \Delta_1{}') \\ \cup (\delta_1 \cup \delta_1{}'),$$
$$Int(D) = (\Gamma_1 \cup \Gamma_1{}') \cup (\gamma_1 \cup \gamma_1{}') \cup (\Delta_2 \cup \Delta_2{}') \cup (\delta_2 \cup \delta_2{}'),$$

respectively: In the right hyperbolic domain $G_2 = \{(x, y) \in D : 0 < x < 1,\ 0 < y < 1\}$ with boundary $\partial G_2 = (O_1B_1) \cup (O_2B_2) \cup (\Gamma_1 \cup \Gamma_1{}') \cup (\Gamma_2 \cup \Gamma_2{}')$, where O_1B_1, O_2B_2 are two parabolic lines with end points $O_1 = (0, 1)$, $B_1 = (1, 1)$ and $O_2 = (0, 0)$, $B_2 = (1, 0)$ and $\Gamma_1, \Gamma_1{}', \Gamma_2, \Gamma_2{}'$ are four characteristics, so that:

$$\Gamma_1 : \int_0^x \sqrt{M(t)}dt = -\int_1^y \sqrt{-K(t)}dt \ : 0 < x < 1, \frac{1}{2} < y < 1, \text{emanating from } O_1 = (0, 1),$$

$$\Gamma_1{}' : \int_0^x \sqrt{M(t)}dt = \int_0^y \sqrt{-K(t)}dt \ : 0 < x < 1\,, 0 < y < \frac{1}{2}, \text{ emanating from } O_2 = (0, 0),$$

$$\Gamma_2 : \int_1^x \sqrt{M(t)}dt = \int_1^y \sqrt{-K(t)}dt \ : 0 < x < 1\,, \frac{1}{2} < y < 1, \text{ emanating from } B_1 = (1, 1),$$

$$\Gamma_2{}' : \int_1^x \sqrt{M(t)}dt = -\int_0^y \sqrt{-K(t)}dt : 0 < x < 1, 0 < y < \frac{1}{2}, \text{emanating from } B_2 = (1, 0),$$

where $M = M(x) > 0$, $0 < x < 1$ and $K = K(y) < 0$, $0 < y < 1$. In the upper hyperbolic domain $G_2{}' = \{(x, y) \in D : -1 < x < 0, 1 < y < 2\}$ with boundary $\partial G_2{}' = (O_1Z_1) \cup (O_1{}'E_1) \cup (\gamma_1 \cup \gamma_1{}') \cup (\gamma_2 \cup \gamma_2{}')$, where O_1Z_1, $O_1{}'E_1$ are two parabolic lines with end points $O_1 = (0, 1)$, $Z_1 = (0, 2)$ and $O_1{}' = (-1, 1)$, $E_1 = (-1, 2)$ and $\gamma_1, \gamma_1{}', \gamma_2, \gamma_2{}'$ are four characteristics, so that:

$$\gamma_1 : \int_0^x \sqrt{-M(t)}dt = -\int_1^y \sqrt{K(t)}dt \;:\; -\frac{1}{2} < x < 0\,,\; 1 < y < 2,$$
$$\text{emanating from } O_1 = (0,1),$$
$$\gamma_1{}' : \int_{-1}^x \sqrt{-M(t)}dt = \int_1^y \sqrt{K(t)}dt \;:\; -1 < x < -\frac{1}{2},\, 1 < y < 2,$$
$$\text{emanating from } O_1{}' = (-1,1),$$
$$\gamma_2 : \int_0^x \sqrt{-M(t)}dt = \int_2^y \sqrt{K(t)}dt \;:\; -\frac{1}{2} < x < 0,\, 1 < y < 2,$$
$$\text{emanating from } Z_1 = (0,2),$$
$$\gamma_2{}' : \int_{-1}^x \sqrt{-M(t)}dt = -\int_2^y \sqrt{K(t)}dt : -1 < x < -\frac{1}{2}, 1 < y < 2,$$
$$\text{emanating from } E_1 = (-1,2),$$

where $M = M(x) < 0$, $-1 < x < 0$ and $K = K(y) > 0$, $1 < y < 2$. In the left hyperbolic domain $G_2{}'' = \{(x,y) \in D : -2 < x < -1, 0 < y < 1\}$ with boundary $\partial G_2{}'' = (O_1{}'A_1) \cup (O_2{}'A_2) \cup (\Delta_1 \cup \Delta_1{}') \cup (\Delta_2 \cup \Delta_2{}')$, where $O_1{}'A_1$, $O_2{}'A_2$ are two parabolic lines with end points $O_1{}' = (-1,1)$, $A_1 = (-2,1)$ and $O_2{}' = (-1,0)$, $A_2 = (-2,0)$ and $\Delta_1, \Delta_1{}', \Delta_2, \Delta_2{}'$ are four characteristics, so that:

$$\Delta_1 : \int_{-2}^x \sqrt{M(t)}dt = -\int_1^y \sqrt{-K(t)}dt : -2 < x < -1, \frac{1}{2} < y < 1,$$
$$\text{emanating from } A_1 = (-2,1),$$
$$\Delta_1{}' : \int_{-2}^x \sqrt{M(t)}dt = \int_0^y \sqrt{-K(t)}dt \;:\; -2 < x < -1,\, 0 < y < \frac{1}{2},$$
$$\text{emanating from } A_2 = (-2,0),$$
$$\Delta_2 : \int_{-1}^x \sqrt{M(t)}dt = \int_1^y \sqrt{-K(t)}dt \;:\; -2 < x < -1,\, \frac{1}{2} < y < 1,$$
$$\text{emanating from } O_1{}' = (-1,1),$$
$$\Delta_2{}' : \int_{-1}^x \sqrt{M(t)}dt = -\int_0^y \sqrt{-K(t)}dt : -2 < x < -1, 0 < y < \frac{1}{2},$$
$$\text{emanating from } O_2{}' = (-1,0),$$

where $M = M(x) > 0$, $-2 < x < -1$ and $K = K(y) < 0$, $0 < y < 1$. In the lower hyperbolic domain $G_2{}''' = \{(x,y) \in D : -1 < x < 0, -1 < y < 0\}$ with boundary $\partial G_2{}''' = (O_2Z_2) \cup (O_2{}'E_2) \cup (\delta_1 \cup \delta_1{}') \cup (\delta_2 \cup \delta_2{}')$, where O_2Z_2, $O_2{}'E_2$ are two parabolic lines with end points $O_2 = (0,0)$, $Z_2 = (0,-1)$ and $O_2{}' = (-1,0)$, $E_2 = (-1,-1)$ and $\delta_1, \delta_1{}', \delta_2, \delta_2{}'$ are four characteristics, so that:

$$\delta_1 : \int_0^x \sqrt{-M(t)}dt = -\int_{-1}^y \sqrt{K(t)}dt : -\frac{1}{2} < x < 0, \; -1 < y < 0,$$
$$\text{emanating from } Z_2 = (0, -1),$$
$$\delta_1' : \int_{-1}^x \sqrt{-M(t)}dt = \int_{-1}^y \sqrt{K(t)}dt : -1 < x < -\frac{1}{2}, -1 < y < 0,$$
$$\text{emanating from } E_2 = (-1, -1),$$
$$\delta_2 : \int_0^x \sqrt{-M(t)}dt = \int_0^y \sqrt{K(t)}dt : -\frac{1}{2} < x < 0, \quad -1 < y < 0,$$
$$\text{emanating from } O_2 = (0, 0),$$
$$\delta_2' : \int_{-1}^x \sqrt{-M(t)}dt = -\int_0^y \sqrt{K(t)}dt : -1 < x < -\frac{1}{2}, -1 < y < 0,$$
$$\text{starting from } O_2' = (-1, 0),$$

where $M = M(x) < 0$, $-1 < x < 0$ and $K = K(y) > 0$, $1 < y < 2$. In the upper right elliptic domain $G_1 = \{(x, y) \in D : x > 0, y > 1\}$ with boundary $\partial G_1 = (O_1B_1) \cup (O_1Z_1) \cup \Gamma_0$, where O_1B_1, O_1Z_1 are two parabolic lines with end points $O_1 = (0, 1)$, $B_1 = (1, 1)$ and $O_1 = (0, 1)$, $Z_1 = (0, 2)$ and Γ_0 is the upper right elliptic arc connecting points $B_1 = (1, 1)$ and $Z_1 = (0, 2)$. In the lower right elliptic domain $G_1' = \{(x, y) \in D : x > 0, y < 0\}$ with boundary $\partial G_1' = (O_2B_2) \cup (O_2Z_2) \cup \Gamma_0'$, where O_2B_2, O_2Z_2 are two parabolic lines with end points $O_2 = (0, 0)$, $B_2 = (1, 0)$ and $O_2 = (0, 0)$, $Z_2 = (0, -1)$ and Γ_0' is the lower right elliptic arc connecting points $B_2 = (1, 0)$ and $Z_2 = (0, -1)$. In the upper left elliptic domain $G_1'' = \{(x, y) \in D : x < -1, y > 1\}$ with boundary $\partial G_1'' = (O_1'E_1) \cup (O_1'A_1) \cup \Gamma_0''$, where $O_1'E_1$, $O_1'A_1$ are two parabolic lines with end points $O_1' = (-1, 1)$, $E_1 = (-1, 2)$ and $O_1' = (-1, 1)$, $A_1 = (-2, 1)$ and Γ_0'' is the upper left elliptic arc connecting points $A_1 = (-2, 1)$ and $E_1 = (-1, 2)$. In the lower left elliptic domain $G_1''' = \{(x, y) \in D : x < -1, y < 0\}$ with boundary $\partial G_1''' = (O_2'A_2) \cup (O_2'E_2) \cup \Gamma_0'''$, where $O_2'E_2$, $O_2'A_2$ are two parabolic lines with end points $O_2' = (-1, 0), E_2 = (-1, -1)$ and $O_2' = (-1, 0), A_2 = (-2, 0)$ and Γ_0''' is the lower left elliptic arc connecting points $A_2 = (-2, 0)$ and $E_2 = (-1, -1)$.

Let us consider the intersection points of the hyperbolic characteristics: $\Gamma_1 \cap \Gamma_1' = \{P_1\}$, where $P_1 = (x_1, \frac{1}{2})$, $0 < x_1 < 1$; $\Gamma_2 \cap \Gamma_2' = \{P_2\}$, where $P_2 = (x_2, \frac{1}{2})$, $0 < x_1 < \frac{1}{2} < x_2 < 1$; $\Delta_1 \cap \Delta_1' = \{P_1'\}$, where $P_1' = (x_1', \frac{1}{2})$, $-2 < x_1' < -1$; $\Delta_2 \cap \Delta_2' = \{P_2'\}$, where $P_2' = (x_2', \frac{1}{2})$, $-2 < x_1' < -\frac{3}{2} < x_2' < -1$; $\gamma_1 \cap \gamma_1' = \{Q_1\}$, where $Q_1 = (-\frac{1}{2}, y_1)$, $1 < y_1 < 2$; $\gamma_2 \cap \gamma_2' = \{Q_2\}$, where $Q_2 = (-\frac{1}{2}, y_2)$, $1 < y_1 < \frac{3}{2} < y_2 < 2$; $\delta_1 \cap \delta_1' = \{Q_1'\}$, where $Q_1' = (-\frac{1}{2}, y_1')$, $-1 < y_1' < 0$; $\delta_2 \cap \delta_2' = \{Q_2'\}$, where $Q_2' = (-\frac{1}{2}, y_2')$, $-1 < y_1' < -\frac{1}{2} < y_2' < 0$. If we denote $\Theta = \Theta(x) = \sqrt{|M(x)|}$,

$H = H(y) = \sqrt{|K(y)|}$, we set

$$D_1(x) = \int_0^x \Theta(t)dt, \quad D_2(x) = \int_1^x \Theta(t)dt,$$
$$D_3(x) = \int_{-1}^x \Theta(t)dt, \quad D_4(x) = \int_{-2}^x \Theta(t)dt,$$
$$G_1(y) = \int_0^y H(t)dt, \quad G_2(y) = \int_1^y H(t)dt,$$
$$G_3(y) = \int_{-1}^y H(t)dt, \quad G_4(y) = \int_2^y H(t)dt.$$

Domains G_1, G_2 differ in notation from functions $G_1(y), G_2(y)$. Thus, we have the following equations for the hyperbolic characteristics

$$\Gamma_1 \cup \gamma_1 : D_1(x) = -G_2(y),\ \Gamma_1{}' \cup \gamma_1{}' : G_2(y) = \begin{cases} D_1(x) \text{ on } \Gamma_1{}', \\ D_3(x) \text{ on } \gamma_1{}', \end{cases}$$

$$\Gamma_2 \cup \gamma_2 : \begin{cases} D_2(x) = G_2(y) & \text{on} \quad \Gamma_2, \\ D_1(x) = G_4(y) & \text{on} \quad \gamma_2, \end{cases}$$

$$\Gamma_2{}' \cup \gamma_2{}' : \begin{cases} D_2(x) = -G_1(y) & \text{on} \quad \Gamma_2{}', \\ D_3(x) = -G_4(y) & \text{on} \quad \gamma_2{}', \end{cases}$$

$$\Delta_1 \cup \delta_1 : \begin{cases} D_4(x) = -G_2(y) & \text{on} \quad \Delta_1, \\ D_1(x) = -G_3(y) & \text{on} \quad \delta_1, \end{cases}$$

$$\Delta_1{}' \cup \delta_1{}' : \begin{cases} D_4(x) = G_1(y) & \text{on} \quad \Delta_1{}', \\ D_3(x) = G_3(y) & \text{on} \quad \delta_1{}', \end{cases}$$

$$\Delta_2 \cup \delta_2 : \begin{cases} D_3(x) = G_2(y) \text{ on } \Delta_2, \\ D_1(x) = G_1(y) \text{ on } \delta_2, \end{cases} \quad \Delta_2{}' \cup \delta_2{}' : D_3(x) = |G_1(y)|.$$

Note that: 1) The boundary ∂D is assumed to be a piecewise continuously differentiable arc. The elliptic arcs are "star-shaped" (counterclockwise).

2) We consider continuous solutions u of the quaterelliptic-quaterhyperbolic equation (1.1) with eight parabolic lines, which have the property that u_x, u_y are continuous in the closure of D. These continuity conditions may be weakened at the following eight points $A_1, A_2, B_1, B_2, O_1, O_2, O_1{}', O_2{}'$, by considering u_x, u_y continuous on the boundary ∂D except at these points. By "quaterelliptic" and "quaterhyperbolic" we mean that equation (1.1) is elliptic in four different subdomains and hyperbolic in four other subdomains of the whole domain D. In fact, equation (1.1) is elliptic and hyperbolic in $G_1 \cup G_1{}' \cup G_1{}'' \cup G_1{}'''$ and $G_2 \cup G_2{}' \cup G_2{}'' \cup G_2{}'''$, respectively.

The Exterior Tricomi Problem or Problem (ET): Consists of finding a solution u of the quaterelliptic -quaterhyperbolic equation (1.1) with eight parabolic lines in D and which assumes continuous prescribed values (2.1).

Definition 2.1. A function $u = u(x, y)$ is a *quasi-regular solution* ([7, 8, 10–16]) of Problem (ET), if i) $u \in C^2(D) \cap C(\overline{D})$, $\overline{D} = D \cup \partial D$;

ii) the Green's theorem is applicable to the integrals $\iint\limits_D u_x Lu dx dy$ and $\iint\limits_D u_y Lu dx dy$;

iii) the boundary and region integrals, which arise, exist; and

iv) u satisfies the mixed type equation (1.1) in D and the following boundary condition on the exterior boundary $Ext(D)$:

$$u = \begin{cases} \varphi_1(s) & \text{on} \quad \Gamma_0; \quad \varphi_2(s) \quad \text{on} \quad {\Gamma_0}'; \\ \varphi_3(s) & \text{on} \quad {\Gamma_0}''; \quad \varphi_4(s) \quad \text{on} \quad {\Gamma_0}'''; \\ \psi_1(x) & \text{on} \quad \Gamma_2; \quad \psi_2(x) \quad \text{on} \quad {\Gamma_2}'; \\ \psi_3(x) & \text{on} \quad \gamma_2; \quad \psi_4(x) \quad \text{on} \quad {\gamma_2}'; \\ \psi_5(x) & \text{on} \quad \Delta_1; \quad \psi_6(x) \quad \text{on} \quad {\Delta_1}'; \\ \psi_7(x) & \text{on} \quad \delta_1; \quad \psi_8(x) \quad \text{on} \quad {\delta_1}' \end{cases} \tag{2.1}$$

with continuous prescribed values.

Uniqueness Theorem 2.2. *Consider the quaterelliptic-quaterhyperbolic equation* (1.1) *with eight parabolic lines and the boundary condition* (2.1). *Assume the above mixed doubly connected domain D and the following conditions:*

(R_1) $r \le 0$ *on the interior boundary* $Int(D)$,

$$(R_2) \begin{cases} x dy - (y-1) dx \ge 0 & on \quad \Gamma_0, \\ x dy - y dx \ge 0 & on \quad {\Gamma_0}', \\ (x+1) dy - (y-1) dx \ge 0 & on \quad {\Gamma_0}'', \\ (x+1) dy - y dx \ge 0 & on \quad {\Gamma_0}''', \end{cases}$$

$$(R_3) \begin{cases} 2r + x r_x + (y-1) r_y \le 0 & in \quad G_1, \\ 2r + x r_x + y r_y \le 0 & in \quad {G_1}', \\ 2r + (x+1) r_x + (y-1) r_y \le 0 & in \quad {G_1}'', \\ 2r + (x+1) r_x + y r_y \le 0 & in \quad {G_1}''', \\ r + x r_x \le 0 & in \quad G_2, \\ r + (y-1) r_y \le 0 & in \quad {G_2}', \\ r + (x+1) r_x \le 0 & in \quad {G_2}'', \\ r + y r_y \le 0 & in \quad {G_2}''', \end{cases}$$

(R_4) $K_i > 0, M_i > 0$ *in* $G_1 \cup {G_1}' \cup {G_1}'' \cup {G_1}'''$, $i = 1, 2$,

$$(R_5)\begin{cases} K_1 < 0, & M_1 > 0 \quad in \quad G_2 \cup G_2'', \\ K_1 > 0, & M_1 < 0 \quad in \quad G_2' \cup G_2''', \end{cases}$$

$$(R_6)\begin{cases} \dot{M}_1 \ge 0, & \dot{M}_2 \ge 0; \quad K_1' \ge 0, \quad K_2' \ge 0 \quad in \quad G_1, \\ \dot{M}_1 \ge 0, & \dot{M}_2 \ge 0; \quad K_1' \le 0, \quad K_2' \le 0 \quad in \quad G_1', \\ \dot{M}_1 \le 0, & \dot{M}_2 \le 0; \quad K_1' \ge 0, \quad K_2' \ge 0 \quad in \quad G_1'', \\ \dot{M}_1 \le 0, & \dot{M}_2 \le 0; \quad K_1' \le 0, \quad K_2' \le 0 \quad in \quad G_1''', \end{cases}$$

(R_7) $K_2 > 0$, $M_2 > 0$ *in* D,

$$(R_8)\begin{cases} \dot{M}_1 \ge 0, & \dot{M}_2 \le 0 \quad in \quad G_2, \\ K_1' \ge 0, & K_2' \le 0 \quad in \quad G_2', \\ \dot{M}_1 \le 0, & \dot{M}_2 \ge 0 \quad in \quad G_2'', \\ K_1' \le 0, & K_2' \ge 0 \quad in \quad G_2'''. \end{cases}$$

Let $()_x = \partial()/\partial x$, $()^{\cdot} = d()/dx$, $()_y = \partial()/\partial y$, $()' = d()/dy$, *where* $f = f(x,y)$ *is continuous in* D, $r = r(x,y)$ *is once-continuously differentiable in* D, $K_i = K_i(y)\,(i = 1,2)$ *are once-continuously differentiable for* $y \in [-k_1, k_2]$ *with* $-k_1 = inf\{y : (x,y) \in D\}$ *and* $k_2 = sup\{y : (x,y) \in D\}$, *and* $M_i = M_i(x)\,(i = 1,2)$ *are once-continuously differentiable for* $x \in [-m_1, m_2]$ *with* $-m_1 = inf\{x : (x,y) \in D\}$ *and* $m_2 = sup\{x : (x,y) \in D\}$. *Then the Problem (ET) has at most one quasi-regular solution in D.*

Proof. We apply the well-known *a*-*b*-*c* energy integral method with $a = 0$, and use the above mixed type equation (1.1) as well as the boundary condition (2.1). First, we assume two quasi-regular solutions u_1, u_2 of the Problem (ET). Then we claim that $u = u_1 - u_2 = 0$ holds in the domain D. In fact, we investigate

$$0 = J = 2 < lu,\ Lu >_0 = \iint_D 2luLudxdy, \tag{2.2}$$

where $lu = bu_x + cu_y$, and $Lu = L(u_1 - u_2) = Lu_1 - Lu_2 = f - f = 0$ with choices

$$b = b(x) = \begin{cases} x & \text{in} \quad G_1 \cup G_1' \cup G_2, \\ x+1 & \text{in} \quad G_1'' \cup G_1''' \cup G_2'', \\ 0 & \text{in} \quad G_2' \cup G_2''', \end{cases}$$

$$c = c(y) = \begin{cases} y & \text{in} \quad G_1' \cup G_1''' \cup G_2''', \\ y-1 & \text{in} \quad G_1 \cup G_1'' \cup G_2', \\ 0 & \text{in} \quad G_2 \cup G_2''. \end{cases}$$

We consider the new differential identities

$$2bK_1M_2u_xu_{xx} = \left(bK_1M_2{u_x}^2\right)_x - (bM_2)_xK_1{u_x}^2,$$

$$2bK_2M_1u_xu_{yy} = \left(2bK_2M_1u_xu_y\right)_y - 2bM_1K_2'u_xu_y - \left(bK_2M_1{u_y}^2\right)_x + (bM_1)_xK_2{u_y}^2,$$

$$2cK_1M_2u_yu_{xx} = \left(2cK_1M_2u_xu_y\right)_x - 2cK_1\dot{M}_2u_xu_y - \left(cK_1M_2{u_x}^2\right)_y + (cK_1)'M_2{u_x}^2,$$

$$2cK_2M_1u_yu_{yy} = \left(cK_2M_1{u_y}^2\right)_y - (cK_2)'M_1{u_y}^2,$$

$$2bruu_x = (bru^2)_x - (br)_xu^2,\ 2cruu_y = (cru^2)_y - (cr)_yu^2,$$

as well as t_1 is the coefficient of u_x in Lu, or $t_1 = t_1(x,y) = K_1(y)\dot{M}_2(x)$, and t_2 is the coefficient of u_y in Lu, or $t_2 = t_2(x,y) = K_2'(y)M_1(x)$. Employing these identities and the classical Green's theorem of the integral calculus we obtain that

$$\begin{aligned} 0 = J = &\iint_D 2(bu_x + cu_y)\left[K_1(M_2u_x)_x + M_1(K_2u_y)_y + ru\right]dxdy \\ = &\iint_D 2(bu_x + cu_y)\left[K_1M_2u_{xx} + K_2M_1u_{yy} + t_1u_x + t_2u_y + ru\right]dxdy \\ = &\ I_D + I_{\partial D}, \end{aligned} \tag{2.3}$$

where

$$I_D = \iint_D Q(u_x,u_y)dxdy = \iint_D (A{u_x}^2 + B{u_y}^2 + \Gamma u^2 + 2\Delta u_xu_y)dxdy,$$

$$I_{\partial D} = \int_{\partial D} \tilde{Q}(u_x,u_y)ds = \int_{\partial D} (\tilde{A}{u_x}^2 + \tilde{B}{u_y}^2 + \tilde{\Gamma}u^2 + 2\tilde{\Delta}u_xu_y)ds,$$

with

$$A = -K_1(bM_2)_x + (cK_1)'M_2 + 2bt_1,\ B = K_2(bM_1)_x - (cK_2)'M_1 + 2ct_2,$$

$$\Gamma = -\left[(br)_x + (cr)_y\right],$$

$$\Delta = -\left[bK_2'M_1 + cK_1\dot{M}_2 - bt_2 - ct_1\right] = -\left[b(K_2'M_1 - t_2) + c(K_1\dot{M}_2 - t_1)\right] = 0$$

in D, and $\tilde{A} = (bv_1 - cv_2)K_1M_2$, $\tilde{B} = (-bv_1 + cv_2)K_2M_1$, $\tilde{\Gamma} = (bv_1 + cv_2)r$, $\tilde{\Delta} = bK_2M_1v_2 + cK_1M_2v_1$ on ∂D, where $v = (v_1, v_2) = (dy/ds, -dx/ds)$ is the outer unit normal vector on the boundary ∂D of the domain D such that

$ds^2 = dx^2 + dy^2 > 0$, $|v| = 1$; $\iint\limits_D ()_x dxdy = \int\limits_{\partial D} ()v_1 ds$, $\iint\limits_D ()_y dxdy = \int\limits_{\partial D} ()v_2 ds$ are the Green's integral formulas. From the above conditions, we obtain

$$0 \le A = \begin{cases} xK_1\dot{M}_2 + (y-1)K_1{}'M_2 & \text{in} \quad G_1, \\ xK_1\dot{M}_2 + yK_1{}'M_2 & \text{in} \quad G_1{}', \\ (x+1)K_1\dot{M}_2 + (y-1)K_1{}'M_2 & \text{in} \quad G_1{}'', \\ (x+1)K_1\dot{M}_2 + yK_1{}'M_2 & \text{in} \quad G_1{}'''; \\ -K_1(M_2 - x\dot{M}_2) & \text{in} \quad G_2, \\ M_2\big(K_1 + (y-1)K_1{}'\big) & \text{in} \quad G_2{}', \\ -K_1\big(M_2 - (x+1)\dot{M}_2\big) & \text{in} \quad G_2{}'', \\ M_2\big(K_1 + yK_1{}'\big) & \text{in} \quad G_2{}'''. \end{cases}$$

Similarly we get

$$0 \le B = \begin{cases} xK_2\dot{M}_1 + (y-1)K_2{}'M_1 & \text{in} \quad G_1, \\ xK_2\dot{M}_1 + yK_2{}'M_1 & \text{in} \quad G_1{}', \\ (x+1)K_2\dot{M}_1 + (y-1)K_2{}'M_1 & \text{in} \quad G_1{}'', \\ (x+1)K_2\dot{M}_1 + yK_2{}'M_1 & \text{in} \quad G_1{}'''; \\ K_2(M_1 + x\dot{M}_1) & \text{in} \quad G_2, \\ -M_1\big(K_2 - (y-1)K_2{}'\big) & \text{in} \quad G_2{}', \\ K_2\big(M_1 + (x+1)\dot{M}_1\big) & \text{in} \quad G_2{}'', \\ -M_1\big(K_2 - yK_2{}'\big) & \text{in} \quad G_2{}'''. \end{cases}$$

Also

$$0 \le \Gamma = -\begin{cases} 2r + xr_x + (y-1)r_y & \text{in} \quad G_1, \\ 2r + xr_x + yr_y & \text{in} \quad G_1{}', \\ 2r + (x+1)r_x + (y-1)r_y & \text{in} \quad G_1{}'', \\ 2r + (x+1)r_x + yr_y & \text{in} \quad G_1{}'''; \\ r + xr_x & \text{in} \quad G_2, \\ r + (y-1)r_y & \text{in} \quad G_2{}', \\ r + (x+1)r_x & \text{in} \quad G_2{}'', \\ r + yr_y & \text{in} \quad G_2{}'''. \end{cases}$$

Therefore,

$$I_D = \int\int_D = \int\int_{G_1 \cup G_1{}' \cup G_1{}'' \cup G_1{}'''} + \int\int_{G_2 \cup G_2{}' \cup G_2{}'' \cup G_2{}'''}$$

$$= \Big(\iint\limits_{G_1} + \iint\limits_{G_1{}'} + \iint\limits_{G_1{}''} + \iint\limits_{G_1{}'''}\Big) + \Big(\iint\limits_{G_2} + \iint\limits_{G_2{}'} + \iint\limits_{G_2{}''} + \iint\limits_{G_2{}'''}\Big) \ge 0.$$

Similarly we prove $I_{\partial D} = I_{Ext(D) \bigcup Int(D)} = I_{Ext(D)} + I_{Int(D)} \geq 0$, where

$$I_{Ext(D)} = (\int_{\Gamma_0} + \int_{\Gamma_0{}'} + \int_{\Gamma_0{}''} + \int_{\Gamma_0{}'''}) + (\int_{\Gamma_2} + \int_{\Gamma_2{}'})$$
$$+(\int_{\gamma_2} + \int_{\gamma_2{}'}) + (\int_{\Delta_1} + \int_{\Delta_1{}'}) + (\int_{\delta_1} + \int_{\delta_1{}'}) \geq 0,$$

and

$$I_{Int(D)} = (\int_{\Gamma_1} + \int_{\Gamma_1{}'}) + (\int_{\gamma_1} + \int_{\gamma_1{}'}) + (\int_{\Delta_2} + \int_{\Delta_2{}'}) + (\int_{\delta_2} + \int_{\delta_2{}'}) \geq 0.$$

There is the following general uniqueness approach:

First, from the maximum principle, if $u|_{\overline{G}_2 \bigcup \overline{G}'_2 \bigcup \overline{G}''_2 \bigcup \overline{G}'''_2} = 0$, it follows that $u|_{\overline{G}_1 \bigcup \overline{G}'_1 \bigcup \overline{G}''_1 \bigcup \overline{G}'''_1} = 0$. Second, from the uniqueness of the solution of the Cauchy problem, if $u|_{\overline{G}_1 \bigcup \overline{G}'_1 \bigcup \overline{G}''_1 \bigcup \overline{G}'''_1} = 0$, it follows $u|_{\overline{G}_2 \bigcup \overline{G}'_2 \bigcup \overline{G}''_2 \bigcup \overline{G}'''_2} = 0$. Thus, $u(x, y) \equiv 0$ everywhere in D, completing the proof of the uniqueness theorem.

Remarks 2.3. 2.3.1. For the existence of weak solutions we follow the pertinent results established by the author in 1999 [15].
2.3.2. Relative results easily follow for the Frankl problem.

Example 2.4. A very simple example with $K_2 = M_2 = 1$ and $K = K_1 = y(y-1)$ and $M = M_1 = x(x+1)$ verifies all the above conditions in our Uniqueness Theorem 2.2. This example may be extended to a general Gellerstedt type example [11].

3. Open Problems

3.1. Extend "quasi-regularity" of solutions to "regularity" by fixing singularities at the following twelve points:

$$O_1 = (0,1),\ O_1{}' = (-1,1),\ O_2 = (0,0), O_2{}' = (-1,0);$$
$$A_1 = (-2,1),\ B_1 = (1,1),\ A_2 = (-2,0),\ B_2 = (1,0);$$
$$E_1 = (-1,2),\ Z_1 = (0,2),\ E_2 = (-1,-1),\ Z_2 = (0,-1).$$

3.2. Investigate the exterior Tricomi and Frankl problems in a multiply connected mixed domain.

3.3. Establish "well-posedness" of solutions for the exterior Tricomi and Frankl problems, in the sense that there is at most one quasi-regular solution and a weak solution exists.

3.4. Solve the n-dimensional Tricomi and Frankl problems in a multiply connected mixed domain.
3.5. Establish the extremum principle for the exterior Tricomi problem: "*A solution of the exterior Tricomi (or Frankl) problem, vanishing on the exterior boundary of the considered mixed domain, achieves neither a positive maximum nor a negative minimum on open arcs of the type-degeneracy curves.*"
3.6. Solve the Tricomi problem for PDE of second order:
3.6.1 $K(y - x^m - x^n)u_{xx} + u_{yy} + r(x, y)u = f(x, y)$;
3.6.2 $u_{xx} + M(x - y^m - y^n)u_{yy} + r(x, y)u = f(x, y)$;
3.6.3 $K(x^m + y^n - 1)u_{xx} + u_{yy} + r(x, y)u = f(x, y)$,
for example $m = n = 2$ or $= 2/3$;
3.6.4 $K\big((y - x^m)(y - x^n)\big)u_{xx} + u_{yy} + r(x, y)u = f(x, y)$;
3.6.5 $K(y - x^n)u_{xx} + M(x - y^m)u_{yy} + r(x, y)u = f(x, y)$;
3.6.6 $K(y^k - x^m \pm x^n)u_{xx} + M(x^k - y^m \pm y^n)u_{yy} + r(x, y)u = f(x, y)$;
3.6.7 $K\big(y^m(y - x^n)\big)u_{xx} + M\big(x^m(x - y^n)\big)u_{yy} + r(x, y)u = f(x, y)$;
3.6.8 $K\big((y-x^m)(y-x^n)\big)u_{xx}+M\big((x-y^\alpha)(x-y^\beta)\big)u_{yy}+r(x, y)u = f(x, y)$.
3.7. Solve the Tricomi problem for PDE of fourth order:

$$\left(\operatorname{sgn}(y - x^l)|y - x^l|^k \frac{\partial^2}{\partial x^2} + \operatorname{sgn}(x - y^n)|x - y^n|^m \frac{\partial^2}{\partial y^2} + r\right)^2 u = f.$$

3.8. Solve the 3-dimensional Tricomi problem for mixed type PDE of second order:

$$\operatorname{sgn}(z)|z|^k(u_{xx} \pm u_{yy}) + \operatorname{sgn}(xy)|x|^m|y|^n u_{zz} + ru = f.$$

Acknowledgment. I am grateful to Professor Guochun Wen for his invitation to the *3rd International Conference,* in Beijing, China.

References

[1] G. Barantsev, On singularities of the Tricomi problem solution by the Fourier method, in *Mixed type equations,* Teubner-Texte zur Mathematik, Vol. 90, Ed. J. M. Rassias, Leipzig, 1986, 47–54.

[2] G. Fichera, Francesco Giacomo Tricomi, in *Mathematical analysis,* Teubner-Texte zur Mathematik, Vol. 79, Ed. J. M. Rassias, Leipzig, 1985, 6–31.

[3] F. I. Frankl, On the problems of Claplygin for mixed subsonic and supersonic flows, *Izv. Akad. Nauk SSSR Ser. Mat.* 9 (1945), 121–143.

[4] M. Kracht and E. Kreyszig, The Tricomi equation and transition problems, in *Mixed type equations,* Teubner-Texte zur Mathematik, Vol. 90, Ed. J. M. Rassias, Leipzig, 1986, 157–165.

[5] E. Kreyszig, *Introductory functional analysis with applications,* Wiley, New York, 1989.

[6] E. Kreyszig, Banach spaces in Bergman operator theory, in *Functional analysis, approximation theory and numerical analysis,* World Scientific, Ed. J. M. Rassias, Singapore, 1994, 155–165.

[7] M. H. Protter, Uniqueness theorems for the Tricomi problem, I, II, *J.Rat. Mech. Anal.* 2 (1953), 107–114; 4 (1955), 721–732.

[8] J. M. Rassias, *Mixed type partial differential equations in* $\mathbb{R}^n$, Ph.D. dissertation, U. C. Berkeley, 1977.

[9] J. M. Rassias, A maximum principle in $\mathbb{R}^{n+1}$, *J. Math. Anal. Appl.*, Acad. Press, New York, 85 (1982), 106–113.

[10] J. M. Rassias, On the Tricomi problem with two parabolic lines of degeneracy, *Bull. Inst. Math., Acad. Sinica* 12 (1983), 62–67.

[11] J. M. Rassias, *Lecture notes on mixed type partial differential equations,* World Scientific, Singapore, 1990.

[12] J. M. Rassias, On the Well-posed Tricomi problem in $\mathbb{R}^2$, *Discuss. Math.* 12 (1992), 85–93.

[13] J. M. Rassias, Uniqueness of quasi-regular solutions for a parabolic elliptic-hyperbolic Tricomi problem, *Bull. Inst. Math., Acad. Sinica* 25 (1997), 277–287.

[14] J. M. Rassias, *Advances in equations and inequalities,* Hadronic Press, Inc., Palm Harbor, FL., U.S.A., 1999.

[15] J. M. Rassias, Existence of weak solutions for a parabolic elliptic-hyperbolic Tricomi problem, *Tsukuba J. Math.* 23 (1999), 37–54.

[16] J. M. Rassias, Uniqueness of quasi-regular solutions for a bi-parabolic elliptic bi-hyperbolic Tricomi problem, *Complex Variables and Elliptic Equations* 47(8) (2002), 707–718.

[17] J. M. Rassias and G. C. Wen, Solvability of the oblique derivative problem for second order equations of mixed type with nonsmooth degenerate curve, *Intern. J. Appl. Math. Stat.* 8(M07) (2007), 96–111.

[18] R. I. Semerdjieva, Uniqueness of regular solutions for a class of non-linear degenerating hyperbolic equations, *Math. Balk.*, 7 (1993), 277–283.

[19] F. G. Tricomi, Sulle equazioni lineari alle parziali di 2^0 ordine di tipo misto, *Atti Accad. Naz. Lincei* 14 (1923), 133–247.

[20] G. C. Wen, The Exterior Tricomi problem for generalized mixed equations with parabolic degeneracy, *Acta Math. Sinica,* English Series 22(5) (2006), 1385–1398.

[21] G. C. Wen, Oblique derivative problems for general Chaplygin-Rassias equations with nonsmooth degenerate line in mixed domains, *Science in China, Series A: Mathematics* 51(1) (2008), 5–36.

[22] G. C. Wen, The Tricomi and Frankl problems for generalized Chaplygin equations in multiply connected domains, *Acta Math. Sinica,* English Series 24(11) (2008), 1759–1774.

[23] G. C. Wen, Oblique derivative problems for generalized Rassias equations of mixed type with several characteristic boundaries, *Electr. J. Diff. Equations* 65 (2009), 1–16.

[24] G. C. Wen, *Elliptic, hyperbolic and mixed complex equations with parabolic degeneracy including Tricomi-Bers and Tricomi-Frankl-Rassias problems,* World Scientific Co. Pte. Ltd., Singapore: Peking University, Series in Mathematics, Vol. 4, 2008.

[25] G. C. Wen and H. Begehr, *Boundary value problems for elliptic equations and systems,* Longman Scientific and Technical Company, Harlow, 1990.

[26] G. C. Wen, D. C. Chen and X. Cheng, General Tricomi-Rassias problem and oblique derivative problem for generalized Chaplygin equation, *J. Math. Anal.* 333 (2007), 679–694.

[27] G. C. Wen and D. C. Chen, Discontinuous Riemann-Hilbert problems for degenerate elliptic complex equations of first order, *Complex Variables* 50 (2005), 707–718.

[28] G. C. Wen and Z. T. Ma, Discontinuous oblique derivative problem for second order equations of mixed type in general domains, *Complex Variables* 48(2) (2003), 119–130.

THE RIEMANN-HILBERT PROBLEM FOR DEGENERATE ELLIPTIC COMPLEX EQUATIONS OF FIRST ORDER[1]

FENG TIAN[a] and GUO-CHUN WEN[b]

[a] *School of Applied Mathematics, Guangdong University of Technology, Guangdong 510940, China*

E-mail: tfeng0309@sina.com

[b] *School of Mathematical Sciences, Peking University, Beijing 100871, China*

E-mail: wengc@math.pku.edu.cn

In this article, we discuss the Riemann-Hilbert Problem for degenerate elliptic systems of first order linear equations in a simply connected domain. We first give the representation of solutions of the boundary value problem for the systems, and then prove the existence and uniqueness of solutions for the problem.

Keywords: Riemann-Hilbert problem, degenerate elliptic complex equations.

AMS No: 35J70, 35J55.

1. Formulation of the Riemann-Hilbert Problem for Degenerate Elliptic Complex Equations

Let D be a domain in the upper half-plane with the boundary ∂D, which consists of $\gamma = \{-1 < x < 1, \hat{y} = y - x^2 = 0\}$ and a curve $\Gamma(\in C^1_\alpha, 0 < \alpha < 1)$ with the end points $-1, 1$ in the upper half-plane: $-\infty < x < \infty$, $\hat{y} = y - x^2 > 0\}$. We consider the linear degenerate elliptic equation of first order

$$\begin{cases} H(\hat{y})u_x - v_y = a_1 u + b_1 v + c_1 \\ H(\hat{y})v_x + u_y = a_2 u + b_2 v + c_2 \end{cases} \text{in } D, \tag{1}$$

where $\hat{y} = y - x^2$, $H(\hat{y}) = \sqrt{K(\hat{y})}$, $G(\hat{y}) = \int_0^{\hat{y}} H(t)dt$, $G'(\hat{y}) = H(\hat{y})$, $K(\hat{y}) = \hat{y}^m h(\hat{y})$ is continuous in $\overline{D}$, here m is a positive number and $h(\hat{y})$ is a continuously differentiable positive function in $\overline{D}$, and $a_j, b_j, c_j\ (j = 1, 2)$ are functions of $z\,(\in D)$. The following degenerate elliptic system is a special case of system (1) with $H(\hat{y}) = \hat{y}^{m/2}$:

$$\begin{cases} \hat{y}^{m/2}u_x - v_y = a_1 u + b_1 v + c_1 \\ \hat{y}^{m/2}v_x + u_y = a_2 u + b_2 v + c_2 \end{cases} \text{in } D. \tag{2}$$

For convenience, we mainly discuss equation (2), and equation (1) can be similarly discussed. From the elliptic condition of system (1), namely

$$J = 4K_1K_4 - (K_2 + K_3)^2 = 4H^2(\hat{y}) > 0 \text{ in } \overline{D}\backslash\gamma, \tag{3}$$

[1]This research is supported by NSFC (No.10971224)

in which $K_j(j=1,...,4)$ are as stated in Chapter II, [1] and Chapter I, [3], and $J=0$ on $\gamma=\{-1<x<1,\hat{y}=y-x^2=0\}$, hence system (1) or (2) is elliptic system of first order equations in D with the parabolic degenerate line $\gamma=(-1,1)$ on the x-axis. Setting $Y=G(\hat{y})=\int_0^{\hat{y}}H(t)dt$, $Z=x+iY$ in $\overline{D}$, if $H(\hat{y})=\hat{y}^{m/2}$, $Y=\int_0^{\hat{y}}H(t)dt=2\hat{y}^{(m+2)/2}/(m+2)$, then its inverse function is $\hat{y}=[(m+2)Y/2]^{2/(m+2)}=JY^{2/(m+2)}$. Denote

$$\begin{aligned} w(z)=u+iv,\ \ w_{\bar{z}}=\frac{1}{2}[H(\hat{y})w_x+iw_y]=\\ =\frac{H(\hat{y})}{2}[w_x+iw_Y]=H(\hat{y})w_{x-iY}=H(\hat{y})w_{\overline{Z}}, \end{aligned} \tag{4}$$

then the system (1.1) can be written in the complex form

$$\begin{aligned} w_{\bar{z}}=H(\hat{y})w_{\overline{Z}}=A_1(z)w+A_2(z)\overline{w}+A_3(z)=g(Z)\ \text{ in } D,\\ A_1=\frac{1}{4}[a_1+ia_2-ib_1+b_2],A_2=\frac{1}{4}[a_1+ia_2+ib_1-b_2],A_3=\frac{1}{2}[c_1+ic_2], \end{aligned} \tag{5}$$

in which D_Z is the image domain of D with respect to the mapping $Z=Z(z)=x+iY=x+iG(\hat{y})$ in D. For the equation (5), we can give a conformal mapping $\zeta=\zeta(Z)$, which maps the domain D_Z onto D_ζ, such that γ and boundary points $-1,1$ in $Z=x+i\hat{y}$-plane are mapped onto themselves, and the boundary $\partial D_\zeta\backslash\gamma(\in C_\alpha^1)$ is a curve with the form $\text{Re}\,\zeta=G(\text{Im}\zeta)-1$ and $\text{Re}\,\zeta=1-G(\text{Im}\zeta)$ near the points $\zeta=-1,1$ respectively. Denote by $Z=Z(\zeta)$ the inverse function of $\zeta=\zeta(Z)$, thus equation (5) is reduced to

$$\begin{aligned} w_{\bar{\zeta}}=g[Z(\zeta)]\overline{Z'(\zeta)}/H(\hat{y}),\ \text{i.e.}\\ w_{\bar{\zeta}}=[A_1(z)w+A_2(z)\overline{w}+A_3(z)]\overline{Z'(\zeta)}/H(\hat{y})\ \text{ in } \overline{D_\zeta}. \end{aligned} \tag{6}$$

In this article, there is no harm in assuming that the boundary Γ is a curve with the form $x=G(\hat{y})-1$ and $x=1-G(\hat{y})$ near the points $\hat{z}=x+i\hat{y}=-1,1$ respectively.

Suppose that equation (5) satisfies the following conditions:

Condition C. The coefficients $A_j[z(Z)]\,(j=1,2,3)$ in (5) satisfy

$$L_\infty[A_j(z(Z)),\overline{D_Z}]\le k_0,\ j=1,2,\ L_\infty[A_3(z(Z)),\overline{D_Z}]\le k_1, \tag{7}$$

where $z(Z)$ is the inverse function of $Z(z)$, and k_0, k_1 are non-negative constants.

Now we formulate the Riemann-Hilbert boundary value problem as follows:

Problem A. Find a solution $w(z)$ of (5) in D, which is continuous in $D^* = \overline{D}\backslash\{-1,1\}$ and satisfies the boundary conditions

$$\operatorname{Re}[\overline{\lambda(z)}w(z)]=r(z) \text{ on } \partial D^*=\partial D\backslash\{-1,1\},\ \operatorname{Im}[\overline{\lambda(z_0)}w(z_0)]=b_0, \tag{8}$$

where $\lambda(z) = a(x) + ib(x)\,(|\lambda(z)| = 1)$, b_0 is a real constants, $z_0\,(\in \Gamma\backslash\{-1,1\})$ is a point, and $\lambda(z)$, $r(z)$, b_0 satisfy the conditions

$$\begin{gathered} C_\alpha[\lambda(z),\Gamma] \le k_0,\ C_\alpha[\lambda(z),\gamma] \le k_0, \\ C_\alpha[r(z),\Gamma]\le k_2,\ C_\alpha[r(z),\gamma]\le k_2,\ |b_0|\le k_2, \end{gathered} \tag{9}$$

where $\alpha\,(0 < \alpha < 1), k_0, k_2$ are non-negative constants. In particular, if $\lambda(z) = a(x) + ib(x) = 1$, then Problem A is the Dirichlet boundary value problem, which will be called Problem D. Denote by $\lambda(z_j-0)$ and $\lambda(z_j+0)$ the left limit and right limit of $\lambda(z)$ as $z \to z_j (j = 1, 2)$ on ∂D^*, and

$$\begin{gathered} e^{i\phi_j}=\frac{\lambda(z_j-0)}{\lambda(z_j+0)},\gamma_j=\frac{1}{\pi i}\ln\left[\frac{\lambda(z_j-0)}{\lambda(z_j+0)}\right]=\frac{\phi_j}{\pi}-K_j, \\ K_j=\left[\frac{\phi_j}{\pi}\right]+J_j,\ J_j=0 \text{ or } 1,\ j=1,2, \end{gathered} \tag{10}$$

in which $z_1 = \hat{z}_1 = -1, z_2 = \hat{z}_2 = 1$, $0 \le \gamma_j < 1$ when $J_j = 0$, and $-1 < \gamma_j < 0$ when $J_j = 1$, $j = 1, 2$, and

$$K=\frac{1}{2}(K_1+K_2)=\frac{1}{2}\sum_{j=1}^{2}[\frac{\phi_j}{\pi}-\gamma_j]$$

is called the index of Problem A. If $\lambda(z)$ on ∂D is continuous, then $K = \Delta_\Gamma \arg\lambda(z)/2\pi$ is a unique integer. If the function $\lambda(z)$ on ∂D is not continuous, we can choose $J_j = 0$ or $1\,$(j = 1,2), hence the index K is not unique. We shall only discuss the case $K = 0$ later on, and the other cases for instance $K = -1/2$, the last point condition in (8) should be cancelled, we can similarly discussed.

2. Representations and Estimates of Solutions of the Riemann-Hilbert Problem for Elliptic Complex Equations

It is clear that the complex equation

$$w_{\overline{Z}} = 0 \text{ in } \overline{D_Z} \tag{11}$$

is a special case of equation (5). On the basis of Theorem 1.3, Chapter I, [3], we can find a unique solution of Problem A for equation (11) in $\overline{D_Z}$.

Now we consider the function $g(Z) \in L_\infty(D_Z)$, and first extend the function $g(Z)$ to the exterior of $\overline{D_Z}$ in $\mathbb{C}$, i.e. set $g(Z)=0$ in $\mathbb{C}\backslash\overline{D_Z}$, hence we can only discuss the domain $D_0 = \{|x|<R_0\} \cap \{\mathrm{Im}Y \geq 0\} \supset \overline{D_Z}$, here $Z = x + iY$, R_0 is a positive number. In the following we shall verify that the integral

$$\Psi(Z)=Tg/H=-\frac{1}{\pi}\iint_{D_0}\frac{g(t)/H(\mathrm{Im}t)}{t-Z}d\sigma_t \text{ in } D_0, \qquad L_\infty[g(Z),D_0] \leq k_3, \tag{12}$$

satisfies the estimate (13) below, where $H(\hat{y}) = \hat{y}^{m/2}$, m ia a positive number. It is clear that the function $g(Z)/H(\hat{y})$ belongs to the space $L_1(D_0)$ and in general is not belonging to the space $L_p(D_0)\,(p > 2,\, m \geq 2)$, and the integral $\Psi(Z_0)$ is definite when $\mathrm{Im}Z_0 > 0$. If $Z_0 \in D_0$ and $\mathrm{Im}Z_0 = 0$, we can define the integral $\Psi(Z_0)$ as the limit of the corresponding integral over $D_0 \cap \{|\mathrm{Re}t - \mathrm{Re}Z_0| \geq \varepsilon\} \cap \{|\mathrm{Im}t - \mathrm{Im}Z_0| \geq \varepsilon\}$ as $\varepsilon \to 0$, where ε is a sufficiently small positive number. The Hölder continuity of the integral can be proved by the method similar to Lemma 2.1, Chapter I, [3].

Lemma 1. *If the function $g(Z)$ in D_Z satisfies the condition in* (12), *and $H(\hat{y}) = \hat{y}^{m/2}$, where m is a positive number, then the integral in* (12) *satisfies the estimate*

$$C_\beta[\Psi(Z),\overline{D_Z}] \leq M_1, \tag{13}$$

where $\beta = 2/(m+2) - \delta$, δ is a sufficiently small positive constant, and $M_1 = M_1(\beta,\, k_3, H, D_Z)$ is a positive constant.

Remark 1. If the condition $H(\hat{y}) = \hat{y}^{m/2}$ in Lemma 1 is replaced by $H(\hat{y}) = \hat{y}^\eta$, herein η is a positive constant satisfying the inequality $\eta < (m+2)/2$, then by the same method we can prove that the integral $\Psi(Z) = T(g/H)$ satisfies the estimate

$$C_\beta[\Psi(Z), D_Z] \leq M_1,$$

where $\beta = 1-2\eta/(m+2)-\delta$, δ is a sufficiently small positive constant, and $M_1 = M_1(\beta, k_3, H, D_Z)$ is a positive constant. In particular if $H(\hat{y}) = \hat{y}$, i.e. $\eta = 1$, then we can choose $\beta = m/(m+2) - \delta$, δ is a sufficiently small positive constant.

Now we give two representation theorems of solutions of Problem A for system (2) or equation (5).

Theorem 2. *Suppose that the equation* (5) *satisfies Condition* C. *Then any solution of Problem* A *for* (5) *can be expressed as*

$$w[z(Z)] = [\tilde{\Phi}(Z) + \tilde{\psi}(Z)]e^{\tilde{\phi}(Z)} \text{ in } D_Z, \tag{14}$$

where $\tilde{\psi}(Z)$, $\tilde{\phi}(Z)$ *possess the form*

$$\tilde{\phi}(Z)=T\tilde{h}=-\frac{1}{\pi}\iint_{D_0}\frac{\tilde{h}(t)}{t-Z}d\sigma_t \text{ in } D_Z,$$

$$\tilde{h}(Z)=\begin{cases}\dfrac{1}{H(\hat{y})}\{A_1[z(Z)]+A_2[z(Z)]\dfrac{\overline{w[z(Z)]}}{w[z(Z)]}\} \text{ if } w[z(Z)]\neq 0, Z\in D_Z,\\ 0 \text{ if } w[z(Z)]=0,\ Z\in D_Z, \text{ or } Z\in D_0\backslash D_Z,\end{cases}$$

$$\tilde{\psi}(Z)=T\tilde{f}=-\frac{1}{\pi}\iint_{D_0}\frac{\tilde{f}(t)}{t-Z}d\sigma_t,\ \tilde{f}(Z)=\frac{A_3[z(Z)]}{H(\hat{y})}e^{-\tilde{\phi}(Z)},$$

in which D_0 *is as stated before,* $\tilde{\phi}(Z)$, $\tilde{\psi}(Z)$ *satisfy the estimate similar to that in* (13), $Z=x+iY=x+iG(\hat{y})$, *and* $\tilde{\Phi}(Z)$ *is an analytic function in* D_Z *satisfying the estimate*

$$C_\delta[X(Z)\tilde{\Phi}(Z),\overline{D_Z}]\leq M_2, \tag{15}$$

where $X(Z)=|Z-t_1|^{\eta_1}|Z-t_2|^{\eta_2}$, *here* $\eta_j=\max(-4\gamma_j,0)+8\delta$, $j=1,2$, $\gamma_j\ (j=1,2)$ *are as stated in* (10), *and* $t_1=-1, t_2=1$, δ *is a sufficiently small positive constant,* $k=(k_0,k_1,k_2)$, *and* $M_2=M_2(\delta,k,H,D_Z)$ *is a non-negative constant.*

Proof. On the basis of Lemma 1, we see that $\tilde{\phi}(Z),\tilde{\psi}(Z)$ in $\overline{D_Z}$ satisfy the similar estimate as in (13). Next it is easy to derive that

$$\tilde{\Phi}_{\overline{Z}}=[w_{\overline{Z}}-w(A_1+A_2\overline{w}/w)/H-A_3/H]e^{-\tilde{\phi}(Z)}=0 \text{ in } D_Z,$$

namely $\tilde{\Phi}(Z)$ is an analytic function in D_Z, which satisfies the boundary conditions

$$\begin{aligned}\text{Re}[\overline{\lambda(z(Z))}e^{\tilde{\phi}(Z)}\Phi(Z)]&=r[z(Z)]-\text{Re}[\overline{\lambda(z(Z))}e^{\tilde{\phi}(Z)}\tilde{\psi}(Z)] \text{ on } \partial D_Z^*,\\ \text{Im}[\overline{\lambda(z_0)}e^{\tilde{\phi}(Z_0)}\tilde{\Phi}(Z_0)]&=b_0-\text{Im}[\overline{\lambda(z_0)}e^{\tilde{\phi}(Z_0)}\tilde{\psi}(Z_0)],\end{aligned} \tag{16}$$

in which $z(Z)$ is the inverse function of $Z(z)$, $Z_0=Z(z_0)$, $\partial D_Z^*=\partial D_Z\backslash\{-1,1\}$, and the index of $\lambda[z(Z)]\exp[\overline{\tilde{\phi}(Z)}]$ on ∂D_Z is $K=0$. Hence according to the proof of Theorems 1.1 and 1.8, Chapter IV, [1], we can derive that $\tilde{\Phi}(Z)$ in $\overline{D_Z}$ satisfies the estimate (15). This completes the proof.

Theorem 3. *Suppose that the equation* (5) *satisfies Condition* C. *Then any solution of Problem* A *for* (5) *can be expressed as*

$$w[z(Z)]=\Phi(Z)e^{\phi(Z)}+\psi(Z) \text{ in } D_Z, \tag{17}$$

where $\psi(Z)$, $\phi(Z)$ *possess the form*

$$\psi(Z)=Tf=-\frac{1}{\pi}\iint_{D_0}\frac{f(t)}{t-Z}d\sigma_t, L_\infty[f(Z)H(\hat{y}),D_Z]<\infty$$

$$\phi(Z)=Th=-\frac{1}{\pi}\iint_{D_0}\frac{h(t)}{t-Z}d\sigma_t \text{ in } D_Z,$$

$$h(Z)=\begin{cases}\dfrac{1}{H(\hat{y})}\{A_1[z(Z)]+A_2[z(Z)]\dfrac{\overline{W(Z)}}{W(Z)}\} \text{ if } W(Z)\neq 0, Z\in D_Z,\\ 0 \text{ if } W(Z)=0,\ Z\in D_Z\cup\{D_0\backslash D_Z\},\end{cases}$$

in which $\psi(Z)$, $\phi(Z)$ *satisfy the estimate* (13), $W(Z)=w[z(Z)]-\psi(Z)$, $Z=x+iY=x+iG(\hat{y})$, *and* $\Phi[z(Z)]$ *is an analytic function in* D_Z.

Proof. First of all, by using the method of parameter extension as stated in the proof of Theorem 5 below, or Theorem 3.3, Chapter II, [1], we can find a solution of equation (5) in the form

$$\psi(Z)=-\frac{1}{\pi}\int_{D_0}\frac{f(t)}{t-Z}d\sigma_t,\ H(\hat{y})f(Z)\in L_\infty(D_Z).$$

On the basis of Theorem 2, the solution of (5) in D_Z can be expressed by $\psi(Z)=\tilde{\psi}(Z)e^{\tilde{\phi}(Z)}$, where

$$\tilde{\phi}(Z)=T\tilde{h}=-\frac{1}{\pi}\iint_{D_0}\frac{\tilde{h}(t)}{t-Z}d\sigma_t \text{ in } D_Z,$$

$$\tilde{h}(Z)=\begin{cases}\dfrac{1}{H(y)}\{A_1[z(Z)]+A_2[z(Z)]\dfrac{\overline{\psi(Z)}}{\psi(Z)}\} \text{ if } \psi(Z)\neq 0, Z\in D_0,\\ 0 \text{ if } \psi(Z)=0,\ Z\in D_0,\end{cases}$$

$$\tilde{\psi}(Z)=T\tilde{f}=-\frac{1}{\pi}\iint_{D_0}\frac{\tilde{f}(t)}{t-Z}d\sigma_t, \tilde{f}(Z)=A_3[z(Z)]e^{-\tilde{\phi}(Z)},$$

it is clear that the functions $\tilde{\phi}(Z),\tilde{\psi}(Z)$ satisfy the estimate similar to (13).

Next let $w(z)$ be a solution of Problem A for equation (5), it is clear that $W(Z)=\Phi(Z)e^{\phi(Z)}=w[z(Z)]-\psi(Z)$ is a solution of the complex equation

$$W_{\overline{Z}}=A_1W(Z)+A_2\overline{W(Z)} \text{ in } D_Z,$$

where $\psi(Z)$ is as stated in (17), and we can verify that the function $\Phi(Z)$ is an analytic function in D_Z. Finally applying the result as in [2], we can find an analytic function $\Phi(Z)$ in D_Z satisfying the boundary conditions

$$\begin{aligned}&\text{Re}[\overline{\lambda(z(Z))}e^{i\text{Im}\phi(Z)}\Phi(Z)]=\{r[z(Z)]-\text{Re}[\overline{\lambda(z(Z))}\psi(Z)]\}e^{-\text{Re}\phi(Z)}\\&\text{on } \partial D^*_,\ \text{Im}[\overline{\lambda(z_0)}e^{i\text{Im}\phi(Z_0)}\Phi(Z_0)]=\{b_0-\text{Im}[\overline{\lambda(z_0)}\psi(Z_0)]\}e^{-\text{Re}\phi(Z_0)},\end{aligned} \tag{18}$$

herein $Z_0 = Z(z_0)$, hence the function $w[z(Z)] = \Phi(Z)e^{\phi(Z)} + \psi(Z)$ in (17) is just the solution of Problem A in D_Z for equation (5).

On the basis of Lemma 1 and the above discussion, we can obtain the estimates of solutions of Problem A for equation (5), namely

Theorem 4. *Any solution $w[z(Z)]$ of Problem* A *for equation* (5) *satisfies the estimates*

$$\hat{C}_\delta[w(z),\overline{D}]=C_\delta[X(Z)w(z(Z)),\overline{D_Z}]\le M_3, \hat{C}_\delta[w(z),\overline{D}]\le M_4(k_1+k_2), \tag{19}$$

in which $X(Z) = |Z-t_1|^{\eta_1}|Z-t_2|^{\eta_2}$, here $\eta_j = \max(-4\gamma_j, 0)+8\delta$, $j = 1, 2$, $\gamma_j (j = 1, 2)$ are as stated in (10), *and $t_1 = -1, t_2 = 1$, δ is a sufficiently small positive constant, $k = (k_0, k_1, k_2)$, and $M_3 = M_3(\delta, k,\ H, D)$, $M_4 = M_4(\delta, k_0, H, D)$ are non-negative constants.*

Proof. Noticing the condition (7), and using Lemma 1 and Theorem 3, we see that the functions $\psi(Z)$, $\phi(Z)$ in (17) satisfy the estimates

$$C_\beta[\psi(Z), \overline{D_Z}] \le M_5,\ C_\beta[\phi(Z), \overline{D_Z}] \le M_5, \tag{20}$$

where $\beta = 2/(m+2)-\varepsilon$, ε is a sufficiently small positive constant, and $M_5 = M_5(\beta, k, H, D)$ is a non-negative constant. Due to the analytic function $\Phi(Z)$ satisfies the boundary condition (18), and from (15) and Theorem 3, we can get the representation and estimate of the analytic function $\Phi(Z)$ in D_Z similar to those in (17) and (15), thus the first estimate of (19) is derived. Moreover we verify the second estimate in (19). If $k = k_1+k_2 > 0$, then the function $w^*(z) = u^*(z) + iv^*(z) = u(z)/k + iv(z)/k$ is a solution of Problem A for equation

$$w^*_{\overline{Z}}=g^*(Z), g^*(Z)=\frac{g(Z)}{kH(\hat{y})}=\frac{1}{H(\hat{y})}[A_1w^*+A_2\overline{w^*}+\frac{A_3}{k}] \text{ in } D_Z, \tag{21}$$

By the proof of the first estimate in (19), we can derive the estimate of the solution $w^*(z)$:

$$\hat{C}_\beta[w^*(z), \overline{D}] \le M_4 = M_4(\beta, k_0, H, D). \tag{22}$$

From the above estimate it follows that the second estimate of (19) holds with $k > 0$. If $k = 0$, we can choose any positive number ε to replace $k = 0$. By using the same proof as before, we have

$$\hat{C}_\beta[w(z), \overline{D}] \le M_4\varepsilon.$$

Let $\varepsilon \to 0$, it is obvious that the second estimate in (19) with $k = 0$ is derived.

3. Solvability of the Riemann-Hilbert Problem for Degenerate Elliptic Complex Equations

Theorem 5. *Suppose that equation* (2) *satisfies Condition* C. *Then Problem* A *for* (5) *has a unique solution in* D.

Proof. We first verify the uniqueness of the solution of Problem A for system (2) or equation (5). Let $w_1(z), w_2(z)$ be any two solutions of Problem A for (5). It is easy to see that $w(z) = w_1(z) - w_2(z)$ satisfies the homogeneous equation and boundary conditions

$$\begin{gathered} w_{\overline{Z}} = [A_1 w + A_2 \overline{w}]/H(\hat{y}) \text{ in } D_Z, \\ \mathrm{Re}[\overline{\lambda((Z))} w(z(Z))] = 0 \text{ in } \partial D^*, \ \mathrm{Im}[\overline{\lambda(z_0)} \Phi(Z_0)] = 0. \end{gathered} \tag{23}$$

Due to the solution $w[z(Z)]$ possesses the expression (17), but $\psi(Z) = 0$ in D_Z, and the index $K = 0$ of $\lambda[z(Z)]$ on ∂D_Z, from the second estimate in (19) with $k_1 = k_2 = 0$, it is easy to derive that $w(z) = w_1(z) - w_2(z) = 0$ in D.

As for the existence of solutions of Problem A for equation (5), we can prove by using the method of parameter extension. In fact, the complex equation (5) can be rewritten as

$$\begin{gathered} w_{\overline{Z}} = F(Z, w), \\ F(Z, w) = \frac{1}{H(\hat{y})} \{A_1[z(Z)] w + A_2[z(Z)] \overline{w} + A_3[z(Z)]\} \text{ in } D_Z. \end{gathered} \tag{24}$$

In order to find a solution $w(z)$ of Problem A in D, we can express $w(z)$ in the form (17), and consider the equation with the parameter $t \in [0, 1]$:

$$w_{\overline{Z}} - tF(z, w) = S(z) \text{ in } \overline{D_Z}, \tag{25}$$

in which the function $S(z)$ satisfies the condition

$$H(y) X(Z) S(z) \in L_\infty(\overline{D_Z}), \tag{26}$$

where $X(Z)$ is as stated in (15). This problem is called Problem A_t.

When $t = 0$, the complex equation (25) becomes the equation

$$w_{\overline{Z}} = S(z) \text{ in } \overline{D_Z}. \tag{27}$$

It is clear that the unique solution of Problem A_0, i.e. Problem A for $w_{\overline{Z}} = S(z)$ can be found, namely $X(Z) w[z(Z)] = \Phi(Z) + TXS$. Suppose that when $t = t_0$ $(0 \le t_0 < 1)$, Problem A_{t_0} is solvable, i.e. Problem A_{t_0} for (25) has a solution $w_0(z)$ ($w_0(z) \in \hat{C}(\overline{D})$, hence $X[Z(z)] w_0(z) \in C(\overline{D})$).

We can find a neighborhood $T_\varepsilon = \{|t - t_0| \le \varepsilon, 0 \le t \le 1\}(0 < \varepsilon < 1)$ of t_0 such that for every $t \in T_\varepsilon$, Problem A_t is solvable. In fact, Problem A_t can be written in the form

$$w_{\overline{Z}} - t_0 F(z, w) = (t_0 - t)F(z, w) + S(z) \text{ in } \overline{D_Z}. \tag{28}$$

Replacing $w_0(z)$ into the right-hand side of (28) by a function $w_0(z) \in \hat{C}(\overline{D})$), especially, we select $w_0(z) = 0$ and substitute it into the right-hand side of (28), it is obvious that the boundary value problem for such equation in (28) then has a solution $w_1(z) \in \hat{C}(\overline{D})$. Using successive iteration, we obtain a sequence of solutions $w_n(z)\,(w_n(z) \in \hat{C}(\overline{D}),\ n = 1, 2, ...)$, which satisfy the equations

$$w_{n+1\overline{Z}} - t_0 F(z, w_{n+1}) = (t - t_0)F(z, w_n) + S(z) \text{ in } \overline{D},$$
$$\mathrm{Re}[\overline{\lambda(z)} w_{n+1}(z)] = r(z) \text{ on } \partial D^*, \mathrm{Im}[\overline{\lambda(z_0)} w_{n+1}(z_0)] = b_0.$$

From the above formulas, it follows that

$$[w_{n+1} - w_n]_{\overline{Z}} - t_0[F(z, w_{n+1}) - F(z, w_n)]$$
$$= (t - t_0)[F(z, w_n) - F(z, w_{n-1})] \text{ in } D,$$
$$\mathrm{Re}[\overline{\lambda(z)}(w_{n+1}(z) - w_n(z))] = 0 \text{ on } \partial D^*,$$
$$\mathrm{Im}[\overline{\lambda(z_0)}(w_{n+1}(z_0) - w_n(z_0))] = 0.$$

Noting that

$$L_\infty[H(\hat{y})X(Z)(F(z, w_n) - F(z, w_{n-1})), \overline{D_Z}] \le 2k_0 \hat{C}[w_n - w_{n-1}, \overline{D_Z}],$$

and then by Theorem 4, we can derive

$$\hat{C}[w_{n+1} - w_n, \overline{D_Z}] \le 2|t - t_0| M_4 \hat{C}[w_n - w_{n-1}, \overline{D_Z}],$$

where the constant $M_4 = M_4(\beta, k_0, H, D)$ is as stated in (19). Choosing the constant ε so small such that $2\varepsilon M_4 \le 1/2$ and $|t - t_0| \le \varepsilon$, it follows that

$$\hat{C}[w_{n+1} - w_n, \overline{D_Z}] \le 2\varepsilon M_4 \hat{C}[w_n - w_{n-1}, \overline{D_Z}] \le \frac{1}{2}\hat{C}[w_n - w_{n-1}, \overline{D_Z}],$$

and when $n, m \ge N_0 + 1$ (N_0 is a positive integer),

$$\hat{C}[w_{n+1} - w_n, \overline{D_Z}] \le 2^{-N_0} \sum_{j=0}^{\infty} 2^{-j} \hat{C}[w_1 - w_0, \overline{D_Z}] \le 2^{-N_0+1} \hat{C}[w_1 - w_0, \overline{D_Z}].$$

Hence $\{w_n(z)\}$ is a Cauchy sequence. According to the completeness of the Banach space $\hat{C}(\overline{D_Z})$, there exists a function $w_*(z) \in \hat{C}(\overline{D_Z})$, so that $\hat{C}[w_n - w_*, \overline{D_Z}] \to 0$ as $n \to \infty$, we can see that $w_*(z)$ is a solution of Problem A_t for every $t \in T_\varepsilon = \{|t - t_0| \le \varepsilon\}$. Because the constant ε is independent of t_0 $(0 \le t_0 < 1)$, therefore from the solvability of Problem A_{t_0} when $t_0 = 0$, we can derive the solvability of Problem A_t when $t = \varepsilon, 2\varepsilon, ..., [1/\varepsilon]\varepsilon, 1$, where $[1/\varepsilon]$ means the integer part of $1/\varepsilon$. In particular, when $t = 1$ and $S(z) = 0$, Problem A_1, namely Problem A for (5) in D has a solution $w(z)$.

References

[1] G. C. Wen and H. Begehr, *Boundary value problems for elliptic equations and systems,* Longman Scientific and Technical, Harlow, 1990.

[2] G. C. Wen, *Conformal mappings and boundary value problems,* Amer. Math. Soc., Providence, RI, 1992.

[3] G. C. Wen, *Elliptic, hyperbolic and mixed complex equations with parabolic degeneracy,* World Scientific Publishing Co., Singapore, 2008.

SOME BOUNDARY VALUE PROBLEMS FOR SECOND ORDER MIXED EQUATIONS WITH DEGENERATE LINES

KE-MING XU, GUANG-WU YANG and SHENG-XUN LI

Dept. of Math., Hebei Univ. of Sci. and Tech., Shijiangzhuang 050018, China

E-mail: xiaobing1_111@sina.com

In [1–4], the authors posed and discussed the Tricomi problem of second order equations of mixed type, and in [6], the author discussed the oblique derivative problems of second order equations of mixed type without parabolic degeneracy. The present paper deals with some discontinuous oblique derivative problems for second order linear equations of mixed (elliptic-hyperbolic) type with parabolic degeneracy. Firstly, we give the formulation and representation of solutions of the above boundary value problems, and then prove the existence of solutions for the problems, in which we use the complex analytic method, namely the complex functions in the elliptic domain and the hyperbolic complex functions in hyperbolic domain are used (see [5,6]).

Keywords: Discontinuous oblique derivative problems, mixed equations, parabolic degeneracy.

AMS No: 35M05, 35J70, 35L80

1. Formulation of Discontinuous Oblique Derivative Problems for Second Order Mixed Equations

Let D be a simply connected bounded domain in the complex plane $\mathbb{C}$ with the boundary $\partial D = \Gamma \cup L$, where $\Gamma\,(\subset \{y>0\}) \in C^2_\alpha\,(0<\alpha<1)$ is a curve with the end points $z=0,2$, without loss of generality, we may assume that the boundary Γ possesses the form $x=-1+\tilde{G}(y)$ and $x=1-\tilde{G}(y)$ including line segments $x=0$ and $x=2$ near 0 and 2 respectively, and $L=L_1\cup L_2$, where

$$\begin{aligned} L_1 &= \{x+\int_0^y \sqrt{|K(t)|}dt = 0,\ x\in[0,1]\}, \\ L_2 &= \{x-\int_0^y \sqrt{|K(t)|}dt = 2,\ x\in[1,2]\}, \end{aligned} \tag{1}$$

where $K(y)=\text{sgn}y|y|^m h(y)$ is continuous in $\overline{D}$, possesses the derivative $K'(y)$ and $yK(y)>0$ on $y\neq 0$, $K(0)=0$, here m is a positive number, $h(y)$ is a continuously differentiable function on $|y|\geq 0$, and z_0 is the intersection point of L_1 and L_2. Denote $D^+=D\cap\{y>0\}$, $D^-=D\cap\{y<0\}$. Consider second order linear equation of mixed type with parabolic degeneracy

$$Lu=K(y)u_{xx}+u_{yy}+a(x,y)u_x+b(x,y)u_y+c(x,y)u=-d(x,y) \text{ in } D, \tag{2}$$

where $a = b = c = d = 0$ in D, then (2) is so-called Chaplygin equation in gas dynamics. Suppose that the coefficients of (2) satisfy the following conditions, namely

Condition C. The coefficients a, b, c, d satisfy

$$\begin{gathered}\tilde{L}_\infty[\eta,\overline{D^+}]=L_\infty[\eta,\overline{D^+}]+L_\infty[\eta_x,\overline{D^+}]\le k_0,\eta=a,b,c,\tilde{L}_\infty[d,\overline{D^+}]\le k_1,\\ \tilde{C}[\eta,\overline{D^-}]=[\eta,\overline{D^-}]+C[\eta_x,\overline{D^-}]\le k_0,\eta=a,b,c,\tilde{C}[d,\overline{D^-}]\le k_1,\\ c\le 0 \text{ in } \overline{D^+},\ |a(x,y)||y|^{1-m/2}=o(1) \text{ in } D^-,\ m\ge 2,\end{gathered}\tag{3}$$

where $k_0\,(\ge \max[2\sqrt{h(y)},1/\sqrt{h(y)},1])$, $k_1\,(\ge k_0)$ are two positive constants. There exist n points $z_1 = a_1 = 0$, $z_2 = a_2$, ..., $z_{n-1} = a_{n-1}, z_n = a_n = 2$ on Γ arranged according to the oppositive direction successively.

The discontinuous oblique derivative problems for equation (2) may be formulated as follows:

Problem P. Find a continuous solution $u(z)$ of equation (2) in $\overline{D}$, where $H(y)u_x$, u_y satisfy the boundary conditions

$$\begin{gathered}\frac{1}{2}\frac{\partial u}{\partial l}=\frac{1}{H(y)}\mathrm{Re}\,[\overline{\lambda(z)}u_{\bar z}]=\mathrm{Re}\,[\overline{\Lambda(z)}u_z]=r(z),\ \ z\in\Gamma^*\cup L_1,\\ \frac{1}{H(y)}\mathrm{Im}[\overline{\lambda(z)}u_{\bar z}]|_{z=z_0}=c_0,\ u(a_1)=c_1,\ u(a_n)=c_2,\end{gathered}\tag{4}$$

where l is the vector of $\Gamma^*\cup L_1$, $\Gamma^*=\Gamma\backslash\{a_1,a_2,...,a_n\}$, l is a vector at every point on Γ, $\lambda(z)=\cos(l,x)-i\cos(l,y)$, $z\in\Gamma^*$, $\lambda(z)=\cos(l,x)+j\cos(l,y)$, $z\in L_1$, c_0,c_1,c_2 are real constants, and $\lambda(z)$, $r(z)$, c_0,c_1,c_2 satisfy the conditions

$$\begin{gathered}C^1_\alpha[\lambda(z),\Gamma^*]\le k_0, C^1_\alpha[r(z),\Gamma^*]\le k_2, C^1_\alpha[\lambda(x),L_1]\le k_0,\\ C^1_\alpha[r(x),L_1]\le k_2,\ \cos(l,n)\ge 0 \text{ on } \Gamma^*\cup L_1,\\ |c_0|,|c_1|,|c_2|\le k_2,\ \max_{z\in L_1}\frac{1}{|a(x)-b(x)|}\le k_0,\end{gathered}\tag{5}$$

where n is the outward normal vector on Γ, $\alpha\,(0<\alpha<1)$, k_0, k_2 are positive constants. Here we mention that if $c = 0$ in equation (1), then we can cancel the condition $\cos(l,n)\ge 0$ on Γ^*, and the last condition in (4) can be replaced by

$$lw(z_{n+1})=\mathrm{Im}[\overline{\lambda(z)}u_{\bar z}]|_{z=z_{n+1}}=H(\mathrm{Im}z_{n+1})c_2=c_2',\tag{6}$$

where z_{n+1} is a fixed point at Γ^*, such that c_2 is a real constant satisfying the condition $|c_2|\le k_2$. Later on we shall use the notation $lw(z_{n+1})$ in some formulas below. The above boundary value problem is called Problem Q.

In particular, for the Tricomi problem with the boundary conditions

$$u(z) = \phi(z) \text{ on } \Gamma, \ u(z) = \psi(z) \text{ on } L_1, \tag{7}$$

in which $\phi(z)$, $\psi(x)$ satisfy the conditions

$$C^2_\alpha[\phi(z), \Gamma] \le k_2, \ C^2_\alpha[\psi(z), L_1] \le k_2, \tag{8}$$

where k_2 is as stated before.

If the boundary Γ near $z = 0, 2$ possesses the form $x = G(y)$ or $x = 2 - G(y)$ respectively, we find the derivative for (7) according to the parameter $s = \operatorname{Re} z = y$ on Γ, and obtain

$$u_s = u_x x_y + u_y = \tilde{H}(y) u_x + u_y = \phi'(y) \text{ on } \Gamma \text{ near } x = 0,$$

$$u_s = u_x x_y + u_y = -\tilde{H}(y) u_x + u_y / \tilde{H}(y) = \phi'(y) \text{ on } \Gamma \text{ near } x = 2,$$

$$u_s = u_x x_y + u_y = -H(y) u_x + u_y = \psi'(y) \text{ on } L_1,$$

in which $\tilde{H}(y) = \tilde{G}'(y)$ on Γ, $H(y) = |K(y)|^{1/2}$, if $H(y) = |y|^{m/2}$, then $x = G(y) = 2|y|^{(m+2)/2}/(m+2)$, and its inverse function is $|y| = G^{-1}(x) = [(m+2)x/2]^{2/(m+2)} = Jx^{2/(m+2)}$ and $H(y) = [(m+2)x/2]^{m/(m+2)} = J^{m/2} x^{m/(m+2)}$ on L_1. It is clear that the complex form of above conditions is as follows

$$\operatorname{Re}[\overline{\lambda(z)}(U + iV)] = \operatorname{Re}[\overline{\lambda(z)}(H(y)u_x - iu_y)]/2 = R(z) \text{ on } \Gamma,$$

$$\operatorname{Re}[\overline{\lambda(z)}(U + jV)] = \operatorname{Re}[\overline{\lambda(z)}(H(y)u_x - ju_y)]/2 = R(z) \text{ on } L_1,$$

$$\operatorname{Im}[\overline{\lambda(z)} u_{\bar{z}}]|_{z_0} = H(\operatorname{Im} z_0) c_0 = c_0', \ u(0) = c_1, u(2) = c_2,$$

where $U = H(y)U_x/2, V = -u_y/2$, $c_1 = \phi(0)$, $c_2 = \phi(2)$, $\lambda(z) = a + ib$ on Γ, $\lambda(z) = a + jb$ on L_1, and

$$\lambda(z) = \begin{cases} -i \text{ on } \Gamma, \text{ if } x_y = \tilde{H}(y) \text{ at } 0, \\ i \text{ on } \Gamma, \text{ if } x_y = -\tilde{H}(y) \text{ at } 2, \\ i \text{ on } L_0, \\ 1 - j \text{ on } L_1, \text{ if } x_y = -H(y), \end{cases} \quad R(z) = \begin{cases} \phi_y/2 \text{ on } \Gamma \text{ at } 0, \\ -\phi_y/2 \text{ on } \Gamma \text{ at } 2, \\ -u_y/2 \text{ on } L_0, \\ -\psi_y/2 \text{ on } L_1, \end{cases}$$

and $b_1 = \operatorname{Im}[(1 + j)u_{\bar{z}}(0)] = -\psi_y/2|_{z=0}$, where $L_0 = \{0 < x < 2, y = 0\}$, $x_y|_{z=0} = \tilde{G}'(y)|_{z=0} = \tilde{H}(y)|_{z=0} = 0$, and $x_y|_{z=2} = -\tilde{G}'(y)|_{z=2} = -\tilde{H}(y)|_{z=2} = 0$. Thus the boundary condition of Problem T in D^+ can be rewritten in the complex form

$$\text{Re}[\overline{\lambda(z)}W(z)] = R(z), \ \ z \in \Gamma^* \cup L_1,$$
$$\text{Im}[\overline{\lambda(z_0)}W(z_0)] = c_1', \ u(0) = c_0, \ u(2) = c_2,$$

in which $R_1(x) = J^{m/2}x^{m/(m+2)}\psi'(x)/2\sqrt{2}$, and $R_4(x)$ is an undetermined function. We have

$$e^{i\phi_1} = \frac{\lambda(t_1-0)}{\lambda(t_1+0)} = e^{-\pi i/2-\pi i/2} = e^{-\pi i}, \gamma_1 = \frac{-\pi}{\pi} - K_1 = 0, \ K_1 = -1,$$
$$e^{i\phi_n} = \frac{\lambda(t_n-0)}{\lambda(t_n+0)} = e^{\pi i/2-\pi i/2} = e^{0\pi i}, \ \gamma_n = \frac{0\pi i}{\pi} - K_n = 0, \ K_n = 0,$$
$$e^{i\phi_l} = \frac{\lambda(t_l-0)}{\lambda(t_l+0)}, \ \gamma_l = \frac{\phi_l}{\pi} - K_l, \ K_l = 0 \ \text{ or } \ 1, \ l = 2, ..., n-1,$$

where $t_1 = a_1 = 0, t_2 = a_2, ..., t_n = a_n$, $\lambda(t) = e^{\pi i/2}$ on L_0, $\lambda(t_1+0) = \lambda(t_n - 0) = \exp(i\pi/2)$, $-1 < \gamma_l < 1$, $l = 1, ..., n-1$, and $K_1, K_2, ..., K_n$ are chosen, such that the index of Problem P in D^+ is $K = (K_1 + \cdots + K_n)/2 = 0$ or $-1/2$ and the other requirements of Problem P.

Noting that the conditions in (5), we find two twice continuously differentiable functions $u_0^\pm(z)$ in $D^\pm$, for instance, which are the solutions of the discontinuous oblique derivative problem with the boundary condition on $\Gamma^* \cup L_1$ in (4) for harmonic equations in $D^\pm$, thus the functions $v(z) = v^\pm(z) = u(z) - u_0^\pm(z)$ in $\overline{D^\pm}$ is the solution of the following equation

$$K(y)v_{xx} + v_{yy} + a(x,y)v_x + b(x,y)v_y + c(x,y)v + \tilde{d}(x,y) = 0 \ \text{ in } D \tag{9}$$

satisfying the corresponding boundary conditions

$$\begin{gathered} \text{Re}[\overline{\lambda(z)}W(z)] = R(z) \ \text{ on } \Gamma_* \cup L_1, \ \text{Im}[\overline{\lambda(z)}W(z)]|_{z_0} = c_0', \\ v(0) = c_1, \ v(2) = c_2 \ \text{ or } \ lw(z_{n+1}) = c_2', \end{gathered} \tag{10}$$

where $\tilde{d} = d + Lu_0^\pm(z)$ in $D^\pm$, $W(z) = U + iV = v_{\bar{z}}^+$ in $\overline{D^+}$, $W(z) = U + jV = v_{\bar{z}}^-$ in $\overline{D^-}$, $R(z) = 0$ on $\Gamma^* \cup L_1$ and $c_0 = c_2 = 0$, and under certain conditions, the coefficients of (9) satisfy the conditions similar to Condition C, hence later on we only discuss the homogeneous boundary value problem (Problem $\tilde{P}$) for equation (2) with the boundary condition (10) and the case of index $K = 0$. Other cases can be similarly discussed. From $v(z) = v^\pm(z) = u(z) - u_0^\pm(z)$ in $\overline{D^\pm}$, we have $u(z) = v^-(z) + u_0^-(z)$ in $\overline{D^-}$, $u(z) = v^+(z) + u_0^+(z)$ in $\overline{D^+}$, $v^+(z) = v^-(z) - u_0^+(z) + u_0^-(z)$, $u_y = v_y^\pm + u_{0y}^\pm$, $v_y^+ = v_y^- - u_{0y}^+ + u_{0y}^- = 2\hat{R}_0(x)$, and $v_y^- = 2\tilde{R}_0(x)$ on $L_0 = D \cap \{y = 0\}$, where $\hat{R}_0(x)$, $\tilde{R}(x)$ are undetermined functions.

2. Representation of Solutions for Discontinuous Oblique Derivative Problems for Second Order Mixed Equations

Denote

$$W(z)=U+iV=\frac{1}{2}[H(y)u_x-iu_y]=u_{\tilde{z}}=\frac{H(y)}{2}[u_x-iu_Y]=H(y)u_Z,$$

$$W_{\bar{\tilde{z}}}=[H(y)W_x+iW_y]/2=H(y)[W_x+iW_Y]/2=H(y)W_{\overline{Z}} \text{ in } \overline{D^+},$$

$$W(z)=U+jV=\frac{1}{2}[H(y)u_x-ju_y]=u_{\tilde{z}}=\frac{H(y)}{2}[u_x-ju_Y]=H(y)u_Z,$$

$$W_{\bar{\tilde{z}}}=[H(y)W_x+jW_y]/2=H(y)[W_x+jW_Y]/2=H(y)W_{\overline{Z}} \text{ in } \overline{D^-},$$

we have

$$H(y)W_{\overline{Z}}=W_{\bar{\tilde{z}}}=He_1(U+V)_\mu+He_2(U-V)_\nu=$$
$$=\frac{e_1}{4}\{[\frac{a}{H}+\frac{H_y}{H}-b](U+V)+[\frac{a}{H}+\frac{H_y}{H}+b](U-V)+cu+d\}$$
$$+\frac{e_2}{4}\{[\frac{a}{H}-\frac{H_y}{H}-b](U+V)+[\frac{a}{H}-\frac{H_y}{H}+b](U-V)+cu+d\},$$
$$H(U+V)_\mu=\frac{1}{4}\{[\frac{a}{H}+\frac{H_y}{H}-b](U+V)+[\frac{a}{H}+\frac{H_y}{H}+b](U-V)+cu+d\},$$
$$H(U-V)_\nu=\frac{1}{4}\{[\frac{a}{H}-\frac{H_y}{H}-b](U+V)+[\frac{a}{H}-\frac{H_y}{H}+b](U-V)+cu+d\} \text{ in } \overline{D^-}, \tag{11}$$

in which $Z=x+iG(y)$ in $\overline{D^+}$, and $Z=x+jG(y)$ in $\overline{D^-}$, thus from (11) and the formula in [7], we have

$$W_{\bar{\tilde{z}}}=A_1(z)W+A_2(z)\overline{W}+A_3(z)u+A_4(z) \text{ in } D, \tag{12}$$

where

$$A_1=\begin{cases}\frac{1}{4}[-\frac{a}{H}+\frac{iH_y}{H}-ib],\\ \frac{1}{4}[\frac{a}{H}+\frac{jH_y}{H}-jb],\end{cases} \quad A_2=\begin{cases}\frac{1}{4}[-\frac{a}{H}+\frac{iH_y}{H}+ib],\\ \frac{1}{4}[\frac{a}{H}+\frac{jH_y}{H}+jb],\end{cases}$$

$$A_3=\begin{cases}-\frac{c}{4},\\ \frac{c}{4},\end{cases} \quad A_4=\begin{cases}-\frac{d}{4} & \text{in } \overline{D^+},\\ \frac{d}{4} & \text{in } \overline{D^-}.\end{cases}$$

The boundary value problem for equations (12) with the boundary condition (4), (5) or (7) ($W(z) = u_{\bar{z}}$) and the relation

$$u(z) = \begin{cases} 2\mathrm{Re}\displaystyle\int_0^z [\frac{\mathrm{Re}W(z)}{H(y)} + i\mathrm{Im}W(z)]dz + b_0 \text{ in } \overline{D^+}, \\ 2\mathrm{Re}\displaystyle\int_0^z [\frac{\mathrm{Re}W(z)}{H(y)} - j\mathrm{Im}W(z)]dz + b_0 \text{ in } \overline{D^-}, \end{cases} \tag{13}$$

will be called Problem B.

The representation of solutions of Problem P for equation (2) is as follows.

Theorem 1. *Suppose that equation* (2) *satisfies Condition* C *and* $u(z)$ *is any solution of Problem* P *for equation* (2) *in* D. *Then the solution* $u(z)$ *can be expressed as follows*

$$\begin{aligned}
&u(z) = -2\int_0^y V(z)dy + u(x) = 2\mathrm{Re}\int_0^z [\frac{\mathrm{Re}W}{H} + \begin{pmatrix} i \\ -j \end{pmatrix} \mathrm{Im}W]dz + b_0 \text{ in } \begin{pmatrix} \overline{D^+} \\ \overline{D^-} \end{pmatrix}, \\
&W(z) = \Phi(Z) + \Psi(Z) = \hat{\Phi}(Z) + \hat{\Psi}(Z),\ \Phi(Z) = T(Z) + \overline{T(\overline{Z})}, \\
&\hat{\Psi}(Z) = T(Z) - \overline{T(\overline{Z})},\ T(Z) = -\frac{1}{\pi}\iint_{D_t^+} \frac{f(t)}{t - Z} d\sigma_t \text{ in } \overline{D_Z^+}, \\
&W(z) = \phi(z) + \psi(z) = \xi(z)e_1 + \eta(z)e_2 \text{ in } \overline{D^-}, \\
&\xi(z) = \zeta(z) + \int_0^y g_1(z)dy = \int_{S_1} g_1(z)dy + \int_0^y g_1(z)dy, \\
&g_1(z) = \tilde{A}_1(U+V) + \tilde{B}_1(U-V) + 2\tilde{C}_1 U + \tilde{D}_1 u + \tilde{E}_1,\ z \in s_1, \\
&\eta(z) = \theta(z) + \int_0^y g_2(z)dy = \int_{S_2} g_2(z)dy + \int_0^y g_2(z)dy, \\
&g_2(z) = \tilde{A}_2(U+V) + \tilde{B}_2(U-V) + 2\tilde{C}_2 U + \tilde{D}_2 u + \tilde{E}_2,\ z \in s_2,
\end{aligned} \tag{14}$$

in which

$$f(Z) = g(Z)/H,\ U = Hu_x/2, V = -u_y/2,$$

and $\tilde{\xi}(z) = \int_{S_1} g_1(z)dy$, $\theta(z) = -\zeta(x + G(y))$ *on* D^-, *and* s_1, s_2 *are two families of characteristics in* D^-:

$$s_1:\ \frac{dx}{dy} = \sqrt{|K(y)|} = H(y),\ s_2:\ \frac{dx}{dy} = -\sqrt{|K(y)|} = -H(y) \tag{15}$$

passing through $z = x + jy \in D^-$, S_1 *is the characteristic curves from the points on* L_1 *to the corresponding points on* L_0 *respectively, and*

$$W(z) = U(z) + jV(z) = \frac{1}{2}Hu_x - \frac{j}{2}u_y,$$

$$\xi(z)=\mathrm{Re}\psi(z)+\mathrm{Im}\psi(z), \eta(z)=\mathrm{Re}\psi(z)-\mathrm{Im}\psi(z),$$

$$\tilde{A}_1=\tilde{B}_2=\frac{1}{2}(\frac{h_y}{2h}-b), \tilde{A}_2=\tilde{B}_1=\frac{1}{2}(\frac{h_y}{2h}+b), \tilde{C}_1=\frac{a}{2H}+\frac{m}{4y},$$

$$\tilde{c}_2 = -\frac{a}{2H} + \frac{m}{4y}, \tilde{D}_1 = -\tilde{D}_2 = \frac{c}{2}, \tilde{E}_1 = -\tilde{E}_2 = \frac{d}{2},$$

in which we choose $H(y) = [|y|^m h(y)]^{1/2}$, $h(y)$ *is a positive continuously differentiable function and*

$$d\mu=d[x + G(y)]=2H(y)dy \text{ on } s_1,\ d\nu=d[x-G(y)]=-2H(y)dy \text{ on } s_2.$$

Proof. From (11) we see that $[Tf]_{\overline{Z}} = f(Z)$, $[\overline{Tf}]_{\overline{Z}} = 0$, $[\Psi]_{\overline{Z}} = 0$, $[\hat{\Phi}]_{\overline{Z}} = 0$ in D_Z^+, and equation (2) in D^- can be reduced to the system of integral equations (14).

3. Solvability of Discontinuous Oblique Derivative Problems for Second Order Mixed Equations

In this section, we prove the existence of solutions of Problems $\tilde{P}$ for equation (12). Firstly we discuss the complex equation

$$W_{\bar{\bar{z}}} = A_1(z)W + A_2(z)\overline{W} + A_3(z)u + A_4(z) \text{ in } D, \tag{16}$$

with the relation (13) and the boundary conditions

$$\begin{gathered}\mathrm{Re}[\overline{\lambda(z)}W(z)]=R(z) \text{ on } \Gamma^*\cup L_1,\ \mathrm{Im}[\overline{\lambda(z)}W(z)]|_{z_0} = c_0,\\ v(0) = c_1,\ v(2) = c_2 \text{ or } lw(z_{n+1}) = c_2',\end{gathered} \tag{17}$$

is called Problem B, where $\lambda(z), R(z), c_l\ (l = 0, 1, 2)$ are as stated in (4)–(6) and (9)–(10). The above boundary value problem about $W(z) = v_{\bar{z}}$ is called Problem $\tilde{B}$. It is not difficult to see that Problem $\tilde{B}$ can be divided into Problem B_1 for equation (16), (13) in D^+ with the boundary conditions

$$\mathrm{Re}[\overline{\lambda(z)}W(z)] = R(z) = 0 \text{ on } \Gamma,\ \mathrm{Re}[-iW(x)] = -\hat{R}_0(x) \text{ on } L_0, \tag{18}$$

and Problem B_2 for equation (16),(13) in D^- with the boundary conditions

$$\begin{aligned}&\mathrm{Re}[\overline{\lambda(z)}W(z)] = R(z) = 0 \text{ on } L_1,\\ &\mathrm{Re}[(-j)W(x)] = \tilde{R}_0(x) \text{ on } L_0,\\ &\mathrm{Im}[\overline{\lambda(z_0)}W(z_0)] = c_0' = 0,\end{aligned} \tag{19}$$

noting Re$W(x) = H(y)v_x = 0$ on L_0. From the result in [7], we know that Problem B_1 for equation (16), (13) has a unique solution $W(z)$. Hence we only prove the unique solvability of Problem B_2 for (16), (13) in D^-.

Theorem 2. *If equation* (2) *satisfies Condition C, then there exists a solution* $[W(z), v(z)]$ *of Problem* B_2 *for* (16), (13).

Proof. Denote $D_0 = \{\delta_0 \le x \le 2-\delta_0, -\delta \le y \le 0\}$, and s_1, s_2 are the characteristics of families (15) emanating from any points $(b_0, 0)$, $(b_1, 0)$ $(\delta_0 = b_0 < b_1 = 2-\delta_0)$, respectively, which intersect at a point $(x', y') \in D^-$, where δ_0, δ are sufficiently small positive numbers.

We may only discuss the case of $K(y) = -|y|^m h(y)$, because otherwise we can similarly discuss. In order to find a solution of the system of integral equations (14), from (2), we have the condition

$$\frac{|y|a(x,y)}{|y|^{m/2}} = o(1), \text{ i.e. } |a(x,y)| = \varepsilon(y)|y|^{m/2-1},\ m \ge 2, \tag{20}$$

in which $\varepsilon(y) \to 0$ as $y\,(\in D^-) \to 0$. It is clear that for two characteristics s_1, s_2 passing through a point $z = x + jy \in D^-$ and x_1, x_2 are the intersection points with the axis $y = 0$ respectively, for any two points $\tilde{z}_1 = \tilde{x}_1 + j\tilde{y} \in s_1, \tilde{z}_2 = \tilde{x}_2 + j\tilde{y} \in s_2$, we have

$$\begin{aligned} &|\tilde{x}_1 - \tilde{x}_2| \le |x_1 - x_2| = 2\Big|\int_0^y \sqrt{-K(t)}dt\Big| \le \frac{2k_0}{m+2}|y|^{1+m/2} \\ &\quad \le M|y|^{m/2+1},\ |y|^{m/2+1} \le \frac{m+2}{4k_0}|x_1 - x_2|, \end{aligned} \tag{21}$$

in which $M\ (> \max[4\sqrt{h(y)}/(m+2), 1])$ is a positive constant. From (3), we can assume that the coefficients of (14) possess continuously differentiable with respect to $x \in L_0$ and satisfy the conditions

$$\begin{aligned} &|\tilde{A}_l|, |\tilde{A}_{lx}|, |\tilde{B}_l|, |\tilde{B}_{lx}|, |\tilde{D}_l|, |\tilde{D}_{lx}|, |\tilde{E}_l|, |\tilde{E}_{lx}|, 2\sqrt{h}, \left|\frac{1}{\sqrt{h}}\right|, \left|\frac{h_y}{h}\right| \le M, \\ &\qquad z \in \overline{D^-},\ l = 1, 2. \end{aligned} \tag{22}$$

According to the method in [6], it is sufficient to find a solution of Problem B_2 for arbitrary segment $-\delta \le y \le 0$, where δ is a sufficiently small positive number. We can choose $v_0 = 0, \xi_0 = 0, \eta_0 = 0$ and substitute them into the corresponding positions of v, ξ, η in the right-hand sides of (14), and by the successive iteration, we can find the sequences of functions $\{v_k\}$,

$\{\xi_k\}, \{\eta_k\}$, which satisfy the relations

$$v_{k+1}(z) = v_{k+1}(x) - 2\int_0^y V_k(z)dy = v_{k+1}(x) + \int_0^y (\eta_k - \xi_k)dy,$$

$$\xi_{k+1}(z) = \zeta_{k+1}(z) + \int_0^y [\tilde{A}_1\xi_k + \tilde{B}_1\eta_k + \tilde{C}_1(\xi_k+\eta_k) + \tilde{D}_1 u_k + \tilde{E}_1]dy, z \in s_1,$$

$$\eta_{k+1}(z) = \theta_{k+1}(z) + \int_0^y [\tilde{A}_2\xi_k + \tilde{B}_2\eta_k + \tilde{C}_2(\xi_k+\eta_k) + \tilde{D}_2 u_k + \tilde{E}_1]dy, z \in s_2,$$

$$k = 0, 1, 2, ... \tag{23}$$

Setting that

$$\tilde{y} = \hat{y} - \hat{y}_1,\ \tilde{t} = \hat{t} - \hat{y}_1,\ \tilde{v}_{k+1}(z) = v_{k+1}(z) - v_k(z),$$
$$\tilde{\xi}_{k+1}(z) = \xi_{k+1}(z) - \xi_k(z),\ \tilde{\eta}_{k+1}(z) = \eta_{k+1}(z) - \eta_k(z),$$
$$\tilde{\zeta}_{k+1}(z) = \zeta_{k+1}(z) - \zeta_k(z),\ \tilde{\theta}_{k+1}(z) = \theta_{k+1}(z) - \theta_k(z),$$

we shall prove that $\{\tilde{v}_k\}, \{\tilde{\xi}_k\}, \{\tilde{\eta}_k\}, \{\tilde{\zeta}_k\}, \{\tilde{\theta}_k\}$ in D_0 satisfy the estimates

$$|\tilde{v}_k(z) - \tilde{v}_k(x)|, |\tilde{\xi}_k(z) - \tilde{\zeta}_k(z)|, |\tilde{\eta}_k(z) - \tilde{\theta}_k(z)| \le M'\gamma^{k-1}|\hat{y}|^{1-\beta}, |\hat{y}| \le \delta,$$
$$|\tilde{\xi}_k(z)|, |\tilde{\eta}_k(z)| \le M(M_0|\tilde{y}|)^{k-1}/(k-1)!,\ \hat{y} \le -\delta, \text{ or } M'\gamma^{k-1}, |\hat{y}| \le \delta,$$
$$|\tilde{\xi}_k(z_1) - \tilde{\xi}_k(z_2) - \tilde{\zeta}_k(z_1) - \tilde{\zeta}_k(z_2)|, |\tilde{\eta}_k(z_1) - \tilde{\eta}_k(z_2) - \tilde{\theta}_k(z_1) - \tilde{\theta}_k(z_2)|$$
$$\le M'\gamma^{k-1}[|x_1 - x_2|^{1-\beta} + |x_1 - x_2|^{\beta}|\hat{y}|^{\beta'}],\ |\hat{y}| \le \delta, |\tilde{v}_k(z_1) - \tilde{v}_k(z_2)|,$$
$$|\tilde{\xi}_k(z_1) - \tilde{\xi}_k(z_2)|, |\tilde{\eta}_k(z_1) - \tilde{\eta}_k(z_2)| \le M(M_0|\tilde{t}|)^{k-1}|x_1 - x_2|^{1-\beta}/(k-1)!,$$
$$\hat{y} \le -\delta, \text{ or } M'\gamma^{k-1}[|x_1 - x_2|^{1-\beta} + |x_1 - x_2|^{\beta}|t|^{\beta'}], |\tilde{\xi}_k(z) + \tilde{\eta}_k(z)$$
$$-\tilde{\zeta}_k(z) - \tilde{\theta}_k(z)|,\ |\tilde{\xi}_k(z) + \tilde{\eta}_k(z)| \le M'\gamma^{k-1}|x_1 - x_2|^{\beta}|\hat{y}|^{\beta'},\ |\hat{y}| \le \delta,$$
$$|\tilde{\xi}_k(z) + \tilde{\eta}_k(z)| \le M(M_0|\tilde{y}|)^{k-1}|x_1 - x_2|^{1-\beta}/(k-1)!,\ \hat{y} \le -\delta, \tag{24}$$

where $z = x + j\hat{y}$, $z = x + jt$ is the intersection point of s_1, s_2 in (15) passing through z_1, z_2, $\beta' = (1 + m/2)(1 - 3\beta)$, β is a sufficiently small positive constant, such that $(2 + m)\beta < 1$, and M_0, M, M' are sufficiently large positive constants. On the basis of the above estimates, we can derive that $\{v_k\}, \{\xi_k\}, \{\eta_k\}$ in D_0 uniformly converge to v_*, ξ_*, η_* satisfying the system of integral equations

$$v_*(z)=v_*(x)-2\int_0^y V_* dy=v_*(x)+\int_0^y(\eta_*-\xi_*)dy,$$

$$\xi_*(z)=\tilde{\xi}_*(z)+\int_0^y[\tilde{A}_1\xi_*+\tilde{B}_1\eta_*+\tilde{C}_1(\xi_*+\eta_*)+\tilde{D}_1u_*+\tilde{E}_1]dy, z\in s_1,$$

$$\eta_*(z)=\tilde{\eta}_*(z)+\int_0^y[\tilde{A}_2\xi_*+\tilde{B}_2\eta_*+\tilde{C}_2(\xi_*+\eta_*)+\tilde{D}_2u_*+\tilde{E}_2]dy, z\in s_2,$$

and the function $[v_*(z)]_{\bar{z}} = W^*(z)$ satisfies equation (16) and boundary condition (19), this shows that Problem B_2 in D_0 has a solution for equation (16). Hence a solution of Problem B_2 for (16) in D^- is obtained. From the above discussion, we can see that the solution of Problem B_2 for (16) in D^- is unique.

From the above result, we have the following theorem.

Theorem 3. *Let equation* (2) *satisfy Condition C. Then the discontinuous oblique derivative problems* (*Problem P*) *for* (2) *has a unique solution.*

References

[1] L. Bers, *Mathematical Aspects of Subsonic and Transonic Gas Dynamics,* Wiley, New York, 1958.

[2] A. V. Bitsadze, *Some Classes of Partial Differential Equations,* Gordon and Breach, New York, 1988.

[3] J. M. Rassias, *Lecture notes on mixed type partial differential equations,* World Scientific, Singapore, 1990.

[4] H. S. Sun, Tricomi problem for nonlinear equation of mixed type, *Sci. in China,* Series A, 35 (1992), 14–20.

[5] G. C. Wen and H. Begehr, *Boundary value problems for elliptic equations and systems,* Longman Scientific and Technical Company, Harlow, 1990.

[6] G. C. Wen, *Linear and quasilinear complex equations of hyperbolic and mixed Type,* Taylor & Francis, London, 2002.

[7] G. C. Wen, *Elliptic, hyperbolic and mixed complex equations with parabolic degeneracy,* World Scientific, Singapore, 2008.

LOCAL WELL-POSEDNESS FOR THE FOURTH ORDER SCHRÖDINGER EQUATION[1]

HUA ZHANG and HONG-FENG WU

College of Sciences, North China University of Technology, Beijing 100144, China

E-mail: zhanghuamaths@163.com,whfmath@gmail.com

In this paper, we prove a local well-posedness of the Cauchy problem for the fourth order nonlinear Schrödinger equation in one demension.

Keywords: Fourth order nonlinear Schrödinger equation, local well-posedness.

AMS No: 35Q55.

1. Introduction

In this paper, we mainly consider the following equation (4NLS)

$$i\partial_t u = \partial_x^4 u + \partial_x(u^3), \quad u(0,x) = \varphi(x), \quad t, x \in \mathbb{R}, \tag{1.1}$$

(1.1) is invariant under the scaling $u \to u_\lambda = \lambda^{3/2} u(\lambda^4 t, \lambda x)$ and moreover,

$$\|u_\lambda\|_{\dot{H}^s(\mathbb{R}^n)} = \lambda^{s+1} \|u(0,\cdot)\|_{\dot{H}^s(\mathbb{R}^n)}. \tag{1.2}$$

From this point of view, we say that $s = -1$ is the critical regularity index of (1.1).

A large amount of interesting work has been devoted to the study of the Cauchy problem for this equations. One can see [1,2,5–7] and references cited therein. Our main result is as following:

Theorem 1.1. *For any* $s \geq 0$, $\varphi(x) \in H^s$, *there exists a* $T(||\varphi(x)||_{H^s})$, *such that* (1.1) *has an unique solution in* $C([0,T]; H^s)$.

2. Notation and Definitions

For $x, y \in \mathbb{R}$, $x \lesssim y$ means that there exist $C > 0$ such that $|x| \leq C|y|$ and $x \sim y$ means that there exist $C_1, C_2 > 0$ such that $C_1|x| \leq |y| \leq C_2|x|$. For $f \in \mathcal{S}'$ we denote by $\widehat{f}$ or $\mathcal{F}(f)$ the Fourier transform of f for both spatial and time variables. Denote by $\mathcal{F}_x$ the the Fourier transform on spatial variable and if there is no confusion, we still write $\mathcal{F} = \mathcal{F}_x$. Let $\mathbf{Z}$ and $\mathbf{N}$ be the sets of integers and natural numbers, respectively. For convenience, let $\mathbf{Z}_+ = \mathbf{N} \bigcup \{0\}$. For $k \in \mathbf{Z}$, let $I_k = \{\xi : |\xi| \in [2^{k-1}, 2^{k+1}]\}$. For $k \in \mathbf{Z}_+$, let $\widetilde{I}_k = [-2, 2]$ if $k = 0$ and $\widetilde{I}_k = I_k$ if $k \geq 1$. We use $f * g$ will stand for the convolution on time and spatial variables.

[1]This research is supported by the Science Research Startup Foundation of North China University of Technology.The authors would like to express their thanks to Doctor Zihua Guo for his valuable suggestions.

Let $\eta_0 : \mathbb{R} \to [0,1]$ denote an even smooth function supported in $[-8/5, 8/5]$ and equal to 1 in $[-5/4, 5/4]$. For $k \in \mathbf{N}$, let $\eta_k(\xi) = \eta_0(\xi/2^k) - \eta_0(\xi/2^{k-1})$. For $k \in \mathbf{Z}$, let $\chi_k(\xi) = \eta_0(\xi/2^k) - \eta_0(\xi/2^{k-1})$.

We introduce the Banach spaces used in [3]. For $k \in \mathbf{Z}_+$, we define the dyadic $X^{b,s}$-type spaces $X_k = X_k(\mathbb{R}^2)$,

$$X_k = \{f \in L^2(\mathbb{R}^2):$$

$$f(\xi,\tau) \text{ is supported in } I_k \times \mathbb{R} \text{ and } ||f||_{X_k} = \sum_{j=0}^{\infty} 2^{j/2} ||\eta_j(\tau - \xi^4) \cdot f||_{L^2}\}.$$

This kind of spaces were introduced, for instance, in [8] and [9] for the BO equation. For $s \geq 0$, we define the following spaces:

$$F^s = \{u \in \mathcal{S}'(\mathbb{R}^2) : ||u||_{F^s}^2 = \sum_{k \in \mathbf{Z}_+} 2^{2sk} ||\eta_k(\xi)\mathcal{F}(u)||_{X_k}^2 < \infty\},$$

$$N^s = \{u \in \mathcal{S}'(\mathbb{R}^2) : ||u||_{N^s}^2 = \sum_{k \in \mathbf{Z}_+} 2^{2sk} ||(i + \tau - \xi^4)^{-1} \eta_k(\xi)\mathcal{F}(u)||_{X_k}^2 < \infty\}.$$

For $T \geq 0$, we can also define the time-localized spaces $F^s(T)$ and $N^s(T)$.

Proposition 2.1. ([3]) *If $s \in \mathbb{R}$, $T \in (0,1]$, and $u \in F^s(T)$, then*

$$\sup_{t \in [0,T]} ||u(t)||_{H^s} \lesssim ||u||_{F^s(T)}.$$

3. Proof of the Trilinear Estimate

According to the standard fixed point argument, we will need the following trilinear estimate.

Lemma 3.1. *If $s \geq 0$, then exists $C > 0$, such that for any $u, v, w \in F^s$,*

$$||\partial_x(uvw)||_{N^s} \leq C(||u||_{F^s}||v||_{F^0}||w||_{F^0}$$
$$+||v||_{F^s}||u||_{F^0}||w||_{F^0} + ||w||_{F^s}||v||_{F^0}||u||_{F^0}).$$

For $\xi_1, \xi_2, \xi_3 \in \mathbb{R}$ and $\omega : \mathbb{R} \to \mathbb{R}$ defined as $\omega(\xi) = \xi^4$. Let

$$\Omega(\xi_1, \xi_2, \xi_3) = \omega(\xi_1) + \omega(\xi_2) + \omega(\xi_3) - \omega(\xi_1 + \xi_2 + \xi_3). \tag{3.1}$$

For compactly supported functions $f, g, h, u \in L^2(\mathbb{R} \times \mathbb{R})$, let

$$J(f,g,h,u) = \int_{\mathbb{R}^6} f(\xi_1,\mu_1) g(\xi_2,\mu_2) h(\xi_3,\mu_3)$$
$$u(\xi_1+\xi_2+\xi_3, \mu_1+\mu_2+\mu_3+\Omega(\xi_1,\xi_2,\xi_3)) d\xi_1 d\xi_2 d\xi_3 d\mu_1 d\mu_2 d\mu_3.$$

Simple changes of variables in the integration and the observation that the function ω is an even function, we have

$$|J(f,g,h,u)| = |J(g,f,h,u)| = |J(f,h,g,u)|.$$

Therefore, we can only assume $k_1 \leq k_2 \leq k_3$ in this paper. Comparing with ω is odd function case (see [4,9]), we have less symmetry. Therefore, we need to consider the magnitude of quantity k_4 compare with k_1, k_2 and k_3 in later proof. According to the methods in [4] and [11], it suffices to prove the following symmetric estimate. Similar ideas for the bilinear estimates can be found in [9].

Lemma 3.2. *Assume $k_1, k_2, k_3, k_4 \in \mathbf{Z}$, $k_1 \leq k_2 \leq k_3$, $j_1, j_2, j_3, j_4 \in \mathbf{Z}_+$ and $f_{k_i,j_i} \in L^2(\mathbb{R}\times\mathbb{R})$ are nonnegative functions supported in $I_{k_i} \times \widetilde{I}_{j_i}$ $(i = 1,2,3,4)$. For simplicity we write $J = |J(f_{k_1,j_1}, f_{k_2,j_2}, f_{k_3,j_3}, f_{k_4,j_4})|$.*

(a) *For any $k_1 \leq k_2 \leq k_3$ and $j_1, j_2, j_3, j_4 \in \mathbf{Z}_+$,*

$$J \leq C2^{(j_{\min}+j_{thd})/2}2^{(k_{\min}+k_{thd})/2}\prod_{i=1}^{4}||f_{k_i,j_i}||_{L^2}. \tag{3.2}$$

(b) *If $k_{thd} \leq k_{sec} - 5$ and $\exists\ i \in \{1,2,3,4\}$, such that $(k_i, j_i) = (k_{thd}, j_{\max})$,*

$$J \leq C2^{(j_1+j_2+j_3+j_4)/2}2^{-j_{\max}/2}2^{-\frac{3k_{\max}}{2}}2^{k_{thd}/2}\prod_{i=1}^{4}||f_{k_i,j_i}||_{L^2}; \tag{3.3}$$

else, we have

$$J \leq C2^{(j_1+j_2+j_3+j_4)/2}2^{-j_{\max}/2}2^{-\frac{3k_{\max}}{2}}2^{k_{\min}/2}\prod_{i=1}^{4}||f_{k_i,j_i}||_{L^2}. \tag{3.4}$$

(c) *For any $k_1, k_2, k_3, k_4 \in \mathbf{Z}$ and $j_1, j_2, j_3, j_4 \in \mathbf{Z}_+$,*

$$J \leq C2^{(j_1+j_2+j_3+j_4)/2}2^{-j_{\max}/2}2^{-(k_{\min}+k_{thd}+k_{sec})/3}\prod_{i=1}^{4}||f_{k_i,j_i}||_{L^2}. \tag{3.5}$$

(d) *If $k_{\min} \leq k_{\max} - 10$, then*

$$J \leq C2^{(j_1+j_2+j_3+j_4)/2}2^{-\frac{8k_{\max}}{3}}\prod_{i=1}^{4}||f_{k_i,j_i}||_{L^2}. \tag{3.6}$$

Here we use $k_{\max}, k_{sec}, k_{thd}$ and $k_{\min}$ denote the maximum, the second maximum, the third maximum number and the minimum of numbers k_1, k_2, k_3 and k_4. The notations $j_{\max}, j_{sec}, j_{thd}$ and $j_{\min}$ are similar.

Proof. Let $A_{k_i}(\xi) = [\int_{\mathbb{R}} |f_{k_i,j_i}(\xi,\mu)|^2 d\mu]^{\frac{1}{2}}$, $i = 1,2,3,4$. Then $||A_{k_i}||_{L^2_\xi} = ||f_{k_i,j_i}||_{L^2_{\xi,\mu}}$. Using the Cauchy-Schwartz inequality and the support properties of the functions f_{k_i,j_i},

$$\begin{aligned}
&|J(f_{k_1,j_1}, f_{k_2,j_2}, f_{k_3,j_3}, f_{k_4,j_4})| \\
&\leq C2^{(j_{\min}+j_{thd})/2} \int_{\mathbb{R}^3} A_{k_4}(\xi_1+\xi_2+\xi_3) \prod_{i=1}^{3} A_{k_i}(\xi_i) d\xi_i \\
&\leq C2^{(k_{\min}+k_{thd})/2} 2^{(j_{\min}+j_{thd})/2} \prod_{i=1}^{4} ||A_{k_i}||_{L^2},
\end{aligned}$$

which is part (a), as desired.

For part (b), by examining the supports of the functions

$$J(f_{k_1,j_1}, f_{k_2,j_2}, f_{k_3,j_3}, f_{k_4,j_4}) \equiv 0,$$

unless

$$k_{\max} \leq k_{sec} + 10. \tag{3.7}$$

We give the proof according to magnitude of quantity k_4 compare with k_1, k_2 and k_3. Firstly, we assume that $k_4 = k_{max}$. Under this assumption, we first consider that $j_2 \neq j_{max}$ and $j_4 = j_{max}$, then we will prove that if $g_i : \mathbb{R} \to \mathbb{R}_+$ are L^2 nonnegative functions supported in I_{k_i} $(i = 1,2,3)$ and $g : \mathbb{R}^2 \to \mathbb{R}_+$ is an L^2 function supported in $I_{k_4} \times \widetilde{I}_{j_4}$, then

$$\begin{aligned}
&\int_{\mathbb{R}^3} g_1(\xi_1) g_2(\xi_2) g_3(\xi_3) g(\xi_1+\xi_2+\xi_3, \Omega(\xi_1,\xi_2,\xi_3)) d\xi_1 d\xi_2 d\xi_3 \\
&\qquad \lesssim 2^{-\frac{3k_{max}}{2}} 2^{k_{min}/2} \prod_{i=1}^{3} ||g_i||_{L^2} ||g||_{L^2}.
\end{aligned} \tag{3.8}$$

This suffices for (3.3). To prove (3.8), we have $|\xi_3 + \xi_2| \sim |\xi_3|$, since $k_2 \leq k_3 - 5$. By change of variables $\xi_1' = \xi_1$, $\xi_2' = \xi_2$, $\xi_3' = \xi_2 + \xi_3$, we get that the left side of (3.8) is dominated by

$$\int_{|\xi_1'|\sim 2^{k_1}, |\xi_2'|\sim 2^{k_2}, |\xi_3'|\sim 2^{k_3}} g_1(\xi_1') g_2(\xi_2') g_3(\xi_3'-\xi_2') g(\xi_1'+\xi_3', \Omega(\xi_1',\xi_2',\xi_3'-\xi_2')) d\xi_1' d\xi_2' d\xi_3'. \tag{3.9}$$

Note that in the integration area, we have

$$\left|\frac{\partial}{\partial_{\xi_2'}} [\Omega(\xi_1',\xi_2',\xi_3'-\xi_2')]\right| = |\omega'(\xi_2') - \omega'(\xi_3'-\xi_2')| \sim 2^{3k_{\max}}, \tag{3.10}$$

where we use the fact $\omega'(\xi) \sim |\xi|^3$ and $k_2 \leq k_3 - 5$. So, we have

$$||g(\xi_1'+\xi_3', \Omega(\xi_1',\xi_2',\xi_3'-\xi_2'))||_{L^2_{\xi_2'}} = 2^{-\frac{3k_3}{2}} ||g(\xi_1'+\xi_3', \mu_2)||_{L^2_{\mu_2}}.$$

By change of variable $\mu_2 = \Omega(\xi_1', \xi_2', \xi_3' - \xi_2')$, we get that (3.9) is dominated by

$$2^{-\frac{3k_3}{2}} \int_{|\xi_1'| \sim 2^{k_1}} g_1(\xi_1') ||g_2||_{L^2} ||g_3||_{L^2} ||g||_{L^2} d\xi_1' \lesssim 2^{-\frac{3k_3}{2}} 2^{k_{\min}/2} ||g||_{L^2} \prod_{i=1}^{3} ||g_i||_{L^2}. \tag{3.11}$$

If $j_3 = j_{\max}$, it suffices to prove $g_i : \mathbb{R}^2 \to \mathbb{R}_+$ are L^2 nonnegative functions supported in $I_{k_i} \times \widetilde{I}_{j_i}$ $(i = 1, 2, 3, 4)$, then

$$\int_{\mathbb{R}^6} g_4(\xi_1 + \xi_2 + \xi_3, \tau_1 + \tau_2 + \tau_3 + \Omega(\xi_1, \xi_2, \xi_3)) \prod_{i=1}^{3} g_i(\xi_i, \tau_i) d\xi_i d\tau_i \lesssim 2^{-\frac{3k_{\max}}{2}} 2^{k_{\min}/2} \prod_{i=1}^{4} ||g_i||_{L^2}. \tag{3.12}$$

By change of variables $\xi_1' = -\xi_1, \tau_1' = -\tau_1, \xi_2' = -\xi_2, , \tau_2' = -\tau_2, \xi_3' = \xi_1 + \xi_2 + \xi_3, \tau_3' = \tau_1 + \tau_2 + \tau_3 + \Omega(\xi_1, \xi_2, \xi_3)$, the left side of (3.12) is

$$\int_{\mathbb{R}^6} g_3(\xi_1' + \xi_2' + \xi_3', \tau_1' + \tau_2' + \tau_3' - \Omega(-\xi_1', -\xi_2', \xi_1' + \xi_2' + \xi_3')) \times g_4(\xi_3', \tau_3') \prod_{i=1,2} g_i(-\xi_i', -\tau_i') d\xi_1' d\xi_2' d\xi_3' d\tau_1' d\tau_2' d\tau_3'.$$

Note that in the integration area, we have

$$\left| \frac{\partial}{\partial_{\xi_2'}} [\Omega(-\xi_1', -\xi_2', \xi_1' + \xi_2' + \xi_3')] \right| = |\omega'(-\xi_2') + \omega'(\xi_1' + \xi_2' + \xi_3')| \sim 2^{3k_{max}}.$$

So, we have

$$||g_3(\xi_1' + \xi_2' + \xi_3', \tau_1' + \tau_2' + \tau_3' - \Omega(-\xi_1', -\xi_2', \xi_1' + \xi_2' + \xi_3'))||_{L^2_{\xi_2'}} = 2^{-\frac{3k_4}{2}} ||g(\xi_1' + \xi_2' + \xi_3', \mu_2)||_{L^2_{\mu_2}}.$$

By change of variable $\mu_2 = \Omega(-\xi_1', -\xi_2', \xi_1' + \xi_2' + \xi_3')$, we get (3.12), as desired. The case $j_1 = j_{\max}$ is similar to $j_3 = j_{\max}$, we omit it.

We assume now that $j_2 = j_{\max}$, it suffices to prove $g_i : \mathbb{R}^2 \to \mathbb{R}_+$ are L^2 nonnegative functions supported in $I_{k_i} \times \widetilde{I}_{j_i}$ $(i = 1, 2, 3, 4)$, then

$$\int_{\mathbb{R}^6} g_4(\xi_1 + \xi_2 + \xi_3, \tau_1 + \tau_2 + \tau_3 + \Omega(\xi_1, \xi_2, \xi_3)) \prod_{i=1}^{3} g_i(\xi_i, \tau_i) d\xi_i d\tau_i \lesssim 2^{-\frac{3k_{\max}}{2}} 2^{k_{thd}/2} \prod_{i=1}^{4} ||g_i||_{L^2}. \tag{3.13}$$

By change of variables $\xi_2' = \xi_1 + \xi_2 + \xi_3$, $\tau_2' = \tau_1 + \tau_2 + \tau_3 + \Omega(\xi_1, \xi_2, \xi_3)$, $\xi_1' = -\xi_1$, $\tau_1' = -\tau_1$, $\xi_3' = -\xi_3$, $\tau_3' = -\tau_3'$ and notice that

$$\left|\frac{\partial}{\partial_{\xi_2'}}[\Omega(-\xi_1', \xi_1' + \xi_2' + \xi_3', -\xi_3')]\right| = |\omega'(\xi_1' + \xi_2' + \xi_3') - \omega'(\xi_2')| \sim 2^{3k_{\max}}$$

in the integration area, we easily get (3.13).

Secondly, we consider the case $k_3 = k_{\max}$ and $k_4 = k_{sec}$. In view of (3.7), we easily get the desired result, as above.

Thirdly, we consider the case $k_4 = k_{thd}$. Under this assumption, we firstly consider the case that $j_4 = j_{\max}$. In this case, we have $(k_4, j_4) = (k_{thd}, j_{\max})$. Therefore, it suffices to prove that (3.13) also hold under this assumption. By change of variables $\xi_1' = \xi_1$, $\tau_1' = \tau_1$, $\xi_2' = \xi_2$, $\tau_2' = \tau_2$, $\xi_3' = \xi_1 + \xi_2 + \xi_3$, $\tau_3' = \tau_1 + \tau_2 + \tau_3 + \Omega(\xi_1, \xi_2, \xi_3)$ and notice that

$$\left|\frac{\partial}{\partial_{\xi_3'}}[\Omega(\xi_1', \xi_2', \xi_3' - \xi_2' - \xi_1')]\right| = |\omega'(\xi_3' - \xi_2' - \xi_1') - \omega'(\xi_3')| \sim 2^{3k_{\max}} \quad (3.14)$$

in the integration area, we easily get (3.13).

The other cases can be treated similarly, we omit them.

For part (c), we first consider the case $k_4 = k_{\max}$ and $j_4 = j_{\max}$. The proof is parallel to [11], we omit it. we can treat other cases just by change of variables. For example, if $j_3 = j_{\max}$, let $\xi_1' = -\xi_1$, $\tau_1' = -\tau_1$, $\xi_2' = -\xi_2$, $\tau_2' = -\tau_2$, $\xi_3' = \xi_1 + \xi_2 + \xi_3$, $\tau_3' = \tau_1 + \tau_2 + \tau_3 + \Omega(\xi_1, \xi_2, \xi_3)$. So, we can treat this case as the first case.

For part (d), we need to consider two cases: $\xi_1 \cdot \xi_2 > 0$ or $\xi_1 \cdot \xi_2 < 0$. The former case is easier to handle. When $\xi_1 \cdot \xi_2 > 0$ and $k_4 = k_{\max}$, $k_{\min} \le k_{\max} - 10$, we have

$$|\Omega(\xi_1, \xi_2, \xi_3)| \ge 2^{k_2} 2^{k_3} 2^{2k_4} \sim 2^{k_2 + 3k_3}. \quad (3.15)$$

If $k_{thd} \le k_{sec} - 5$, similar to (3.7), by examining the supports of functions, we have $j_{\max} \ge k_2 + 3k_3 - 20$, part (d) holds by part (b).

If $k_{thd} \ge k_{sec} - 5$, observing $-\frac{k_1+k_2+k_3}{3} \sim -\frac{k_1+2k_3}{3}$ and owning to $j_{\max} \ge k_2 + 3k_3 - 20$, part (d) holds by part (c).

If $k_4 = k_{sec}$, in view of (3.7), we have $|\Omega(\xi_1, \xi_2, \xi_3)| = |(\xi_1 + \xi_2)(\xi_1 + \xi_2 + 2\xi_3)(\xi_2^2 + \xi_4^2)| \sim 2^{2k_{sec}} 2^{2k_{\max}} \sim 2^{4k_{\max}}$.

Whether $k_4 = k_{thd}$ or $k_4 = k_{\min}$, it is easy to see that $|\Omega(\xi_1, \xi_2, \xi_3)| \sim 2^{4k_{\max}}$. Which goes back to (3.15).

We assume now $\xi_1 \cdot \xi_2 < 0$. Firstly, we assume that $k_4 = k_{\max}$. Now we should consider serval cases according to $j_i = j_{\max}$. If $j_4 = j_{\max}$, it suffices to prove that if A_i is L^2 nonnegative functions supported in I_{k_i} $(i = 1, 2, 3)$

and B is a L^2 nonnegative function supported in $I_{k_4} \times \widetilde{I}_{j_4}$, then

$$\begin{aligned}\int_{\mathbb{R}^3\cap\{\xi_1\cdot\xi_2<0\}} & B(\xi_1+\xi_2+\xi_3,\Omega(\xi_1,\xi_2,\xi_3))\prod_{i=1}^{3}A_i(\xi_i)d\xi_1 d\xi_2 d\xi_3 \\ & \lesssim 2^{j_4/2}2^{-3k_3}\|B\|_{L^2}\prod_{i=1}^{3}\|A_i\|_{L^2}.\end{aligned} \tag{3.16}$$

By localizing $|\xi_1+\xi_2| \sim 2^l$ for $l \in \mathbf{Z}$, we get that the right-hand side of (3.16) is dominated by

$$\sum_l \int_{\mathbb{R}^3} \chi_l(\xi_1+\xi_2)B(\xi_1+\xi_2+\xi_3,\Omega(\xi_1,\xi_2,\xi_3))\prod_{i=1}^{3}A_i(\xi_i)d\xi_i. \tag{3.17}$$

From the support properties of the functions A_i, B and the fact that in the integration area

$$|\Omega(\xi_1,\xi_2,\xi_3)| = |(\xi_1+\xi_2)(\xi_1+\xi_2+2\xi_3)(\xi_4^2+\xi_3^2)| \sim 2^{l+3k_3}.$$

We get that

$$j_{\max} \geq l + 3k_3 - 20. \tag{3.18}$$

By change of variables $\xi_1' = \xi_1+\xi_2$, $\xi_2' = \xi_2$, $\xi_3' = \xi_1+\xi_3$, we obtain that (3.17) is dominated by

$$\begin{aligned}\sum_l \int_{|\xi_1'|\sim 2^l,|\xi_2'|\sim 2^{k_2},|\xi_3'|\sim 2^{k_3}} & \chi_l(\xi_1')A_1(\xi_1'-\xi_2')A_2(\xi_2')A_3(\xi_2'+\xi_3'-\xi_1') \\ & B(\xi_2'+\xi_3',\Omega(\xi_1'-\xi_2',\xi_2',\xi_2'+\xi_3'-\xi_1'))d\xi_1'd\xi_2'd\xi_3'.\end{aligned} \tag{3.19}$$

Since in the integration area

$$\left|\frac{\partial}{\partial_{\xi_1'}}[\Omega(\xi_1'-\xi_2',\xi_2',\xi_2'+\xi_3'-\xi_1')]\right| = |\omega'(\xi_1'-\xi_2')-\omega'(\xi_2'+\xi_3'-\xi_1')| \sim 2^{3k_3}, \tag{3.20}$$

then we get from (3.20) that (3.19) is dominated by

$$\begin{aligned}&\sum_l \int_{|\xi_1'|\sim 2^l} \chi_l(\xi_1')\|A_1\|_{L^2}\|A_3\|_{L^2} \\ &\times\|A_2(\xi_2')B(\xi_2'+\xi_3',\Omega(\xi_1'-\xi_2',\xi_2',\xi_2'+\xi_3'-\xi_1'))\|_{L^2_{\xi_2',\xi_3'}} d\xi_1' \\ &\lesssim \sum_l 2^{l/2}2^{-\frac{k_3}{2}}\|B\|_{L^2}\prod_{i=1}^{3}\|A_i\|_{L^2} \lesssim 2^{j_{\max}/2}2^{-3k_3}\|B\|_{L^2}\prod_{i=1}^{3}\|A_i\|_{L^2},\end{aligned}$$

where we used (3.18) in the last inequality. When $j_3 = j_{\max}$, by change of variables $\xi_1' = -\xi_1$, $\tau_1' = -\tau_1$, $\xi_2' = -\xi_2$, $\tau_2' = -\tau_2$, $\xi_3' = \xi_1 + \xi_2 + \xi_3$, $\tau_3' = \tau_1 + \tau_2 + \tau_3 + \Omega(\xi_1, \xi_2, \xi_3)$ and notice that

$$|\frac{\partial}{\partial_{\xi_1'}}[\Omega(-\xi_1', -\xi_2', \xi_1' + \xi_2' + \xi_3')]| = |\omega'(\xi_1') + \omega'(\xi_1' + \xi_2' + \xi_3')| \sim 2^{3k_{\max}}$$

in the integration area, which can be treated as before. In other cases, we always have a condition like (3.15) or (3.10), so we can treat them similarly.

4. Local Well-Posedness for 4NLS Equation

From Duhamel's principle, (1.1) is equivalent to the following equation

$$u = S(t)\phi + \int_0^t S(t-t')(\partial_x(u^3)(t'))dt'. \tag{4.1}$$

We will mainly work on the following truncated version

$$u = \psi(t)S(t)\phi + \psi(t)\int_0^t S(t-t')(\partial_x[(\psi(t')u)^3](t'))dt', \tag{4.2}$$

where $S(t) = e^{it\Delta^2}$, $\psi(t) = \eta_0(t)$ is a smooth cut-off function. Then we easily see that if u is a solution to (4.2) on $\mathbb{R}$, then u solves (4.1) on $t \in [-1, 1]$.

Our first proposition is on the estimate for the linear solution.

Proposition 4.1. *If $s \geq 0$ and $\phi \in H^s$, then*

$$||\psi(t) \cdot (S(t)\phi)||_{F^s} \leq C||\phi||_{H^s}. \tag{4.3}$$

Proof. Noticing that $\mathcal{F}[\psi(t) \cdot (S(t)\phi)](\xi, \tau) = \widehat{\phi}(\xi)\widehat{\psi}(\tau - \xi^4)$. In view of definition, it suffices to prove that if $k \in \mathbf{Z}_+$, then

$$||\eta_k(\xi)\widehat{\phi}(\xi)\widehat{\psi}(\tau - \xi^4)||_{X_k} \leq C||\eta_k(\xi)\widehat{\phi}(\xi)||_{L^2}. \tag{4.4}$$

Indeed, from definition we have

$$\begin{aligned}||\eta_k(\xi)\widehat{\phi}(\xi)\widehat{\psi}(\tau - \xi^4)||_{X_k} &\leq C\sum_{j=0}^{\infty} 2^j ||\eta_k(\xi)\widehat{\phi}(\xi)||_{L^2}||\eta_j(\tau)\widehat{\psi}(\tau)||_{L^2} \\ &\leq C||\eta_k(\xi)\widehat{\phi}(\xi)||_{L^2},\end{aligned}$$

which is (4.4), as desired.

The second proposition is on the estimate for the retarded linear term. These estimates were also used in [9] and [4].

Proposition 4.2. *If $l, s \geq 0$ and $u \in \mathcal{S}(\mathbb{R} \times \mathbb{R})$, then*

$$||\psi(t) \cdot \int_0^t S(t-s)(u(s))ds||_{F^s} \leq C||u||_{N^s}.$$

Proof. A straightforward computation shows that

$$\mathcal{F}\left[\psi(t) \cdot \int_0^t S(t-s)(u(s))ds\right](\xi, \tau)$$
$$= c\int_{\mathbb{R}} \mathcal{F}(u)(\xi, \tau')\frac{\widehat{\psi}(\tau-\tau') - \widehat{\psi}(\tau-\xi^4)}{\tau'-\xi^4}d\tau'.$$

For $k \in \mathbf{Z}_+$, let $f_k(\xi, \tau') = \mathcal{F}(u)(\xi, \tau')\chi_k(\xi)(\tau' - \xi^4 + i)^{-1}$. For $f_k \in X_k$, let

$$T(f_k)(\xi, \tau) = \int_{\mathbb{R}} f_k(\xi, \tau')\frac{\widehat{\psi}(\tau-\tau') - \widehat{\psi}(\tau-\xi^4)}{\tau'-\xi^4}(\tau' - \xi^4 + i)d\tau'.$$

In view of the definitions, it suffices to prove that

$$||T||_{X_k \to X_k} \leq C \text{ uniformly in } k \in \mathbf{Z}_+,$$

which follows from the proof of Lemma 4.2 in [9].

Now we prove Theorem 1.1. Define the operator

$$\Phi_\phi(u) = \psi(t)S(t)\phi + \psi(t)\int_0^t S(t-t')(\partial_x((\psi(t')u)^3)(t'))dt',$$

and we will show that when $s \geq 0$, $\Phi_\phi(\cdot)$ is a contraction mapping from $\mathcal{D} = \{u \in F^s : ||u||_{F^s} \leq 2cr\}$ into itself.

From (1.2), it suffices to construct u_λ on the time interval $[-1, 1]$. Observing that for $s \geq 0$, $||u_\lambda||_{H^s} = \lambda^{s+1}||\varphi||_{H^s}$, so we can choose $0 < \lambda = \lambda(||\varphi||_{H^s}) \ll 1$, such that $||u_\lambda||_{H^s} < r$. The choice of the parameter r will be made later. For simplicity, we still denote u_λ by u in later without confusion. From Propositions 4.1, 4.2 and Lemma 3.1, we get if $u \in \mathcal{D}$, then

$$\begin{aligned}||\Phi_\phi(u)||_{F^s} &\leq c||\phi||_{H^s} + ||\psi(t)\int_0^t S(t-t')(\partial_x((\psi(t')u)^3)(t'))dt'||_{F^s}\\ &\leq cr + c||\partial_x((\psi(t')u)^3||_{N^s}\\ &\leq cr + c||u||_{F^s}^3 \leq cr + c(2cr)^3 \leq 2cr,\end{aligned}$$

provided that r satisfies $8c^3r^2 \leq 1/2$. Similarly, for $u, v \in \mathcal{D}$

$$\begin{aligned}||\Phi_\phi(u)-\Phi_\phi(v)||_{F^s} &\leq c||\psi(t)\int_0^t S(t-t')\partial_x(\psi^3(\tau)(u^3(\tau)-v^3(\tau)))dt'||_{F^s}\\ &\leq 8c^3r^2||u-v||_{F^s} \leq \tfrac{1}{2}||u-v||_{F^s}.\end{aligned}$$

Thus $\Phi_\phi(\cdot)$ is a contraction. Therefore, there exists a unique $u \in \mathcal{D}$ solves the integral equation (4.2) in the time interval $[0,1]$. Noticing that Proposition 2.1, we have the solution in $C([0,1];H^s)$. Moreover, we can use the same method in [3] extend the uniqueness to the whole space F^s. Unraveling the scaling, we can get the solution exists on the time interval $[0,\lambda^4]$.

References

1] M. Ben-Artzi, H. Koch and J. C. Saut, Dispersion estimates for fourth order Schrödinger equations, *C. R. Acad. Sci. Paris Sér. I Math.*, 330(2) (2000), 87–92.

[2] B. L. Guo and B. X. Wang, The global Cauchy problem and scattering of solutions for nonlinear Schrödinger equations in H^s, *Diff. Integral Eqs.*, 15(9) (2002), 1073–1083.

[3] Z. H. Guo and B. X. Wang, Global well posedness and inviscid limit for the Korteweg-de Vries-Burgers equation, *J. Diff. Eqs.*, 246(10) (2009), 3864-3901.

[4] Z. H. Guo, Local Well-posedness and a priori bounds for the modified Benjamin-Ono equation without using a gauge transformation, *arXiv:* 0807.3764v1 .

[5] C. C. Hao, L. Hsiao and B. X. Wang, Wellposedness for the fourth order nonlinear Schrödinger equations, *J. Math. Anal. Appl.*, 320(1) (2006), 246–265.

[6] C. C. Hao, L. Hsiao and B. X. Wang, Wellposedness of the fourth order nonlinear Schrödinger equations in multi-dimension spaces, *J. Math. Anal. Appl.*, 328(1) (2007), 58–83.

[7] Z. H. Huo and Y. L. Jia, The Cauchy problem for the fourth-order nonlinear Schrödinger equation related to the vortex filament, *J. Diff. Eqns.*, 214 (2005), 1–35.

[8] A. D. Ionescu, C. E. Kenig and D. Tataru, Global well-posedness of KP-I initial-value problem in the energy space, *arXiv*: 0704.4239v1.

[9] A. D. Ionescu and C. E. Kenig, Global well-posedness of the Benjamin-Ono equation in low-regularity spaces, *J. Amer. Math. Soc.*, 20 (2007), 3:753-798.

[10] T. Tao, Multiplinear weighted convolution of L^2 functions and applications to nonlinear dispersive equations, *Amer. J. Math.*, 123(5) (2001), 839-908.

[11] H. Zhang and L. J. Han, Global well-posedness and inviscid limit for the modified Korteweg-de Vries-Burgers equation, To appear in *Nonlinear analysis*.

AN INVERSE PROBLEM OF CALIBRATING VOLATILITY IN JUMP-DIFFUSION OPTION PRICING MODELS[1]

CHANG JIN[a], QING-HUA MA[b] and ZUO-LIANG XU[c]

[a,b,c] *Renmin University of China, Beijing 100872, China*

[a] *E-mail: 45265171@qq.com*

[b] *E-mail: qinghua@ygi.edu.cn*

[c] *E-mail: xuzl@ruc.edu.cn*

This paper mainly concerns calibrating volatility from a jump diffusion model to a finite set of observed option pricing. We proposed a regularization algorithm based on Cont and Tankov's relative entropy regularization to solve this problem. We determine the regularization parameter using quasi-optimality criterion with original data error level unknown. Iteratively Guass-Newton method is developed for solving the unconstrained optimization problem. Finally, the theoretical results are illustrated by numerical experiments.

Keywords: Gauss-Newton method, jump diffusion model, relative entropy, regularization, volatility.

AMS No: 65J15, 65J20, 91B28.

1. Introduction

It is well known that the constant volatility assumption made in Black-Scholes model framework for option pricing is not valid in real market. Obtaining estimation of the volatility is major challenge for market finance. Unlike historical estimates of the volatility, calibration rely on the anticipation of the trading agents reflected in the prices of traded option products derived from the stock price S. We consider in this paper a widely studied inverse problem in mathematical finance, that of calibrating a volatility function from a given set of option prices in a jump-diffusion model.

This calibration problem has received intensive study in the past ten years. In this paper, we shall focus on the regularization method based on minimal relative entropy, following an approach introduced by R. Cont and P. Tankov in 2004 (see [2]). In our paper, inspired by the results in [2], we attempt to present a new algorithm to calibrate implied volatility in jump-diffusion option pricing models.

In a jump-diffusion model with a deterministic volatility function, consider a stock S affected by two sources of uncertainty: a standard Brownian motion W_t and a Poisson counting process $\eta_t^{\mathbb{Q}}$ with the deterministic jump intensity $\lambda^{\mathbb{Q}}$. The risk-neutral evolution of the underlying asset price $S(t)$

[1]This research is supported by NSFC (No.10971224), Beijing Talents Foundation and College of Art & Science of Beijing Union University

is given by

$$\frac{dS(t)}{S(t-)} = (r - q - \kappa^{\mathbb{Q}}\lambda^{\mathbb{Q}})dt + \sigma(S(t-),t)dW_t^{\mathbb{Q}} + (J-1)\eta_t^{\mathbb{Q}}, \tag{1.1}$$

where t-denotes the instant immediately before time t, r is the risk free rate, q is the dividend yield, and $\sigma(S,t)$ is a deterministic volatility function. The superscript $\mathbb{Q}$ denotes the pricing measure. In addition, J is a random variable representing the jump amplitude with $\kappa^{\mathbb{Q}} = E^{\mathbb{Q}}[J-1]$. For simplicity, $\log J$ is assumed to be normally distributed with constant mean $\mu^{\mathbb{Q}}$ and standard deviation $\gamma^{\mathbb{Q}}$. We will refer to the process (1.1), with a constant volatility σ and a lognormal jump density as Merton's jump-diffusion model.

In our terminology a jump-diffusion is a lévy process with finite jump activity. In this paper, we use the term "jump-diffusion" to denote a lévy process with a finite activity of jumps, that is, a linear combination of a Brownian motion and a compound Poisson jump process. We assume that dynamic of log price under risk-neutral measure $\mathbb{Q}$ is a lévy process.

2. Volatility Calibration of Exponential Iévy Model Using a Relative Entropy Regularization Method

General reference works on lévy process are by J. Bertoin(1996) and Applobanm(2003) (see Chapter V, [1]). Here we give some necessary theoretical background of lévy process.

A lévy process is a stochastic process $(X_t)_{t\geq 0}$ with statianary independent increments satisfying $X_0 = 0$. The characteristic function of X_t satisfies the following lévy-Khintchine formula $E[e^{izX_t}] = \Phi_t(z) = \exp[t\Psi(z)]$, and

$$\Psi(z) = i\gamma z - \frac{1}{2}\sigma^2 z^2 + \int_{-\infty}^{+\infty} (e^{izx} - \mathbf{1} - izx\mathbf{1}_{\{|x|<1\}})\nu(dx), \tag{2.1}$$

where $\gamma \in \mathbb{R}$, $\sigma \geq 0$ and ν is a positive measure on $\mathbb{R}$. If the measure $\nu(dx)$ admit a density with respect to the Lebesgue measure, we will call it lévy density of X and denote it by $\nu(x)$. In the case where $\lambda = \int v(dx) < +\infty$, the lévy process is said to be of finite activity, and the measure ν can then be normalized to define a probability measure μ on $\mathbb{R}$, which can be interpreted as the distribution of jump sizes: $\mu(x) = \frac{\nu(dx)}{\lambda}$.

The exponential of lévy process: $S_t = e^{rt+X_t}$, where X_t is a lévy process with characteristic triplet (σ, ν, γ) and the interest rate r. Since the discounted price process $e^{-rt}S_t = e^{X_t}$ is a martingale, this gives a constraint on the triplet (σ, ν, γ):

$$\gamma = \gamma(\sigma, \nu) = -\frac{\sigma^2}{2} - \int (e^y - \mathbf{1} - y\mathbf{1}_{|y|\le 1})v(dy).$$

Different exponential lévy models (i.e. compound Poisson models and infinite activity models) proposed in the financial modeling literature simply correspond to different parameterization of the lévy process.

The calibration problem consist of identifying the lévy measure ν and a volatility σ from a set of observations of call option prices described as below:

Calibration Problem 1: $S_t = \exp X_t$ where X_t is a lévy process defined by the characteristic function (σ, ν). Given the (observed) market prices $C_0^i(T_i, K_i)$ $(i = 1, \dots n)$ for a set of call options, find a constant $\sigma > 0$ and a lévy measure ν such that $C^{\sigma,\nu}(T_i, K_i) = C_0^i(T_i, K_i)$, where $C^{\sigma,\nu}$ is the option price computed for the lévy process with triplet$(\sigma, \nu, \gamma(\sigma, \nu))$.

If we know call option prices for all maturities and all strikes, we could deduce σ and ν in the following way:

1. Compute the option price. We can compute call option price in exponential lévy model using fast Fourier transform (Carr and Madan, 1998, [3]). For a European call option with log strike $k = \log K$,

$$C_T^0(K) = e^{-rT}E^Q[(s_T - K)^+]$$
$$= e^{-rT}E^Q[(e^{sT} - e^k)^+] = e^{-rT}\int_{-\infty}^{+\infty}(e^s - e^k)q_T(s)ds,$$

where s_T is the terminal log price with density $q_T(s)$, $q_T(k) = e^{-k}\{C^{''}(k) - C^{'}(k)\}$. The characteristic function of this density is defined by $\phi_T(u) = \int_{-\infty}^{+\infty} e^{ius}q_T(s)$. On the other hand, the characteristic function of the log price is given by lévy-Khintchine formula:

$$\Phi_T(u) = \exp\{T(-\frac{1}{2}\sigma^2u^2 + i\gamma(v)u + \int_{-\infty}^{+\infty}(e^{iux} - 1)\nu(x)dx)\}.$$

The key idea of FFT is to instead compute the Fourier transform of the (modified) time value of the option, that is, the function

$$z_T(k) = e^{-rT}E[(e^{sT} - e^k)^+] - (1 - e^{k-rT})^+, \tag{2.2}$$

$\xi_T(\nu)$ denote the Fourier transform of the time value

$$\xi_T(\nu) = \int_{-\infty}^{+\infty} e^{i\nu k} z_T(k)dk = \frac{e^{-rT}\phi_T(\nu - i) - e^{i\nu rT}}{i\nu(1 + i\nu)}. \tag{2.3}$$

2. Deduce σ and ν from Φ_T. First, the volatility of Gaussian component σ can be found as follows $\sigma^2 = \lim_{u\to\infty} -\frac{2\ln\Phi_T(u)}{Tu^2}$. Denote

$$\Psi(u) = \ln\frac{\Phi_T(u)}{T} + \frac{1}{2}\sigma^2u^2,$$

it has been proved that ν can be uniquely determined from Ψ by Fourier inversion (see [7]).

Despite its simple form, Problem 1 is an ill-posed inverse problem: there may exist no solution at all or an infinite number of solutions. To obtain a unique solution of this ill-posed problem in a stable manner, we must use a regularization method.

Many choices are possible for the penalization term. In this paper, following Cont and Tankov (2004), we use a calibration regularized based on the relative entropy with respect to a prior model. The relative entropy of probability measure $\mathbb{Q}$ with respect to $\mathbb{Q}_0$ is defined as:

$$\varepsilon(\mathbb{Q}|\mathbb{Q}_0) = E^{\mathbb{Q}}\left[\ln \frac{d\mathbb{Q}}{d\mathbb{Q}_0}\right] = E^{\mathbb{Q}_0}\left[\frac{d\mathbb{Q}}{d\mathbb{Q}_0} \ln \frac{d\mathbb{Q}}{d\mathbb{Q}_0}\right].$$

If $\mathbb{Q}$ and $\mathbb{Q}_0$ correspond to exponential lévy model, the relative entropy can be expressed in terms of the corresponding lévy measure. Namely,

$$\varepsilon(\mathbb{Q}|\mathbb{Q}_0) = \frac{T}{2\sigma^2}\{\int_{-\infty}^{+\infty} (e^x - 1)(\nu - \nu_0)(dx)\}^2$$

$$+T\int_{-\infty}^{+\infty} (\frac{d\nu}{d\nu_0} \ln(\frac{d\nu}{d\nu_0}) + 1 - \frac{d\nu}{d\nu_0})\nu_0(dx) = H(\nu, \nu_0) = H(\nu).$$

And then Problem 1 becomes:

Calibration Problem 2: Given a prior the exponential lévy model $\mathbb{Q}_0$ with characteristics (σ_0, ν_0), find a parameter vector ν which minimizes

$$J(\nu) = \sum_{i=1}^{N}(C_0^\nu(S_0, T_i, K_i) - C_0^i(T_i, K_i))^2 + 2\alpha H(\nu). \tag{2.4}$$

The function (2.4) consists two parts: the relative entropy function, which is convex in its argument ν, and the quadratic pricing error, which measures the precision of calibration.

For solving Problem 2, we represent the calibrated lévy measure ν by discretizing it on a grid. The grid must be uniform for the FFT algorithm to be used for option pricing. We localize ν one some bounded interval $[-M, M]$ and then choosing a partition, $\pi = (-M = x_1 < \ldots < x_N = M)$. Define L_π as a set of lévy measure with support in π:

$$L_\pi = \{\sum_{x\in\pi} a(x)\delta_x, a \in (R^+)^\pi\},$$

where δ_x is a measure that affects unit mass to point x, π is a finite set of points, we implicitly assume ν is finite.

The discretized calibration problem becomes

$$\min_{\nu \in L_\pi} J(\nu), \tag{2.5}$$

where ν corresponds to the entire lévy measure discretized on a grid, (2.5) always has a finite solution that depends continuously on the input prices and converges to the least square solution with minimal entropy when α goes to zero. Moreover, the use of entropy penalization could make our (discretized) problem well-posed which has been proved in [2].

Now we need present a numerical implementation algorithm for solving the discretized (2.5). First, we make the additional hypothesis that both the prior and the calibrated lévy process have finite jump activity. Below we propose a numerical algorithm for solving (2.5) different from Cont and Tankov's.

3. Choice of the Regularization Parameter

The regularization parameter α of (2.5) determines the tradeoff between the accuracy of calibration and the numerical stability of the results with respect to the input option prices. The principle of choose α make that a posterior error (calibration error) has the same level as the a prior error (error on input prices). Parameter choice methods can roughly be divided into two classes depending on their assumptions about data error level. The two classes can be characterized as follows:

1. Methods based on knowledge, or a good estimate of data error. When data error level is known, then it is crucial to make use of this information. The most widespread method is the discrepancy principle, usually attributed to Morozov (1966), Cont and Tankov used this method.

2. Methods that do not require data error level, but instead seek to exact this information from the given right-hand side.

Due to it is difficult to estimate original data error level in finance, we consider this class method to determine the regularization parameter. In this paper, we choose the parameter by using *quasi-optimality criterion* [4]. Now let (σ, ν_α) be the solution of (2.5) for a given regularization parameter $\alpha > 0$. The fundamental idea of this approach is to find a good balance between perturbation error and regularization errors in ν_α. This method is, strictly speaking, only defined for regularization parameter α and according

$$\alpha_{opt} = \min_{\alpha > 0} \left\{ \left\| \alpha \frac{d\nu_\alpha}{d\alpha} \right\| \right\} \tag{3.1}$$

to determine the parameter α.

Denote $Q(\alpha) = \|\alpha\frac{dx_\alpha}{d\alpha}\|^2$, $\alpha > 0$, in the linear case $Ax = b$, $Q(\alpha)$ could be calculated from formula

$$\alpha\frac{dx_\alpha}{d\alpha} = -\alpha[A^*A + \alpha I]^{-1}x_\alpha. \tag{3.2}$$

For nonlinear problems, we will now shown in a more heuristic way how the parameter choice may be generalized to the nonlinear case. We will assume that C is twice continuously Fréchet-differential. Denote

$$\eta(\nu) = \sum_{i=1}^{N}(C(\nu) - C_0),$$

then (2.5) becomes

$$J(\nu) = 2\alpha H(\nu) + (\eta(\nu))^2. \tag{3.3}$$

Since ν_α is the minimizer of (3.3), it satisfies the first order necessary condition

$$\eta'(\nu_\alpha)^*\eta(\nu_\alpha) + \alpha H'(\nu_\alpha) = 0.$$

A formal differentiation of this equation with respect to α yields

$$\eta''(\nu_\alpha)^*\eta(\nu_\alpha))\frac{d\nu_\alpha}{d\alpha} + \eta'(\nu_\alpha)^*\eta'(\nu_\alpha)\frac{d\nu_\alpha}{d\alpha} + \alpha H''(\nu_\alpha)\frac{d\nu_\alpha}{d\alpha} = -H'(\nu_\alpha).$$

If we neglect the second derivative term in this equation, we obtain the approximation

$$\alpha\frac{d\nu_\alpha}{d\alpha} \approx -\alpha[\eta'(\nu_\alpha)^*\eta'(\nu_\alpha) + \alpha H''(\nu_\alpha)]^{-1}H'(\nu_\alpha). \tag{3.4}$$

The right-hand side of (3.4) is a differentiable function of α, so the solution can be obtained with few iterations, for example, by gradient descent method. In this case, all integrals in $H(\nu)$ become finite sums and the relative entropy taking the following form:

$$H(\nu)=\frac{T}{2\sigma^2}\{\sum_{i=1}^{N}(e^{x_i}-1)(\nu_i-\nu_0(x_i))\}^2+T\sum_{i=1}^{N}\{\nu_i\ln\frac{\nu_i}{\nu_0(x_i)}+\nu_0(x_i)-\nu_i\},$$

where we denote $\nu_i = \nu(x_i)$. The first and the second term of $H(\nu)$ are all continuous for $\nu_i \geq 0, \forall i$. Of course we can compute the derivative of $H(\nu)$ with respect to ν simply. The method for calculating the derivative $C'(\nu_\alpha)$ will be described in the next section. We notice that in finite dimensional cases there always have $Q(0) = 0$, hence in practice we should make initial value α_0 slightly bigger.

4. Method for Minimizing $J(\nu)$

In order to solve the minimal problem (2.5), Cont and Tankov choose a L-BFGS-B gradient descent method. This method use a BFGS matrix to approximate the Hessian of objective function $J(\nu)$. In this paper, we consider another idea to solve this nonlinear optimization problems.

Review the Tikhonov regularization methods for nonlinear system $F(x) = y$, one of them is iteratively regularized Guass-Newton method, iterative process is

$$x_{k+1}=x_k-\tau_k[G'(x_k)^*G'(x_k)+\alpha_k I]^{-1}G'(x_k)^*G(x_k)+\alpha_k(x_k-\xi), \quad (4.1)$$

where $\xi = x_0$ is an initial guess for the true solution, τ_k and α_k are a sequence of positive number, $G(x) = F(x) - y$. Then the approximate solution x_{k+1} minimizes the Tikhonov functional

$$\|F(x) - y\|^2 + \alpha_k\|x - x_0\|^2, \quad (4.2)$$

where the nonlinear function F is linearized around x_k.

Back to our problem $\min J(\nu)$. We rewrite $J(\nu)$ of (2.5) into (3.3) in Section 3. According to the definition of relative entropy, $\varepsilon(\mathbb{Q}|\mathbb{Q}_0)$ is a convex non-negative functional of $\mathbb{Q}$ for fixed $\mathbb{Q}_0$, equal to zero if and only if $\frac{d\mathbb{Q}}{d\mathbb{Q}_0} = 1$. Namely, $H(\nu) = H(\nu, \nu_0) = 0$ if and only if $\nu = \nu_0$. We could find that feature of function $H(\nu)$ is similar to the second term of Tikhonov functional (4.2). Namely, the minimization problem $J(\nu)$ could be regard as minimization of Tikhonov functional (4.2). So the methods for solving (4.2) could be used to solve (2.5). We present a iteratively Gauss-Newton method, arising from A. Bakushinsky and A. Goncharsky (see [5]) for solving $\min J(\nu)$.

For solving (2.5), we give the iterative process as below:

$$\nu_{k+1}=\nu_k-[\eta'(\nu_k)^*\eta'(\nu_k)+\alpha_k H''(\nu)]^{-1}\eta'(\nu_k)^*\eta(\nu_k)+\alpha_k H'(\nu_k,\nu_0), \quad (4.3)$$

in which α_k is the regularization parameter obtained by quasi-optimal criterion $\eta(\nu) = \sum_{i=1}^N (C(\nu) - C_0)$.

In order to minimize the function (2.5) using a Gauss-Newton iterative formula (4.3), the essential step is the computation of the gradient of the calibration function with respect to the discretized values of the lévy measure. We represent the calibrated lévy measure ν by discretizing it on a grid $(x_i, i = 1, \dots N)$, where $x_i = x_0 + i \triangle x, \nu = \Sigma_{i=1}^N \nu_i\delta(x_i)$. This means that we effective allow a fixed (but large) number of jump sizes and calibrate the intensities of these jumps. The lévy process is then represented as a sum of independent standard Poisson processes with different intensities. The grid must be uniform in order to use the FFT algorithm for option pricing. This means that we effectively allow a fixed (but large) number of jump sizes and calibrate the intensities of these jumps. The lévy process is then represented as a weighted sum of independent standard Poisson processes with different intensities. The prior lévy measure ν_0 must also be discretized on the same grid, namely, using the formula $\nu_0(x_i) = \int_{x_i-\triangle x}^{x_i+\triangle x} \nu_0(dx)$ for the points $x_i, \ldots, x_{N-1}$.

The main step is to compute the variational derivative of option price, $DC_T(K)[v]$. Let $k = \log K$, computing the derivative of option price is equal to the derivative of time value, $z_T(k)[\nu]$, defined by formula (2.3). The function which maps ν into the time value $z_T(k)[\nu]$ is a superposition of the lévy-Khinchin formula (2.1) and equation (2.2). Direct computation show that

$$\begin{aligned} DZ_T(k)[v] = {} & T(1-e^{x_j})e^{-rT}\frac{1}{2\pi}\int_{-\infty}^{+\infty} e^{i\nu k}\frac{\Phi_T(\nu-i)}{1+i\nu}d\nu \\ & + e^{x_j}[C_T(k-x_j) - C_T(k)]. \end{aligned}$$

Therefore, the gradient may be represented in terms of the option price and one auxiliary function. Since we are using FFT to compute option prices for the whole price sheet, we already know these prices for the whole range of strikes.

5. Numerical Algorithm

As explained in Section 2, we tackle the ill-posedness of the initial calibration problem by transforming it into an optimization problem (2.5). Based on our discussion in sections 3 and 4, we now describe a numerical algorithm for solving this problem. Here is the final algorithm as implemented in the examples below.

1. Obtain an estimate of volatility σ_0 and a candidate for the prior lévy measure ν_0 based on historical estimation or calibration a simple auxiliary jump-diffusion model. Here, the prior does not contain any additional information and is only used to regularize the problem.

2. Use quasi-optimality criterion as explained in Section 3 to determine optimal regularization parameter α_{opt}. Choose an initial value α_0, minimize $J(\nu)$ in step 3 with α_0 and obtain the next lévy measure ν_1. Then back to (3.1) and (3.4), the optimal α_{opt} is found by running the gradient descent method with line search by Armijo's rule.

3. Solve variational problem for $J(\nu)$ with α_{opt} by iteratively Gauss-Newton method which introduced in section 4, we just need to compute the first derivatives of calibration function with respect to the discretized lévy measure ν.

6. Numerical Results

In this section, we describe two computational experiments with our proposed algorithm for calibration from Index call options.

In the first test, option prices were generated using Merton jump-diffusion model (see [8]), and a lévy measure given by

$$\nu(x) = \frac{\lambda}{\sqrt{2\pi}\delta} \exp\left[- \frac{(x-\mu)^2}{2\delta^2} \right]. \tag{6.1}$$

We set $\delta = 0.15$, $\mu = 0.1$, $\lambda = 1$. Estimation based on historical was used as the prior.

In the second test, option prices were generated using double exponential model (Kou, 2002, [7]) and a lévy measure given by

$$\nu(x) = p\alpha_1 e^{-\alpha_1 x}\mathbf{1}_{x>0} + (1-p)\alpha_2 e^{-\alpha_2 |x|}\mathbf{1}_{x<0}, \tag{6.2}$$

$\alpha_1 = \frac{1}{0.07}$, $\alpha_2 = \frac{1}{0.13}$, and $p = 0.35$ was chosen to have the jump intensity equal to 1. The results of Merton jump-diffusion model in the first test was used as the prior in this test. The option prices were computed using fast Fourier transform (FFT, [7]) method in both tests. We will apply our regularized algorithm to empirical data sets of index options and examine the implied lévy measures thus obtained.

The left graph in Figure 1 compares the calibration of lévy measure to the true Merton jump-diffusion measure, which is known to be of the form (6.2). The right compares calibration of measure and true double exponential (Kou) measure. The main features of true measure are successfully calibrated with our algorithm. Figure 2 presents the results of calibration of implied volatility surface in Kou's model.

7. Conclusion

In this paper, we investigate the problem of calibrating the implied volatility surface in jump-diffusion option pricing models. Jump-diffusion is a lévy process with finite jump activity. Based on the relative entropy regularization method (Cont and Tankov), we have presented a numerical algorithm to solve this inverse problem. In particular, according to it is difficult to estimate the error levels in financial data, we developed a quasi-optimality criterion to determine the regularization parameter. And we use a iterative Gauss-Newton method to solve the unconstraint optimization problem. Our numerical tests indicate how this algorithm can be used in an efficient calibration to market quoted option prices. Further theoretical and numerical developments in this problem are left to future research.

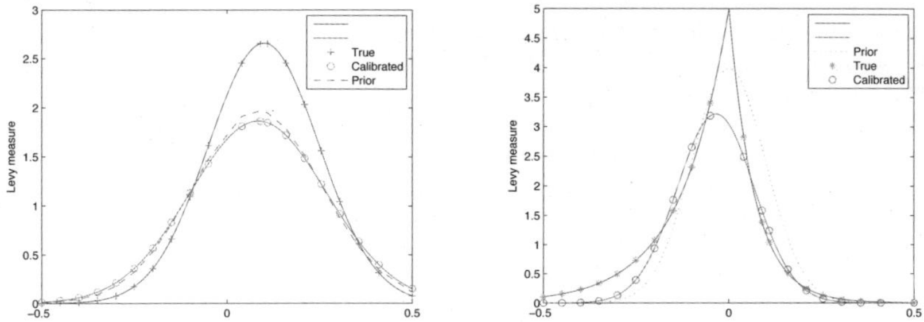

Figure 1: Left: Implied lévy measure calibrated to option prices simulated from Merton's model. Right: Implied lévy measure calibrated to option prices simulated from Kou's model.

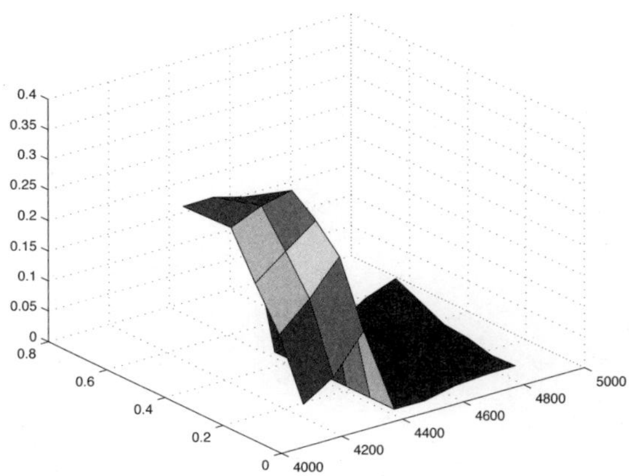

Figure 2: Market implied volatility surface

References

[1] J. Bertoin, *Iévy processes,* Cambridge University Press, Cambridge, 2010.

[2] R. Cont and P. Tankov, Non-parametric calibration of jump-diffusion option pricing models, *Journal of Computational Finance,* 7(2) (2004), 1–49.

[3] P. Carr and D. Madan, Option valuation using the fast Fourier transform, *Journal of Computational Finance,* 2 (1998), 61–73.

[4] M. Avellaneda, C. Andreasen, R. Holmes and D. Samperi, Calibrating volatility surfaces via relative entropy minimization, *Appl. Math. Finance,* 4 (1997), 37–64.

[5] A. Bakushinsky and A. Goncharsky, *Ill-posed problems: Theory and Applications,* Kluwer Academic Publishers, Dordrecht, 1994.

[6] L. Anderson and J. Friedman, Jump diffusion models: volatility smile fitting and numerical methods for pricing, *Review of Derivatives Research,* 4 (2000), 231–262.

[7] K. Sato, *Lévy processes and infinitely divisible distributions,* Cambridge University Press, Cambridge, 1999.

[8] R. Merton, Option pricing when underlying stock returns are discontinuous, *Journal of Financial Economics,* 3 (1976), 125-144.

[9] H. W. Engl, M. Hanke and A. Neubauer, *Regularization of inverse problems,* Kluwer Academic Publishers, Dordrecht, 2000.

INVERSE PROBLEMS OF TEXTILE MATERIAL DESIGN UNDER LOW TEMPERATURE[1]

DING-HUA XU, JIAN-XIN CHENG and YUAN-BO CHEN

Department of Mathematics, College of Sciences, Zhejiang Sci-Tech University, Hangzhou 310018, P. R. China

E-mail: dhxu6708@zstu.edu.cn

Based on the model of steady-state heat and moisture transfer through textiles, we propose inverse problems of single layer textile material design under low temperature, for example the thickness design or type design. Adopting the idea of regularization method, solving the inverse problems can be formulated into function minimization problems. Combining the finite difference method for ordinary differential equations with direct search method of one-dimensional minimization problems, we derive some iteration algorithms of regularized solution for the inverse design problems. Numerical simulation is achieved in order to verify the validity of proposed methods.

Keywords: Textiles, heat and moisture transfer, inverse problems, thickness design, type design, regularization method, numerical solution.

AMS No: 34A55, 34B15, 65L09.

1. Background for Textile Material Design

Simultaneous heat and moisture transfer in porous media is of growing interest in a wide range of science and engineering fields, such as civil engineering, safety analysis of dam, meteorology, energy storage and energy conservation, functional clothing design. As for functional clothing design, there are many requirements on human body comfort, healthier and safer textile and so on. As for the human body comfort textiles, it is hoped that the textiles are of fast decalescence, fast heat radiation, soft or stand-up apparel. Of course, clothing should be light and keeps body warm under low temperature, meanwhile sweat vaporizes fast and body feels cool under high temperature. Specially for industrial textile products, people need special materials which possess radiation-proof and keep the body in stable pressure. People usually seek greener and healthier textiles. According to above requirements, we should choose/determine material texture (such as cotton, tingle, feather, terylene, upgraded materials or composite materials), thickness of textiles (such as light-thin textiles, medium thick textiles or thick-heavy textiles) and physical structure of textiles (such as single layer or multi-layer textiles; parallel pore textiles or pellet-accumulated textiles).

[1]This research is supported by NSFC (No. 11071221 and 10561001)

In practical applications, modeling becomes much more interesting and important, since it provides an efficient way for evaluating new designs or testing new materials. The mathematical modeling and numerical simulation can be helpful to study heat and moisture transfer characteristics in textiles. The researchers usually focus on the characteristics of textile materials (such as moisture absorption, condensation characteristics, etc.) and textile structural features (such as porous media, multi-layer structure, etc.). Some mathematical models [9–12] have given predictions on the properties of heat and moisture transfer through different textiles and have shown effective in clothing design.

But to our knowledge, we have not seen the mathematical formulation of inverse problems of textile materials design on heat and moisture transfer properties. Therefore, in this paper, we propose the formulation of inverse problems of textile material design, such as inverse problem of thickness design, and inverse problem of type design, which are both based on the steady-state model of coupled heat and moisture transfer through parallel pore textiles [1]. The inverse problem of textile material design is of highly theoretical advantages, since it can predict/guide the textile design and clothing equipment design scientifically.

2. Mathematical Model of Heat and Mass Transfer

Textile material design is a kind of inverse problems in mathematical physics fields. In this paper, we consider inverse problems of thickness design and type design based on a new well-posed steady-state model of heat and moisture transfer through parallel pore textiles under low temperature conditions. The model of steady-state heat and moisture transfer through parallel pore textiles can be described as a mixed problem of coupled ordinary differential equations [1]:

$$\frac{k_1\varepsilon(x)r(x)}{\tau(x)} \cdot \frac{p_v}{T^{3/2}} \cdot \frac{dp_v}{dx} + m_v(x) = 0, \tag{1}$$

$$\frac{dm_v}{dx} + \Gamma(x) = 0, \tag{2}$$

$$\kappa\frac{d^2T}{dx^2} + \lambda\Gamma(x) = 0, \tag{3}$$

$$\Gamma(x) = \frac{-k_2\varepsilon(x)r(x)}{\tau(x)} \cdot (p_{sat} - p_v) \cdot \frac{1}{\sqrt{T}}, \tag{4}$$

and

$$\begin{cases} T(0) = T_L, \\ T(L) = T_R, \\ m_v(0) = m_{v,0}, \\ p_v(0) = p_{v,0}, \end{cases} \tag{5}$$

where $0 < x < L$, L represents the thickness of textile material. $T(x)$ is temperature(K); $m_v(x)$ is mass flux of water vapor$(kg/m^2 \cdot s)$; p_v is water vapor pressure(pa); $\Gamma(x)$ is the rate of condensation$(kg/m^3 \cdot s)$. The saturation vapor pressure within the parallel pore is given as follows [2]:

$$p_{sat}(T) = 100 \cdot \exp[18.956 - \frac{4030}{(T - 273.16) + 235}], \tag{6}$$

k_1 and k_2 are both constants which are related with molecular weight and gas constant; $\varepsilon(x)$ is porosity of textile surface; $r(x)$ is radius of cylindrical pore(m); $\tau(x)$ is effective tortuosity of the textile; λ is latent heat of sorption and condensation of water vapor(J/kg); κ is thermal conductivity of textiles$(W/m \cdot K)$; Let $k_3 = \frac{\kappa}{\lambda}$.

$T(0)$ and $T(L)$ are the temperatures of inner fabric and outside fabric respectively; $m_{v,0}$ is mass flux of water vapor of inner side of fabric; $p_{v,0}$ is water vapor pressure of inner side of fabric.

The above mixed problem (2.1)–(2.5) of coupled ordinary differential equations is usually called a direct problem **(DP)**.

3. Inverse Problems of Textile Material Design

In this section, we propose the formulation of inverse problems of textile material design and take the textile thickness design and textile type design as two examples.

4. Thickness Design of Textile Materials Under Low Temperature

We consider an inverse problem of thickness design for single layer textile material [3].

Suppose that the environmental temperature and relative humidity are given as follows:

$$(T, RH) \in [T_{min}, T_{max}] \times [H_{min}, H_{max}],$$

where T_{min} and T_{max} are minimum average temperature and maximum average temperature at a specific place during a specific time period respectively. Similarly H_{min} and H_{max} are minimum average relative humidity and maximum average relative humidity respectively.

Suppose that the structure and type of single layer textile are known. The structure of textile includes the radius of pore, porosity of textile surface and effective tortuosity of the textile.

The literatures on clothing thermal comfort have indicated that the comfort indices in the clothing microclimate, which is located between the skin surface and the inner surface of fabric, are given as follows [8] : temperature $(32 \pm 1)^o C$, relative humidity $(50\% \pm 10\%)$, wind speed (25 ± 15)cm/s.

According to the requirements of clothing thermal comfort, we intend to determine the fabric thickness L. Thus, the inverse problem of thickness design can be formulated as follows:

IP 1 (Inverse Problem 1). Given the environmental temperature and relative humidity and the above comfort indices, according to the boundary value conditions

$$\begin{cases} T(0) = T_L, \\ T(L) = T_R, \\ m_v(0) = m_{v,0}, \\ p_v(L) = p_{v,R}, \end{cases} \tag{7}$$

we need determine the thickness L of fabric through the model of ODEs (2.1)-(2.4), where $p_{v,R}$ is related with the temperature and relative humidity of environment.

5. Regularized Solution of the IP 1

In order to obtain the regularized solution, we discretize the combination of environmental temperature and humidity as (T_i, RH_j) $(i = 1, 2, \cdots, k;$ $j = 1, 2, \cdots, m)$. Let $RH_{i,j,0}(x)$ is relative humidity of inner fabric, which will be solved by coupled ordinary differential equations. Suppose that RH_0^* is experience value of relative humidity in comfortable state.

We can attribute the inverse problem to the following least squared problem:

$$\min \sum_{i=1}^{k} \sum_{j=1}^{m} (RH_{i,j,0}(x) - RH_0^*)^2.$$

Since above least squared problem doesn't exist unique solution or the solutions are unstable, we adopt regularized idea to improve the least squared method.

In this respect, we define the following function:

$$J(x) = \alpha \cdot x^2 + \sum_{i=1}^{k} \sum_{j=1}^{m} (RH_{i,j,0}(x) - RH_0^*)^2.$$

This function is different from the least squared function, as it is added a penalty term on the least squared function, where $\alpha\ (>0)$ is a regularization parameter. Set $M=[0,L]$, which is called the permissible solution set. If x_{reg} satisfies

$$J(x_{reg})=\min_{x\in M} J(x),$$

then it is called the regularized solution of the inverse problem, or the generalized solution.

6. Iteration Algorithms of the Regularized Solution

Step 1. Numerical computation of the DP

We decouple the ODES(2.1)-(2.4), and use the finite difference method to discretize the differential equation as follows:

$$\sqrt{T_i}\cdot\frac{T_i-2T_{i-1}+T_{i-2}}{h^2}=\frac{k_2}{k_3}A(x_i)$$

$$\times\left[p_{sat}(T_i)-\sqrt{p_v^2(x_n)+2\sum_{j=i}^{N-1}\frac{1}{k_1\cdot A(x_{j+1})}\cdot T_{j+1}^{3/2}\cdot[k_3\cdot\frac{T_{j+1}-T_j}{h}+C_1^*]\cdot h}\right],$$

$$i=2,\cdots,N-1,$$

$$\sqrt{T_N}\cdot\frac{T_N-2T_{N-1}+T_{N-2}}{h^2}=\frac{k_2}{k_3}A(x_N)\left[p_{sat}(T_N)-p_v(x_n)\right],$$

$$p_{v,i}=\sqrt{p_{v,R}^2+2\sum_{j=i}^{N-1}\frac{1}{k_1\cdot A(x_{j+1})}\cdot T_{j+1}^{3/2}\cdot[k_3\cdot\frac{T_{j+1}-T_j}{h}+C_1^*]\cdot h},$$

$$i=0,1,2,\cdots,N-1,$$

where $h=\frac{x_n}{N}$, $C_1^*=m_{v,0}-k_3\cdot\frac{T_1-T_0}{h}$.

Hence, we can obtain the relative humidity of inner side of fabric:

$$p_{v,0}=\sqrt{p_{v,R}^2+2\sum_{j=0}^{N-1}\frac{1}{k_1\cdot A(x_{j+1})}\cdot T_{j+1}^{3/2}\cdot[k_3\cdot\frac{T_{j+1}-T_j}{h}+C_1^*]\cdot\frac{x_n}{N}},$$

$$RH_{i,j,0}(x_n)=\frac{p_{v,0}}{p_{sat}(T_0)}$$

$$=\frac{\sqrt{p_{v,R}^2+2\sum_{j=0}^{N-1}\frac{1}{k_1\cdot A(x_{j+1})}\cdot T_{j+1}^{3/2}\cdot[k_3\cdot\frac{T_{j+1}-T_j}{h}+C_1^*]\cdot\frac{x_n}{N}}}{100\cdot exp[18.956-\frac{4030}{(T_0-273.16)+235}]}.$$

Step 2. Search method of one-dimensional minimization problems

The optimization problem involved in this paper is a single variable problem. As we know, $RH_{i,j,0}(x)$ is relative humidity of inner fabric which is a numerical solution calculated by coupled ordinary differential equations, and it is difficult to obtain the derivative of $RH_{i,j,0}(x)$, hence we must use the direct search method. Taking this actual situation into account, we can use Hooke-Jeeves pattern search algorithm [4,8], direct search algorithm by Cai [5] and 0.618 method [6] to solve the above optimization problem. We use Hooke-Jeeves pattern search algorithm to solve above optimization problem as an example.

Hooke-Jeeves pattern search algorithm (1961)

Step 1. x_1 is given. Set initial step $\Delta_1 > 0$, acceleration factor $\gamma \geq 1$, reduced rate $\beta \in (0,1)$, permissible error $\varepsilon > 0$, search direction $e_1 = 1$, $e_2 = -1$. set $y_1 = x_1$, $i = 1$.

Step 2. If $J(y_1 + \Delta_1 \cdot e_1) < J(y_1)$, then

$$y_2 = y_1 + \Delta_1 \cdot e_1$$

carry out Step 4; otherwise, carry out Step 3.

Step 3. If $J(y_1 + \Delta_1 \cdot e_2) < J(y_1)$, then

$$y_2 = y_1 + \Delta_1 \cdot e_2$$

carry out Step 4; otherwise, if $J(y_1 + \Delta_1 \cdot e_2) \geq J(y_1)$, then

$$y_2 = y_1,$$

carry out Step 4.

Step 4. If $J(y_2) < J(x_i)$, carry out Step 5; otherwise, if $J(y_2) \geq J(x_i)$, carry out Step 6.

Step 5. $x_{i+1} = x_i; y_1 = x_{i+1} + \gamma(x_{i+1} - x_i); i = i + 1$; go to Step 2.

Step 6. If $\Delta_1 \leq \varepsilon$, then stop,$x^* = x_i$; otherwise, $\Delta_1 = \beta \cdot \Delta_1; y_1 = x_i; x_{i+1} = x_i; i = i + 1$; go to Step 2.

7. Type Design of Textile Materials Under Low Temperature

In this section, we consider an inverse problem of type design for single layer textile material [7].

IP 2 (Inverse Problem 2). Given the combinations of environmental temperature and relative humidity and the above comfort indices in 3.1, according to the boundary value conditions

$$\begin{cases} T(0) = T_L, \\ T(L) = T_R, \\ m_v(0) = m_{v,0}, \\ p_v(L) = p_{v,R}, \end{cases} \tag{8}$$

we need to determine the thermal conductivity of textile κ through the model of ODEs (2.1)–(2.4), as it represents the type of textiles. Where the thickness of textile L is known, and $p_{v,R}$ is related with the temperature and relative humidity of environment.

8. Regularized Solution of the Inverse Problem

We define the following function:

$$J(\kappa) = \alpha \cdot \kappa^2 + \sum_{i=1}^{k} \sum_{j=1}^{m} (RH_{i,j,0}(\kappa) - RH_0^*)^2.$$

This function is different from the least squared function, as it is added a penalty term on the least squared function, where $\alpha > 0$ is a regularization parameter. Set $M = [0, K]$, which is called the permissible solution set. If κ^* satisfies

$$J(\kappa^*) = \min_{\kappa \in M} J(\kappa),$$

then it is called the regularized solution of the inverse problem, or the generalized solution.

9. Iteration Algorithms of the Regularized Solution

Step 1. Numerical computation of the DP

Decoupling the ODES (2.1)–(2.4), and adopting the finite difference method to discretize the differential equation, we obtain the relative humidity of inner side of fabric:

$$RH_{i,j,0}(\kappa_n) = \frac{p_{v,0}}{p_{sat}(T_0)}$$

$$= \frac{\sqrt{p_{v,R}^2 + 2\sum_{j=0}^{N-1} \frac{1}{k_1 \cdot A(x_{j+1})} \cdot T_{j+1}^{3/2} \cdot [k_3 \cdot \frac{T_{j+1} - T_j}{h} + C_1^*] \cdot \frac{L}{N}}}{100 \cdot exp[18.956 - \frac{4030}{(T_0 - 273.16) + 235}]},$$

where $h = \frac{L}{N}, C_1^* = m_{v,0} - k_3 \cdot \frac{T_1 - T_0}{h}, k_3 = \frac{\kappa}{\lambda}$.

Step 2. Search method of one-dimensional minimization problems

We can also use direct search methods of one-dimensional minimization problem mentioned in 3.1.

10. Numerical Solution

Numerical simulations are carried out to verify the validity of above numerical method for inverse problems of thickness design and type design. Suppose that the initial mass flux of water vapor is $m_v(0) = 3.3084 \times 10^{-5} kg/m^2 \cdot s$. The temperature of the inner side of fabric is assumed to be 32^oC to guarantee that temperature in microclimate is in the comfort index interval, that is $T(0) = 305.16K$. In the model, $k_1 = 0.00006$, $k_2 = 0.00007$.

We choose four different environmental conditions under low temperature for simulation. Environmental condition 1: $T \in [-10^oC, 0^oC]$, $RH \in [40\%, 90\%]$; Environmental condition 2: $T \in [0^oC, 10^oC]$, $RH \in [30\%, 85\%]$; Environmental condition 3: $T \in [-15^oC, 0^oC]$, $RH \in [40\%, 90\%]$; Environmental condition 4: $T \in [0^oC, 15^oC]$, $RH \in [30\%, 85\%]$.

Example 1. The inverse problem of thickness design

Table 1. Numerical results of thickness design for wool under environmental condition 1

Initial values (m)	Thickness of wool (cm)
0.0005	0.7875
0.001	0.7888
0.007	0.7938
0.01	0.795

Table 2. Numerical results of thickness design for wool under environmental condition 2

Initial values (m)	Thickness of wool (cm)
0.0005	0.7525
0.001	0.74625
0.007	0.74625
0.01	0.74625

Example 2. The inverse problem of type design

Table 3. Numerical solution of thermal conductivity of textile under environmental condition 3

thickness of textile (mm)	Numerical heat conductivity
5.5	0.017273
6.5	0.019396

Table 4. Numerical solution of thermal conductivity of textile under environmental condition 4

thickness of textile (mm)	Numerical heat conductivity
6.5	0.027926
7.5	0.044988
8.0	0.051447

According to the above numerical results, we conclude that the numerical results under low temperature conditions are reasonable and acceptable, as the thickness of textile under low temperature is between $0.5mm$ and $10mm$ and the heat conductivity of textiles lies in reasonable empirical measurement interval. Hence the formulation of mathematical model is correct and the proposed numerical algorithm is efficient. Subsequently we have provided theoretical explanation for textile material design in engineering.

References

[1] Dinghua Xu, Jianxin Cheng and Xiaohong Zhou, A model of heat and moisture transfer through the parallel pore textiles, in *Proc. Textile Bioengineering and Informatics Symposium*, Shanghai, Track 10 (2010), 1151-1156.

[2] Jianhua Huang, *Clothing comfort*, Science Press, Beijing, 2008.

[3] Dinghua Xu, Jianxin Cheng and Xiaohong Zhou, An inverse problem of thickness design for single layer textile material under low temperature, To appear in *Journal of Mathematics-for-Industry,* 2010B.

[4] Virginia Torczon, On the convergence of pattern search algorithms, *SIAM. J. Optim.*, 7 (1997), 1-25.

[5] Zhijie Cai and Deqiang Chen, The direct search algorithm of nonlinear optimization and its convergence proof, *Journal of Fudan University* (*Natural Science Edition*), 45 (2006), 396-403.

[6] Baolin Chen, *Optimization problems and algorithms*, Tsinghua University, Beijing, 2005.

[7] Dinghua Xu, Yuanbo Chen and Jianxin Cheng, An inverse problem of type design for single layer textile material under low temperature, Submitted to *Chinese Journal of Textile Research.*

[8] R. Hooke and T. A. Jeeves, "Direct search" solution of numerical and statistical problems, *J. Assoc. Comput. Mach.*, 8 (1961), 212-229.

[9] Jintu Fan, Zhongxuan Luo and Yi Li, Heat and moisture transfer with sorption and condensation in porous clothing assemblies and numerical simulation, *Int. J. Heat Mass Transfer*, 43 (2000), 2989-3000.

[10] Jintu Fan and Xinghuo Wen, Modeling heat and moisture transfer through fibrous insulation with phase change and mobile condensates, *Int. J. Heat Mass Transfer*, 45 (2002), 4045-4055.

[11] Jintu Fan, Xiaoyin Cheng, Xinhuo Wen and Weiwei Sun, An improved model of heat and moisture transfer with phase change and mobile condensates in fibrous insulation and comparison with experimental results, *Int. J. Heat Mass Transfer*, 47 (2004), 2343-2352.

[12] Huijun Wu and Jintu Fan, Study of heat and moisture transfer within multi-layer clothing assemblies consisting of different types of battings, *International Journal of Thermal Sciences*, 47 (2008), 641-647.

A NEW METHOD FOR CONSTRUCTING COEFFICIENTS OF ELLIPTIC COMPLEX EQUATIONS WITH RIEMANN-HILBERT TYPE MAP[1]

GUO CHUN WEN

School of Mathematical Sciences, Peking University, Beijing 100871, China

E-mail: wengc@math.pku.eu.cn

The present paper mainly concerns the inverse problem for linear elliptic complex equations of first order with Riemann-Hilbert type map in simply connected domains. Firstly the formulation and the complex form of the problem for the equations are given, and then the existence of solutions for the above problem is proved by a new complex analytic method, where the advantage of the other methods is absorbed, and the used method in this paper is more simple and the obtained result is more general. As an application of the above results, we can derive the corresponding results of the inverse problem for second order elliptic equations from Dirichlet to Neumann map.

Keywords. Inverse problem, elliptic complex equations, Riemann-Hilbert type map.

AMS Classification. 35R30, 35J45.

1. Formulation of Inverse Problem for Elliptic Complex Equations of First Order

In [1–4], the authors posed and discussed the inverse problem of second order elliptic equations. In this paper, the existence of solutions of the inverse problem for linear elliptic systems of first order equations with Riemann-Hilbert type map is considered. By using the complex analytic method, we first formulate the inverse problem for elliptic complex equations of first order in simply connected domains and give some properties of its solutions, and then discuss the existence of solutions of the inverse problem for the above elliptic complex equations of first order.

Let D be a simply connected domain in the complex plane $\mathbb{C}$ with the boundary $\partial D = \Gamma \in C^1_\mu(0 < \mu < 1)$. There is no harm in assuming that a point $a_0 \in \Gamma$ with $\arg a_0 = 0$ and $z = 0 \in D$. Consider the linear elliptic system of first order equations

$$u_x - v_y + au + bv = 0,\ v_x + u_y + cu + dv = 0 \ \text{in}\ D, \tag{1.1}$$

[1]This research is supported by NSFC (No.10971224)

where $a = a(z), b = b(z), c = c(z), d = d(z)$ are real measurable functions of $z = x + iy\,(\in D)$ with the conditions $a, b, c, d \in L_p(\overline{D})$, $p\,(> 2)$ is a positive constant. The above conditions will be called Condition C.

Denote

$$W(z)=u+iv,\ W_{\bar z}=\frac{1}{2}[W_x+iW_y]=\frac{1}{2}[u_x-v_y+i(v_x+u_y)]\ \text{in}\ D, \tag{1.2}$$

where $z = x + iy$, we can get

$$\begin{aligned} W_{\bar z} &= \frac{1}{2}[W_x + iW_y] = -\frac{1}{2}[au + bv] - \frac{i}{2}[cu + dv] \\ &= -\frac{1}{4}[a(W+\overline{W})+ib(\overline{W}-W)] - \frac{1}{4}[ic(W+\overline{W})-d(\overline{W}-W)] \\ &= -A(z)W - B(z)\overline{W}\ \text{in}\ D. \end{aligned} \tag{1.3}$$

Here $A(\zeta) = [a+d-ib+ic]/4$, $B(\zeta) = [a-d+ib+ic]/4$, and assume that the coefficients $A(z)$, $B(z) \in L_p(\overline{D})$, $p\,(> 2)$ is a positive number. In this paper the notations are the same as those in [5] or [7].

Introduce the Riemann-Hilbert boundary condition for the complex equation (1.3) as follows:

$$\begin{cases} \operatorname{Re}[\overline{\lambda(z)}W(z)] = r(z) + g(z) = h_1(z),\ z \in \Gamma, \\ \operatorname{Im}[\overline{\lambda(a_j)}W(a_j)] = b_j,\ j = 1, ..., 2K+1,\ K \geq 0, \end{cases} \tag{1.4}$$

where

$$g(z) = \begin{cases} 0,\ K \geq 0, \\ g_0 + \operatorname{Re}\displaystyle\sum_{m=1}^{-K-1} (g_m^+ + ig_m^-)[\zeta(z)]^m,\ K < 0. \end{cases} \tag{1.5}$$

Herein $\lambda(z)\,(\neq 0)$, $r(z) \in C_\alpha(\Gamma)$, $\alpha\,(\leq (p-2)/p)$ is a positive constant, g_0, $g_m^{\pm}\,(m = 1, ..., -K-1, K < 0)$ are unknown real constants to be determined appropriately, $a_j(\in \Gamma,\ j = 1, ...2K+1,\ K \geq 0)$ are distinct points, and $b_j(j = 1, ..., 2K+1)$ are all real constants, in which $K = \frac{1}{2\pi}\Delta_L \arg\lambda(z)$ is called the index of $\lambda(z)$ on Γ, and $z = z(\zeta)$ is a conformal mapping from the unit disk $|\zeta| < 1$ onto D. The above Riemann-Hilbert boundary value problem is called Problem RH for equation (1.3). From (5.114) and (5.115), Chapter VI, [5], we see that Problem RH of equation (1.3) possesses the important application to the shell and elasticity. Under Condition C, the solution $W(z)$ of Problem RH for (1.3) in D can be found. It is clear that the above solution $W(z)$ satisfies the following Riemann-Hilbert type boundary condition for the equation (1.4):

$$\operatorname{Im}[\overline{\lambda(z)}W(z)] = h_2(z)\ \text{on}\ \Gamma, \tag{1.6}$$

and then the boundary conditions of Riemann-Hilbert to Riemann-Hilbert type map can be written as follows

$$\begin{aligned}&\overline{\lambda(z)}W(z)=h_1(z)+ih_2(z)=h(z)\ \text{ on } \Gamma,\ \text{i.e.}\\ &W(z))=H(z)=h(z)/\overline{\lambda(z)}\ \text{on}\ \Gamma,\end{aligned}\tag{1.7}$$

which will be called Problem RR for the complex equation (1.3) (or (1.1)), where $h(z)\,(\in C_\alpha(\Gamma))$ is a complex function. Thus we can define the Riemann-Hilbert to Riemann-Hilbert type map $\Lambda: C_\alpha(\Gamma)\to C_\alpha(\Gamma)$, i.e. $h_1(z)\to h_2(z)$ by $\Lambda h_1 = h_2$.

Our inverse problem is to determine the real coefficients $a(z)$, $b(z)$, $c(z)$, $d(z)$ of the equation (1.1) (or the complex coefficients $A(z)$, $B(z)$ of the complex equation (1.3)) from the map Λ. Obviously the function $h(z)$ is corresponding to the function $H(z)$ one by one. Denote by R_h and R_H the sets of $\{h(z)\}$ and $\{H(z)\}$ respectively. It is clear that for any function $h_1(z)$ of the set $C_\alpha(\Gamma)$ in the Riemann-Hilbert boundary condition (1.4), there is a set $\{h_2(z)\}$ of the functions of Riemann-Hilbert type boundary condition (1.6), where $R_h=\{h(z)\}$ is corresponding to the complex equation (1.3). Inversely from the set $R_h=\{h(z)\}$ or $R_H=\{H(z)\}$, one complex equation in (1.3) can be determined, which will be verified later on.

2. Existence of Solutions of Inverse Problem for Elliptic Complex Equations of First Order

According to [5], introduce the notations

$$Tf(z)=-\frac{1}{\pi}\iint_{\mathbf{C}}\frac{f(\zeta)}{\zeta-z}d\sigma_\zeta,$$

in which $f(z)\in L_p(\overline{D})$, $p>2$. Obviously $(Tf)_{\bar z}=f(z)$ in $\overline{D}$. We consider the complex equation

$$g_{\bar z}+Ag+B\overline{g}=0,\ \text{i.e.}\ W_{\bar z}+AW+B\overline{W}=0\ \text{ in }\ \overline{D},\tag{2.1}$$

where $g(z)=W(z)$. On the basis of the Pompeiu formula (see Chapters I and III, [5]), the corresponding integral equation of the complex equation (2.1) is as follows

$$g(z)-T[Ag+B\overline{g}]=\frac{1}{2\pi i}\int_\Gamma\frac{g(\zeta)}{\zeta-z}d\zeta\ \text{ in } D.\tag{2.2}$$

For simplicity we can only consider the following integral equation

$$g(z)-T[Ag+B\overline{g}]=1\ \text{ or }\ i\ \text{ in }\ D\tag{2.3}$$

later on.

We first prove the following lemma by the similar way as in [2,3] and [6].

Lemma 2.1. The function $g(z)$ is a unique solution of one of the integral equations

$$g(z)+TAg+TB\overline{g}=\begin{cases}1,\\ i,\end{cases}\text{ in }\overline{D},\ g(z)=\begin{cases}H_1(z),\\ H_2(z),\end{cases}\text{ on }\Gamma,\tag{2.4}$$

with the condition $g(z)=H_j(z)\,(H_j(z)\in R_H, z\in\Gamma, j=1,2)$ if and only if $H_1(z)$, $H_2(z)$ are the solutions of the integral equations

$$\begin{cases}\dfrac{1}{2}g(z)+\dfrac{1}{2\pi i}\displaystyle\int_\Gamma\frac{g(\zeta)}{\zeta-z}d\zeta=\begin{cases}1,\\ i,\end{cases}\ g(\zeta)=\begin{cases}H_1(\zeta),\\ H_2(\zeta),\end{cases}\ \text{i.e.}\\ \dfrac{H_1(z)}{2}+\dfrac{1}{2\pi i}\displaystyle\int_\Gamma\frac{H_1(\zeta)}{\zeta-z}d\zeta=1,\ \dfrac{H_2(z)}{2}+\dfrac{1}{2\pi i}\displaystyle\int_\Gamma\frac{H_2(\zeta)}{\zeta-z}d\zeta=i\text{ on }\Gamma,\end{cases}\tag{2.5}$$

respectively.

Proof. It is obvious that we can only discuss the case of H_1. If $g(z)$ is a solution of the first equation in (2.4), then $g_{\bar z}=-AJg-BJ\overline{g}$. On the basis of the Pompeiu formula

$$g(z)=\frac{1}{2\pi i}\int_\Gamma\frac{g(\zeta)}{\zeta-z}d\zeta-T[g(\zeta)]_{\bar\zeta}=\frac{1}{2\pi i}\int_\Gamma\frac{g(\zeta)}{\zeta-z}d\zeta-T[Ag+B\overline{g}]\text{ in }D\tag{2.6}$$

(see Chapters I and III, [5]), we have

$$g(z)+TAg+TB\overline{g}=1=\frac{1}{2\pi i}\int_\Gamma\frac{g(\zeta)}{\zeta-z}d\zeta\text{ in }D,\tag{2.7}$$

where $g(\zeta)=H_1(\zeta)$ on Γ. Moreover by using the Plemelj-Sokhotzki formula for Cauchy type integrals (see [5,7])

$$1=\frac{1}{2\pi i}\int_\Gamma\frac{g(\zeta)}{\zeta-z}d\zeta+\frac{1}{2}g(z),\ g(\zeta)=H_1(\zeta)\text{ on }\Gamma,$$

this is the first formula in (2.5).

Inversely if the first formula in (2.5) is true, noting that there exists a unique solution of the equation $g_{\bar z}=-AJg-BJ\overline{g}$ in $\overline{D}$ with the boundary value $g(\zeta)=H_1(\zeta)$ on Γ, we have

$$g(z)+TAg+TB\overline{g}=\frac{1}{2\pi i}\int_\Gamma\frac{g(\zeta)}{\zeta-z}d\zeta\text{ in }D,$$

where the integral $\frac{1}{2\pi i}\int_\Gamma \frac{g(\zeta)}{\zeta-z}d\zeta$ in D is analytic, whose boundary value on Γ is

$$\lim_{z'(\in D)\to z(\in\Gamma)} \frac{1}{2\pi i}\int_\Gamma \frac{g(\zeta)}{\zeta-z'}d\zeta = \frac{1}{2}g(z) + \frac{1}{2\pi i}\int_\Gamma \frac{g(\zeta)}{\zeta-z}d\zeta = 1,$$

hence

$$\frac{1}{2\pi i}\int_\Gamma \frac{g(\zeta)}{\zeta-z}d\zeta = 1 \text{ in } D,$$

and the first equation in (2.4) is true. Now we furthermore explain the uniqueness of the solutions of the first equation in (2.4), and can assume that the equation possesses the form $g_{\bar z} = R(z)\,(\in L_\infty(D))$ in D, it is easy to see that there exists a solution $g_1(z)$ of the equation in $\overline{D}$ with the boundary condition $g_1(z) = H_1(z)$ on Γ. If we have the another solution $g_2(z)$ of the equation in $\overline{D}$ with the same boundary condition, then $g_1(z) - g_2(z)$ satisfies the equation $[g_1(z) - g_2(z)]_{\bar z} = 0$ with the homogeneous boundary condition $g_1(z) - g_2(z) = 0$ on Γ, and then $g_1(z) - g_2(z) = 0$, i.e. $g_1(z) = g_2(z)$ in $\overline{D}$.

Lemma 2.2. Under the above conditions, the functions $H_1(z)$, $H_2(z)$ as stated in (1.7) are the solutions of the system of integral equations

$$\begin{cases} \frac{1}{2}(1-iS)H_1 = 1, \ \frac{1}{2}(1-iS)H_2 = i, \\ SH_1 = \frac{1}{\pi}\int_\Gamma \frac{H_1(\zeta)}{\zeta-z}d\zeta, \ SH_2 = \frac{1}{\pi}\int_\Gamma \frac{H_2(\zeta)}{\zeta-z}d\zeta. \end{cases} \tag{2.8}$$

Proof. From the theory of integral equations (see [6,8]), we can derive the solutions $H_1(z)$ and $H_2(z)$ of (2.4). In fact, on the basis of Lemma 2.1 we can find the solutions of the following integral equations

$$\begin{aligned} W_1(z) &= 1 + \frac{1}{\pi}\iint_D \frac{J[AW_1 + B\overline{W_1}]}{\zeta - z}d\sigma_\zeta, \\ W_2(z) &= i + \frac{1}{\pi}\iint_D \frac{J[AW_2 + B\overline{W_2}]}{\zeta - z}d\sigma_\zeta, \end{aligned} \quad z \in \overline{D}. \tag{2.9}$$

By using the Pompeiu formula, the above equations can be rewritten as

$$\begin{aligned} W_1(z) &= \frac{1}{2\pi i}\int_\Gamma \frac{W_1(t)}{t-z}dt + \frac{1}{\pi}\iint_D \frac{J[AW_1 + B\overline{W_1}]}{\zeta - z}d\sigma_\zeta, \\ W_2(z) &= \frac{1}{2\pi i}\int_\Gamma \frac{W_2(t)}{t-z}dt + \frac{1}{\pi}\iint_D \frac{J[AW_2 + B\overline{W_2}]}{\zeta - z}d\sigma_\zeta, \end{aligned} \quad z \in \overline{D}, \tag{2.10}$$

and $W_1(z) = H_1(z)$ and $W_2(z) = H_2(z)$ on Γ, the functions

$$\frac{1}{2\pi i}\int_\Gamma \frac{H_j(t)}{t-z}dt\,(j = 1, 2)$$

are analytic in D. According to the Plemelj-Sokhotzki formula for Cauchy type integrals, we immediately obtain the formulas in (2.8).

Theorem 2.3. For the inverse problem of Problem RR for equation (1.3) with Condition C, we can reconstruct the convective coefficients $a(z), b(z), c(z)$ and $d(z)$ of the system (1.1).

Proof. On the basis of Lemma 2.2, we can find two solutions $\phi_1(z) = W_1(z)$ and $\phi_2(z) = W_2(z)$ of complex equation

$$[\phi]_{\bar{z}} + A(z)\phi + B(z)\overline{\phi} = 0 \text{ in } \overline{D}, \tag{2.11}$$

and the above solutions $\phi_1(z) = F(z)$, $\phi_2(z) = G(z)$ are also the solutions of integral equations

$$\left\{ \begin{array}{l} F(z) + T\{AF + B\overline{F}\} = 1, \\ G(z) + T\{AG + B\overline{G}\} = i, \end{array} \right. \text{in } \overline{D}, \tag{2.12}$$

and we can require that the above solutions satisfy the boundary conditions

$$F(z) = H_1(z),\ G(z) = H_2(z) \text{ on } \Gamma, \tag{2.13}$$

where $H_1(z)$, $H_2(z) \in R_H$. From Lemma 2.4 below, we can verify that

$$\operatorname{Im}[\overline{F(z)}G(z)] = [F(z)\overline{G(z)} - \overline{F(z)}G(z)]/2i \neq 0 \text{ in } D.$$

In addition, noting that $F(z)$, $G(z)$ satisfy the complex equations

$$\left\{ \begin{array}{l} F_{\bar{z}} + AF + B\overline{F} = 0, \\ G_{\bar{z}} + AG + B\overline{G} = 0, \end{array} \right. \text{in } D, \tag{2.14}$$

we can determine the coefficients A and B as follows

$$A = -\frac{F_{\bar{z}}\overline{G} - G_{\bar{z}}\overline{F}}{F\overline{G} - \overline{F}G},\ B = \frac{F_{\bar{z}}G - G_{\bar{z}}F}{F\overline{G} - \overline{F}G} \text{ in } \overline{D}. \tag{2.15}$$

From the above formulas, the coefficients $a(z), b(z), c(z)$ and $d(z)$ of the system (1.1) can be obtained, namely

$$a(z) + ic(z) = 2[A(z) + B(z)],\ d(z) - ib(z) = 2[A(z) - B(z)] \text{ in } D. \tag{2.16}$$

Theorem 2.4. For the solutions $[F(z), G(z)]$ of equations in (2.12), we have

$$2i\operatorname{Im}[F(z)\overline{G(z)}] = F(z)\overline{G(z)} - \overline{F(z)}G(z) \neq 0 \text{ in } D. \tag{2.17}$$

Proof. Suppose that (2.17) is not true, then there exists a point $z_0 \in D$ such that $\mathrm{Im}[\overline{F(z_0)}G(z_0)] = 0$, namely

$$\begin{vmatrix} \mathrm{Re}F(z_0) & \mathrm{Im}F(z_0) \\ \mathrm{Re}G(z_0) & \mathrm{Im}G(z_0) \end{vmatrix} = 0.$$

Thus we have two real constants c_1, c_2, which are not all equal to 0, such that $c_1F(z_0) + c_2G(z_0) = 0$.

In the following, we prove that the equality of $c_1F(z_0) + c_2G(z_0) = 0$ is not true. If $W(z_0) = c_1F(z_0) + c_2G(z_0) = 0$, then $W(z) = \Phi(z)e^{\phi(z)} = (z - z_0)\Phi_1(z)e^{\phi(z)}$, where $\Phi(z)$, $\Phi_1(z)$ are analytic functions in D, and

$$(z-z_0)\Phi_1(z)e^{\phi(z)} + \frac{1}{\pi}\iint_D \frac{(\zeta-z_0)\Phi_1(\zeta)e^{\phi(z)}[A+B\overline{W(\zeta)}/W(\zeta)]}{\zeta - z}d\sigma_\zeta$$
$$= c_1 + ic_2.$$

Letting $z \to z_0$, one has

$$\frac{1}{\pi}\iint_D \Phi_1(\zeta)e^{\phi(z)}[A + B\overline{W(\zeta)}/W(\zeta)]d\sigma_\zeta = c_1 + ic_2,$$

and then

$$c_1+ic_2 = (z - z_0)\Phi_1(z)e^{\phi(z)}$$
$$+\frac{1}{\pi}\iint_D \frac{(\zeta-z+z-z_0)\Phi_1(\zeta)e^{\phi(z)}[A+B\overline{W(\zeta)}/W(\zeta)]}{\zeta - z}d\sigma_\zeta$$
$$= (z-z_0)[\Phi_1(z)e^{\phi(z)} + \frac{1}{\pi}\iint_D \frac{\Phi_1(\zeta)e^{\phi(z)}[A+B\overline{W(\zeta)}/W(\zeta)]}{\zeta - z}d\sigma_\zeta]$$
$$+\frac{1}{\pi}\iint_D \Phi_1(\zeta)e^{\phi(z)}[A + B\overline{W(\zeta)}/W(\zeta)]d\sigma_\zeta.$$

The above equality implies

$$\Phi_1(z)e^{\phi(z)} + \frac{1}{\pi}\iint_D \frac{\Phi_1(\zeta)e^{\phi(z)}[A + B\overline{W(\zeta)}/W(\zeta)]}{\zeta - z}d\sigma_\zeta = 0 \text{ in } D,$$

and the above homogeneous integral equation only have the trivial solution, namely $\Phi_1(z) = 0$ in D, thus $W(z) = \Phi(z)e^{\phi(z)} = (z - z_0)\Phi_1(z)e^{\phi(z)} \equiv 0$ in D. This is impossible.

From the above discussion, we can see that four real coefficients $a(z)$, $b(z)$, $c(z)$, $d(z)$ of the system (1.1) or two complex coefficients $A(z)$, $B(z)$ of the complex equation (1.3) can be determined by two boundary functions $H_1(z)$, $H_2(z)$ in the set R_H.

3. Inverse Problem for Second Order Elliptic Equations from Dirichlet to Neumann Map

Let D be a simply connected bounded domain in the complex plane $\mathbb{C}$ with the boundary $\partial D = \Gamma\,(\in C^2_\mu,\ 0<\mu<1)$. There is no harm in assuming that a point $a_0 \in \Gamma$ with $\arg a_0 = 0$ and $z = 0 \in D$. Consider the linear elliptic equation of second order

$$u_{xx} + u_{yy} + au_x + bu_y = 0 \ \text{ in } D, \tag{3.1}$$

where $a = a(z)$, $b = b(z)$ are real functions of $z = x + iy(\in D)$ with the conditions $a, b \in L_p(\overline{D})$, where $p\,(>2)$ is a positive constant. The above conditions will be called Condition C.

Similarly to Section 1, denote

$$\begin{aligned} W(z) &= U + iV = \frac{1}{2}[u_x - iu_y] = u_z, \\ W_{\bar z} &= \frac{1}{2}[W_x + iW_y] = u_{z\bar z} = \frac{1}{4}[u_{xx} + u_{yy}] \ \text{ in } D, \end{aligned} \tag{3.2}$$

we can get

$$\begin{aligned} u_{z\bar z} &= W_{\bar z} = \frac{1}{2}[W_x + iW_y] = -\frac{1}{4}[au_x + bu_y] \\ &= -\frac{1}{4}[a(W+\overline{W}) + ib(W-\overline{W})] = -\frac{1}{4}[(a+ib)W + (a-ib)\overline{W}] \\ &= -A(z)W - B(z)\overline{W} = -2\mathrm{Re}[A(z)W] \ \text{ in } D, \end{aligned} \tag{3.3}$$

in which $A = A(z) = \overline{B(z)} = \overline{B} = [a+ib]/4$, and $A(z) \in L_p(\overline{D})$, $p\,(>2)$ is a positive constant.

Introduce the Dirichlet boundary condition for the equation (3.1) as follows:

$$u = f(z) \ \text{ on } \Gamma, \text{ i.e. } u = f(z) \text{ on } \Gamma, \tag{3.4}$$

where $f(z) \in C^1_\alpha(\Gamma)$, $f(z) \in C^1_\alpha(\Gamma)$, $\alpha\,(0 < \alpha \le (p-2)/p)$ is a positive constant, which is called Problem D for equation (3.1). If we find the derivative of positive tangent direction with respect to the unit arc length parameter s of the boundary Γ with $s(0) = \arg(a_0 + 0) = 0$, where $a_0 \in \Gamma_0$, then

$$f_s = \frac{\partial f(z)}{\partial s} = u_z z_s + u_{\bar z}\bar z_s = 2\mathrm{Re}[z_s u_z] \ \text{ on } \Gamma. \tag{3.5}$$

It is clear that the equivalent boundary value problem is found a solution $[W(z), u(z)]$ of the complex equation (3.3) with the boundary conditions

$$\mathrm{Re}[\overline{\lambda(z)}w(z)] = \mathrm{Re}[z_s w(z)] = \frac{f_s}{2},\ z \in \Gamma,\ u(a_0) = f(a_0) = b_0, \tag{3.6}$$

and the relation

$$u(z) = 2\mathrm{Re}\int_{a_0}^{z} w(z)dz + b_0 \text{ in } \overline{D}, \tag{3.7}$$

in which $\lambda(z) = \overline{z_s}$, $z \in \Gamma$. Taking into account the index $K = \Delta_\Gamma \arg[\lambda(z)] = -1$, obviously this is a special case of the Riemann-Hilbert boundary value problem (Problem RM) as stated in Sections 1 and 2.

Under the above condition, the corresponding Neumann boundary condition is

$$u_n = \frac{\partial u}{\partial n} = u_z z_n + u_{\bar z}\bar z_n = u_z z + u_{\bar z}\bar z = 2\mathrm{Im}[z_s u_z] = g(z) \text{ on } \Gamma, \tag{3.8}$$

where n is the unit outwards normal vector of Γ. The boundary value problem (3.1)(or (3.3)), (3.8) will be called Problem N. Hence the boundary conditions of Dirichlet and Neumann problems can be written as follows

$$\begin{aligned} u_s + iu_n &= \mathrm{Re}[z_s u_z] + 2i\mathrm{Im}[z_s u_z] = 2z_s w(z),\ z \in \Gamma, \text{ i.e.} \\ w(z) &= h(z) = h_1(z) + ih_2(z) = [u_s + iu_n]/2z_s,\ z \in \Gamma, \end{aligned} \tag{3.9}$$

which will be called Problem DN for the complex equation (3.3) (or (3.1)) with the relation (3.7), where $h(z) \in C_\alpha(\Gamma)$ is a complex function satisfying the condition $\int_\Gamma \mathrm{Re}[z_s]u_z ds = 0$. For any function $f(z)$ of the set $C^1_\alpha(\Gamma)$ in the Dirichlet boundary condition (3.4), there is a set $\{g(z)\}$ of the functions of Neumann boundary condition (3.8), which is called the Dirichlet to Neumann map. Moreover we see that the set $\{h(z)\}$ is corresponding to the set $\{f_s + iu_n\}$ one by one, which can be determined the coefficients of equation (3.1). We denote the set of functions $\{h(z)\}$ by R_h, and $h(z)$ is as stated in (3.9).

According to the method as stated in Section 2, we can only consider the following integral equations

$$g(z) - T[Ag + B\overline{g}] = 1 \text{ or } i \text{ in } D, \tag{3.10}$$

and can prove a lemma similar to Lemma 2.1. Finally we can obtain the following result.

Theorem 3.1. For the inverse problem of Problem DN for complex equation (3.1) with Condition C, we can reconstruct the convective coefficients $a(z)$ and $b(z)$ of equation (3.1).

Proof. On the basis of Lemma 3.1, we can find two solutions $\phi_1(z) = F(z)$ and $\phi_2(z) = G(z)$ of integral equations

$$\left\{\begin{aligned} F(z) + T\{AF + \overline{AF}\} = 1, \\ G(z) + T\{AG + \overline{AG}\} = i, \end{aligned}\right. \quad \text{in } \overline{D},$$

and we can require that the above solutions satisfy the boundary conditions

$$F(z) = h_1(z),\ G(z) = h_2(z) \text{ on } \Gamma,$$

where $h_1(z)$, $h_2(z) \in R_h$. It is obvious that the above solutions $\phi_1(z) = F(z)$, $\phi_2(z) = G(z)$ are also the solutions of complex equation

$$[\phi]_{\bar{z}} + \{A(z)\phi + \overline{A(z)]\phi}\} = 0 \text{ in } \overline{D}.$$

We can verify that

$$\operatorname{Im}[F(z)\overline{G(z)}] = [F(z)\overline{G(z)} - \overline{F(z)}G(z)]/2i \neq 0 \text{ in } \overline{D},$$

thus the complex coefficients of the equation (3.3):

$$A = -\frac{F_{\bar{z}}\overline{G} - G_{\bar{z}}\overline{F}}{F\overline{G} - \overline{F}G},\ \overline{A} = \frac{F_{\bar{z}}G - G_{\bar{z}}F}{F\overline{G} - \overline{F}G} \text{ in } \overline{D}$$

can be determined. From the above formulas, we immediately get the real coefficients $a(z)$, $b(z)$ of the system (3.1), namely

$$a(z) = 2[A(z) + \overline{A(z)}],\ b(z) = -2i[A(z) - \overline{A(z)}] \text{ in } D.$$

References

[1] J. Cheng and M. Yamamoto, The global uniqueness for determining two convection coefficients from Dirichlet to Neumann map in two dimensions, *Inverse Problems,* 16 (2000), 25–30.

[2] A. Tamasan, On the scattering method for the $\overline{\partial}$-equation and reconstruction of convection coefficients, *Inverse Problems,* 20 (2004), 1807–1817.

[3] Z. L. Tong, J. Cheng, and M. Yamamoto, A method for constructing the convection coefficients of elliptic equations from Dirichlet to Neumann map, *Science in China, Ser. A,* 34 (2004), 752–766 (In Chinese).

[4] J. Cheng and M. Yamamoto, Determination of two convection coefficients from Dirichlet to Neumann map in the two-dimensional case, *SIAM J. Math. Anal.*, 35 (2004), 1371–1393.

[5] I. N. Vekua, *Generalized analytic functions,* Pergamon, Oxford, 1962.

[6] N. I. Mushelishvili, *Singular integral equations,* Noordhoff, Groningen, 1953.

[7] G. C. Wen and H. Begehr, *Boundary value problems for elliptic equations and systems,* Longman Scientific and Technical Company, Horlow, 1990.

[8] G. C. Wen, *Conformal mappings and boundary value problems,* Amer. Math. Soc., Providence, RI, 1992.

NUMERICAL INVERSION OF THE EXPONENTIAL RADON TRANSFORM[1]

YUN-GANG DENG and JIN-PING WANG*

Department of Mathematics, Ningbo University, Ningbo 315211, China

* *Corresponding author. E-mail: wangjinping@nbu.edu.cn(Jin-Ping Wang)*

In this paper we obtain the numerical inversion formula of the exponential Radon transform, with Chebyshev polynomials and the inverse formula of the exponential Radon transform. Furthermore we deduce the algorithm for the numerical inversion formula.

Keywords: Exponential Radon transform, inversion formula, numerical inversion.

AMS No: 42C15, 44A12.

1. Introduction

The Radon transform was derived by Radon in his paper published in 1917. And he discussed the problem how to get a function f from its line integrals. After that, much literature did their research about Radon transform not only in theory but also in application, see [4,9]. A summary result of Radon transform and generalized Radon transform, which has a general weighted function in the integral, was given in [3]. With the first brain scanner developed, people began to investigate the exponential Radon transform which is a special kind generalized Radon transform, for it is the foundation of mathematics of the single-photon emission computer tomograph. They also gave the inverse formula of the exponential Radon transform in two dimension case within [3]. Our results are obtained on the base of this formula, with the Chebyshev polynomials and the character of Hilbert transform.

The exponential Radon transform $T_\mu f(\theta, s)$, $0 \leq \theta < 2\pi$, $-\infty < s < \infty$, of a function $f \in C_0^\infty(\Omega)$ with the variable $x = (x_1, x_2)$, where Ω is the unit disk in the $\mathbb{R}^2$ plane, is defined by

$$T_\mu f(\theta, s) = \int_{-\infty}^{\infty} e^{\mu t} f(su_\theta + tv_\theta)dt, \tag{1}$$

where $u_\theta = (\cos\theta, \sin\theta)$ and $v_\theta = u_{\theta+\pi/2} = (-\sin\theta, \cos\theta)$. An inversion formula is an expression for $f(x)$ in terms of $T_\mu f$, which had been solved by Bellini et al in theory and C. E. Metz together with X. Pan gave the algorithm in practical. There are many such expressions involving various

[1]This research is supported by the NSFC (No. 60872095), and Ningbo Natural Science Foundation (2008A610018, 2009B21003, 2010A610100).

hypotheses on the function f, see [1–3, 8]. One reason such transform is of interest is due to applications in computed tomography; several such applications are described in [5–7].

This paper is organized as follows. First, some preliminaries are given in Section 2. The numerical inversion method is developed in Section 3. In Section 4 the algorithm on the numerical inversion is established.

2. Some Preliminaries

We use the Cartesian coordinate system to denote one point in the $\mathbb{R}^2$ plane by (x_1,x_2), or by the polar system (r,φ). Then we have $(x_1,x_2) = re^{i\varphi}$, $x_1 = r\cos\varphi$, $x_2 = r\sin\varphi$. Then we can express one line in the $\mathbb{R}^2$ plane as

$$s = x_1\cos\theta + x_2\sin\theta = x\cdot u_\theta = r\cos(\varphi-\theta), \tag{2}$$

where u_θ just as above and $x=(x_1,x_2)$, $0<r\leq 1$, with "$\cdot$" denote the inner product of two vectors in the $\mathbb{R}^2$ plane. So we also have $x\cdot v_\theta = r\sin(\varphi-\theta)$. For the sake of discussing simply, we assume $f\in C_0^\infty(\Omega)$, and define the n-dimensions Fourier transform as:

$$\hat{f}(\sigma) = (2\pi)^{-n/2}\int_{\mathbb{R}^n} f(x)e^{-ix\cdot\sigma}dx,\ \sigma\in\mathbb{R}^n,\ n\geq 2. \tag{3}$$

Next we define the Hilbert transform of functions in $C_0^\infty(\mathbb{R}^2)$ as

$$Hf(x) = \frac{1}{\pi}\int_{\mathbb{R}^2}\frac{f(y)}{x-y}dy. \tag{4}$$

Lemma 1. *For $f\in C_0^\infty(\mathbb{R}^2)$, the Hilbert transform defined above, we have $\widehat{Hf(\sigma)} = -i\,\mathrm{sgn}(\sigma)\hat{f}(\sigma)$.*

Proof.

$$\begin{aligned}
\widehat{Hf}(\sigma) &= \frac{1}{\sqrt{2\pi}}\frac{1}{\pi}\int_{\mathbb{R}}\frac{f(s)}{t-s}ds\int_{\mathbb{R}}e^{-it\sigma}dt\frac{-i}{\sqrt{2\pi}}\\
&= \frac{1}{\sqrt{2\pi}}\frac{1}{\pi}\int_{\mathbb{R}}f(s)e^{-it\sigma}e^{-is\sigma}e^{is\sigma}ds\int_{\mathbb{R}}\frac{1}{t-s}dt\\
&= \frac{1}{\sqrt{2\pi}}\frac{1}{\pi}\int_{\mathbb{R}}f(s)e^{-is\sigma}ds\int_{\mathbb{R}}\frac{1}{t-s}e^{-i(t-s)\sigma}dt\\
&= \frac{-i}{\sqrt{2\pi}}\frac{1}{\pi}\int_{\mathbb{R}}f(s)e^{-is\sigma}ds\int_{\mathbb{R}}\frac{\sin(u\sigma)}{u}du\\
&= \frac{-2i}{\pi}\hat{f}(\sigma)\int_0^\infty\frac{\sin(u\sigma)}{u}du = -i\mathrm{sgn}(\sigma)\hat{f}(\sigma),
\end{aligned} \tag{5}$$

where we make use of the formula:

$$\int_0^\infty \frac{\sin(t\sigma)}{t} dt = -\mathrm{sgn}(\sigma) \int_0^\infty \frac{\sin u}{u} du = -\mathrm{sgn}(\sigma)\frac{\pi}{2}.$$

Then we can use the Fourier transform relation between the derivative of function f and itself to obtain

$$\widehat{(Hf)'}(\sigma) = \sigma\,\mathrm{sgn}(\sigma)\hat{f}(\sigma). \tag{6}$$

3. Numerical Inversion

Lemma 2. *If f satisfies the above hypothesis and $g = T_\mu f$ is the exponential Radon transform of f, then the following result holds:*

$$f = \frac{-1}{4\pi^2} \int_0^{2\pi} e^{-\mu r \sin(\varphi-\theta)} \int_{-1}^{1} \frac{g'(s)}{s - r\cos(\varphi-\theta)} ds d\theta. \tag{7}$$

Proof. We can follow the inversion formula of the exponential radon transform ([3])

$$f = \frac{1}{4\pi} T^\sharp_{-\mu} I_\mu^{-1} g,$$

where $g = T_\mu f$ and $T^\sharp_\mu$ is the adjoint operator of T_μ. Here the operator I_μ^{-1} is the Risze potential satisfies:

$$\widehat{I_\mu^{-1} g}(\sigma) = \begin{cases} 0, & \text{if} \quad |\sigma| \le |\mu|, \\ |\sigma|\hat{g}(\sigma), & \text{if} \quad |\sigma| > |\mu|. \end{cases} \tag{8}$$

From the first lemma and (6), we have $\widehat{I_\mu^{-1} g}(\sigma) = \widehat{(Hg)'}(\sigma)$. Then by the inverse Fourier theory, it holds: $I_\mu^{-1} g(\sigma) = (Hg)'(\sigma)$. Following this and $\mathrm{supp} f \subset \Omega$, we obtain

$$\begin{aligned} f &= \frac{1}{4\pi} T^\sharp_{-\mu} I_\mu^{-1} g \\ &= \frac{1}{4\pi} \int_{S^1} e^{-\mu x\cdot u_\theta} (Hg)'(u_\theta, s) du_\theta \\ &= \frac{1}{4\pi^2} \int_0^{2\pi} e^{-\mu r \sin(\varphi-\theta)} \int_{-1}^{1} \frac{g'(s)}{r\cos(\varphi-\theta) - s} ds d\theta, \end{aligned} \tag{9}$$

with alternative of the sign can give the result.

As we all know from the Fourier theory, any function can be expressed by the Fourier series. So we will utilize this fact to derive our result. If we assume the Fourier series of function f is $f(x) = f(r, \varphi) = G_l(r)e^{il\varphi}$ or the linear combination of this form. As the exponential radon transform of f by $g(\theta, s) = g_l(s)e^{il\theta}$ or the linear combination of this form, then the following theorem holds:

Theorem 1. *For the above f and g, we have*

$$G_l(r) = \frac{1}{2\pi^2 r}\int_0^1 g_l{}'(s)W_l(\frac{s}{r})ds, \tag{10}$$

where

$$W_l(\frac{s}{r}) = \int_{-1}^1 (x - \frac{s}{r})^{-1}T_l(x)(1-x^2)^{-\frac{1}{2}}((-1)^l e^{\mu r\sqrt{1-x^2}} + e^{-\mu r\sqrt{1-x^2}})dx,$$

and $T_l(x)$is the first Chebyshev polynomial.

Proof. Take the first order derivative of $g(\theta, s) = g_l(s)e^{il\theta}$ with s and use Lemma 2, we have

$$\begin{aligned} f(x) &= f(r, \varphi) \\ &= \frac{-1}{4\pi^2}\int_0^{2\pi} e^{-\mu r\sin(\varphi-\theta)}\int_{-1}^1 \frac{g'(s)}{s - r\cos(\varphi-\theta)}dsd\theta \\ &= \frac{-1}{4\pi^2}\int_0^{2\pi} e^{-\mu r\sin(\varphi-\theta)}\int_{-1}^1 \frac{e^{il\theta}}{s-t}\frac{\partial g_l}{\partial s}dsd\theta \\ &\quad \times\int_{-r}^r \delta(t - r\cos(\varphi-\theta))dt. \end{aligned} \tag{11}$$

Next we set

$$J = \int_0^{2\pi} e^{-\mu r\sin(\varphi-\theta)}e^{il\theta}\delta(t - r\cos(\varphi-\theta))d\theta, \tag{12}$$

then

$$\begin{aligned} J &= \int_0^{\pi} e^{-\mu r\sin(\varphi-\theta)}e^{il\theta}\delta(t - r\cos(\varphi-\theta))d\theta \\ &\quad + \int_{\pi}^{2\pi} e^{-\mu r\sin(\varphi-\theta)}e^{il\theta}\delta(t - r\cos(\varphi-\theta))d\theta \\ &= \int_0^{\pi} e^{-\mu r\sin(\varphi-\theta)}e^{il\theta}\delta(t - r\cos(\varphi-\theta))d\theta \\ &\quad + (-1)^l\int_0^{\pi} e^{\mu r\sin(\varphi-\theta)}e^{il\theta}\delta(t + r\cos(\varphi-\theta))d\theta, \end{aligned} \tag{13}$$

we make the substitute $x = \cos(\varphi - \theta)$, then $\varphi - \theta = \arccos(x)$, so (13) can be transformed into

$$J = \int_{-\cos\varphi}^{\cos\varphi} \frac{1}{\sqrt{1-x^2}} e^{-\mu r\sqrt{1-x^2}} e^{il(\varphi-\arccos(x))}\delta(t-rx)dx$$
$$+(-1)^l \int_{-\cos\varphi}^{\cos\varphi} \frac{1}{\sqrt{1-x^2}} e^{\mu r\sqrt{1-x^2}} e^{il(\varphi-\arccos(x))}\delta(t+rx)dx, \quad (14)$$

with the character of δ function and $0 < r \leq 1$, $0 < \frac{t}{r} \leq 1$.

We have

$$J = \frac{\exp(il\varphi - il\arccos(\frac{t}{r}))}{\sqrt{r^2-t^2}}((-1)^l e^{\mu\sqrt{r^2-t^2}} + e^{-\mu\sqrt{r^2-t^2}}), \quad (15)$$

following the Euler formula and the concern conclusion of the first or second kind Chebyshev polynomial, we get

$$J = e^{il\varphi}[T_l(\frac{t}{r}) - iU_{l-1}(\frac{t}{r})\sqrt{1-\frac{t^2}{r^2}}]$$
$$\times \frac{1}{\sqrt{r^2-t^2}}((-1)^l e^{\mu}\sqrt{r^2-t^2} + e^{-\mu\sqrt{r^2-t^2}}), \quad (16)$$

where $T_l(x)$ is the first kind Chebyshev polynomial and $U_l(x)$ is the second. Then (11) is changed into

$$f(r,\varphi) = G_l(r)e^{il\varphi} = \frac{-1}{4\pi^2}\int_{-r}^{r}\frac{J}{s-t}dt\int_{-1}^{1} g_l{}'(s)ds. \quad (17)$$

From the expression of J, we know

$$G_l(r) = \frac{-1}{4\pi^2}\int_{-1}^{1} g_l{}'(s)ds \int_{-r}^{r}\frac{1}{s-t}$$
$$\times[T_l(\frac{t}{r})(r^2-t^2)^{-\frac{1}{2}} - iU_{l-1}(\frac{t}{r})r^{-1}]$$
$$\times((-1)^l e^{\mu}\sqrt{r^2-t^2} + e^{-\mu\sqrt{r^2-t^2}})dt, \quad (18)$$

let $t = rx, -1 \leq x \leq 1$,

$$G_l(r) = \frac{-1}{4\pi^2}\int_{-1}^{1} g_l{}'(s)ds \int_{-1}^{1}\frac{1}{s-rx}[T_l(x)(1-x^2)^{-\frac{1}{2}} - iU_{l-1}(x)]$$
$$\times((-1)^l e^{\mu r\sqrt{1-x^2}} + e^{-\mu r\sqrt{1-x^2}})dx$$
$$= \frac{1}{4\pi^2 r}\int_{-1}^{1} g_l{}'(s)ds \int_{-1}^{1}(x-\frac{s}{r})^{-1}[T_l(x)(1-x^2)^{-\frac{1}{2}} - iU_{l-1}(x)] \quad (19)$$
$$\times((-1)^l e^{\mu r\sqrt{1-x^2}} + e^{-\mu r\sqrt{1-x^2}})dx$$
$$= I_1 + I_2,$$

in which I_1 is the first one in the above sum and I_2 is the second. As the character of the second kind Chebyshev polynomial and substitute $-x$, $-s$ into x, s respectively, we know the following formula is identical to zero:

$$\begin{aligned}I_2 = \frac{1}{4\pi^2 r}\int_{-1}^{1} g_l{}'(s)ds\int_{-1}^{1}(x-\frac{s}{r})^{-1}\\ \times[-iU_{l-1}(x)]((-1)^l e^{\mu r\sqrt{1-x^2}}+e^{-\mu r\sqrt{1-x^2}})dx,\end{aligned} \tag{20}$$

then (19) is left I_1 to calculate. We write $W_l(\frac{s}{r})$ in I_1 as follows

$$W_l(\frac{s}{r})=\int_{-1}^{1}(x-\frac{s}{r})^{-1}T_l(x)(1-x^2)^{-\frac{1}{2}}((-1)^l e^{\mu r\sqrt{1-x^2}}+e^{-\mu r\sqrt{1-x^2}})dx. \tag{21}$$

Moreover from (19), we have

$$G_l(r) = \frac{1}{4\pi^2 r}\int_{-1}^{1} g_l{}'(s)W_l(\frac{s}{r})ds. \tag{22}$$

To find (22), we take (21) into account and separate (21) into two parts: one is when $|\frac{s}{r}| \leq 1$, the other is when $|\frac{s}{r}| > 1$. Then (22) becomes

$$G_l(r) = \frac{1}{4\pi^2 r}\{\int_{-1}^{-r}+\int_{-r}^{0}+\int_{0}^{r}+\int_{r}^{1} g_l{}'(s)W_l(\frac{s}{r})\}ds. \tag{23}$$

Substituting $-x$, $-s$ into x, s respectively, we know

$$W_l(-\frac{s}{r}) = (-1)^{l+1}W_l(\frac{s}{r}),$$

and noticing that

$$g_l{}'(s) = \frac{dg_l(s)}{ds},\ g_l{}'(-s) = (-1)^{l+1}g_l{}'(s),$$

then the following equalities hold:

$$\begin{aligned}\int_{-1}^{-r} g_l{}'(s)W_l(\frac{s}{r})ds = \int_{r}^{1} g_l{}'(s)W_l(\frac{s}{r})ds,\\ \int_{-r}^{0} g_l{}'(s)W_l(\frac{s}{r})ds = \int_{0}^{r} g_l{}'(s)W_l(\frac{s}{r})ds.\end{aligned} \tag{24}$$

So we obtain

$$G_l(r) = \frac{1}{2\pi^2 r}\int_{0}^{1} g_l{}'(s)W_l(\frac{s}{r})ds, \tag{25}$$

which is the expected result.

4. Algorithm

From the above result, we have the algorithm as follows:
Step 1. Get the Radon data for each θ_i, $T_\mu f(\theta_i, s_l)$, $l = -p, \cdots p$, where $i = -q, \cdots q$
Step 2. For each θ_i, using interpolation to get the radon data $(T_\mu f)'(\theta_i, s_l)$, with respect to s,
Step 3. Use a quadrature rule in (21) to obtain $W_l(\frac{s}{r})$, and take another quadrature rule in the formula (10) to get the fourier coefficient $G_l(r)$, so as to obtain function f in each φ_i direction.
Step 4. Reconstruct the activity image $f(r, \varphi)$ in polar coordinates with $f(x) = f(r, \varphi) = G_l(r)e^{il\varphi}$.

Remark 1. We use $x_i = \cos(\varphi_i - \theta_i)$ to acquire φ_i according to θ_i with x_i in the quadrature rule of (21) in Step 4. And $(T_\mu f)'(\theta_i, s_l)$ in Step 2 is acquired by Mid-point differential formula (see below) with respect to s

$$(T_\mu f)'(\theta_i, s_l) = \frac{T_\mu f(\theta_i, s_{l+1}) - T_\mu f(\theta_i, s_{l-1})}{2(s_l - s_{l-1})}. \tag{26}$$

5. Numerical Simulation

The left figure is the Shepp-Logan phantom and the right is the exponential radon transform graphic

The left figure is the derivative of the exponential radon transform and the right one is the numerical inversion graphic of the exponential radon transform according to (25).

From above we can see that while the numerical inversion graphic has an error contrast to the phantom, it is more convenient to realize in computer.

Remark 2. All the graphics above are produced by Matlab when the parameter $\mu = 0$.

References

[1] Jinping Wang, Some results on the exponential Radon transform, *Acta Math. Scientia.* 26A (2005), 031-038.
[2] Jinping Wang, The harmonic decomposition of exponential Radon transform, *J. Analyse Math.* 47 (2001), 69-85.
[3] F. Natterer, *The mathematics of computerized tomography,* John Wiley Sons, Stuttgart, 1986.
[4] E. Clarkson, Projections onto the range of the exponential Radon transform and reconstruction algorithms, *Inverse Problems* 15 (1999), 563-571.
[5] Qiulin Tang, Analytical fan-beam and cone-beam reconstruction algorithms with uniform attenuation correction for SPECT, *Phys. Med. Biol.* 50 (2005), 3153-3170.
[6] A. V. Bronnikov, A filtering approach to image reconstruction in 3D SPECT, *Phys. Med. Biol.* 45 (2000), 2639-2651.

[7] C. E. Metz and X. Pan, A unified analysis of exact methods of inverting the 2D exponential Radon transform, with implications for noise control in SPECT, *IEEE Trans. Med. Imag.* 17 (1995), 643-658.

[8] J. Bourgain, Averages in the plane over convex curves and maximal operators, *J. Analyse Math.* 47 (1986), 69-85.

[9] S. R. Deans, *The Radon transform and some of its applications*, Wiley, New York, 1983.

UNIQUENESS AND STABILITY ESTIMATES FOR A SEMILINEARLY PARABOLIC BACKWARD PROBLEM[1]

DING-HUA XU and HAI-LI ZHANG

Department of Mathematics, College of Sciences, Zhejiang Sci-Tech University, Hangzhou 310018, P. R. China

E-mail: dhxu6708@zstu.edu.cn

The problem of finding the temperature u satisfying

$$u_t = u_{xx} + u_{yy} + f(u, \nabla u),\ (x, y, t) \in \mathbb{R}^2 \times [0, T),$$

from the final data $u(x, y, T) \equiv \varphi(x, y)$, is discussed. Due to nonlinear ill-posedness for the inverse problem, the Fourier transforms was used to derive a nonlinear integral equation, from which a regularized solution was deduced by perturbing directly the integral equation. It was also mentioned that one should give the appropriate choice of the time T, after which the solution $u(x, y, t)$ will blow up. For the backward parabolic problem, the uniqueness and stability estimates are achieved for the proposed algorithm.

Keywords: Backward parabolic problem, blow-up time, contraction principle, nonlinearity, ill-posed problem, unique solution, stability estimate.

AMS No: 35R30, 65M32, 42B99.

1. Introduction

We consider the following nonlinear inverse problem (NIP):

NIP: Find the initial distribution $u(x, y, 0) = \phi(x, y)$ or intermediate distribution $u(x, y, t)\mid_{t\in(0,T)}$ such that

$$u_t = u_{xx} + u_{yy} + f(u, \nabla u),\ (x, y, t) \in \mathbb{R}^2 \times [0, T] \tag{1.1}$$

by the final measurements $u(x, y, T) \equiv \varphi(x, y)$, here the function $f(\cdot)$ is known, T is a fixed positive number.

The inverse problem (NIP) is usually called the nonlinearly backward parabolic problem. The function $f(\cdot, \cdot)$ is often nonlinear in practical situations. For the different expressions of $f(u, \nabla u)$, the parabolic equation (1.1) describes different kinds of physical phenomena, which can be formulated mathematically to be the well-known parabolic equations (see, for example, [1–5]). The equation (1.1) is, respectively, the well-known chemical

[1]This research is Supported by the NSFC (No. 10561001 and 11071221), the Science Foundation of Zhejiang Sci-Tech University(ZSTU)(0613263).

reaction equation when $f(u, \nabla u) = u^n (n \in N^+)$; Fisher equation in population model when $f(u, \nabla u) = u - u^2$; Cahan-Allen equation in the biology chemistry when $f(u, \nabla u) = u - u^3$; the simple exothermic reaction model when $f(u, \nabla u) = \lambda e^{\frac{u}{1+\varepsilon u}} \approx \lambda e^u$ (here $\varepsilon > 0$ is very small); the generalized Burger's equation when $f(u, \nabla u) = u^p + u^{q-1} a \cdot \nabla u$ $(p, q > 1, a \in \mathbb{R}^2)$; the population dynamics model when $f(u, \nabla u) = u^p - \mu |\nabla u|^q$ $(p, q > 1, \mu > 0)$.

As well known, the problem is severely ill-posed, more precisely, its solution does not always uniquely exist. Even though in the case of unique existence, the solution does not dependent continuously on the given observation data (see, for example, [6]). Moreover it is significantly important for the inverse problem that the terminal time T should be chosen before the blow-up time T^*. With respect to the blow-up properties for the nonlinear parabolic equations with unbounded domains, we can refer to some articles (see [7–11]). Of course the blow-up properties for the nonlinear parabolic equations with bounded domains can be found in more research papers (see [12–15]).

Once we know the "blow-up time", we may reasonably choose the terminal time T. At this time the measurements $u(x, y, T)$ must be given for the inverse problems. Let $0 < T < T^*$, we consider the problem of finding the temperature $u(x, y, t), (x, y, t) \in \mathbb{R}^2 \times [0, T)$.

There are so many works on the backward problems for linear parabolic equations in the one-dimensional cases (see, for example, [16–18]). But only a little literature has been concerning with the nonlinear case or multidimensional case. Recently some new numerical methods are proposed to solve the nonlinear backward problems both in 1-D case and in 2-D cases, for example in [19–22], but the uniqueness and stability results were not achieved.

In this paper we will adopted the regularization method proposed in [6] for the problem (NIP). We generalized the method for the 1-D case with absence of ∇u in [6] to the 2-D case with the term ∇u, which is more difficult to be theoretically discussed and has much more practical background in physics and engineering. Moreover, comparing with [6], the given assumptions in the paper are more reasonable and natural, and the results are much more applicable to various nonlinear parabolic equations.

From now on, we consider the (NIP) under following assumptions.

(A_1) Let T be less than the blow-up time T^*. So there exists a positive number $H > 0$ such that for all $(x, y, t) \in \mathbb{R}^2 \times [0, T]$,

$$|u(x, y, t)| < H, \ |\frac{\partial f}{\partial u}| < H, \ |\frac{\partial f}{\partial u_x}| < H \text{ and } |\frac{\partial f}{\partial u_y}| < H.$$

(A_2) $f(\cdot) \in L^\infty((-H, H)^3)$. Meanwhile there are constants $k_1, k_2, k_3 >$

0 independent of w, v, such that

$$|f(w, w_x, w_y) - f(v, v_x, v_y)| \leq k_1|w - v| + k_2|w_x - v_x| + k_3|w_y - v_y|.$$

(A_3) $\varphi \in L^2(\mathbb{R}^2)$. Let $\varphi_\varepsilon \in L^2(\mathbb{R}^2)$ be a measured data such that $\|\varphi_\varepsilon - \varphi\| < \varepsilon$, here the error level ε is very small.

2. The Regularization Method and Main Results

Let us denote the Fourier transform $\hat{g}(p_1, p_2)$ for any 2-D functions $g(x, y)$ by

$$\hat{g}(p_1, p_2) \equiv F(g) \equiv \frac{1}{2\pi} \int_{-\infty}^{+\infty} \int_{-\infty}^{+\infty} g(x, y) e^{-i(p_1 x + p_2 y)} dx dy, \tag{2.1}$$

while its inverse can be given as follows

$$g(x, y) \equiv F^{-1}(\hat{g}) \equiv \frac{1}{2\pi} \int_{-\infty}^{+\infty} \int_{-\infty}^{+\infty} \hat{g}(p_1, p_2) e^{i(p_1 x + p_2 y)} dp_1 dp_2. \tag{2.2}$$

Using the Fourier transform we can rewrite the above nonlinear inverse problem (NIP) as the following nonlinear integral equation (NIE):

NIE:

$$\begin{aligned} \hat{u}(p_1, p_2, t) &= e^{(T-t)(p_1^2+p_2^2)} \hat{\varphi}(p_1, p_2) \\ &- \int_t^T e^{-(t-s)(p_1^2+p_2^2)} \hat{f}(u, \nabla u) ds, \ 0 \leq t \leq T. \end{aligned} \tag{2.3}$$

It is well known that the problem (NIP), or the integral equation (NIE), is severely ill-posed. A regularization method should be adopted to derive the solution in a stable way. We approximate the problem (NIE) by the following integral equation with a small perturbation (IEP):

IEP:

$$\begin{aligned} \hat{u^\varepsilon}(p_1, p_2, t) &= \frac{e^{-t(p_1^2+p_2^2)}}{\varepsilon + e^{-T(p_1^2+p_2^2)}} \hat{\varphi}(p_1, p_2) \\ &- \int_t^T \frac{e^{-t(p_1^2+p_2^2)}}{\varepsilon^{s/T} + e^{-s(p_1^2+p_2^2)}} \hat{f}(u^\varepsilon, \nabla u^\varepsilon) ds, \ 0 \leq t \leq T, \end{aligned} \tag{2.4}$$

or

$$\begin{aligned} u^\varepsilon(x, y, t) &= \frac{1}{2\pi} \int_{-\infty}^{+\infty} \int_{-\infty}^{+\infty} \frac{e^{-t(p_1^2+p_2^2)}}{\varepsilon + e^{-T(p_1^2+p_2^2)}} \hat{\varphi}(p_1, p_2) e^{i(p_1 x + p_2 y)} dp_1 dp_2 \\ &- \frac{1}{2\pi} \int_{-\infty}^{+\infty} \int_{-\infty}^{+\infty} \int_t^T \frac{e^{-t(p_1^2+p_2^2)}}{\varepsilon^{s/T} + e^{-s(p_1^2+p_2^2)}} \hat{f}(u^\varepsilon, \nabla u^\varepsilon) e^{i(p_1 x + p_2 y)} ds dp_1 dp_2, \\ &\qquad 0 \leq t \leq T, \end{aligned} \tag{2.5}$$

here the number $\varepsilon > 0$ is very small.

Theorem 1. (The well-posedness of (IEP)) *Let φ, f satisfy the assumptions $A_1 - A_3$. Then the problem* (IEP) *has a unique solution $u^\varepsilon \in C([0,T]; H^1(\mathbb{R}^2))$. The solution depends continuously on φ in $C([0,T]; H^1(\mathbb{R}^2))$.*

Theorem 2. (Uniqueness of (NIP)) *Let φ, f satisfy the assumptions A_1–A_3. Then the problem* (NIP) *has at most one solution $u \in C([0,T]; H^1(\mathbb{R}^2)) \bigcap L^2(0,T; H^2(\mathbb{R}^2)) \bigcap C^1((0,T); H^1(\mathbb{R}^2))$.*

We further assume that

(A_4) $e^{T(p_1^2+p_2^2)}\hat{\varphi}(p_1,p_2) \in L^2(\mathbb{R}^2)$.

(A_5) $\frac{\partial}{\partial t}(e^{t(p_1^2+p_2^2)}\hat{u}(p_1,p_2,t)) \in L^2(\mathbb{R}^2 \times [0,T])$, that is

$$\int_0^T \int_{-\infty}^{+\infty} \int_{-\infty}^{+\infty} |\frac{\partial}{\partial t}(e^{t(p_1^2+p_2^2)}\hat{u}(p_1,p_2,t))|^2 dp_1 dp_2 dt < +\infty.$$

Theorem 3. (Stability Estimate with exact data φ) *Let φ, f satisfy the assumptions A_1–A_4, and the problem* (NIE) *has a solution $u \in C([0,T]; H^1(\mathbb{R}^2))$, satisfying A_5. Then for any $t \in (0,T)$,*

$$\|u(\cdot,\cdot,t) - u^\varepsilon(\cdot,\cdot,t)\|_{H^1}^2 \le (1+c_1)\sqrt{M}\varepsilon^{t/T} + \frac{3}{2}k^2T^2(1+c_1)^2(\frac{\varepsilon^{1-t/T}-1}{\ln\varepsilon}),$$

where

$$\begin{aligned} M = 3\int_{-\infty}^{+\infty}\int_{-\infty}^{+\infty} |e^{T(p_1^2+p_2^2)}\hat{\varphi}(p_1,p_2)|^2 dp_1 dp_2 \\ +3T\int_0^T \int_{-\infty}^{+\infty}\int_{-\infty}^{+\infty} |\frac{\partial}{\partial s}(e^{s(p_1^2+p_2^2)}\hat{u}(p_1,p_2,s))|^2 dp_1 dp_2 ds. \end{aligned} \tag{2.6}$$

Remark 1. Since

$$\lim_{\varepsilon\to 0^+} \varepsilon^{t/T} \cdot (\frac{\varepsilon^{1-t/T} - 1}{\ln\varepsilon})^{-1} = 0,$$

then we have

$$\|u(\cdot,\cdot,t) - u^\varepsilon(\cdot,\cdot,t)\|_{H^1}^2 \le 2A\frac{1-\varepsilon^{1-t/T}}{\ln\frac{1}{\varepsilon}},$$

where u^ε is the unique solution of the problem (IEP), and

$$A = \max\{(1+c_1)\sqrt{M}, \frac{3}{2}k^2T^2(1+c_1)^2\}. \tag{2.7}$$

Theorem 4. *Let φ, f satisfy the assumptions A_1–A_4 and the problem* (NIE) *has a solution $u \in C([0,T];\, H^1(\mathbb{R}^2))$ satisfying $u_t \in L^2([0,T]; H^1(\mathbb{R}^2))$ and A_5. Then for all $\varepsilon \in (0,1)$ there exists a $t_\varepsilon > 0$ such that*

$$\|u(\cdot,\cdot,0) - u^\varepsilon(\cdot,\cdot,t_\varepsilon)\|_{H^1} \leq 2M_2(\ln\frac{1}{\varepsilon})^{-\frac{1}{6}},$$

where

$$M_2 = \max\{2\sqrt{(1+c_1)N}(M_1T)^{\frac{1}{6}}, \frac{\sqrt{6}}{2}kT(1+c_1)\}, \tag{2.8}$$

$$N = \sqrt{\int_0^T \|u_t(\cdot,\cdot,s)\|^2 ds}, \tag{2.9}$$

$$M_1 = \sqrt{M}/N. \tag{2.10}$$

Theorem 5. (Stability Estimate with nonexact data φ_ε) *Let φ, f satisfy the assumptions A_1–A_4 and the problem* (NIE) *has a solution $u \in C([0,T]; H^1(\mathbb{R}^2))$, satisfying $u_t \in H^1([0,T]; L^2(\mathbb{R}^2))$ and A_5. Then from φ_ε we can construct a function u^ε satisfying*

$$\|u(\cdot,\cdot,t) - u^\varepsilon(\cdot,\cdot,t)\|^2_{H^1} \leq 2M_3[\frac{1-\varepsilon^{1-t/T}}{\ln\frac{1}{\varepsilon}}]^{\frac{1}{2}}, \quad 0 < t < T,$$

for every $\varepsilon \in (0,1)$ and

$$\|u^\varepsilon(\cdot,\cdot,0) - u(\cdot,\cdot,0)\|_{H^1} \leq 5M_2(\ln\frac{1}{\varepsilon})^{-\frac{1}{6}}, \quad t = 0,$$

where

$$B = \max\{\sqrt{2}(1+c_1), k^2T^2(1+c_1)^2\},\ M_3 = \max\{\sqrt{2A}, \sqrt{2B}\}. \tag{2.11}$$

3. The Proof of Theorems

Lemma 1. *For any functions $u(x,y) \in H^2(\mathbb{R}^2)$, there exists a positive constant c_1 such that*

$$\| \nabla u \|^2_{L^2(\mathbb{R}^2)} \leq c_1 \| u \|_{L^2(\mathbb{R}^2)},$$

hence when $\| u \|_{L^2(\mathbb{R}^2)} \leq 1$, we have

$$\| u \|^2_{H^1(\mathbb{R}^2)} \leq (1+c_1) \| u \|_{L^2(\mathbb{R}^2)}.$$

The proof of the Lemma 1 can easily be implemented by means of the Green's formula and the Hölder inequality.

Lemma 2. *Let the functions $u(t)$, $v(t)$ be locally integrable, non-negative functions on $[0,T]$, the constant $K > 0$, $t_0 \in [0,T]$. Then the following inequality*

$$u(t) \leq K + \int_{t_0}^{t} v(s)u^{\frac{1}{2}}(s)ds, \quad t \in [0,T]$$

implies that

$$u(t) \leq (K^{\frac{1}{2}} + \frac{1}{2}\int_{t_0}^{t} v(s)ds)^2, \quad t \in [0,T].$$

The Lemma 2 is a special case of the general Gronwall's inequality in [23]. The result can be derived straightforwardly from the Theorem 2 in [23].

Lemma 3. *For any positive integer m, there exist a constant K, which is independent of η, m and $\eta_0 > 0$, such that*

$$K\|\lambda^{-m}(v_{xx}+v_{yy}-v_t)\|_2^2 \geq \|\lambda^{-m}v_x\|_2^2 + \|\lambda^{-m}v_y\|_2^2 + \|\lambda^{-m-1}v\|_2^2$$

for every $v \in p_T \equiv \{u(x,y,t) \mid u(x,y,t) \in C([0,T];H^1(\mathbb{R}^2)) \cap L^2(0,T;H^2(\mathbb{R}^2)) \cap C^1((0,T);H^1(\mathbb{R}^2))$, $u(x,y,t) = 0$ on $\mathbb{R}^2 \times \{t = T\}, \mathbb{R}^2 \times \{t = 0\}\}$, $0 < \eta < \eta_0, \lambda = t - T - \eta$.

The result can be obtained from the Lemma 1 in [6].

Proof of Theorem 1. The proof is divided into two steps.

Step 1. The existence and the uniqueness of solution of (IEP).

Define a nonlinear integral operator G: for any $w \in C([0,T];H^2(\mathbb{R}^2))$,

$$\hat{G}(w)(x,y,t) \equiv \frac{1}{2\pi}\psi(x,y,t)$$
$$-\frac{1}{2\pi}\int_{-\infty}^{+\infty}\int_{-\infty}^{+\infty}\int_{t}^{T} \frac{e^{-t(p_1^2+p_2^2)}}{\varepsilon^{s/T} + e^{-s(p_1^2+p_2^2)}}\hat{f}(w,\nabla w)e^{i(p_1x+p_2y)}ds dp_1 dp_2,$$

where

$$\psi(x,y,t) = \int_{-\infty}^{+\infty}\int_{-\infty}^{+\infty} \frac{e^{-t(p_1^2+p_2^2)}}{\varepsilon + e^{-T(p_1^2+p_2^2)}}\hat{\varphi}(p_1,p_2)e^{i(p_1x+p_2y)}dp_1 dp_2.$$

From the Lipschitz property of $f(w,\nabla w)$ with respect to $w, \nabla w$, it follows that $\hat{G}(w) \in C([0,T];H^2(\mathbb{R}^2))$. Then by the mathematical induction principle, for every $w, v \in C([0,T];H^2(\mathbb{R}^2))$, $m \geq 1$, we have

$$\|G^m(w)(\cdot,\cdot,t) - G^m(v)(\cdot,\cdot,t)\|^2 \leq (\frac{k}{\varepsilon})^{2m}\frac{(T-t)^m c^m}{m!}|||w-v|||^2, \quad (3.1)$$

where $c = \max\{T, 1\}$, $\|\cdot\|$ is the norm in L^2 and $|||\cdot|||$ is the sup norm in $C([0,T]; H^2(\mathbb{R}^2))$. Subsequently, we get

$$\|G^m(w) - G^m(v)\| \leq (\frac{k}{\varepsilon})^m \frac{T^{m/2}}{\sqrt{m!}} \sqrt{c^m} |||w - v|||$$

for all $w, v \in C([0,T]; H^2(\mathbb{R}^2))$.

Since $\lim_{m\to\infty} (\frac{k}{\varepsilon})^m \frac{T^{m/2}}{\sqrt{m!}} \sqrt{c^m} = 0$, there exists a positive integer number m_0 such that G^{m_0} is a contraction in $C([0,T]; H^2(\mathbb{R}^2))$. It follows that the equation $G^{m_0}(w) = w$ has a unique solution $u^\varepsilon \in C([0,T]; H^2(\mathbb{R}^2))$.

We claim that $G(u^\varepsilon) = u^\varepsilon$. In fact, one has $G(G^{m_0}(u^\varepsilon)) = G(u^\varepsilon)$. Hence $G^{m_0}(G(u^\varepsilon)) = G(u^\varepsilon)$ by the uniqueness of the fixed point of G^{m_0}, one has $G(u^\varepsilon) = u^\varepsilon$, i.e, the equation $G(w) = w$ has a unique solution $u^\varepsilon \in C([0,T]; H^2(\mathbb{R}^2))$.

Step 2. The solution of the problem (IEP) depends continuously on $\varphi \in L^2(\mathbb{R}^2)$.

Let u and v be two solutions of (IEP) corresponding to the final function φ and Φ, respectively. From (2.4) and (2.5) we have

$$\begin{aligned}
&\|u(\cdot,\cdot,t) - v(\cdot,\cdot,t)\|^2 \\
\leq & 2\int_{-\infty}^{+\infty}\int_{-\infty}^{+\infty} |\frac{e^{-t(p_1^2+p_2^2)}}{\varepsilon + e^{-T(p_1^2+p_2^2)}} (\hat{\varphi}(p_1,p_2) - \hat{\Phi}(p_1,p_2)|^2 dp_1 dp_2 \\
&+2\int_{-\infty}^{+\infty}\int_{-\infty}^{+\infty} |\int_t^T \frac{e^{-t(p_1^2+p_2^2)}}{\varepsilon^{s/T} + e^{-s(p_1^2+p_2^2)}} (\hat{f}(u, \bigtriangledown u) - \hat{f}(v, \bigtriangledown v)) ds|^2 dp_1 dp_2.
\end{aligned}$$

One has for $s > t$ and $\alpha > 0$

$$\frac{e^{-t(p_1^2+p_2^2)}}{\alpha + e^{-s(p_1^2+p_2^2)}} = \frac{e^{-t(p_1^2+p_2^2)}}{(\alpha + e^{-s(p_1^2+p_2^2)})^{t/s}(\alpha + e^{-s(p_1^2+p_2^2)})^{1-t/s}} \leq \alpha^{t/s-1}.$$

Letting $\alpha = \varepsilon$, $s = T$, we get $\frac{e^{-t(p_1^2+p_2^2)}}{\varepsilon + e^{-T(p_1^2+p_2^2)}} \leq \varepsilon^{t/T-1}$. Letting $\alpha = \varepsilon^{s/T}$, we get $\frac{e^{-t(p_1^2+p_2^2)}}{\varepsilon^{s/T} + e^{-s(p_1^2+p_2^2)}} \leq \varepsilon^{t/T-s/T}$. Hence, it follows that

$$\begin{aligned}
&\|u(\cdot,\cdot,t) - v(\cdot,\cdot,t)\|_{L^2}^2 \leq 2\varepsilon^{2(t/T-1)} \|\hat{\varphi} - \hat{\Phi}\|^2 + 2(T-t)\varepsilon^{2t/T} k^2 \\
&\times \int_t^T [\varepsilon^{-2s/T} \|u(\cdot,\cdot,s) - v(\cdot,\cdot,s)\|_{L^2}^2 + \| \bigtriangledown u(\cdot,\cdot,s) - \bigtriangledown v(\cdot,\cdot,s)\|_{L^2}^2] ds.
\end{aligned}$$

So, we have

$$\begin{aligned}
&\varepsilon^{-6t/T} \|u(\cdot,\cdot,t) - v(\cdot,\cdot,t)\|_{L^2}^2 \leq 2\varepsilon^{-4t/T-2} \|\hat{\varphi} - \hat{\Phi}\|_{L^2}^2 \\
&+ 2k^2(T-t)\varepsilon^{-4t/T} \int_t^T [\varepsilon^{-2s/T} \|u(\cdot,\cdot,s) - v(\cdot,\cdot,s)\|_{H^1}^2] ds.
\end{aligned}$$

Using Lemma 1, we can get

$$\|u(\cdot,\cdot,t)-v(\cdot,\cdot,t)\|_{L^2(\mathbb{R}^2)} \geq \frac{1}{1+c_1}\|u(\cdot,\cdot,t)-v(\cdot,\cdot,t)\|^2_{H^1(R^2)}.$$

Hence

$$\begin{aligned}&(\varepsilon^{-3t/T}\|u(\cdot,\cdot,t)-v(\cdot,\cdot,t)\|^2_{H^1})^2\\ \leq\ &2\varepsilon^{-4t/T-2}(1+c_1)^2\|\hat{\varphi}-\hat{\Phi}\|^2_{L^2}\\ &+2k^2T(1+c_1)^2\varepsilon^{-4t/T}\int_t^T \varepsilon^{s/T}\cdot\varepsilon^{-3s/T}\|u(\cdot,\cdot,s)-v(\cdot,\cdot,s)\|^2_{H^1}ds.\end{aligned}$$

And using Gronwall's inequality, we have the following inequality

$$\|u(\cdot,\cdot,t)-v(\cdot,\cdot,t)\|^2_{H^1} \leq \sqrt{2}(1+c_1)\varepsilon^{t/T-1}\|\varphi-\Phi\|+k^2T^2(1+c_1)^2\frac{\varepsilon^{1-t/T}-1}{\ln\varepsilon},$$

which shows the continuous dependence of the solution once $\|\varphi-\Phi\|$ is sufficiently small. This completes the proof of Theorem 1.

Proof of Theorem 2. Let u_1 and u_2 be two solution of the problem (NIP) such that u_1, u_2 satisfy the assumptions A_1-A_4. Put $w(x,y,t) = u_1(x,y,t)-u_2(x,y,t)$, then w satisfies the equation

$$\begin{aligned}&w_t(x,y,t)-w_{xx}(x,y,t)-w_{yy}(x,y,t)\\ =\ &f(u_1,\nabla u_1)-f(u_2,\nabla u_2)\\ =\ &\frac{\partial f}{\partial u}(\bar{u}(x,y,t))w(x,y,t)+\frac{\partial f}{\partial u_x}(\bar{u}_x(x,y,t))w_x(x,y,t)\\ &+\frac{\partial f}{\partial u_y}(\bar{u}_y(x,y,t))w_y(x,y,t)\\ \leq\ &H(w+w_x+w_y)\end{aligned}$$

for some $\bar{u}(x,y,t)$. It follows that $(w_t(x,y,t)-w_{xx}(x,y,t)-w_{yy}(x,y,t))^2 \leq 3H^2(w^2+w_x^2+w_y^2)$, then $w(x,y,t)\equiv 0$, $(x,y,t)\in\mathbb{R}^2\times[0,T]$. Put

$$\eta_0' = \min\{\eta_0, \frac{1}{H\sqrt{8K}}\}, \mu = \frac{1}{H\sqrt{6K}} - \eta_0',$$

then $\eta_0', \mu > 0$ and $3(\mu+\eta_0')^2KH^2 = \frac{1}{2}$. Now, we shall suppose that $0 < \eta < \eta_0' < \eta_0$, $T \leq \min\{\mu, T^*\}$. Then remainder of the proof is divided into two cases:

Case 1. $T \leq \mu$. Let $0 < t_1 < t_2 < T$ and $\xi \in C^2(\mathbb{R})$ satisfy $\xi(t) = 0$, when $0 < t < t_1$; $\xi(t) = 1$, when $t_2 < t < T$. Then function $v = \xi w$ belongs to P_T and therefore by Lemma 3 and in view of the inequality $(w_t - w_{xx} - w_{yy})^2 \leq 3H^2(w^2 + w_x^2 + w_y^2)$, we have

$$\begin{aligned}
&\int_{t_2}^{T}\int_{-\infty}^{+\infty}\int_{-\infty}^{+\infty}(\lambda^{-m-1}w)^2 dxdydt\\
&\leq \|\lambda^{-m}v_x\|_2^2 + \|\lambda^{-m}v_y\|_2^2 + \|\lambda^{-m-1}v\|_2^2\\
&\leq K\int_{t_1}^{t_2}\int_{-\infty}^{+\infty}\int_{-\infty}^{+\infty}(\lambda^{-m}(v_{xx} + v_{yy} - v_t))^2 dxdydt\\
&\quad +K\int_{t_2}^{T}\int_{-\infty}^{+\infty}\int_{-\infty}^{+\infty}(\lambda^{-m}(w_{xx} + w_{yy} - w_t))^2 dxdydt\\
&\leq \frac{1}{2}\int_{t_1}^{t_2}\int_{-\infty}^{+\infty}\int_{-\infty}^{+\infty}(\lambda^{-m-1})^2(v^2 + v_x^2 + v_y^2) dxdydt\\
&\quad +\frac{1}{2}\int_{t_2}^{T}\int_{-\infty}^{+\infty}\int_{-\infty}^{+\infty}(\lambda^{-m-1})^2(w^2 + w_x^2 + w_y^2) dxdydt.
\end{aligned}$$

(1) If

$$\begin{aligned}
&\frac{1}{2}\int_{t_1}^{t_2}\int_{-\infty}^{+\infty}\int_{-\infty}^{+\infty}(\lambda^{-m-1})^2(v^2 + v_x^2 + v_y^2) dxdydt \neq 0,\\
&\frac{1}{2}\int_{t_2}^{T}\int_{-\infty}^{+\infty}\int_{-\infty}^{+\infty}(\lambda^{-m-1})^2(w^2 + w_x^2 + w_y^2) dxdydt \neq 0,
\end{aligned}$$

we have

$$\begin{aligned}
&K\int_{t_1}^{t_2}\int_{-\infty}^{+\infty}\int_{-\infty}^{+\infty}(\lambda^{-m}(v_{xx} + v_{yy} - v_t))^2 dxdydt\\
&\quad +K\int_{t_2}^{T}\int_{-\infty}^{+\infty}\int_{-\infty}^{+\infty}(\lambda^{-m}(w_{xx} + w_{yy} - w_t))^2 dxdydt\\
&\leq k\int_{t_1}^{t_2}\int_{-\infty}^{+\infty}\int_{-\infty}^{+\infty}(\lambda^{-m-1})^2(v^2 + v_x^2 + v_y^2) dxdydt,
\end{aligned}$$

where

$$\begin{aligned}
k = 1 + [&\frac{1}{2}\int_{t_1}^{t_2}\int_{-\infty}^{+\infty}\int_{-\infty}^{+\infty}(\lambda^{-m-1})^2(v^2 + v_x^2 + v_y^2) dxdydt,\\
&\frac{1}{2}\int_{t_2}^{T}\int_{-\infty}^{+\infty}\int_{-\infty}^{+\infty}(\lambda^{-m-1})^2(w^2 + w_x^2 + w_y^2) dxdydt].
\end{aligned}$$

Hence

$$k\int_{t_1}^{t_2}\int_{-\infty}^{+\infty}\int_{-\infty}^{+\infty}(\lambda^{-m-1})^2(v^2+v_x^2+v_y^2)dxdydt$$
$$\geq \int_{t_2}^{T}\int_{-\infty}^{+\infty}\int_{-\infty}^{+\infty}(\lambda^{-m-1}w)^2dxdydt.$$

This inequality implies, for any $t_2 < t_3 < T$,

$$k\int_{t_1}^{t_2}\int_{-\infty}^{+\infty}\int_{-\infty}^{+\infty}(T+\eta-t_2)^{-2m-2}(v^2+v_x^2+v_y^2)dxdydt$$
$$\geq (T+\eta-t_3)^{-2m-2}\int_{t_3}^{T}\int_{-\infty}^{+\infty}\int_{-\infty}^{+\infty}w^2dxdydt.$$

Because $((T+\eta-t_2)/(T+\eta-t_3))^{-2m} \to 0$ as $m\to\infty$, we have $w\equiv 0$ for $x\in\mathbb{R}$, $t_3<t<T$. Since t_3 can be taken arbitrarily small, $w\equiv 0$ in $\mathbb{R}^2\times[0,T]$.

(2) If

$$\frac{1}{2}\int_{t_2}^{T}\int_{-\infty}^{+\infty}\int_{-\infty}^{+\infty}(\lambda^{-m-1})^2(w^2+w_x^2+w_y^2)dxdydt=0,$$

we get

$$\frac{1}{2}\int_{t_1}^{t_2}\int_{-\infty}^{+\infty}\int_{-\infty}^{+\infty}(\lambda^{-m-1})^2(v^2+v_x^2+v_y^2)dxdydt$$
$$\geq \int_{t_2}^{T}\int_{-\infty}^{+\infty}\int_{-\infty}^{+\infty}(\lambda^{-m-1}w)^2dxdydt.$$

It can be proved as above.

(3) If

$$\frac{1}{2}\int_{t_1}^{t_2}\int_{-\infty}^{+\infty}\int_{-\infty}^{+\infty}(\lambda^{-m-1})^2(v^2+v_x^2+v_y^2)dxdydt=0,$$
$$\frac{1}{2}\int_{t_2}^{T}\int_{-\infty}^{+\infty}\int_{-\infty}^{+\infty}(\lambda^{-m-1})^2(w^2+w_x^2+w_y^2)dxdydt\neq 0.$$

Let $0<t_1<t_2^{'}<T$, which satisfies $\frac{1}{2}\int_{t_1}^{t_2^{'}}\int_{-\infty}^{+\infty}\int_{-\infty}^{+\infty}(\lambda^{-m-1})^2(v^2+v_x^2+v_y^2)dxdydt\neq 0$ ($t_2^{'}$ is always exist) and $\xi\in C^2(\mathbb{R})$ satisfy $\xi(t)=0$, when $0<t<t_1$; $\xi(t)=1$, when $t_2^{'}<t<T$. Then function $v=\xi w$ belongs to P_T. Then the proof can be finished similarly as (1).

Case 2. $T>\mu$. We can also prove that $w=0$ in $\mathbb{R}\times[T-\mu,T]$, then in $\mathbb{R}\times[T-2\mu,T-\mu]$, etc. From above cases, we know that w is identically equal to zero, i.e., $u_1\equiv u_2$. This completes the proof of Theorem 2.

Proof of Theorem 3. First we mention that

$$\begin{aligned}
&\|u(\cdot,\cdot,t)-u^{\varepsilon}(\cdot,\cdot,t)\|^2\\
&=\int_{-\infty}^{+\infty}\int_{-\infty}^{+\infty}|\hat{u}(p_1,p_2,t)-\hat{u^{\varepsilon}}(p_1,p_2,t)|^2dp_1dp_2\\
&=\int_{-\infty}^{+\infty}\int_{-\infty}^{+\infty}|(e^{(T-t)(p_1^2+p_2^2)}-\frac{e^{-t(p_1^2+p_2^2)}}{\varepsilon+e^{-T(p_1^2+p_2^2)}})\hat{\varphi}(p_1,p_2)\\
&\quad-\int_t^T e^{-(t-s)(p_1^2+p_2^2)}\hat{f}(u,\bigtriangledown u)ds\\
&\quad+\int_t^T\frac{e^{-t(p_1^2+p_2^2)}}{\varepsilon^{s/T}+e^{-s(p_1^2+p_2^2)}}\hat{f}(u^{\varepsilon},\bigtriangledown u^{\varepsilon})ds|^2dp_1dp_2\\
&\le 3\varepsilon^{2t/T}\int_{-\infty}^{+\infty}\int_{-\infty}^{+\infty}|e^{T(p_1^2+p_2^2)}\hat{\varphi}(p_1,p_2)|^2dp_1dp_2\\
&\quad+3k^2\varepsilon^{2t/T}T\int_t^T\varepsilon^{-2s/T}\|u^{\varepsilon}(\cdot,\cdot,s)-u(\cdot,\cdot,s)\|^2ds\\
&\quad+3\varepsilon^{2t/T}T\int_0^T\int_{-\infty}^{+\infty}\int_{-\infty}^{+\infty}|\frac{\partial}{\partial s}(e^{s(p_1^2+p_2^2)}\hat{u}(p_1,p_2,s))|^2dp_1dp_2ds.
\end{aligned}$$

Therefore

$$\begin{aligned}
&\varepsilon^{-6t/T}\|u^{\varepsilon}(\cdot,\cdot,t)-u(\cdot,\cdot,t)\|_{L^2}^2\\
&\le M\varepsilon^{-4t/T}+3k^2T\varepsilon^{-4t/T}\int_t^T\varepsilon^{-2s/T}\|u^{\varepsilon}(\cdot,\cdot,s)-u(\cdot,\cdot,s)\|_{H^1}^2ds.
\end{aligned}$$

Using Gronwall's inequality, we get

$$\|u(\cdot,\cdot,t)-u^{\varepsilon}(\cdot,\cdot,t)\|_{H^1}^2\le(1+c_1)\sqrt{M}\varepsilon^{t/T}+\frac{3}{2}k^2T^2(1+c_1)^2(\frac{\varepsilon^{1-t/T}-1}{\ln\varepsilon}),$$

where M is defined in (2.6). This completes the proof of Theorem 3.

Proof of Theorem 4. From $u(x,y,t)-u(x,y,0)=\int_0^t u_t(x,y,s)ds$, we have

$$\|u(\cdot,\cdot,0)-u(\cdot,\cdot,t)\|^2\le t\int_0^t\|u_t(\cdot,\cdot,s)\|^2ds=N^2t.$$

Using Lemma 1, we have

$$\|u(\cdot,\cdot,0)-u(\cdot,\cdot,t)\|_{H^1}\le[(1+c_1)^2N^2t]^{1/4}.$$

Using Theorem 3, we get

$$\begin{aligned}
&\|u(\cdot,\cdot,0) - u^\varepsilon(\cdot,\cdot,t)\|_{H^1} \\
&\le \|u(\cdot,\cdot,0) - u(\cdot,\cdot,t)\|_{H^1} + \|u(\cdot,\cdot,t) - u^\varepsilon(\cdot,\cdot,t)\|_{H^1} \\
&\le [(1+c_1)^2 N^2 t]^{1/4} + [(1+c_1)\sqrt{M}\varepsilon^{t/T}]^{\frac{1}{2}} + [\frac{3}{2}k^2T^2(1+c_1)^2(\frac{1}{\ln\frac{1}{\varepsilon}})]^{1/2}.
\end{aligned}$$

For every $\varepsilon \in (0,1)$, there exists uniquely a positive number t_ε such that

$$[(1+c_1)^2 N^2 t_\varepsilon]^{1/4} = [(1+c_1)\sqrt{M}\varepsilon^{t_\varepsilon/T}]^{\frac{1}{2}}, \tag{3.2}$$

i.e.,

$$\sqrt{t_\varepsilon} = M_1 \varepsilon^{t_\varepsilon/T}.$$

Using inequality $\ln t > -(1/t)$ for every $t > 0$, we get

$$t_\varepsilon < (\frac{M_1 T}{\ln\frac{1}{\varepsilon}})^{\frac{2}{3}}.$$

Hence

$$\begin{aligned}
&\|u(\cdot,\cdot,0) - u^\varepsilon(\cdot,\cdot,t)\|_{H^1} \\
&\le 2[(1+c_1)^2 N^2 t_\varepsilon]^{\frac{1}{4}} + \tfrac{\sqrt{6}}{2}kT(1+c_1)(\ln\tfrac{1}{\varepsilon})^{-\frac{1}{2}} \\
&\le 2\sqrt{(1+c_1)N}(M_1T)^{\frac{1}{6}}(\ln\frac{1}{\varepsilon})^{-\frac{1}{6}} + \frac{\sqrt{6}}{2}kT(1+c_1)(\ln\frac{1}{\varepsilon})^{-\frac{1}{2}} \\
&\le 2M_2(\ln\frac{1}{\varepsilon})^{-\frac{1}{6}},
\end{aligned}$$

where M_1, M_2, N is defined as in (2.10), (2.8) and (2.9). This completes the proof of Theorem 4.

Proof of Theorem 5. Let v^ε be the solution of the problem (IEP) corresponding to φ and let w^ε be the solution of the problem (IEP) corresponding to φ_ε, where φ, φ_ε are in the right-hand side of (2.5). Let t_ε be the unique solution of

$$\sqrt{t_\varepsilon} = M_1 \varepsilon^{t_\varepsilon/T}. \tag{3.3}$$

From Theorem 4, we have

$$\|u(\cdot,\cdot,0) - u^\varepsilon(\cdot,\cdot,t)\|_{H^1} \le 2M_2(\ln\frac{1}{\varepsilon})^{-\frac{1}{6}}. \tag{3.4}$$

Put

$$u^\varepsilon(\cdot,\cdot,t) = \begin{cases} w^\varepsilon(\cdot,\cdot,t), & 0 < t < T, \\ w^\varepsilon(\cdot,\cdot,t_\varepsilon), & t = 0. \end{cases}$$

Using Theorem 3 and Step 2 in Theorem 1, we get

$$\begin{aligned}
&\|u^\varepsilon(\cdot,\cdot,t) - u(\cdot,\cdot,t)\|_{H^1} \\
&\le \|w^\varepsilon(\cdot,\cdot,t) - v^\varepsilon(\cdot,\cdot,t)\| + \|v^\varepsilon(\cdot,\cdot,t) - u(\cdot,\cdot,t)\| \\
&\le [\sqrt{2}(1+c_1)\varepsilon^{t/T} + k^2T^2(1+c_1)^2\tfrac{\varepsilon^{1-t/T}-1}{\ln\varepsilon} \\
&\le 2M_3[\frac{\varepsilon^{1-t/T}-1}{\ln\varepsilon}]^{\frac{1}{2}}
\end{aligned}$$

for every $t \in (0,T)$. From (3.3) and (3.4), and Step 2 in Theorem 1, we have

$$\begin{aligned}
&\|u^\varepsilon(\cdot,\cdot,0) - u(\cdot,\cdot,0)\|_{H^1} \\
&\le \|w^\varepsilon(\cdot,\cdot,t_\varepsilon) - v^\varepsilon(\cdot,\cdot,t_\varepsilon)\| + \|v^\varepsilon(\cdot,\cdot,t_\varepsilon) - u(\cdot,\cdot,0)\| \\
&\le 5M_2(\ln\frac{1}{\varepsilon})^{-\frac{1}{6}},
\end{aligned}$$

where A, B, M_3 is defined as in (2.7) and (2.11). This completes the proof of Theorem 5.

References

[1] J. R. Ockendon, S. D. Howison, A. A. Lacey et al, *Applied partial differential equations (revised edition)*, Oxford Unversity Press, Oxford, 2003.

[2] I. Ndayirinde I and W. Malfliet, New special solutions of the Brusselator reaction model, *Phys: A Math Gen.*, 30 (1997), 5151-5157.

[3] M. J. Ablowitz and P. S. Clarkson, *Nonlinear evolution and inverse scatting*, Cambridge University Press, Cambridge, 1991.

[4] P. Constantin, C. Foias and B. Nicolaenko, *Integral manifolds and inertial manifolds for dissipative PDEs*, Springer-Verlag, Berlin, 1989.

[5] J. M. Marban and C. Palencia, A new numerical method for backward parabolic problems in the maximum-norm setting, *SIAM J. Numer. Ana.*, 140 (2002), 1405-1420.

[6] P. H. Quan and D. D. Trong, A nonlinearly backward heat problem: uniqueness, regularization and error estimate, *Applicable Analysis*, 85 (2006), 6-7:641-657.

[7] N. Mizoguchi and E. Yanagida, Critical exponents for the blow-up of solutions with sign changes in a semilinear parabolic equation, *Math. Ann.*, 307 (1997), 663-675.

[8] K. Mochizuki and R. Suzuki, Critical exponent and critical blow up for quasilinear parabolic equations, *Israel J. Math.*, 98 (1997), 141-156.

[9] C. Bandl, H. A. Levine and Q. S. Zhang, Critical exponents of Fujita type for inhomogeneous parabolic equations and systems, *J. Math. Anal. Appl.*, 251 (2000), 624-648.

[10] Q. S. Zhang, A critical behavior for semilinear parabolic equations involving sign changing solutions, *Nonlinear Analysis*, 50 (2002), 967-980.

[11] P. Souplet and F. B. Weissler, Self-similar subsolutions and blow-up for nonlinear parabolic equations, *Nonlinear Analysis*, 30 (1997), 7:4637-4641.

[12] J. Ding, Blow-up solutions for a class of nonlinear parabolic equations with Dirichlet boundary conditions, *Nonlinear Analysis*, 52 (2003), 1645-1654.

[13] K. Deng, On The role of critical exponents in blow-up theorems: the sequel, *Journal of Mathematical Analysis and Applications*, 243 (2000), 85-126.

[14] C. Cortazar, M. Elgueta and J. D. Rossi, The blow-up problem for a semi-linear parabolic equation with a potential, *J. Math. Anal. Appl.*, 335 (2007), 418-427.

[15] L. E. Payne and P. W. Schaefer, Lower bounds for blow-up time in parabolic problems under Dirichlet conditions, *J. Math. Anal. Appl.*, 328 (2007), 1196-1205.

[16] C. R. Ye and Z. Z. Sun, On the stability and convergence of a difference scheme for an one-dimensional parabolic inverse problem, *Applied Mathematics and Computation*, 188 (2007), 1:214-225.

[17] Q. Chen and J. J. Liu, Solving an inverse parabolic problem by optimization from final measurement data, *Journal of Computational and Applied Mathematics*, 193 (2006), 183-203.

[18] A. G. Fatullayev, Numerical procedure for the simultaneous determination of unknown coefficients in a parabolic equation, *Applied Mathematics and Computation*, 162 (2005), 1367-1375.

[19] S. Zeng and D. Xu, On the simultaneous inversion of diffusivity coefficient and initial function of a class of nonlinear parabolic equations with the variational adjoint method, *Jiangxi Science*, 25 (2007), 598-602.

[20] J. He, D. Xu and W. Zhang, A new numerical method for inverse problems of nonlinear parabolic equations, *Journal of East China Institute of Technology*, 32 (2007), 196-200.

[21] M. Ge and D. Xu, Numerical method of an inverse problem for two-dimensional advection-dispersion-reaction equations, *Journal of Zhejiang Sci-Tech University*, 24 (2007), 577-582.

[22] B. Wang and D. Xu, The research on the variational adjoint method for percolation coefficient inversion for one dimensional flowing Boussinesq equation, *Mathematics in Practice and Theory*, 38 (2008), 194-200.

[23] D. Willett and J. S. W. Wong, On the discrete analogues of some generalizations of Gronwall's inequality, *Monatshefte fur Mathematic*, 69 1965), 4:362-367.

ON PROJECTIVE GRADIENT METHOD FOR THE INVERSE PROBLEM OF OPTION PRICING[1]

QING-HUA MA[ab], ZUO-LIANG XU[a] and CHANG JIN[a]

[a] *Renmin University of China, Beijing 100872, China*

E-mail: qinghua@ygi.edu.cn, xuzl@ruc.edu.cn

[b] *Beijing Union University, Beijing 100191, China*

E-mail: qinghua@ygi.edu.cn

This paper investigates the inverse problem of calibrating the volatility function from given option price data. This is an ill-posed problem because of at least one of three well-posed conditions violating. We start with a simplified model of pure price-dependent volatility to gather insight on the nature of ill-posedness of the problem. We formulate the problem by the operator equations and establish a Tikhonov regularization model. Projected gradient methods are developed for solving the regularizing problem.

Key Words: Option pricing problems, volatility function, ill-posed, regularization, projective gradient methods.

AMS Subject Classification: 65J15, 65J20, 91B28.

1. Introduction

Option pricing models are used in practice to price derivative securities given knowledge of the volatility and other market variables. Volatility is a measure of the amount of fluctuation in the asset prices; a measure of the randomness. Volatility is subject to a tremendous amount of uncertainty due not only to the quality of available data, but also on the modeling of the financial instrument and the underlying assets under consideration. The constant-volatility Black-Scholes model (see [1]) is the most often used option pricing model in practice. Based on the assumption of constant volatility, Black-Scholes formula can be used to evaluate European options simply and quickly by using the estimated or forecast volatility constant as an input. In many situations, we invert the Black-Scholes formula to determine the volatility from the market option price, this is implied volatility. Implied Black-Scholes volatilities vary with strikes and time to maturity, which are respectively known as the smile effect and the term structure.

In this survey, we are dealt with a specific ill-posed inverse problem that arises in the course of determining the volatility from quoted data by using the Black-Scholes formula. We assume that the relative changes in stock prices (the returns) follow Brownian motion with drift. We suppose that in an infinitesimal time dt, the stock price S changes $S + dS$, S follows a

[1] This research is supported by NSFC (No.10971224), Beijing Talents Foundation and College of Art & Science of Beijing Union University

general diffusion of the form $\frac{dS}{S} = \mu dt + \sigma dZ$, where μ is the drift rate, σ is the stock volatility, and dZ is the increment of a standard Wiener process. The European call option premium $u = u(S, t; K, T)$ satisfies the following pricing model, the Black-Scholes partial differential equation

$$\begin{aligned} &\frac{\partial u}{\partial t} + \frac{1}{2}S^2\sigma^2\frac{\partial^2 u}{\partial S^2} - (r-q)S\frac{\partial u}{\partial S} - ru = 0, \quad (S,t) \in R^+ \times [0,T), \\ &u(0,t) = 0, \\ &u(S,T) = (S-K)^+ := \max\{S-K, 0\}, \end{aligned} \tag{1}$$

here $:=$ means "defined as", $K > 0$ is the strike, T is the maturity, $r \geq 0$ is the risk-free interest rate, q is the dividend yield on the stock and $\sigma(S,t)$ is the only parameter in the pricing model that is not directly observable. Similar to the implied volatility in the constant volatility model, one possible idea is to imply this local volatility function from the market option price data. The inverse problem was first considered by Dupire in [2]. He showed that option prices given for all possible strikes and maturities completely determine the local volatility and Bouchouev and Isakov in [5,6] obtained a nonlinear Fredholm equation for volatility and solved the approximate problem iteratively. Unfortunately, the market European option prices are typically limited to a relatively few different strikes and maturities. Therefore the problem of determining the local volatility function can be regarded as a function approximation problem from a finite data set with an observation functional. Due to insufficient market option price data, this is a well known ill-posed problem. With the derivation of a dual problem, the problem (1) can be reduced to a standard parabolic equation with new variables, it was found by Dupire [2] and rigorously justified

$$\begin{aligned} &\frac{\partial u}{\partial \tau} - \frac{1}{2}K^2\sigma^2\frac{\partial^2 u}{\partial K^2} + (r-q)K\frac{\partial u}{\partial K} + qu = 0, \ (K,\tau) \in R^+ \times R^+, \\ &u|_{K=0} = e^{-q\tau}S, \\ &u|_{\tau=0} = (S-K)^+, \end{aligned} \tag{2}$$

where $\tau = T - t$ is times remaining to maturity, which is varying between zero and the upper time limit T. Here, we focus on the case of volatility that independent of time and use only option prices with different strikes and a fixed maturity date. That is to say, we find $\sigma(K)$ such that the solution of (2) satisfies

$$u(S^*, t^*, K, \tau)|_{\tau=\tau^*} = u^*(K),$$

where $\tau^* = T - t^*$ and $u^*(K)$ is the current market price of options for different $K > 0$ at current time t^* with stock price S^*.

To remove the singularity at $K = 0$, we make the logarithmic substitution $y = \log \frac{K}{S}$, $v = \frac{1}{S}e^{q\tau}u$, which transforms (see [3,4]) the problem and the additional market data into the following Cauchy problem

$$\begin{aligned} &\frac{\partial v}{\partial \tau} - a(y)\left(\frac{\partial^2 v}{\partial y^2} - \frac{\partial v}{\partial y}\right) + (r-q)\frac{\partial v}{\partial y} = 0,\ (y,\tau) \in R \times (0, \tau^*], \\ &v|_{\tau=0} = (1-e^y)^+, \end{aligned} \tag{3}$$

where $v(y,\tau) = \frac{1}{S}e^{q\tau}u(K,T)$, $a(y) = \frac{1}{2}\sigma^2(K)$.

Regularization is a must for solving inverse problems. The essence of the regularization lies in that approximate the original ill-posed problems by a family of well-posed problems, and solve the well-posed problems to get the approximate solution. After using regularization methods, the ill-posed problem is replaced by a stabilized problem. In this paper, we use Tikhonov regularization model to find $a(y)$ such that the solution of (3) satisfies

$$v(y,\tau)|_{\tau=\tau^*} = v^*(y),$$

where $v^*(y) = \frac{1}{S^*}e^{q\tau^*}u^*(K)$. The structure of the paper is organized as follows: Section 2 addresses the regularization method, which incorporates both regularization parameter and smoothness constraint. Both the regularization parameter and the smoothness constraint can be considered as *a priori* information. Then, we solve the regularization model by an efficient gradient method. Details about implementation of the method are addressed. Numerical results for the market data are given in Section 3. In Section 4, some concluding remarks are given.

2. The Gradient Regularization Method

Both the discrete and continuous calibration problems are ill-posed. This is the case in the continuous calibration problem, because the solution depends upon the data in an unstable manner, and in the discrete calibration problem because the full surface $a(y) = \frac{1}{2}\sigma^2(K)$ is simply under determined by the discrete data. It is then necessary to introduce stabilizing procedures in the reconstruction method for the local volatility function. One of these is known as the Tikhonov regularization method (see [5–8]). The idea is to tackle the calibration problem as a minimization problem see [9]), where the cost criterion to be minimized is

$$\min J(a) := \frac{1}{2}\|v(y,\tau^*,a) - v^*\|^2_{L^2(R)} + \frac{\alpha}{2}\|La\|^2_{L^2(R)}. \tag{4}$$

In which, $a \in A := \{a \in L^2(R) |\ 0 < a_0 \le a(y) \le a_1\}$, a_0, a_1 are the lower and upper bounds of $a(y)$, α is the regularization parameter, and L is the weight of the function a. We assume that $\int_R |v^*(y) - H(-y)|^2 dy < +\infty$, $H(.)$ is the Heaviside function, for any given $a \in A$, there is a unique solution $v(y,\tau)$ to the Cauchy problem (3) by the known theory for parabolic equations (see [10]).

Since the nonlinearity of the function J and the gradient can be easily obtained, it is convenient to use gradient methods for the minimization problem (4). In fact, the gradient of J can be evaluated as

$$g(a) = \frac{1}{2}\frac{\mathrm{d}}{\mathrm{d}a}\|v(y,\tau^*,a) - v^*\|^2_{L^2(R)} + \alpha(L^*La, a).$$

In computer calculation, only the matrix-vector multiplication is performed.

There have a special gradient method, which is known as the Landweber iteration

$$a_{k+1} = a_k + \omega s_k, \tag{5}$$

where $s_k = -g_k$, $g_k = g(a_k)$, the steplengths $\omega > 0$ is a constant in each iteration. However, as is reported that this method is quite slow in convergence and can not be used for practical problems (see [8]). We consider nonmonotone gradient methods for solving (4), and projection techniques for box constrained problems with adaptive stepsizes are also developed in [9]. It is reported that the Barzilai and Borwein's method is a simple and efficient nonmonotone method (see [11]) and performs well for ill-posed problems (see [12]). The key point of Barzilai and Borwein's method is the two choices of the stepsize ω_k

$$\omega_k^{BB1} = \frac{(s_{k-1}, s_{k-1})}{(s_{k-1}, j_{k-1})}, \tag{6}$$

and

$$\omega_k^{BB2} = \frac{(s_{k-1}, j_{k-1})}{(j_{k-1}, j_{k-1})}, \tag{7}$$

where $j_k = g_{k+1} - g_k$, $s_k = a_{k+1} - a_k$. We adopt the recently developed gradient method with adaptive stepsizes

$$\omega_k = \begin{cases} \omega_k^{BB1}, & \text{if } r_k \geq r, \\ \omega_k^{BB2}, & \text{otherwise}, \end{cases} \tag{8}$$

where $r \in (0,1)$ is a parameter close to 0.5 and $r_k = \frac{\omega_k^{BB2}}{\omega_k^{BB1}}$, which reflects the deviation of the gradient search direction g_k to the transformed search direction.

3. Projection

In iterative process, negative point values may occur. To tackle this problem, we develop the projection technique in this paper. Note that the set $Q := \{a : 0 \leq a < \infty\}$ is bounded below and convex, therefore, there exist an orthogonal projection operator P_S onto S such that

$$P_Q : \mathbb{R}^N \to Q,$$

and

$$P_Q^* = P_Q, \ P_Q^2 = P_Q.$$

Therefore, we actually solve a constrained regularizing model

$$\min J[a] := \frac{1}{2}\|v(y, \tau^*, a) - v^*\|^2 + \frac{\alpha}{2}\|La\|^2, \ \text{s.t. } a \in Q, \tag{9}$$

our choice of the projection P_Q is that

$$P_Q(x) = \text{argmin}_z \|x - z\|,$$

the projection operator associated with a bounded domain Q is defined by

$$(P_Q(x))(t) = \chi_Q(t)x(t),$$

and $\chi_Q(t)$ is the characteristic function of the domain Q. This operator projects onto the subspace of all functions, which are zero outside the domain Q. The i-th component of $P_Q(x)$ is

$$[P_Q(x)]_i = \max(x_i, 0) = \begin{cases} x_i, & \text{if } x_i \geq 0, \\ 0, & \text{otherwise.} \end{cases}$$

Assume that the current iterate a_k is feasible, then the next iteration point can be obtained by

$$a_{k+1} = P_Q(a_k - \omega_k g_k).$$

For computing method, the proper termination of the iterative process is important. In our algorithm, the initial search direction is the negative gradient direction. To be sure that the first step is a decreasing step, we use the Wolfe inexact line search strategy. And, we use the following stopping condition (see [12]) in our numerical tests:

$$\|\tilde{g}_k\|/\|g_1\| \leq \varepsilon, \tag{10}$$

where ε is a preassigned tolerance, and $\tilde{g}_k$ is defined as

$$(\tilde{g}_k)_i = \begin{cases} (g_k)_i, & \text{if } (a_k)_i > 0, \\ \min\{(g_k)_i, 0\}, & \text{if } (a_k)_i = 0. \end{cases}$$

For choosing the regularization parameter α, it is a delicate thing. It is preassigned by users. Currently, we apply an *a priori* choice rule, i.e., we set α to be a constant in each iteration but be limited in $(0,1)$. It is clear that α can not be large or typically small. Otherwise, over-regularization or instability will occur.

We argue that other methods for supplying nonnegative constraints, such as negative entropy (see [13]) and active set method (see [14]), may be also applied. However, it is not so straightforward.

4. Numerical Experiments

To verify the feasibility of our inversion method, we have tested it by computer simulations. The simulation consists of three steps. First, given the market price of option $u^*(k)$ and the underlying S^* at time t^*, we would then determine $v^*(y)$ by its definition $v^*(y) = \frac{1}{S^*}e^{q(T-t^*)}u^*(K)$. Our task is solving for $a(y)$ from the equations (3). Finally, we would be able to get the volatility function by $\sigma(S) = \sqrt{2 * a(\ln(\frac{S}{S^*}))}$.

Now, given τ^*, $v^*(y)$, we solve $a(y)$ using our developed gradient regularization method. In our simulation, we choose three cases of implicit volatility functions: the constant volatility ($\sigma = 0.25$), the smile volatility ($\sigma = \frac{(\ln S - \ln 10)^2}{40} + 0.2$) and the skew volatility ($\sigma = -\frac{(\ln S - \ln 10)^3}{80} + 0.2$). The computational results of the three cases are shown in Figures 1, 2 and 3, respectively. It is evident that our algorithm works stably for all tests.

5. Conclusion

In this paper, we investigate establishing regularization model and gradient methods for retrieval of the local volatility function. The simulation results show that the new developed methods are simple and suitable for recovering the volatility function from the option pricing model. Hence it can be used for practical applications.

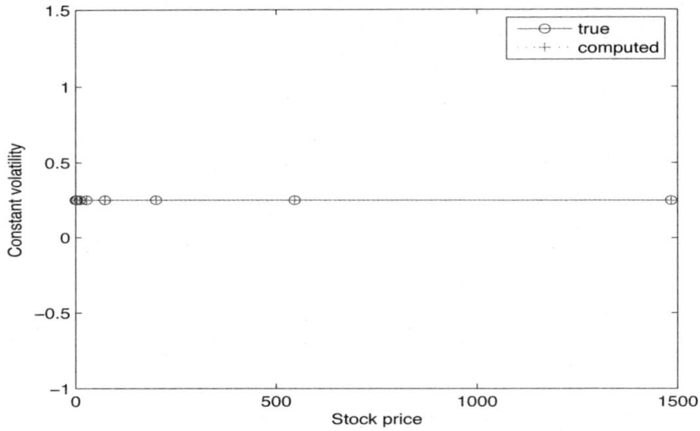

Figure 1: Retrieved results with our inversion method for the constant volatility. The true volatility is $\sigma = 0.25$.

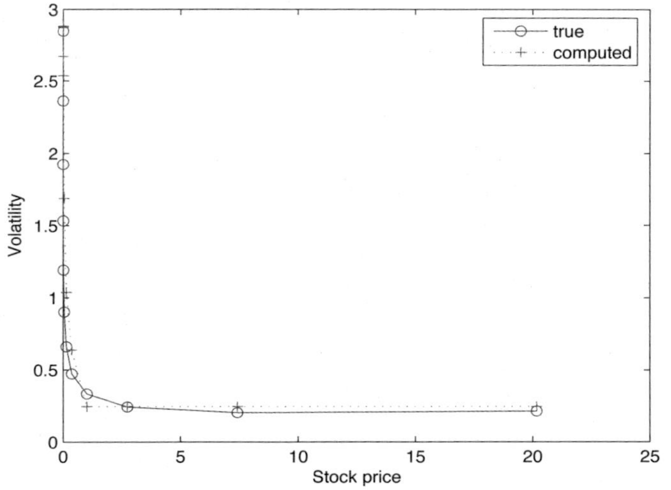

Figure 2: Retrieved results with our inversion method for the smile volatility. The true volatility is $\sigma = \frac{(\ln S - \ln 10)^2}{40} + 0.2$.

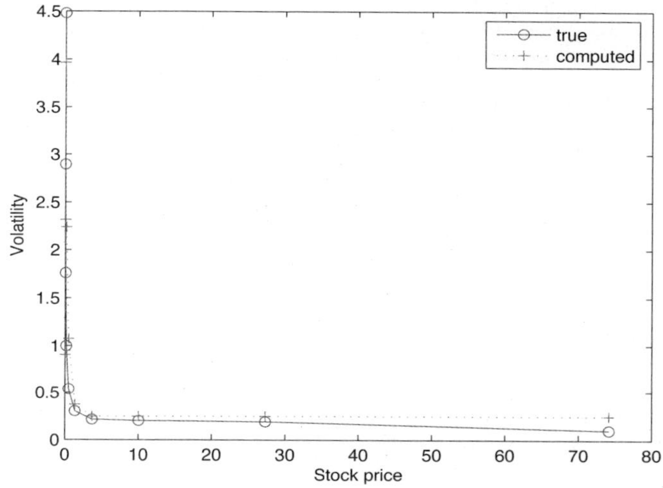

Figure 3: Retrieved results with our inversion method for the skew volatility. The true volatility is $\sigma = -\frac{(\ln S - \ln 10)^3}{80} + 0.2$.

References

[1] F. Black and M.Scholes, The pricing of options and corporate liabilities, *J. Political Econ.,* 81 (1973), 637–659.

[2] B. Dupire, Pricing with a smile, *Risk,* 7 (1994), 18–20.

[3] L. S. Jiang and Y. S. Tao, Identifyibng the volatility of underlying assets from option prices, *Inverse Problems,* 17 (2001), 137–155.

[4] L. S. Jiang, Q. H. Chen, L. J. Wang and J. E. Zhang, A new well-posed algorithm to recover implied local volatility, *Quantitative Finance,* 3 (2003), 451–457.

[5] I. Bouchouev and V. Isakov, The inverse problem of option pricing, *Inverse Problems,* 13 (1997), 11–17.

[6] I. Bouchouev and V. Isakov, Uniqueness, stability and numerical methods for the inverse problem that arises in financial markets, *Inverse Problems,* 15 (1999), R95–R116.

[7] T. Hein and B. Hofmann, On the nature of ill-posedness of an inverse problem arising in option pricing, *Inverse Problems,* 19 (2003), 1319C1338.

[8] Y. F. Wang, *Computational methods for inverse problems and their applications,* Higher Education Press, Beijing, 2007.

[9] Y. F. Wang and Q. H. Ma, A gradient method for regularizing retrieval of aerosol particle size distribution Function, *Journal of Industrial and Management Optimization,* 5 (2009), 115–126.

[10] O. Ladyzenskaya, V. Solinnikov and N. N. Ural'ceva, *Linear and quasilinear equations of parabolic type,* American Mathematical Society, Providence, RI, 1968.

[11] J. Barzilai and J. Borwein, Two-point step size gradient methods, *IMA Journal of Numerical Analysis,* 8 (1988), 141–148.

[12] Y. F. Wang and S. Q. Ma, Projected Barzilai-Borwein methods for large scale non-negative image restorations, *Inverse Problems in Science and Engineering,* 15 (2007), 559–583.

[13] Y. F. Wang, An efficient gradient method for maximum entropy regularizing retrieval of atmospheric aerosol particle size distribution function, *Journal of Aerosol Science,* 39 (2008), 305–322.

[14] Y. F. Wang and C. C. Yang, A regularizing active set method for retrieval of atmospheric aerosol particle size distribution function, *Journal of Optical Society of America A,* 25 (2008), 348–356.

LOCAL REGULARITY FOR SOLUTIONS TO OBSTACLE PROBLEMS WITH WEIGHT[1]

HONG-YA GAO[a], JING GUO[b], JIN-JING QIAO[c] and SHUANG LIANG[d]
[a,b] *College of Mathematics and Computer Science, Hebei University, Baoding 071002, China*
[a] *E-mail: hongya-gao@sohu.com*
[b] *E-mail: guojing_201@yahoo.com.cn*
[c] *College of Mathematics and Computer, Hunan Normal University, Changsha 410081, China*
E-mail: qiaojinjing1982@163.com
[d] *School of Science, Agricultural University of Hebei, Baoding 071000, China*
E-mail: 944204016@qq.com

This paper gives a local regularity result for solutions to obstacle problems of the $\mathcal{A}$-harmonic equation $\text{div}\mathcal{A}(x, \nabla u) = 0$ with $|\mathcal{A}(x, \xi)| \approx w(x)|\xi|^{p-1}$, where $w(x)$ is a Muckenhoupt A_p weight with $1 < p < \infty$.

Keywords: Local regularity, obstacle problem, Muckenhoupt weight, $\mathcal{A}$-harmonic equation.
AMS No: 35J60.

1. Introduction and Statement of Result

Let w be a locally integrable nonnegative function in $\mathbb{R}^n$. Then a Radon measure μ is canonically associated with the weight w,

$$\mu(E) = \int_E w(x)dx. \tag{1.1}$$

Thus $d\mu(x) = w(x)dx$, where dx is the n-dimensional Lebesgue measure. In what follows the weight w and the measure μ are identified via (1.1).

Definition 1[1]. Given a nonnegative locally integrable function w, we say that w belongs to the A_p class of Muckenhoupt, $1 < p < \infty$, if

$$\sup_B \left(\frac{1}{|B|}\int_B w dx\right)\left(\frac{1}{|B|}\int_B w^{1/(1-p)}dx\right)^{p-1} = A_p(w) < \infty, \tag{1.2}$$

where supremum is taken over all balls B of $\mathbb{R}^n$. When $p = 1$, replace the inequality (1.2) with

$$\mathcal{M}w(x) \leq cw(x)$$

for some fixed constant c and a.e. $x \in \mathbb{R}^n$, in which $\mathcal{M}$ is the Hardy-Littlewood maximal operator.

[1]This research is supported by NSFC (No.10971224)

It is well-known that $A_1 \subset A_p$ whenever $p > 1$, see [1]. We say that a weight w is doubling, if there is a constant $C > 0$ such that

$$\mu(2B) \le C\mu(B),$$

whenever $B \subset 2B$ are concentric balls in $\mathbb{R}^n$, where $2B$ is the ball with the same center as B and with radius twice that of B. Given a measurable subset E of $\mathbb{R}^n$, we will denote by $L^p(E, w)$, $1 < p < \infty$, the Banach space of all measurable functions f defined on E for which

$$\|f\|_{L^p(E,w)} = \left(\int_E |f(x)|^p w(x)dx\right)^{\frac{1}{p}} < \infty.$$

The weighted Sobolev class $W^{1,p}(E, w)$ consists of all functions f and its first generalized derivatives belong to $L^p(E, w)$. The symbols $L^p_{loc}(E, w)$ and $W^{1,p}_{loc}(E, w)$ are self-explanatory.

If $x_0 \in \Omega$ and $t > 0$, then B_t denotes the ball of radius t centered at x_0. For a function $u(x)$ and $k > 0$, set $A_k = \{x \in \Omega : |u(x)| > k\}$, $A_{k,t} = A_k \cap B_t$. Let $T_k(u)$ be the usual truncation of u at level $k > 0$, that is

$$T_k(u) = \max\{-k, \min\{k, u\}\}.$$

Let Ω be a bounded domain in $\mathbb{R}^n, n \ge 2$. We consider the second order degenerate elliptic equation (also called $\mathcal{A}$-harmonic equation or Leray-Lions equation)

$$\text{div}\mathcal{A}(x, \nabla u) = 0, \tag{1.3}$$

in which $\mathcal{A}(x, \xi) : \Omega \times \mathbb{R}^n \to \mathbb{R}^n$ is a Carathéodory function satisfying the following assumptions:

1. $\langle \mathcal{A}(x, \xi), \xi\rangle \ge \alpha w(x)|\xi|^p$,
2. $|\mathcal{A}(x, \xi)| \le \beta w(x)|\xi|^{p-1}$,

where $0 < \alpha \le \beta < \infty$, $w \in A_p$, $1 < p < \infty$ and $w > 0$. Suppose ψ is any function in Ω with values in the extended reals $[-\infty, +\infty]$ and that $\theta \in W^{1,p}(\Omega, w)$. Let

$$\mathcal{K}^p_{\psi,\theta} = \mathcal{K}^p_{\psi,\theta}(\Omega, w)$$
$$= \{v \in W^{1,p}(\Omega, w) : v \ge \psi, \text{ a.e. } x \in \Omega \text{ and } v - \theta \in W^{1,p}_0(\Omega, w)\}.$$

The function ψ is an obstacle and θ determines the boundary values.

Definition 2[1]**.** A weak solution to the $\mathcal{K}^p_{\psi,\theta}$-obstacle problem is a function $u \in \mathcal{K}^p_{\psi,\theta}(\Omega, w)$ such that

$$\int_\Omega \langle \mathcal{A}(x, \nabla u), \nabla(v - u)\rangle dx \ge 0, \tag{1.4}$$

whenever $v \in \mathcal{K}^p_{\psi,\theta}(\Omega, w)$.

The local and global higher integrability of the derivatives in obstacle problems with $w(x) \equiv 1$ was first considered by Li and Martio [2] in 1994, by using the so-called reverse Hölder inequality. Gao and Tian [3] gave a local regularity result for weak solutions to obstacle problems in 2004. Regularity theory for very weak solutions of the $\mathcal{A}$-harmonic equations with $w(x) \equiv 1$ have been considered in [4] by Iwaniec and Sbordone, and the regularity theory for very weak solutions to obstacle problems with $w(x) \equiv 1$ have been explored in [5] by Li and Gao. In this paper, we continue to consider obstacle problems, and obtain a local regularity result for weak solutions to obstacle problems with weight. The main result of this paper is the following theorem.

Theorem 1. *Suppose that $\psi \in W^{1,s}_{loc}(\Omega, w)$, $s > p$. A solution u to the $\mathcal{K}^p_{\psi,\theta}$-obstacle problem belongs to $L^t_{loc}(\Omega, w)$, where*

$$t = \begin{cases} \dfrac{pK}{2-K}, & \text{if } pK < s < \dfrac{pK}{K-1} \text{ and } 1 < K < 2, \\ \dfrac{spK}{s - K(s-p)}, & \text{otherwise,} \end{cases}$$

where the value of K comes from Lemma 1, see Section 2 below.

2. Preliminary Lemmas

The following lemma comes from [6]. Chiarenza and Frasca gave a simplified proof in [7].

Lemma 1. *Let B be any ball in $\mathbb{R}^n$, $w \in A_p$, $1 < p < \infty$ and $u \in C_0^\infty(B)$. Then there exist constants c and $\delta^* > 0$ such that for all $1 \le k \le K = \frac{n}{n-1} + \delta^*$, we have*

$$\left(\frac{1}{\mu(B_R)} \int_{B_R} |u|^{kp} d\mu\right)^{\frac{1}{kp}} \le cR \left(\frac{1}{\mu(B_R)} \int_{B_R} |\nabla u|^p d\mu\right)^{\frac{1}{p}}. \tag{2.1}$$

Obviously (2.1) can be extended to functions $u \in W_0^{1,p}(B_R, w)$ by an approximation argument.

The following lemma comes from [8], see also [9].

Lemma 2. *Let $f(t)$ be a nonnegative bounded function defined for $0 \le T_0 \le t \le T_1$. Suppose that for $T_0 \le t < s \le T_1$, we have*

$$f(t) \le A(s-t)^{-\alpha} + B + \theta f(s),$$

where A, B, α, θ are non-negative constants and $\theta < 1$. Then there exist a constant c, depending only on α and θ, such that for every $\rho, R, T_0 \le \rho <$

$R \le T_1$, we have

$$f(\rho) \le c[A(R-\rho)^{-\alpha} + B].$$

We now state a key lemma, which will be used in the proof of the main theorem.

Lemma 3. *Let $u \in W^{1,p}_{loc}(\Omega, w)$, $\phi_0 \in L^r_{loc}(\Omega, w)$ and $w \in A_p, w(x) > 0$, where $1 < p < \infty$ and r satisfies*

$$1 < r < \frac{K}{K-1},$$

where K comes from Lemma 1. Assume that the following integral estimates holds

$$\int_{A_{k,\tau}} |Du|^p d\mu \le c_0 \left[\int_{A_{k,t}} |\phi_0| d\mu + (t-\tau)^{-\alpha} \int_{A_{k,t}} |u|^p d\mu \right] \tag{2.2}$$

for every $k \in N$ and $R_0 \le \tau < t \le R_1$, where c_0 is a positive constant that depends only on $n, p, r, R_0, R_1, \mu(B_{R_0}), K$ and $|\Omega|$, α is a real positive constant. Then $u \in L^s_{loc}(\Omega, w)$ with

$$s = \begin{cases} \dfrac{Kp}{2-K}, & \text{if } K < r < \dfrac{K}{K-1} \text{ and } 1 < K < 2, \\ \dfrac{rpK}{r - K(r-1)}, & \text{otherwise}. \end{cases}$$

Proof. The proof of this lemma is a little change of which in [10]. We omit the details.

3. Proof of Theorem 1

Let u be a weak solution to the $\mathcal{K}^p_{\psi,\theta}$-obstacle problem. By Lemma 3, it is sufficient to prove that u satisfies the inequality (2.2) with $\alpha = p$. Let $B_{R_1} \subset\subset \Omega$ and $0 < R_0 \le \tau < t \le R_1$ be arbitrarily fixed. Fix a cut-off function $\phi \in C^\infty_0(B_t)$ such that

$$\text{supp}\phi \subset B_t, 0 \le \phi \le 1, \phi = 1 \text{ in } B_\tau \text{ and } |\nabla\phi| \le 2(t-\tau)^{-1}.$$

Consider the function

$$v = u - T_k(u) - \phi^p(u - \psi_k^+),$$

where $T_k(u)$ is the usual truncation of u at level $k \ge 0$ defined in Section 1 and $\psi_k^+ = \max\{\psi, T_k(u)\}$. Now $v \in \mathcal{K}^p_{\psi - T_k(u), \theta - T_k(u)}$. Indeed,

$$v - (\psi - T_k(u)) = u - \theta - \phi^r(u - \psi_k^+) \in W^{1,p}_0(\Omega, w),$$

since $\phi \in C_0^\infty(\Omega)$ and

$$v - (\psi - T_k(u)) = (u - \psi) - \phi^r(u - \psi_k^+) \geq (1 - \phi^r)(u - \psi) \geq 0,$$

a.e. in Ω. Since

$$\nabla v = \nabla(u - T_k(u)) - \phi^p(\nabla u - \nabla\psi_k^+) - p\phi^{p-1}\nabla\phi(u - \psi_k^+),$$

and $u - T_k(u)$ is a solution to the $\mathcal{K}^p_{\psi - T_k(u), \theta - T_k(u)}$-obstacle problem, we have by definition that

$$\int_{A_{k,t}} \langle \mathcal{A}(x, \nabla u), \phi^p(\nabla u - \nabla\psi_k^+) + p\phi^{p-1}\nabla\phi(u - \psi_k^+)\rangle dx \leq 0,$$

that is

$$\begin{aligned} &\int_{A_{k,t}} \langle \mathcal{A}(x, \nabla u), \phi^p \nabla u\rangle dx \\ \leq &\int_{A_{k,t}} \langle \mathcal{A}(x, \nabla u), \phi^p \nabla\psi_k^+\rangle dx + \int_{A_{k,t}} \langle \mathcal{A}(x, \nabla u), p\phi^{p-1}\nabla\phi(u - \psi_k^+)\rangle dx \\ \leq &\ I_1 + I_2. \end{aligned} \tag{3.1}$$

Now, we estimate the left-hand side and the right-hand side of (3.1) respectively. Firstly

$$\int_{A_{k,t}} \langle \mathcal{A}(x, \nabla u), \phi^p \nabla u\rangle dx \geq \int_{A_{k,\tau}} \langle \mathcal{A}(x, \nabla u), \nabla u\rangle dx \geq \alpha \int_{A_{k,\tau}} |\nabla u|^p d\mu. \tag{3.2}$$

Secondly, by Young's inequality

$$ab \leq \varepsilon a^{p'} + c_2(\varepsilon, p) b^p,\ \frac{1}{p'} + \frac{1}{p} = 1,\ a, b \geq 0,\ \varepsilon \geq 0,\ p \geq 1,$$

we can derive that

$$\begin{aligned} |I_1| &= \left| \int_{A_{k,t}} \langle A(x, \nabla u), \phi^p \nabla\psi_k^+\rangle dx \right| \leq \beta \int_{A_{k,t}} |\nabla u|^{p-1} |\nabla\psi_k^+| d\mu \\ &\leq \beta\, \varepsilon \int_{A_{k,t}} |\nabla u|^p d\mu + \beta c(\varepsilon, p) \int_{A_{k,t}} |\nabla\psi|^p d\mu. \end{aligned} \tag{3.3}$$

Finally, using Young's inequality again we have

$$\begin{aligned}|I_2| &= \left|\int_{A_{k,t}} \langle A(x,\nabla u), p\phi^{p-1}\nabla\phi(u-\psi_k^+)\rangle dx\right| \\ &\le \frac{2\beta p}{t-\tau}\int_{A_{k,t}} |\nabla u|^{p-1}|u-\psi_k^+|d\mu \\ &\le \frac{2\beta p}{t-\tau}\int_{A_{k,t}} |\nabla u|^{p-1}|u|d\mu \\ &\le 2\beta p\varepsilon\int_{A_{k,t}} |\nabla u|^p d\mu + 2\beta pc(\varepsilon,p)\int_{A_{k,t}} \frac{|u|^p}{(t-\tau)^p}d\mu.\end{aligned} \tag{3.4}$$

Combining the inequalities (3.1)–(3.4) all together, we have

$$\begin{aligned}\int_{A_{k,\tau}} |\nabla u|^p d\mu \le \frac{\beta(2p+1)\varepsilon}{\alpha}\int_{A_{k,t}} |\nabla u|^p d\mu + \frac{\beta c(\varepsilon,p)}{\alpha}\int_{A_{k,t}} |\nabla\psi|^p d\mu \\ + \frac{2\beta pc(\varepsilon,p)}{\alpha(t-\tau)^p}\int_{A_{k,t}} |u|^p d\mu.\end{aligned} \tag{3.5}$$

Now we want to eliminate the first term on the right-hand side including ∇u. Choosing ε small enough such that $\eta = \frac{\beta(2p+1)\varepsilon}{\alpha} < 1$, and let ρ, R be arbitrarily fixed with $R_0 \le \rho < R \le R_1$. Thus from (3.5) we deduce that for every t and τ such that $\rho \le \tau < t \le R$, it results

$$\begin{aligned}\int_{A_{k,\tau}} |\nabla u|^p d\mu \le \eta\int_{A_{k,t}} |\nabla u|^p d\mu + \frac{\beta c(\varepsilon,p)}{\alpha}\int_{A_{k,R}} |\nabla\psi|^p d\mu \\ + \frac{2\beta pc(\varepsilon,p)}{\alpha(t-\tau)^p}\int_{A_{k,R}} |u|^p d\mu.\end{aligned} \tag{3.6}$$

Applying Lemma 3 in (3.6), we conclude that

$$\int_{A_{k,\rho}} |\nabla u|^p d\mu \le \frac{C\beta c(\varepsilon,p)}{\alpha}\int_{A_{k,R}} |\nabla\psi|^p d\mu + \frac{C\beta pc(\varepsilon,p)}{\alpha(R-\rho)^p}\int_{A_{k,R}} |u|^p d\mu.$$

Thus u satisfies inequality (2.2). The theorem follows from Lemma 3.

References

[1] J. Heinonen, T. Kilpeläinen and O. Martio, *Nonlinear potential theory of degenerate elliptic equations,* The Clarendon Press, Oxford University Press, New York, 1993.

[2] G. B. Li and O. Martio, Local and global integrability of gradients in obstacle problems, *Ann. Acad. Sci. Fenn. Ser. A. I. Math.*, 19 (1994), 25–34.

[3] H. Y. Gao and H. Y. Tian, Local regularity result for solutions of obstacle problems, *Acta Math. Sci.*, 24B (2004), 71–74.

[4] T. Iwaniec and C. Sbordone, Weak minima of variational integrals, *J. Reine. Angew. Math.*, 454 (1994), 143–161.

[5] J. Li and H. Y. Gao, Local regularity result for very weak solutions of obstacle problems, *Radovi Mathematički*, 12 (2003), 19–26.

[6] E. Fabes, C. Kenig and R. Serapioni, The local regularity of solutions of degenerate elliptic equations, *Comm. Part. Diff. Eqns.*, 7 (1982), 77–116.

[7] F. Chiarenza and M. Frasca, A note on weighted Sobolev inequality, *Proc. Amer. Math. Soc.*, 93 (1985), 703–704.

[8] M. Giaquinta and E. Giusti, On the regularity of the minima of variational integrals, *Acta Math.*, 148 (1982), 31–46.

[9] M. Giaquinta, *Multiple integrals in the calculus of variations and nonlinear elliptic systems*, Ann. of Math. Studies, Princeton University Press, Princeton, NJ, 1983.

[10] D. Giachetti and M. M. Porzio, Local regularity results for minima of functionals of the Calculus of Variation, *Nonlinear Analysis, TMA,* 39 (2000), 463–482.

BELTRAMI SYSTEM WITH TWO CHARACTERISTIC MATRICES AND VARIABLE COEFFICIENTS[1]

HONG-YA GAO[a], SHEN-ZHOU ZHENG[b] and YING-QIANG YUE[c]

[a] *College of Mathematics and Computer Science, Hebei University, Baoding 071002, China*

E-mail: hongya-gao@sohu.com

[b] *College of Sciences, Beijing Jiaotong University, Beijing 100044, China*

E-mail: zhengsz@vip.sina.com

[c] *Department of Mathematics and Computer, Hebei Normal College for Nationalities, Chengde 067000, China*

E-mail: yuehongda8888@126.com

This paper deals with the Beltrami system with two characteristic matrices and variable coefficients

$$D^t f(x) H(x) Df(x) = J(x,f)^{2/n} G(x),$$

where the matrices $H(x), G(x) \in S(n)$ satisfy some conditions. A homogeneous elliptic equation of divergence type

$$\mathrm{Div}\mathcal{A}(x, Df(x)) = 0$$

is derived from the Beltrami system by using the energy and variational methods. A regularity property is obtained by using the Div-Curl fields.

Keywords: Beltrami system, elliptic equation of divergence type, energy functional, regularity, div-curl field.

AMS No: 30C65, 35J20.

1. Introduction and Statement of Result

Throughout this paper we let Ω be a cube in $\mathbb{R}^n$, $n \geq 2$, and $\sigma\Omega$ a cube of the same center as Ω but σ times smaller than Ω, for $0 < \sigma \leq 1$. Consider a mapping $f = (f^1, f^2, \cdots, f^n) : \Omega \to \mathbb{R}^n$. The Jacobian matrix $Df = \left(\frac{\partial f^i}{\partial x_j}\right)_{i,j=1,\cdots,n} \in \mathbb{R}^{n\times n}$ and it's determinant $J(x,f)$ are defined almost everywhere in Ω. Denote by $D^t f(x)$ and $D^{-1} f(x)$ the transpose and inverse matrices of $Df(x)$, respectively. For $X, Y \in \mathrm{R}^{n\times n}$, the inner product of X and Y is defined by $\langle X, Y\rangle = \mathrm{Trace}(X^t Y) = \mathrm{Trace}(Y^t X)$. The Hilbert-Schmidt norm of the matrix X is then $|X| = \langle X, X\rangle^{1/2} = (\mathrm{Trace} X^t X)^{1/2}$. A basic assumption in this paper is $J(x,f) > 0$, a.e. Ω, that is, f is sense-preserving.

We consider the Beltrami system with two characteristic matrices and

[1]The first author is partially supported by NSFC (No.10971224) and the second author is partially supported by NSFC (No.11071012).

variable coefficients

$$D^t f(x)H(x)Df(x) = J(x,f)^{2/n}G(x), \tag{1.1}$$

where $H(x), G(x)$ are measurable functions valued in the space $S(n)$ of positive definite, symmetric matrices of determinant 1, and satisfy the following elliptic conditions

$$\alpha_1(x)|\xi|^2 \leq \langle H(x)\xi,\, \xi\rangle \leq \beta_1(x)|\xi|^2, \tag{1.2}$$

$$\alpha_2(x)|\eta|^2 \leq \langle G(x)\eta,\, \eta\rangle \leq \beta_2(x)|\eta|^2, \tag{1.3}$$

for all $\xi, \eta \in \mathbb{R}^n$, in which $0 < \alpha_1(x) \leq 1 \leq \beta_1(x) < \infty$ and $0 < \alpha_2(x) \leq 1 \leq \beta_2(x) < \infty$, a.e. Ω.

If f is a generalized solution of (1.1), then it transforms the infinitesimal ellipsoid at x with the characteristic $G(x)$ into the infinitesimal ellipsoid at $f(x)$ with the characteristic $H(x)$.

If $H(x) = \mathrm{Id}$, the identity matrix, then (1.1) becomes to the following Beltrami system

$$D^t f(x)Df(x) = J(x,f)^{2/n}G(x). \tag{1.4}$$

Its generalized solutions are called *mappings with finite distortion.* If, moreover, $\alpha_2(x) \geq \alpha_2 > 0$ and $\beta_2(x) \leq \beta_2 < \infty$, a.e. Ω, then the generalized solutions of (1.4) are called *quasiregular mappings.* Quasiregular mappings were first introduced and studied by Reshetnyak in a series of articles that began to appear in 1966. He used the phrase *mappings with bounded distortion* for *quasiregular mappings.* See also the monograph [1] for details. Quasiregular mappings are interesting not only because of the results obtained about them, but also because of the many new ideas generated in the course of the development of their theory. For other related works on quasiregular mappings, see [2–5].

If $n = 2$, then set $z = x_1 + ix_2$ and $f(z) = f^1(x_1,x_2) + if^2(x_1,x_2)$, (1.4) and (1.1) reduce to the following Beltrami equation with one and two characteristic functions in the plane

$$f_{\bar{z}} = \mu(z)f_z, \tag{1.5}$$

and

$$f_{\bar{z}} = \mu_1(z)f_z + \mu_2(z)\overline{f_z}, \tag{1.6}$$

respectively, where the complex coefficients μ, μ_1 and μ_2 are given by

$$\mu(z) = \frac{G_{11} - G_{22} + 2iG_{12}}{G_{11} + G_{22} + 2},$$

$$\mu_1(z) = \frac{G_{11} - G_{22} + 2iG_{12}}{G_{11} + G_{22} + H_{11} + H_{22}},\ \mu_2(z) = \frac{H_{22} - H_{11} - 2iH_{12}}{G_{11} + G_{22} + H_{11} + H_{22}}.$$

By the conditions (1.2) and (1.3), one can derive

$$|\mu(z)| \leq k_1 < 1, \tag{1.7}$$

and

$$|\mu_1(z)| + |\mu_2(z)| \leq k_2 < 1. \tag{1.8}$$

The equations (1.5) and (1.6) which satisfy the conditions (1.7) and (1.8) respectively have been studied widely, see [6–8].

Beltrami system is important in modern geometric function theory and nonlinear analysis. In recent years, the properties and some applications to other fields of the generalized solutions of (1.1) and (1.4) have been widely studied, see, for example, [9–12]. The study for the properties of generalized solutions of (1.1) and (1.4) is difficult, the main reason for which is that they are non-linear, non-uniform and over-determined.

In the present paper, we will have a further study of the generalized Beltrami system (1.1). The main result is the following regularity property.

Theorem 1.1. *If $f = (f^1, \cdots, f^n) \in W^{1,n}(\Omega, \mathbb{R}^n)$ is a generalized solution of (1.1) with $Df \in L^n \log^{1/\gamma} L(\Omega, \mathbb{R}^{n\times n})$ and*

$$\frac{\beta_1(x)\beta_2(x)}{\alpha_1(x)\alpha_2(x)} \in \mathrm{Exp}_{\frac{n\gamma}{2}}(\Omega), \tag{1.9}$$

for some $\gamma > 1$, then $Df \in L^n \log^{\alpha} L(\sigma\Omega, \mathbb{R}^{n\times n})$ for any $\alpha > 0$ and $0 < \sigma < 1$.

For the notations used in Theorem 1.1, see Section 2.

2. Some Notations

In this section, we give some notations used in this paper.

If $B : \Omega \to \mathbb{R}^{n\times n}$ and $E : \Omega \to \mathbb{R}^{n\times n}$ are matrix fields on Ω such that

$$\mathrm{Div}B = (\mathrm{div}B_1, \cdots, \mathrm{div}B_n) = 0,$$

and

$$\mathrm{Curl}E = (\mathrm{curl}E_1, \cdots, \mathrm{curl}E_n) = 0$$

in the sense of distributions, then the scalar product $\langle B, E\rangle$ is referred to as a Div-Curl product. Where, as usual, for the vector fields $B_i = (B_i^1, \cdots, B_i^n)$ and $E_i = (E_i^1, \cdots, E_i^n)$, $i = 1, \cdots, n$, the "div" and "curl" operators are defined by

$$\mathrm{div}B_i = \sum_{j=1}^{n} \frac{\partial B_i^j}{\partial x_j}, \quad \mathrm{curl}E_i = \left(\frac{\partial E_i^j}{\partial x_k} - \frac{\partial E_i^k}{\partial x_j}\right)_{j,k=1,\cdots,n}.$$

We investigate a class of Div-Curl fields $\mathcal{F} = (B, E)$, which is coupled by the distortion inequality

$$\frac{|B|^p}{p} + \frac{|E|^q}{q} \le k(x)\langle B, E\rangle, \quad \text{a.e. in } \Omega, \tag{2.1}$$

where $1 \le k(x) < \infty$ is a measurable function in Ω and $1 < p,\, q < \infty$ are conjugate Hölder exponents, $\frac{1}{p} + \frac{1}{q} = 1$. We shall also assume that

$$\langle B, E\rangle \in L^1(\Omega). \tag{2.2}$$

A continuous strictly increasing function $\Psi\colon [0,\infty] \to [0,\infty]$ with $\Psi(0) = 0$ and $\Psi(\infty) = \infty$ is called an Orlicz function. If, moreover, Ψ is convex, then it is called a Young function. The Orlicz space $L^\Psi(\Omega)$ consists of all measurable functions h on Ω such that

$$\int_\Omega \Psi(\lambda^{-1}|h(x)|)dx < \infty$$

for some $\lambda = \lambda(h) > 0$. The Luxemburg functional

$$\|h\|_\Psi = \|h\|_{L^\Psi} = \inf\left\{\lambda > 0 : \fint_\Omega \Psi(\lambda^{-1}|h|) \le \Psi(1)\right\},$$

where $\fint_\Omega$ stands for the integral mean over Ω, is a norm provided that Ψ is a Young function, but it need not to be if Ψ is only an Orlicz function.

For $\Phi(t) = t^p \log^\alpha(e+t)$, where $1 \le p < \infty$ and α is a real number, the spaces L^Φ are called Zygmund spaces and denoted by $L^p \log^\alpha L$. For $p \ge 1$ and $\alpha \ge 0$ the non-linear functional

$$[h]_{L^p \log^\alpha L} = \left[\int_\Omega |h|^p \log^\alpha\left(e + \frac{|h|}{\|h\|_p}\right)\right]^{1/p}$$

is comparable with the Luxemburg norm via constants depending only on p and α and not on Ω. Denote by $\mathrm{Exp}_\gamma(\Omega)$ the Orlicz space defined by the function $\Phi(t) = \exp(t^\gamma) - 1$, where $\gamma > 0$ is a real number. A basic result is the complementary to $t\log^{1/\gamma}(e+t)$, $\gamma > 0$, is the function $\exp(t^\gamma) - 1$.

3. Some Lemmas

In this section we give some preliminary lemmas used in the proof of Theorem 1.1. The following lemma comes from Lemma 2.1 in [13].

Lemma 3.1. *Suppose $H, G \in S(n)$. For any $A \in R^{n\times n}$, we have*

$$n^{n/2}|\det A| \le \langle AG^{-1},\, HA\rangle^{n/2}.$$

Equality occurs if and only if $A^t HA = |\det A|^{2/n} G$.

The next lemma can be found in [Lemma 4.7.2, 3].

Lemma 3.2. *If* $f, g \in W^{1,n}(\Omega, \mathbb{R}^n)$ *and* $f = g$, $x \in \partial\Omega$ *in the Sobolev sense (abbr.* $g \in f + W_0^{1,n}(\Omega, \mathbb{R}^n)$*), then*

$$\int_\Omega J(x, g)\, dx = \int_\Omega J(x, f)\, dx.$$

Lemma 3.3. *Suppose* $f = (f^1, f^2, \cdots, f^n) \in W^{1,n}(\Omega, \mathbb{R}^n)$ *be a generalized solution of* (1.1) *with* $Df \in L^n \log^{1/\gamma} L(\Omega, \mathbb{R}^{n\times n})$ *and* (1.9) *holds for some* $\gamma > 1$*, then*

$$\mathrm{div}\mathcal{A}(x, Df(x)) = 0 \tag{3.1}$$

in the distributional sense, where $\mathcal{A}(x, A) : \Omega \times \mathbb{R}^{n\times n} \to \mathbb{R}^{n\times n}$ *is given by*

$$\mathcal{A}(x, A) = \langle AG^{-1}, HA\rangle^{(n-2)/2} HAG^{-1},$$

and the operator $\mathcal{A}$ *satisfies*

(i) *Lipschitz type condition*

$$|\mathcal{A}(x, A) - \mathcal{A}(x, B)| \le (n-1) \left(\frac{\beta_1(x)}{\alpha_2(x)}\right)^{n/2} (|A| + |B|)^{n-2} |A - B|.$$

(ii) *Monotonicity condition*

$$\langle \mathcal{A}(x, A) - \mathcal{A}(x, B), A - B\rangle \ge 2^{2-n} \left(\frac{\alpha_1(x)}{\beta_2(x)}\right)^{n/2} (|A| + |B|)^{n-2} |A - B|^2.$$

(iii) *Homogeneous condition*

$$\mathcal{A}(x, tA) = |t|^{n-2} t\mathcal{A}(x, A),\ \forall t \in \mathbb{R}.$$

Proof. For any $g \in f + W_0^{1,n}(\Omega, \mathbb{R}^n)$ with $J(x, g) \ge 0$ and $Dg \in L^n \log^{1/\gamma}(\Omega, \mathbb{R}^{n\times n})$, we define the stored energy function $E(x, Dg)$, where $E(x, A)\colon \Omega \times \mathbb{R}^{n\times n} \to \mathbb{R}$ is given by

$$E(x, A) = \langle AG^{-1}, HA\rangle^{n/2}.$$

Our nearest goal is to show that $E(x, Dg)$ is integrable. To this end, notice from the assumptions (1.2) and (1.3) that

$$|E(x, Dg)| = \left|\langle DgG^{-1}, HDg\rangle^{n/2}\right| \le \left(\frac{\beta_1(x)}{\alpha_2(x)}\right)^{n/2} |Dg|^n.$$

Since

$$\begin{aligned}[|Dg|^n]_{L\log^{1/\gamma}L} &= \int_\Omega |Dg|^n \log^{1/\gamma}\left(e+\frac{|Dg|^n}{\||Dg|^n\|_1}\right)dx \\ &= \int_\Omega |Dg|^n \log^{1/\gamma}\left(e+\left(\frac{|Dg|}{\|Dg\|_n}\right)^n\right)dx \\ &\le n^{1/\gamma}\int_\Omega |Dg|^n \log^{1/\gamma}\left(e+\frac{|Dg|}{\|Dg\|_n}\right)dx \\ &= n^{1/\gamma}\,[|Dg|]_{L^n\log^{1/\gamma}L}<\infty,\end{aligned}$$

then $|Dg|^n \in L\log^{1/\gamma}L(\Omega,\mathbb{R})$. This implies that $E(x,Dg)$ is integrable due to the facts that $\left(\frac{\beta_2(x)}{\alpha_1(x)}\right)^{n/2} \le \left(\frac{\beta_1(x)\beta_2(x)}{\alpha_1(x)\alpha_2(x)}\right)^{n/2} \in \mathrm{Exp}_\gamma(\Omega)$ and the complementary to $t\log^{1/\gamma}(e+t)$ is the function $\exp(t^\gamma)-1$.

We define the energy functional of g by

$$\mathcal{E}[g]=\int_\Omega E(x,Dg)\,dx.$$

By Lemma 3.1, the energy $\mathcal{E}[g]$ of g is non-negative. In the following, we prove that the generalized solution f of (1.1) takes the minimum value under the same boundary conditions. In fact, (1.1) implies

$$\langle D^t f(x)H(x)Df(x)G^{-1}(x),\,\mathrm{Id}\rangle = J(x,f)^{2/n}\langle \mathrm{Id},\mathrm{Id}\rangle.$$

Thus

$$\langle Df(x)G^{-1}(x),\,H(x)Df(x)\rangle^{n/2}=n^{n/2}J(x,f). \tag{3.2}$$

By (3.2), Lemmas 3.2 and 3.1, we have

$$\begin{aligned}\mathcal{E}[f] &= \int_\Omega \langle DfG^{-1},HDf\rangle^{n/2}\,dx = n^{n/2}\int_\Omega J(x,f)\,dx \\ &= n^{n/2}\int_\Omega J(x,g)\,dx \le \int_\Omega \langle DgG^{-1},HDg\rangle^{n/2}\,dx=\mathcal{E}[g].\end{aligned}$$

For any $\varphi\in W_0^{1,n}(\Omega,\mathbb{R}^n)$ and $t\in\mathbb{R}$, one has

$$\mathcal{E}[f+t\varphi]\ge\mathcal{E}[f].$$

Thus $t\mapsto \mathcal{E}[f+t\varphi]$ take its minimum value at $t=0$. Therefore

$$\left.\frac{d\mathcal{E}[f+t\varphi]}{dt}\right|_{t=0}=0.$$

An easy calculation yields

$$\frac{d\mathcal{E}[f+t\varphi]}{dt}\bigg|_{t=0}=\int_\Omega \langle\langle DfG^{-1},HDf\rangle^{(n-2)/2}HDfG^{-1},\ D\varphi\rangle\,dx=0. \quad (3.3)$$

This is nothing but the definition of solutions of (3.1) in the distributional sense.

In the following, we derive the conditions for the operator $\mathcal{A}(x,A)$. By $H=O_1^t\Gamma_1^2O_1$, $G=O_2^t\Gamma_2O_2$, $\Gamma_1=\mathrm{diag}(\gamma^1(x),\gamma^2(x),\cdots,\gamma^n(x))$, $\Gamma_2=\mathrm{diag}(\gamma_1(x),\gamma_2(x),\cdots,\gamma_n(x))$, (1.2) and (1.3), we know $\sqrt{\alpha_1(x)}\le\gamma^i(x)\le\sqrt{\beta_1(x)}$, $\sqrt{1/\beta_2(x)}\le\gamma_i(x)^{-1}\le\sqrt{1/\alpha_2(x)}$, $i=1,2,\cdots,n$. By a basic inequality, see [(35), 12]

$$||\xi|^{p-2}\xi-|\zeta|^{p-2}\zeta|\le(p-1)(|\xi+|\zeta|)^{p-2}|\xi-\zeta|,\ \forall\xi,\zeta\in\mathbb{R}^{n\times n},\ p\ge 2,$$

we can derive that

$$\begin{aligned}
&\Big|\mathcal{A}(x,A)-\mathcal{A}(x,B)\Big|\\
&=\Big|\langle AG^{-1},HA\rangle^{(n-2)/2}HAG^{-1}-\langle BG^{-1},HB\rangle^{(n-2)/2}HBG^{-1}\Big|\\
&=\Big|\sqrt{H}\Big(|\sqrt{H}A\sqrt{G^{-1}}|^{n-2}\sqrt{H}A\sqrt{G^{-1}}\\
&\quad-|\sqrt{H}B\sqrt{G^{-1}}|^{n-2}\sqrt{H}B\sqrt{G^{-1}}\Big)\sqrt{G^{-1}}\Big|\\
&\le(n-1)\sqrt{\frac{\beta_1(x)}{\alpha_2(x)}}\Big(|\sqrt{H}A\sqrt{G^{-1}}|+|\sqrt{H}B\sqrt{G^{-1}}|\Big)^{n-2}\Big|\sqrt{H}(A-B)\sqrt{G^{-1}}\Big|\\
&\le(n-1)\Big(\beta_1(x)\alpha_2(x)\Big)^{n/2}(|A|+|B|)^{n-2}|A-B|.
\end{aligned}$$

This is the Lipschitz type condition. By another basic inequality

$$\langle|\xi|^a\xi-|\zeta|^a\zeta,\ \xi-\zeta\rangle\ge\frac{1}{2}(|\xi|^a+|\zeta|^a)|\xi-\zeta|^2,\ \forall\, a>0,\ \xi,\zeta\in\mathbb{R}^{n\times n},$$

one can derive that

$$\begin{aligned}
&\langle\mathcal{A}(x,A)-\mathcal{A}(x,B),\ A-B\rangle\\
&=\Big\langle\langle AG^{-1},HA\rangle^{(n-2)/2}HAG^{-1}-\langle BG^{-1},HB\rangle^{(n-2)/2}HBG^{-1},A-B\Big\rangle
\end{aligned}$$

$$
\begin{aligned}
&= \Big\langle |\sqrt{H}A\sqrt{G^{-1}}|^{n-2}\sqrt{H}A\sqrt{G^{-1}} \\
&\quad -|\sqrt{H}B\sqrt{G^{-1}}|^{n-2}\sqrt{H}B\sqrt{G^{-1}},\ \sqrt{H}(A-B)\sqrt{G^{-1}}\Big\rangle \\
&\geq \tfrac{1}{2}\left(|\sqrt{H}A\sqrt{G^{-1}}|^{n-2}+|\sqrt{H}B\sqrt{G^{-1}}|^{n-2}\right)|\sqrt{H}(A-B)\sqrt{G^{-1}}|^2 \\
&\geq \frac{1}{2}\left(\frac{\alpha_1(x)}{\beta_2(x)}\right)^{n/2}\left(|A|^{n-2}+|B|^{n-2}\right)|A-B|^2 \\
&\geq 2^{2-n}\left(\frac{\alpha_1(x)}{\beta_2(x)}\right)^{n/2}(|A|+|B|)^{n-2}|A-B|^2.
\end{aligned}
$$

This is the monotonicity condition. By the definition of $\mathcal{A}(x,A)$, one can easily derive the homogeneous condition. The proof of Lemma 3.3 has been completed.

Lemma 3.4. *Let $\mathcal{F} = (B,E)$ be a Div-Curl field verifying* (2.1). *If $k(x) \in Exp_\gamma(\Omega)$ for some $\gamma > 1$, then $B \in L^p \log^\alpha L(\sigma\Omega, \mathbb{R}^{n\times n})$ and $E \in L^q \log^\alpha L(\sigma\Omega, \mathbb{R}^{n\times n})$ for any $\alpha > 0$ and $0 < \sigma < 1$. Moreover for any $\alpha > 1$,*

$$
\||B|^p + |E|^q\|_{L\log^{\alpha-1/\gamma}L(\sigma\Omega)} \leq c\|\langle B,E\rangle\|_{L\log^{\alpha-1}L(\Omega)},
$$

where $c = c(\sigma,p,\alpha,n,\|k\|_{Exp_\gamma(\Omega)})$.

The difference between Lemma 3.4 and Theorem 1.1 in [14] is that the vector fields $B, E \in \mathbb{R}^n$ are replaced by matrix fields $B, E \in \mathbb{R}^{n\times n}$. The proof of Lemma 3.4 is almost the same as in [14], thus we omit the details.

4. Proof of Theorem 1.1

In the following, we will derive that the conditions in Lemma 3.4 are satisfied, thus Theorem 1.1 follows from Lemma 3.4. Let $E = Df$ and $B = \langle DfG^{-1}, HDf\rangle^{\frac{n-2}{2}} HDfG^{-1}$. It is obvious that Curl$E = 0$, and the fact Div$B = 0$ follows from Lemma 3.3. Thus $\mathcal{F} = (B,E)$ be a Div-Curl field. By Lemma 3.3 again, we have

$$
|E|^n = |Df|^n,
$$

$$
|B|^{\frac{n}{n-1}} = |\langle DfG^{-1}, HDf\rangle^{\frac{n-2}{2}} HDfG^{-1}|^{\frac{n}{n-1}} \leq \left(\frac{\beta_1(x)}{\alpha_2(x)}\right)^{\frac{n}{2}} |Df|^n,
$$

and since

$$
\langle B,E\rangle = E(x,Df) \geq \left(\frac{\alpha_1(x)}{\beta_2(x)}\right)^{\frac{n}{2}} |Df|^n,
$$

then

$$\frac{|B|^{n/(n-1)}}{n/(n-1)} + \frac{|E|^n}{n} \le k(x)\langle B, E\rangle,$$

where

$$k(x) = \frac{(n-1)\left(\frac{\beta_1(x)}{\alpha_2(x)}\right)^{\frac{n}{2}} + 1}{n\left(\frac{\alpha_1(x)}{\beta_2(x)}\right)^{\frac{n}{2}}} \le \left(\frac{\beta_1(x)\beta_2(x)}{\alpha_1(x)\alpha_2(x)}\right)^{n/2} \in \mathrm{Exp}_\gamma(\Omega).$$

Theorem 1.1 follows from Lemma 3.4.

References

[1] Yu. G. Reshetnyak, *Space mappings with bounded distortion*, Trans. Math. Monographs, Vol.73, Amer. Math. Soc., Providence, RI, 1989.

[2] S. Rickman, *Quasiregular mappings*, Spriner-Verlag, Berlin, 1993.

[3] T. Iwaniec and G. Martin, *Geometric function theory and non-linear analysis*, Clarendon Press, Oxford, 2001.

[4] T. Iwaniec and G. Martin, Quasiregular mappings in even dimensions, *Acta Math.*, 170 (1993), 29-81.

[5] T. Iwaniec, p-harmonic tensors and quasiregular mappings, *Ann. of Math.*, 136 (1992), 589-624.

[6] I. N. Vekua, *Generalized analytic functions*, Pergamon, Oxford, 1962.

[7] L. Bers, *Theory of pseudo-analytic functions*, New York University, New York, 1953.

[8] G. C. Wen, *Linear and non-linear elliptic complex equations*, Shanghai Sci. & Tech. Press, Shanghai, 1986.

[9] S. Z. Zheng and A. N. Fang, Two characteristic Beltrami system and quasiregular mappings, *Acta Math. Sin.*, 40(5) (1997), 745-750 (in Chinese).

[10] J. F. Cheng and A. N. Fang, Generalized Beltrami system in even dimensions and Beltrami system in high dimensional spaces, *Chinese Ann. of Math.*, 18A(6) (1997), 789-798 (in Chinese).

[11] J. F. Cheng and A. N. Fang, (G, H)-quasiregular mappings and B-harmonic equations, *Acta Math. Sin.*, 42(5) (1999), 883-888 (in Chinese).

[12] H. Y. Gao and Z. M. Wu, On Beltrami system with two characteristic matrices, *Acta Math. Sci.*, 22(4) (2002), 433-440 (in Chinese).

[13] H. Y. Gao and L. F. Zhao, On generalized Beltrami system, *Acta Math, Sci.*, Ser. A, 30(4) (2010), 953-958.

[14] L. Migliaccio and G. Moscariello, Higher integrability of div-curl products, *Ricerche di Matematica*, XLIX (2000), 151-161.

CACCIOPPOLI TYPE ESTIMATE FOR VERY WEAK SOLUTIONS OF NONHOMOGENEOUS A-HARMONIC TYPE EQUATIONS[1]

JIAN-TAO GU[a], JUAN LI[b] and HONG LIU[c]

[a] *College of Science, Hebei Polytechnic University, Tangshan 063009, China*
E-mail: gujiantaolg@126.com

[b] *Department of Mathematics, Ningbo University, Ningbo 315211, China*
E-mail: juanjuan-li@163.com

[c] *College of Mathematics and Computer Science, Hebei University, Baoding 071002, China*
E-mail: liuhongmath@163.com

Using the estimates of so-called Hodge decomposition of disturbed vector fields, a Caccioppoli type estimate is established for very weak Solutions of a class of nonlinear equations involved nonhomogeneous items.

Keywords: Hodge decomposition, A-harmonic type equation, Caccioppoli type estimate, very weak solutions.

AMS No: 35B65, 35J60.

1. Introduction

Recently, that Caccioppoli type estimate were established for the so-called nonlinear A-harmonic equations of homogeneous items (see [1–8]). In this paper, based on using the Hodge decomposition of disturbed vector fields, we shall consider a class of nonlinear equations of divergent type involved nonhomogeneous items.

Before embarking on the discussion we refer to some notations we shall use. Throughout this paper, Ω will denote an open, connected subset of $\mathbb{R}^n$, and E is a closed set of zero Lebesgue measure in $\mathbb{R}^n$. We denote Sobolev space of weakly differentiable functions defined locally on Ω as $W^{1,p}_{loc}(\Omega)$, and its differential $\nabla u \in L^p_{loc}(\Omega)$. In order to avoid some technical difficulties related to the imbedding theorem, we shall illustrate our approach only for p smaller than the spatial dimension of Ω.

Definition 1. A compact set $E \subset \mathbb{R}^n$ is said to have zero r-capacity for $1 < r \le n$, if for some bounded domain Ω containing E, there exists a sequence $\{\varphi_k(x)\}$, $k = 1, 2, ...$, of functions $\varphi_k(x) \in C_0^\infty(\Omega)$, such that

(1) $0 \le \varphi_k \le 1$;

(2) each φ_k equals to 1 on its own neighborhood of E;

(3) $\lim_{k\to\infty} \|\nabla\varphi_k\|_r = 0$; $\lim_{k\to\infty} \varphi_k = 0$ for every $x \in \Omega \setminus E$.

[1]The research supported by Natural Science Foundation of Hebei Province (A2010000910), Projects of Hebei Province Education Department (Z2010261) and Tangshan Science and Technology Projects (09130206c).

A closed set $E \subset \mathbb{R}^n$ has zero r-capacity, if every compact subset of E has zero r-capacity.

This definition of a set of zero r-capacity coincides with the customary one. Notice that for $r = p - \varepsilon$, $0 < \varepsilon < 1$, a closed set $E \subset \mathbb{R}^n$ of Hausdorff dimension $\dim_H(E) < \varepsilon$ has zero r-capacity.

We shall study this problem for very weak solutions of PDEs of divergent type with nonhomogeneous items, which is also called A-harmonic type equations. It is necessary to impose some assumpations near the singular set for the very weak solution $u(x)$ and its gradient ∇u is as follows.

$$Du = -\text{div}A(x, u, \nabla u) + f(x) = 0. \tag{1.1}$$

Let the operators $A : \mathbb{R}^n \times \mathbb{R}^n \mapsto \mathbb{R}^n$ be a measurable mapping satisfying

$$\begin{aligned} &\text{(i) } \langle A(x, u, \nabla u), \nabla u\rangle \geq a|\nabla u|^p; \\ &\text{(ii) } |A(x, u, \nabla u)| \leq b|\nabla u|^{p-1} + c|u|^s + g(x); \end{aligned} \tag{1.2}$$

where $a, b, c\,(0 < a, b, c < \infty)$ are constants, $g(x)$ is a nonnegative function satisfying $g(x) \in L^{\frac{r}{p-1}}(\Omega)$, and $f(x) \in L^{\frac{nr}{n(p-1)+r}}(\Omega)$. The nonnegative constant s satisfies $0 < s < \frac{n(p-1)}{n-r}$. A function u of Sobolev class $W^{1,r}_{loc}(\Omega)$ for $1 < r < p$ is said to be a very weak solution, if

$$\int_\Omega \langle A(x, u, \nabla u), \nabla\eta(x)\rangle dx = \int_\Omega f(x)\eta(x)dx \tag{1.3}$$

for all test functions $\eta(x) \in W_0^{1-\frac{r}{r-p+1}}(\Omega)$.

Note that the **very weak** means that the Sobolev integrable exponent r of u can be smaller than p.

In order to investigate removable singular sets, it is important to consider very weak solutions of A-harmonic type equation (1.1). This makes the problem more difficult, because the very weak solution and its modifications cannot be used test functions (1.1). In the following, we shall make use of the estimates of so-called Hodge decomposition of disturbed vector fields, see [4–5]. To this effect our first result is the following weaker estimate of Caccioppoli type.

Theorem 1.1. (Caccioppoli type estimate) *There exists* $p_0 = p_0(n, p, a, b) \in \left[\frac{n(p-1)}{n(p-1)-1}, p\right)$, *such that every very weak solution of A-harmonic equations* (1.1) *in Sobolev class* $W^{1,r}_{loc}(\Omega)$ *with* $p_0 \leq r < p$ *satisfies*

$$\begin{aligned} \|\varphi\nabla u\|_r dx \leq C(n, p, a, b)\Big(&\|\nabla\varphi \otimes (u - \bar{u}_R)\|_r + \|g(x)\|^{\frac{1}{p-1}}_{\frac{r}{p-1}} \\ &+ \|f(x)\|^{\frac{1}{p-1}}_{\frac{nr}{n(p-1)+r}} + \|\nabla u\|^{\frac{s}{p-1}}_{\tilde{p}}\Big), \end{aligned} \tag{1.4}$$

for all test functions $\varphi \in C_0^\infty(\Omega)$*, the exponent* $\tilde{p} = \frac{nsr}{np-n+sr} < r$*. Furthermore,* $u \in W^{1,p}_{loc}(\Omega)$ *is classical weak solution in the usual sense.*

2. Proof of Theorem 1.1

Proof. Fixed $R_0 : R_0 \leq d = \text{dist}(x_0, \partial\Omega)$ for all $x_0 \in \Omega$, we denote

$$B_R = B_R(x_0) = \{x||x - x_0| < R\},$$

and

$$u_R = \int_{B_R} u(x)dx = \frac{1}{|BR|}\int_{B_R} u(x)$$

for any $0 < R < R_0$. Let $u \in W^{1,r}(B_R)$, $r = p - \varepsilon$ for some $0 < \varepsilon < 1$, and $\varphi(x) \in C_0^\infty(B_R)$. Let C denote a constant (not necessarily the same at different settings). By making use of so-called Hodge decomposition acted on a perturbed vector field

$$|\nabla(\varphi(u - \bar{u}_R))|^{-\varepsilon}\nabla(\varphi(u - \bar{u}_R)),$$

we have

$$|\nabla(\varphi(u - \bar{u}_R))|^{-\varepsilon}\nabla(\varphi(u - \bar{u}_R)) = \nabla\Phi + H, \tag{2.1}$$

where $H \in L^{\frac{r}{1-\varepsilon}}(B_R)$ and $\Phi \in W_0^{1,\frac{r}{1-\varepsilon}}(B_R)$. Moreover, the following estimates are valid

$$\|H\|_{\frac{r}{1-\varepsilon}} \leq C\varepsilon\|\nabla(\varphi(u - \bar{u}_R))\|_r^{1-\varepsilon}, \tag{2.2}$$

and

$$\|\nabla\Phi\|_{\frac{r}{1-\varepsilon}} \leq C\|\nabla(\varphi(u - \bar{u}_R))\|_r^{1-\varepsilon}, \tag{2.3}$$

where $C = C(n, r, R_0)$ are independent of ε, also see [2–3]. Write

$$E(\varphi, \nabla u) = |\varphi\nabla u|^{-\varepsilon}\varphi\nabla u - |\nabla(\varphi(u - \bar{u}_R))|^{-\varepsilon}\nabla(\varphi(u - \bar{u}_R)),$$

then we derive

$$|E(\varphi, \nabla u)| \leq \frac{2^\varepsilon(1+\varepsilon)}{1-\varepsilon}|\nabla\varphi \otimes (u - \bar{u}_R)|^{1-\varepsilon}. \tag{2.4}$$

A rather remarkable technique is that Φ in the Hodge decomposition (2.1) will be used as a test function. Then, the A-harmonic type equation (1.1) becomes

$$\begin{aligned}
&\int_{B_R} \langle A(x,\varphi\nabla u), |\varphi\nabla u|^{-\varepsilon}\varphi\nabla u\rangle dx \\
&= \int_{B_R} \langle A(x,\varphi\nabla u), |\nabla(\varphi(u-\bar{u}_R))|^{-\varepsilon}\nabla(\varphi(u-\bar{u}_R))\rangle dx \\
&\quad + \int_{B_R} \langle A(x,\varphi\nabla u), E(\varphi,\nabla u)\rangle dx \\
&= \int_{B_R} \langle A(x,\varphi\nabla u), \nabla\Phi\rangle dx + \int_{B_R} \langle A(x,\varphi\nabla u), H\rangle dx \\
&\quad + \int_{B_R} \langle A(x,\varphi\nabla u), E(\varphi,\nabla u)\rangle dx \\
&= \int_{B_R} f(x)\Phi dx + \int_{B_R} \langle A(x,\varphi\nabla u), H + E(\varphi,\nabla u)\rangle dx.
\end{aligned}$$

By the conditions (1.2) imposed on the operators A, we have

$$\begin{aligned}
a\int_{B_R} |\varphi\nabla u|^{p-\varepsilon} dx &\le \int_{B_R} |\varphi\nabla u|^{-\varepsilon}\langle A(x,\varphi\nabla u), \varphi\nabla u\rangle dx \\
&= \int_{B_R} \langle A(x,\varphi\nabla u), |\varphi\nabla u|^{-\varepsilon}\varphi\nabla u\rangle dx \\
&\le \int_{B_R} |f(x)||\Phi| dx \\
&\quad + \int_{B_R} |A(x,\varphi\nabla u)|(|H| + |E(\varphi,\nabla u)|) dx \qquad (2.5)\\
&\le \int_{B_R} |f(x)||\Phi| dx + \int_{B_R} g(x)(|H| + |E(\varphi,\nabla u)|) dx \\
&\quad + b\int_{B_R} |\varphi\nabla u|^{p-1}(|H| + |E(\varphi,\nabla u)|) dx \\
&\quad + c\int_{B_R} |u|^s(|H| + |E(\varphi,\nabla u)|) dx.
\end{aligned}$$

On the basis of the Hölder inequality and the inequality (2.4), we can obtain

$$\begin{aligned}
a\int_{B_R} |\varphi\nabla u|^{p-\varepsilon} dx &\le \|f(x)\|_{\frac{t}{p-1}} \|\Phi\|_q + b\|\varphi\nabla u\|_r^{p-1} \|H\|_{\frac{r}{1-\varepsilon}} \\
&\quad + \frac{2^\varepsilon(1+\varepsilon)}{1-\varepsilon} b \int_{B_R} |\varphi\nabla u|^{p-1} |\nabla\varphi \otimes (u-\bar{u}_R)|^{1-\varepsilon} dx
\end{aligned}$$

$$
\begin{aligned}
&+\frac{2^{\varepsilon}(1+\varepsilon)}{1-\varepsilon}b\int_{B_R} g(x)|\nabla\varphi\otimes(u-\bar{u}_R)|^{1-\varepsilon}dx\\
&+b\|g(x)\|_{\frac{r}{p-1}}\|H\|_{\frac{r}{1-\varepsilon}}+c\|u\|^{s}_{\frac{sr}{p-1}}\|H\|_{\frac{r}{1-\varepsilon}} \qquad (2.6)\\
&+\frac{2^{\varepsilon}(1+\varepsilon)}{1-\varepsilon}c\int_{B_R} |u|^s|\nabla\varphi\otimes(u-\bar{u}_R)|^{1-\varepsilon}dx,
\end{aligned}
$$

where $q=\dfrac{n\frac{r}{1-\varepsilon}}{n-\frac{r}{1-\varepsilon}}$ and $t=\frac{(p-1)q}{q-1}>1$ satisfying $\frac{1}{q}+\frac{p-1}{t}=1$.

Applying Poincaré-Sobolev's inequality to Φ, u, and the estimate items of the Hodge decomposition in (2.2) and (2.3), we have

$$
\begin{aligned}
a\int_{B_R}|\varphi\nabla u|^{p-\varepsilon}dx &\le C\|f(x)\|_{\frac{t}{p-1}}\|\nabla\Phi\|_{\frac{r}{1-\varepsilon}}+C\|\varphi\nabla u\|_r^{p-1}\|H\|_{\frac{r}{1-\varepsilon}}\\
&\quad+C\|\varphi\nabla u\|_r^{p-1}\|\nabla\varphi\otimes(u-\bar{u}_R)\|_r^{1-\varepsilon}\\
&\quad+C\|g(x)\|_{\frac{r}{p-1}}\|H\|_{\frac{r}{1-\varepsilon}}+C\|u\|^{s}_{\frac{sr}{p-1}}\|H\|_{\frac{r}{1-\varepsilon}}\\
&\quad+C\|g(x)\|_{\frac{r}{p-1}}\|\nabla\varphi\otimes(u-\bar{u}_R)\|_r^{1-\varepsilon}\\
&\quad+C\|u\|^{s}_{\frac{sr}{p-1}}\|\nabla\varphi\otimes(u-\bar{u}_R)\|_r^{1-\varepsilon}\\
&\quad+C\|u\|^{s}_{\frac{sr}{p-1}}\|\nabla\varphi\otimes(u-\bar{u}_R)\|_r^{1-\varepsilon}\\
&\le C\|f(x)\|_{\frac{t}{p-1}}\|\nabla(\varphi(u-\bar{u}_R))\|_r^{1-\varepsilon} \qquad (2.7)\\
&\quad+C\varepsilon\|\varphi\nabla u\|_r^{p-1}\|\nabla(\varphi(u-\bar{u}_R))\|_r^{1-\varepsilon}\\
&\quad+C\|\varphi\nabla u\|_r^{p-1}\|\nabla\varphi\otimes(u-\bar{u}_R)\|_r^{1-\varepsilon}\\
&\quad+C\varepsilon\|g(x)\|_{\frac{r}{p-1}}\|\nabla(\varphi(u-\bar{u}_R))\|_r^{1-\varepsilon}\\
&\quad+C\|g(x)\|_{\frac{r}{p-1}}\|\nabla\varphi\otimes(u-\bar{u}_R)\|_r^{1-\varepsilon}\\
&\quad+C\varepsilon\|\nabla u\|^{s}_{\tilde{p}}\|\nabla(\varphi(u-\bar{u}_R))\|_r^{1-\varepsilon}\\
&\quad+C\|\nabla u\|^{s}_{\tilde{p}}\|\nabla\varphi\otimes(u-\bar{u}_R)\|_r^{1-\varepsilon},
\end{aligned}
$$

where $\tilde{p}=\dfrac{n\frac{sr}{p-1}}{n+\frac{sr}{p-1}}=\dfrac{nsr}{np-n+sr}$. For $0<s<\frac{n(p-1)}{n-r}$, then $\tilde{p}<r$. Since

$$
\begin{aligned}
\|\nabla(\varphi(u-\bar{u}_R))\|_r^{1-\varepsilon} &= \|\nabla\varphi\otimes(u-\bar{u}_R)+\varphi\nabla u\|_r^{1-\varepsilon}\\
&\le \|\nabla\varphi\otimes(u-\bar{u}_R)\|_r^{1-\varepsilon}+\|\varphi\nabla u\|_r^{1-\varepsilon} \qquad (2.8)
\end{aligned}
$$

for any $0<\varepsilon<1$. Hence

$$
\begin{aligned}
a\|\varphi\nabla u\|_r^r \le{} & C\|f(x)\|_{\frac{t}{p-1}}\left(\|\nabla\varphi\otimes(u-\bar{u}_R)\|_r^{1-\varepsilon}+\|\varphi\nabla u\|_r^{1-\varepsilon}\right)\\
&+C\varepsilon\|\varphi\nabla u\|_r^{p-1}\left(\|\nabla\varphi\otimes(u-\bar{u}_R)\|_r^{1-\varepsilon}+\|\varphi\nabla u\|_r^{1-\varepsilon}\right)\\
&+C\|\varphi\nabla u\|_r^{p-1}\|\nabla\varphi\otimes(u-\bar{u}_R)\|_r^{1-\varepsilon}\\
&+C\varepsilon\|g(x)\|_{\frac{r}{p-1}}\left(\|\nabla\varphi\otimes(u-\bar{u}_R)\|_r^{1-\varepsilon}+\|\varphi\nabla u\|_r^{1-\varepsilon}\right)\\
&+C\|g(x)\|_{\frac{r}{p-1}}\|\nabla\varphi\otimes(u-\bar{u}_R)\|_r^{1-\varepsilon}\\
&+C\varepsilon\|\nabla u\|_{\tilde{p}}^{s}\left(\|\nabla\varphi\otimes(u-\bar{u}_R)\|_r^{1-\varepsilon}+\|\varphi\nabla u\|_r^{1-\varepsilon}\right)\\
&+C\|\nabla u\|_{\tilde{p}}^{s}\|\nabla\varphi\otimes(u-\bar{u}_R)\|_r^{1-\varepsilon}.
\end{aligned}
$$

Applying Young's inequality $|ab|\le\delta|a|^{\frac{r}{p-1}}+\delta^{-\frac{p-1}{1-\varepsilon}}|b|^{\frac{r}{1-\varepsilon}}$ for any $\delta>0$ to the right hand side of the above inequality, we derive

$$
\begin{aligned}
& a\|\varphi\nabla u\|_r^r\\
\le{} & C\delta_1\|f(x)\|_{\frac{t}{p-1}}^{\frac{r}{p-1}}+C\|\nabla\varphi\otimes(u-\bar{u}_R)\|_r^r+C\|f(x)\|_{\frac{s}{p-1}}^{\frac{r}{p-1}}\\
&+C\delta_2\|\varphi\nabla u\|_r^r+C\varepsilon\delta_3\|\varphi\nabla u\|_r^r+C\varepsilon\|\nabla\varphi\otimes(u-\bar{u}_R)\|_r^r\\
&+C\varepsilon\|\varphi\nabla u\|_r^r+C\delta_4\|\varphi\nabla u\|_r^r+C\|\nabla\varphi\otimes(u-\bar{u}_R)\|_r^r\\
&+C(\varepsilon+1)\delta_5\|g(x)\|_{\frac{r}{p-1}}^{\frac{r}{p-1}}+C(\varepsilon+1)\|\nabla\varphi\otimes(u-\bar{u}_R)\|_r^r\\
&+C\varepsilon\|g(x)\|_{\frac{r}{p-1}}^{\frac{r}{p-1}}+C\varepsilon\delta_6\|\varphi\nabla u\|_r^r+C(\varepsilon+1)\delta_7\|\nabla u\|_{\tilde{p}}^{\frac{sr}{p-1}}\\
&+C(\varepsilon+1)\|\nabla\varphi\otimes(u-\bar{u}_R)\|_r^r+C\|\nabla u\|_{\tilde{p}}^{\frac{sr}{p-1}}+C\varepsilon\delta_8\|\varphi\nabla u\|_r^r\\
\le{} & C\left(\|f(x)\|_{\frac{t}{p-1}}^{\frac{r}{p-1}}+\|g(x)\|_{\frac{r}{p-1}}^{\frac{r}{p-1}}\right)+C\|\nabla\varphi\otimes(u-\bar{u}_R)\|_r^r\\
&+C\|\nabla u\|_{\tilde{p}}^{\frac{sr}{p-1}}+C(\delta_2+\varepsilon\delta_3+\varepsilon+\delta_4+\varepsilon\delta_6+\varepsilon\delta_8)\|\varphi\nabla u\|_r^r.
\end{aligned}
$$

Now, let us take $\delta_2,\delta_3,\delta_4,\delta_6,\delta_8$ and ε to be small enough, such that

$$C(\delta_2+\varepsilon\delta_3+\varepsilon+\delta_4+\varepsilon\delta_6+\varepsilon\delta_8)<a,$$

then we get

$$
\begin{aligned}
\|\varphi\nabla u\|_r^r \le{} & C\left(\|f(x)\|_{\frac{t}{p-1}}^{\frac{r}{p-1}}+\|g(x)\|_{\frac{r}{p-1}}^{\frac{r}{p-1}}\right)+C\|\nabla\varphi\otimes(u-\bar{u}_R)\|_r^r\\
&+C\|\nabla u\|_{\tilde{p}}^{\frac{sr}{p-1}}
\end{aligned}
$$

for all $\varphi\in C_0^\infty(B_R)$, where $t=\frac{(p-1)q}{q-1}=\frac{(p-1)nr}{n(p-1)+r}=\frac{nr}{n+r/(p-1)}>1$, $\tilde{p}=\frac{n\frac{sr}{p-1}}{n+\frac{sr}{p-1}}=\dfrac{nsr}{np-n+sr}<r$. This completes the proof of theorem.

References

[1] L. Budney and T. Iwaniec, Removability of singularities of A-harmonic functions. *Differential and integral equations*, 12(2) (1999), 261-274.

[2] T. Iwaniec, P-harmonic tensor and quasi-regular mappings, *Ann. Math.*, 136 (1992), 589-624.

[3] T. Iwaniec and C. Sbordone, Weak minima of variational integrals. *J. Reine Angew Math.*, 454 (1994), 143-161.

[4] P. Koskela and O. Martio, Removability theorems for solution of degenerate elliptic, *PDE. Ark. Mat.*, 31 (1993), 339-353.

[5] J. Lewis, On very weak solutions of certain elliptic systems. Comm. *Part. Diff. Equ.*, 18 (1993), 1515-1537.

[6] Hongya Gao, Jinjing Qiao and Yuming Chu, Local regularity and local boundedness results for very weak solutions of obstacle problems, *J. Inequal. Appl.* 2010, Art. ID 878769, 12 pp.

[7] Hongya Gao and Huiying Tian, Local regularity result for solutions of obstacle problems, *Acta Mathematica Scientia B,* 24 (2004), 71-74.

[8] T. Iwaniec, L. Migliaccio, L. Nania and C. Sbordone, Integrability and removability results for quasiregular mappings in high dimensions, *Math. Scand.*, 75 (1994), 263-279.

[9] Shenzhou Zheng, Removable singularities of solutions of A-harmonic type equations, *Acta Mathematicae Applicatae Sinica, English Series,* 20 (2004), 115-122.

[10] Hongya Gao, Jing Guo et al, Local regularity result in obstacle problems, *Acta Mathematica Scientia B,* 30 (2010), 208-214.

DIFFERENTIAL FORMS AND QUASIREGULAR MAPPINGS ON RIEMANNIAN MANIFOLDS

ZHI-HUA GU and XIU-YIN SHANG

College of Sciences, Agricultural University of Hebei, Boading 071001, P.R.China

E-mail: guzhihuahbu@163.com

E-mail: sxshangxiuyin@hebau.edu.cn

The purpose of this paper is to discuss the relations between (K_1, K_2)-quasiregular mappings on Riemannian manifolds and differential forms. Two classes of differential forms are introduced and it is shown that some differential expressions connected in a natural way to (K_1, K_2)-quasiregular mapping.

Keywords: Weakly closed, differential form, Riemannian manifold, (K_1, K_2)-quasiregular mapping.

AMS No: 35J60.

1. Introduction

We first introduce some notations and symbols used in this paper. Most of them can be found in [3]. We list them all here for completion of this paper. Let $\mathcal{M}$ be an n-dimensional Riemannian manifold with boundary or without boundary. Throughout this paper we assume that the manifold $\mathcal{M}$ is orientable and of class C^3. By $T(\mathcal{M})$ we denote the tangent bundle and by $T_m(\mathcal{M})$ the tangent space at the point $m \in \mathcal{M}$. For each pair of vectors $x, y \in T_m(\mathcal{M})$, the symbol $\langle x, y\rangle$ denotes their scalar product. Below we shall use standard notation for function classes on manifolds. Thus, for example, the symbol $L^p_{loc}(D)$ stands for the set of all Lebesgue measurable functions on an open set $D \subset \mathcal{M}$, locally integrable to the power $p\,(1 \leq p \leq \infty)$ on D. The symbol $W^{1,p}_{loc}(D)$ stands for the set of functions that have generalized partial derivatives in the sense of Sobolev of class $L^p_{loc}(D)$. Let $\mathcal{M}$ and $\mathcal{N}$ be Riemannian manifolds of class $C^k, k \geq 3$, and $F : D \to \mathcal{N}, D \subset \mathcal{M}$, a mapping. We shall say that $F \in L^p_{loc}(D)$, if for an arbitrary function $\phi \in C^0(\mathcal{N})$ we have $\phi \circ F \in L^p_{loc}(D)$. The mapping F is in the class $W^{1,p}_{loc}(D)$, if $\phi \circ F \in W^{1,p}_{loc}(D)$ for every $\phi \in C^1(\mathcal{N})$. Let $V(\mathcal{M})$ be a vector bundle on $\mathcal{M}$. Let in the elements of this bundle be given an Euclidean scalar product and let the linear connection on $V(\mathcal{M})$ preserve this scalar product. In this case we may say that $V(\mathcal{M})$ is a Riemannian vector bundle over $\mathcal{M}$.

By $\bigwedge^k(\mathcal{M})$ and $\bigwedge_k(\mathcal{M})$ we denote Riemannian vector bundles $\bigwedge^k(T_m(\mathcal{M}))$ and $\bigwedge_k(T_m(\mathcal{M}))$. The sections of these bundles are the fields of k-covectors (k-forms) and k-vectors.

Let $x^1, x^2, \cdots, x^n$ be local coordinates in the neighborhood of point $m \in \mathcal{M}$. The square of a line element on $\mathcal{M}$ has the following expression in terms of the local coordinates $x^1, x^2, \cdots, x^n$:

$$ds^2 = \sum_{i,j=1}^{n} g_{ij} dx^i dx^j.$$

By the symbol g^{ij} we shall denote the contravariant tensor defined by the equality $g^{ik})(g_{kj}) = (\delta^i_j)$, $i, j = 1, \cdots, n$, where δ^i_j is the Kronecker symbol.

Each section α of the bundle $\bigwedge^k(\mathcal{M})$ (that is a differential form) can be writen in terms of the local coordinates $x^1, x^2, \cdots, x^n$ as the linear combination

$$\alpha = \sum_{I \in \wedge(k,n)} \alpha_I dx_I = \sum_{1 \le i_1 < \cdots < i_k \le n} \alpha_{i_1 \cdots i_k} dx^{i_1} \wedge \cdots \wedge dx^{i_k},$$

where we have denoted by $\bigwedge(k, n)$ the set of all ordered multi-indices $I = (i_1, i_2, \cdots, i_k)$ of integers $1 \le i_1 < \cdots < i_k \le n$.

Let α be a differential form defined on an open set $D \subset \mathcal{M}$. If $\mathcal{F}(D)$ is a class of functions defined on D, then we say that the differential from α is in this class provided that $\alpha_I \in \mathcal{F}(D)$. For instance, the differential form α is in the class $L^p(D)$, if all its coefficients are in this class.

The operator $\star : \bigwedge^k(\mathcal{M}) \to \bigwedge^{n-k}(\mathcal{M})$, called Hodge star operator has the following properties: If $\alpha, \beta \in \bigwedge^k(\mathcal{M})$, and $a, b \in \mathrm{R}$, then

$$\star(a\alpha + b\beta) = a \star \alpha + b \star \beta.$$

For every w with $\deg w = k$, we have

$$\star(\star w) = (-1)^{k(n-k)} w.$$

Let w be a differential form of degree k, we set

$$\star^{-1} w = (-1)^{k(n-k)} \star w.$$

The operator $\star^{-1}$ is an inverse to $\star$ in the sense that $\star^{-1}(\star w) = \star(\star^{-1} w) = w$. The inner or scalar product has the usual properties of the scalar product. We set

$$\langle \alpha, \beta \rangle = \star^{-1} \langle \alpha, \star\beta \rangle = \star(\alpha \wedge \star\beta).$$

A differential form w of degree k is called simple, if there are differential forms $\alpha_1, \ldots, \alpha_1$ of degree 1 such that

$$w = \alpha_1 \wedge \ldots \wedge \alpha_k.$$

We note the following useful property of the Euclidean norm: If $\alpha, \beta \in \wedge^\star(R^n)$, then

$$|\alpha \wedge \beta| \leq |\alpha||\beta|,$$

if at least one of the differential forms α, β is simple.

If α ($\deg \alpha = k$, $0 \leq k \leq n$) is a differential from, whose coefficients are in $C^1(\mathcal{M})$, then $d\alpha$, $\deg(d\alpha) = k+1$, denotes its differential defined by

$$d\alpha = \sum_{I \in \wedge(k,n)} d\alpha_I \wedge dx_I.$$

The differentiation is a linear operation for which the following properties hold:

If α and β are arbitrary differential forms that are differentiable in a domain $U \subset \mathcal{M}$, then

(i) $d(\alpha \wedge \beta) = d\alpha \wedge \beta + (-1)^k \alpha \wedge d\beta$,

(ii) $d(d\alpha) = 0$,

where $k = \deg(w)$ is the degree of the differential form α.

The operator $\star$ and the exterior differentiation d define the codiffernetial operator d^* by the formula

$$d^*\alpha = (-1)^k \star^{-1} d \star \alpha$$

for a differential form α of degree k. Clearly, $d^*\alpha$ is a differential form of degree $k-1$.

2. Differential Forms on Riemannian Manifolds

Definition 2.1. A differential form α of degree k on the manifold $\mathcal{M}$ with coefficients $\alpha_{i_1 \cdots i_k} \in L^p_{loc}(\mathcal{M})$ is called weakly closed, if for each differential form β, $\deg\beta = k+1$, with

$$\operatorname{supp} \beta \cap \partial\mathcal{M} = \phi, \ \operatorname{supp} \beta = \overline{\{m \in \mathcal{M} : \beta \neq 0\}} \subset \mathcal{M},$$

and with coefficients in the class $W^{1,q}_{loc}(\mathcal{M})$, $(\frac{1}{p} + \frac{1}{q} = 1,\ 1 \leq p,\ q \leq \infty)$, *we have*

$$\int_{\mathcal{M}} \langle \alpha, d^*\beta \rangle dv = 0. \tag{2.1}$$

We next introduce two classes of differential forms with generalized derivatives. These classes can be used to study the associated classes of (K_1, K_2)-quasiregular mappings.

Definition 2.2. A weakly closed differential form

$$w \in L^p_{loc}(\mathcal{M}),\ \deg w = k,\ 0 \leq k \leq n,\ p > 1 \tag{2.2}$$

is said to be of the class $\mathcal{WT}_1$ on $\mathcal{M}$, if there exists a weakly closed differential form

$$\theta \in L^q_{loc}(\mathcal{M}),\ \deg\theta = n-k,\ \frac{1}{p}+\frac{1}{q}=1, \tag{2.3}$$

such that almost everywhere on $\mathcal{M}$, we have

$$|w|^p \le \nu_1\langle w, *\theta\rangle + \nu_2, \tag{2.4}$$

and

$$\nu_3|\theta| \le |w|^{p-1} \tag{2.5}$$

is satisfied, with constants $\nu_1, \nu_2, \nu_3 > 0$.

Definition 2.3. A simple differential form

$$w = w_1 \wedge \cdots \wedge w_k,\ \ w_i \in L^p_{loc}(\mathcal{M}),\ 1 \le i \le k \tag{2.6}$$

is said to be of the class $\mathcal{WT}_2$ on $\mathcal{M}$, if there exists a weakly closed differential form (2.3) such that almost everywhere on $\mathcal{M}$ the inequality (2.5) holds and

$$\|w\|^{kp} \le \nu_4\langle w, \star\theta\rangle + \nu_5 \tag{2.7}$$

is satisfied, with constants $\nu_4, \nu_5 > 0$.

For an arbitrary simple differential form of degree k

$$w = w_1 \wedge \ldots \wedge w_k,$$

we set

$$\|w\| = (\sum_{i=1}^k |w_i|^2)^{1/2}.$$

For a simple differential form w, we have Hadamard's inequality

$$|w| \le \prod_{i=1}^k |w_i|.$$

Taking these and using the inequality between geometric and arithmetic means

$$(\prod_{i=1}^k |w_i|)^{1/k} \le \frac{1}{k}\sum_{i=1}^k |w_i| \le (\frac{1}{k}\sum_{i=1}^k |w_i|^2)^{1/2},$$

we obtain

$$|w| \le k^{-k/2}\|w\|^k. \tag{2.8}$$

Theorem 2.1. *The following inclusion holds between these $\mathcal{WT}$-classes*

$$\mathcal{WT}_2 \subset \mathcal{WT}_1.$$

Proof. From (2.8), we easily obtain

$$|w|^p \leq kp^{-kp/2}\|w\|^{kp} \leq kp^{-kp/2}\nu_4\langle w, \star\theta\rangle + kp^{-kp/2}\nu_5.$$

3. (K_1, K_2)-Quasiregular Mappings

Let $\mathcal{M}$ and $\mathcal{N}$ be a Riemannian manifolds of dimension n. A mapping $F: \mathcal{M} \to \mathcal{N}$ of the class $W^1_{n,loc}(\mathcal{M})$ is called a (K_1, K_2)-quasirgular mapping, if F satisfies

$$|F^{'}(m)|^n \leq K_1 J_F(m) + K_2, \tag{3.1}$$

almost everywhere on $\mathcal{M}$. Here $F^{'}(m): T_m(\mathcal{M}) \to T_n(\mathcal{N})$ is the formal derivative of $F(m)$, further, $|F^{'}(m)| = \max_{|h|=1} |F^{'}(m)h|$. We denote by $J_F(m)$ the Jacobian of F at the point $m \in M$, i.e., the determinant of $F^{'}(m)$.

If $F: \mathcal{M} \to \mathcal{N}$ is a quasiregular homeomorphism, then the mapping F is called quasiconformal.

Let $\mathcal{A}$ and $\mathcal{B}$ be Riemannian manifolds of dimensions $\dim \mathcal{A} = k$, $\dim \mathcal{B} = n - k\,(1 \leq k \leq n)$, and with scalar products$\langle,\rangle_{\mathcal{A}}, \langle,\rangle_{\mathcal{B}}$, respectively. On the Cartesian product $\mathcal{N} = \mathcal{A} \times \mathcal{B}$, we introduce the natural structure of a Riemannian maniflod with the scalar product

$$\langle,\rangle = \langle,\rangle_{\mathcal{A}} + \langle,\rangle_{\mathcal{B}}.$$

We denote by $\pi: \mathcal{A} \times \mathcal{B} \to \mathcal{A}$ and $\eta: \mathcal{A} \times \mathcal{B} \to \mathcal{B}$ the natural projections of the manifold $\mathcal{N}$ onto submanifolds.

If w_A and w_B are volume forms on $\mathcal{A}$ and $\mathcal{B}$, respectively, the differential form $w_{\mathcal{N}} = \pi^\star w_{\mathcal{A}} \wedge \eta^\star w_{\mathcal{B}}$ is a volume form on $\mathcal{N}$.

Theorem 3.1. *Let $F: \mathcal{M} \to \mathcal{N}$ be a (K_1, K_2)-quasiregular mapping and let $f = \pi \circ F: \mathcal{M} \to \mathcal{A}$. The differential form $f^\star w_{\mathcal{A}}$ is of the class $\mathcal{WT}_1$ on $\mathcal{M}$ with the structure constants $p = n/k$, $\nu_1 = \nu_1(n, k, K_1)$, $\nu_2 = \nu_2(n, k, K_2)$ and $\nu_3 = \nu_3(n, k, K_1, K_2)$.*

Remark 3.1. From the proof of the theorem, it will be clear that structure constants can be chosen to be

$$\nu_1 = (k + \frac{n-k}{\overline{c}^2})^{-n/2} n^{n/2} K_1, \nu_2 = (k + \frac{n-k}{\overline{c}^2})^{-n/2} n^{n/2} K_2, \nu_3 = \underline{c}^{n-k},$$

where $\overline{c} = \overline{c}(k, n, K_1, K_2)$, and $\underline{c} = \underline{c}(k, n, K_1, K_2)$ are, respectively, the greatest and least positive roots of the equation

$$(k\xi^2 + (n-k))^{n/2} - n^{n/2} K_1 \xi^2 - n^{n/2} K_2 |g^\star w_{\mathcal{B}}|^{-n/(n-k)} = 0. \tag{3.2}$$

Proof. Setting $g = \eta \circ F \colon \mathcal{M} \to \mathcal{B}$, we choose $\theta = g^\star w_{\mathcal{B}}$. The volume form $w_{\mathcal{B}}$ is weakly closed. In fact, if the mapping g is sufficiently regular, then

$$d\theta = dg^\star w_{\mathcal{B}} = g^\star dw_{\mathcal{B}} = 0.$$

In the general case for the verification of condition (2.1), we approximate the mapping $g : \mathcal{M} \to \mathcal{B}$ in the norm of W_n^1 by smooth maps $g_l\,(l = 1, 2, \cdots)$. Because condition (2.1) holds for each of the differential form $g_l^\star w_{\mathcal{B}}$, it must hold also for the differential form $g^\star w_{\mathcal{B}}$.

The weak closedness of the differential form $f^\star w_{\mathcal{A}}$ follows similarly.

Fix a point $m \in \mathcal{M}$, in which the relation (3.1) holds. Set $a = f(m), b = g(m)$. Then

$$T_{F(m)}(\mathcal{N}) = T_a(\mathcal{A}) \times T_b(\mathcal{B}).$$

The computations can be conveniently carried out as follows. We first rewrite condition (3.1) in the form

$$|F^{'}(m)|^n \leq K_1 |F^\star w_{\mathcal{N}}| + K_2, \tag{3.3}$$

where $w_{\mathcal{N}}$ is a volume form on $\mathcal{N}$.

For the point $a \in \mathcal{A}$, $b \in \mathcal{B}$, we choose neighborhoods and local systems of coordinates $y^1, \cdots y^k$, and $y^{k+1}, \cdots y^n$, orthinormal at a and b, respectively. We have

$$\begin{aligned} f^\star w_{\mathcal{A}} &= f^\star(dy^1 \wedge \cdots \wedge dy^k) = f^\star dy^1 \wedge \cdots \wedge f^\star dy^k \\ &= df^1 \wedge \cdots \wedge df^k, \; f^i = y^i \circ f, \;\; i = 1, \cdots, k. \end{aligned}$$

Because the differential form $w_{\mathcal{A}}$ is a simple, we obtain by the inequality between the geometric and arithmetic means

$$|df^1 \wedge \cdots \wedge df^k|^{1/k} \leq (\prod_{i=1}^k |df^i|)^{1/k} \leq \frac{1}{k} \sum_{i=1}^k |df^i| \leq (\frac{1}{k} \sum_{i=1}^k |df^i|^2)^{1/2}. \tag{3.4}$$

Similarly

$$|dg^{k+1} \wedge \cdots \wedge dg^n|^{1/(n-k)} \leq (\frac{1}{n-k} \sum_{i=k+1}^n |dg^i|^2)^{1/2}. \tag{3.5}$$

It is not difficult to see that

$$F^\star w_{\mathcal{N}} = F^\star(\pi^\star w_{\mathcal{A}} \wedge \eta^\star w_{\mathcal{B}}) = f^\star w_{\mathcal{A}} \wedge g^\star w_{\mathcal{B}} = f^\star w_{\mathcal{A}} \wedge \theta,$$

and further that

$$|F^\star w_{\mathcal{N}}| = |f^\star w_{\mathcal{A}} \wedge g^\star w_{\mathcal{B}}| \leq |df^1 \wedge \cdots \wedge df^k||dg^{k+1} \wedge \cdots \wedge dg^n|.$$

We have

$$|dF|^2 = \sum_{i=1}^{k} |df^i|^2 + \sum_{i=k+1}^{n} |dg^i|^2 \leq n|F'|^2.$$

Therefore we get from (3.3), (3.4) and (3.5) that

$$\begin{aligned}&(k|f^\star w_{\mathcal{A}}|^{2/k} + (n-k)|g^\star w_{\mathcal{B}}|^{2/(n-k)})^{n/2}\\ &= (\sum_{i=1}^{k} |df^i|^2 + \sum_{i=k+1}^{n} |dg^i|^2)^{n/2}\\ &\leq (n|F'|^2)^{n/2} = n^{n/2}|F'|^n \leq n^{n/2}(K_1|F^\star w_{\mathcal{N}}| + K_2)\\ &= n^{n/2}K_1|F^\star w_{\mathcal{N}}| + n^{n/2}K_2 = n^{n/2}K_1\langle f^\star w_{\mathcal{A}}, \star\theta\rangle + n^{n/2}K_2.\end{aligned}$$

Set

$$\xi = \frac{|f^\star w_{\mathcal{A}}|^{1/k}}{|g^\star w_{\mathcal{B}}|^{1/(n-k)}}.$$

The preceding relations takes the form

$$\begin{aligned}&(k\xi^2 + (n-k))^{n/2}\\ &\leq n^{n/2}K_1\langle f^\star w_{\mathcal{A}}, \star\theta\rangle|g^\star w_{\mathcal{B}}|^{-n/(n-k)} + n^{n/2}K_2|g^\star w_{\mathcal{B}}|^{-n/(n-k)},\end{aligned} \tag{3.6}$$

or takes the form as

$$(k\xi^2 + (n-k))^{n/2} \leq n^{n/2}K_1\xi^k + n^{n/2}K_2|g^\star w_{\mathcal{B}}|^{-n/(n-k)}. \tag{3.7}$$

Using the notations $\underline{c}$ and $\overline{c}$ for the least and greatest positive roots of equation (3.2), we have $\underline{c} \leq \xi \leq \overline{c}$, and

$$\underline{c}\,|g^\star w_{\mathcal{B}}|^{1/(n-k)} \leq |f^\star w_{\mathcal{A}}|^{1/k} \leq \overline{c}\,|g^\star w_{\mathcal{B}}|^{1/(n-k)}. \tag{3.8}$$

From (3.8) it follows that

$$\underline{c}^{n-k}|\theta| \leq |f^\star w_{\mathcal{A}}|^{(n-k)/k}.$$

As above, from (3.6) it follows that

$$|f^\star w_{\mathcal{A}}|^{n/k} \leq (k + \frac{n-k}{\overline{c}^2})^{-n/2}(n^{n/2}K_1\langle f^\star w_{\mathcal{A}}, \star\theta\rangle + n^{n/2}K_2).$$

That is

$$\begin{aligned}|f^\star w_{\mathcal{A}}|^{n/k} \leq\ &(k+\frac{n-k}{\overline{c}^2})^{-n/2}n^{n/2}K_1\langle f^\star w_{\mathcal{A}}, \star\theta\rangle\\ &+(k + \frac{n-k}{\overline{c}^2})^{-n/2}n^{n/2}K_2.\end{aligned} \tag{3.9}$$

Thus condition (2.4) for the membership of the differential form $f^\star w_{\mathcal{A}}$ of degree k in the class $\mathcal{WT}_1$ is indeed satisfied. This ends the proof of Theorem 3.1.

Let $y^1, \cdots y^k$ be an orthonormal system of coordinates in R^k, $1 \leq k \leq n$. Let $\mathcal{A}$ be a domain in R^k and let $\mathcal{B}$ be an $(n-k)$-dimensional Riemannian manifold. We consider the maniflod $\mathcal{N} = \mathcal{A} \times \mathcal{B}$.

Let $F = (f^1, f^2, \cdots, f^k, g) : \mathcal{M} \to \mathcal{N}$ be a mapping of the class $W^p_{n,loc}(\mathcal{M})$ and $g = \eta \circ F$ as defined above. We have $f^\star w_{\mathcal{A}} = df^1 \wedge \cdots \wedge df^k$.

Theorem 3.2. *Let F be a (K_1, K_2)-quasiregular mapping. Then the differential form $f^\star w_{\mathcal{A}}$ is of the class $\mathcal{WT}_2$ on $\mathcal{M}$ with the structure constants $p = n/k, \nu_4 = \nu_4(n, k, K_1, K_2)$ and $\nu_5 = \nu_5(n, k, K_1, K_2)$.*

Remark 3.2. We can choose the constants ν_3, ν_4, ν_5 to be

$$\nu_3 = \underline{c}_1^{n-k},\ \nu_4 = \left(\frac{1}{\overline{c}_1^2}+1\right)^{-n/2} n^{n/2} K_1,\ \nu_5 = \left(\frac{1}{\overline{c}_1^2}+1\right)^{-n/2} n^{n/2} K_2,$$

where $\underline{c}_1$ is the least and $\overline{c}_1$ the greatest positive root of the equation

$$\begin{aligned}(\xi^2+1)^{n/2} - k^{-k/2}(n-k)^{-(n-k)/2} n^{n/2} K_1 \xi^k \\ - n^{n/2} K_2 \Big(\sum_{i=k+1}^{n} |dg^i|^2\Big)^{-n/2} = 0.\end{aligned} \tag{3.10}$$

Proof. In contrast to the previous case, the k-form $f^\star w_{\mathcal{A}}$ has now a global coordinate representation. Because the earlier arguments had local character, they are applicable to the present case,too. As in the previous case we can choose $\theta = g^\star w_{\mathcal{B}}$. Condition (2.5) holds with the same constant. We now proceed to verify condition(2.7).

Combining (3.3), (3.4) and (3.5), we get

$$\begin{aligned}
&\Big(\sum_{i=1}^{k} |df^i|^2 + \sum_{i=k+1}^{n} |dg^i|^2\Big)^{n/2} \leq (n|F'|^2)^{n/2} = n^{n/2}|F'|^n \\
&\leq n^{n/2}(K_1|F^\star w_{\mathcal{N}}| + K_2) = n^{n/2}K_1|F^\star w_{\mathcal{N}}| + n^{n/2}K_2 \\
&\leq n^{n/2}K_1|df^1 \wedge \cdots \wedge df^k||dg^{k+1} \wedge \cdots \wedge dg^n| + n^{n/2}K_2 \\
&\leq n^{n/2}K_1\Big(\frac{1}{k}\sum_{i=1}^{k}|df^i|^2\Big)^{k/2}\Big(\frac{1}{n-k}\sum_{i=k+1}^{n}|dg^i|^2\Big)^{(n-k)/2} + n^{n/2}K_2 \\
&= k^{-k/2}(n-k)^{-(n-k)/2}n^{n/2}K_1\Big(\sum_{i=1}^{k}|df^i|^2\Big)^{k/2}\Big(\sum_{i=k+1}^{n}|dg^i|^2\Big)^{(n-k)/2} + n^{n/2}K_2.
\end{aligned}$$

Hence we have

$$\xi = (\frac{\sum_{i=1}^{k} |df^i|^2}{\sum_{i=k+1}^{n} |dg^i|^2)})^{1/2}.$$

Moreover we can get

$$(\xi^2+1)^{n/2} \leq k^{-k/2}(n-k)^{-(n-k)/2} n^{n/2} K_1 \xi^k + n^{n/2} K_2 (\sum_{i=k+1}^{n} |dg^i|^2)^{-n/2}.$$

If $\underline{c}_1$ and $\overline{c}_1$ are, respectively, the least and greatest of the positive roots of (3.10), then

$$\underline{c}_1 (\sum_{i=k+1}^{n} |dg^i|^2)^{1/2} \leq (\sum_{i=1}^{k} |df^i|^2)^{1/2} \leq \overline{c}_1 (\sum_{i=k+1}^{n} |dg^i|^2)^{1/2}. \quad (3.11)$$

From the relation (3.3), it follows that

$$(\frac{1}{\overline{c}_1^2} + 1)^{n/2} (\sum_{i=1}^{k} |df^i|^2)^{n/2} \leq ((\frac{1}{\xi^2} + 1)(\sum_{i=1}^{k} |df^i|^2))^{n/2}$$
$$= (\sum_{i=1}^{k} |df^i|^2 + \sum_{i=k+1}^{n} |dg^i|^2)^{n/2} \leq n^{n/2} |F^{'}|^n$$
$$\leq n^{n/2}(K_1 |F^\star w_{\mathcal{N}}| + K_2) = n^{n/2}(K_1 \langle f^\star w_{\mathcal{A}}, \star\theta \rangle + n^{n/2} K_2.$$

That is

$$\|f^\star w_{\mathcal{A}}\|^n \leq (\frac{1}{\overline{c}_1^2} + 1)^{-n/2} n^{n/2} K_1 \langle f^\star w_{\mathcal{A}}, \star\theta \rangle + (\frac{1}{\overline{c}_1^2} + 1)^{-n/2} n^{n/2} K_2,$$

which guarantees the truth of (2.7).

References

[1] J. M. Ball, Convexity conditions and existence theorems in nonlinear elasticity, *Arch. Rat. Mech. Anal.* 63 (1977), 337–403.

[2] B. Bojarski and T. Iwaniec, Analytical foundations of the theory of quasiconformal mappings in R^n ,*Ann. Acad. Sci. Fenn. Ser. A. I. Math.* 8 (1983), 257–324.

[3] D. Franke, O. Martio, V. M. Miklyukov, M. Vuorinen and R. Wisk, Quasiregular mappings and $\mathcal{WT}$-classes of differential forms on Riemannian manifolds, *Pacific J. of Math.* 1 (202) (2002), 73–92.

[4] J. Heinonen, T. Kilpeläinen and O. Martio, *Nonlinear potential theory of degenerate elliptic equations,* Clarendon press, Oxford, 1993.

[5] T. Iwaniec and G. Martin, *Geometric function theory and nonlinear analysis,* Clarendon Press, Oxford, 2001.

BELTRAMI SYSTEM WITH THREE CHARACTERISTIC MATRICES IN EVEN DIMENSIONS

CHUN-XIA GAO[a] and HONG LIU[b]

[a,b] *College of Electronic and Information Engineering, Hebei University, Baoding 071002, China*

[a] *E-mail: chunxia-gao@163.com*

[b] *E-mail: liuhongmath@163.com*

This paper deals with the Beltrami system with three characteristic matrices in even dimensions

$$D^t f(x)H(x)Df(x) = J_f^{\frac{2}{n}}(x)G(x) + K(x)D^t f(x)Df(x). \qquad (*)$$

An elliptic equation of divergence type

$$\mathrm{div}A(x, \nabla u) = \mathrm{div}B(x, Df)$$

is derived from $(*)$ under the uniformly elliptic conditions on the matrices $H(x), G(x) \in S(n)$ and $K(x)$ of diagonal and positive.

Keywords: Beltrami system with three characteristic matrices, elliptic system of divergence form, quasiregular mapping.

AMS No: 30C65.

1. Introduction

Let Ω be a bounded domain in $\mathbb{R}^n$, $n \geq 2$. Consider a mapping $f = (f^1, f^2, \cdots, f^n) \in W^{1,n}_{loc}(\Omega, \mathbb{R}^n)$. Denote by $Df(x) = (\frac{\partial f^i}{\partial x_j})_{1 \leq i,j \leq n}$ and $J_f(x) = \det Df(x)$ the Jacobian matrix and the Jacobian determinant of f, respectively. Denote by $D^t f(x)$ and $|Df(x)|$ the transpose and the norm of $Df(x)$, in which $|Df(x)|^2 = tr(D^t f(x)Df(x))$. In this paper, we also need another norm of $Df(x)$, denoted by $|Df(x)|_2$, which ia defined by

$$|Df(x)|_2 = \sup_{|h| \in S^n} |Df(x)h|, \qquad (1.1)$$

where S^n denotes the unit sphere in $\mathbb{R}^n$. The two matrix norms satisfy

$$|Df(x)|_2 \leq |Df(x)| \leq n^{\frac{1}{2}} |Df(x)|_2. \qquad (1.2)$$

In this paper, we always assume that f is orientation-preserving, i.e. $J_f(x) \geq 0$, a.e. Ω.

Definition 1.1. A mapping $f \colon \Omega \to \mathbb{R}^n$ is called K-quasiregular mapping, $1 \leq K < \infty$, if $f(x)$ satisfies

(1) $f \in W^{1,n}_{loc}(\Omega, \mathbb{R}^n)$;

(2) $|Df|^n \leq KJ_f(x)$, *a.e.* $x \in \Omega$.

If f is also homeomorphous, then f is called quasiconformal mapping (see [1]).

Beltrami system with one characteristic matrix

$$D^t f(x)Df(x) = J_f^{\frac{2}{n}}(x)G(x), \tag{1.3}$$

and Beltrami system with two characteristic matrices

$$D^t f(x)H(x)Df(x) = J_f^{\frac{2}{n}}(x)G(x) \tag{1.4}$$

are relevant with quasiregular mappings, in which $G(x), H(x) \in GL(n)$ are $n \times n$ matrices: positive, symmetric with determinant 1, and satisfy the following conditions

$$\alpha_1|\,\xi|^2 \leq \langle G(x)\xi, \xi\rangle \leq \beta_1|\,\xi|^2,\ 0 \leq \alpha_1 \leq \beta_1 < \infty,\ 0 \neq \xi \in \mathbb{R}^n, \tag{1.5}$$

$$\alpha_2|\,\eta|^2 \leq \langle H(x)\eta, \eta\rangle \leq \beta_2|\,\eta|^2,\ 0 \leq \alpha_2 \leq \beta_2 < \infty,\ 0 \neq \eta \in \mathbb{R}^n. \tag{1.6}$$

From calculation, if $f(x) \in W^{1,n}_{loc}(\Omega, \mathbb{R}^n)$ is a generalized solution of (1.4), then $f(x)$ is $(\frac{\beta_1}{\alpha_1})^n$-quasiregular mapping (see [2]).

In complex plane, the study of the property of the solutions of Beltrami systems with one characteristic matrix and two characteristic matrices is very important and we have derived embedded and systematic results. How to generalize the results in two dimensions to high dimensions, the allied sufficient conditions and the regularity of solutions are problems which is the mathematicians are studying all along (see [3–5]). In this paper, we consider the Beltrami system with three characteristic matrices in even dimensions

$$D^t f(x)H(x)Df(x) = J_f^{\frac{2}{n}}(x)G(x) + K(x)D^t f(x)Df(x). \tag{1.7}$$

In the following, we will translate (1.7) into an elliptic equation of divergence form

$$\mathrm{div}A(x, \nabla u) = \mathrm{div}B(x, Df). \tag{1.8}$$

Because divergence form elliptic equation is important for the study of quasiregular mappings, this paper constructs a bridge of (1.7) and quasiregular mappings, such that we study the theory of quasiregular mappings with the way of partial differential equations.

In (1.7), $G(x), K(x) \in GL(n)$, $H(x)$ is positive and diagonal matrix, and satisfy

(i) $\alpha_1|\,\xi|^2 \leq \langle G(x)\xi, \xi\rangle \leq \beta_1|\,\xi|^2,\ 0 \leq \alpha_1 \leq \beta_1 < \infty$;

(ii) $\alpha_2|\,\eta|^2 \leq \langle H(x)\eta, \eta\rangle \leq \beta_2|\,\eta|^2,\ 0 \leq \alpha_2 \leq \beta_2 < \infty$;

(iii) $\alpha_3|\,\langle \xi, \zeta\rangle| \leq \langle K(x)\xi, \zeta\rangle \leq \beta_3|\,\langle \xi, \zeta\rangle|\ ,\ 0 \leq \alpha_3 \leq \beta_3 < \infty$,

where $0 \neq \xi, \eta, \zeta \in \mathbb{R}^{\mathbf{n}}$.

From direct calculation, if $f(x) \in W^{1,n}_{loc}(\Omega, \mathbb{R}^n)$ is a generalized solution of (1.7) which is satisfy (i), (ii) and (iii), then $f(x)$ is a $\frac{n^{\frac{1}{2}} 2^{\frac{n-2}{2}} \beta_1^{\frac{n}{2}}}{\alpha_2^{\frac{n}{2}} - 2^{\frac{n-2}{2}} \beta_3^{\frac{n}{2}}}$-quasiregular mapping.

Definition 1.2. If for every testing function $\phi \in W^{1,n}_0(\Omega, \mathbb{R}^n)$, which has compact supports, we have

$$\int_\Omega \langle A(x, \nabla u), \nabla \phi \rangle dx = \int_\Omega \langle B(x, Df), \nabla \phi \rangle dx, \tag{1.9}$$

then $u = f^l$ is the weak solution of the elliptic equation (1.8).

The main result is the following theorem.

Theorem 1.1. *Assume that* $f(x)\, (\in W^{1,n}_{loc}(\Omega, \mathbb{R}^n), n = 2k, k = 1, 2, \ldots)$ *is a generalized solution of* (1.7), *which is satisfy* (i), (ii) *and* (iii). *Then* $u = f^l\, (l = 1, 2, ...n)$ *are weak solutions of* (1.8)

$$\text{div} A(x, \nabla u) = \text{div} B(x, Df),$$

in which

$$\begin{aligned} A(x, \nabla u) &= \left(\frac{\langle G^{-1}(x)\nabla u, \nabla u\rangle}{H^{ll}(x)}\right)^{\frac{n-2}{2}} G^{-1}(x)\nabla u, \\ B(x, Df) &= B_1(x, Df) + B_2(x, \nabla u) + B_3(x, \nabla u), \\ B_1(x, Df) &= (H^{ll}(x) - H^{ll}(x_0)) J_f(x) D^{-1} f(x) e^l, \\ B_2(x, \nabla u) &= H^{ll}(x) J_f^{\frac{n-2}{n}}(x) G^{-1}(x) K(x) \nabla u(x), \\ B_3(x, \nabla u) &= \sum_{p=1}^{k-1} (-1)^{p+1} C^p_{k-1} \left(\frac{\langle G^{-1}(x)\nabla u, \nabla u\rangle}{H^{ll}(x)}\right)^{k-1-p} \\ &\quad \times \langle G^{-1}(x) K(x) \nabla u, \nabla u\rangle^p G^{-1}(x) \nabla u, \end{aligned}$$

where $H^{ll}(x)$ *denotes the* l*th element of main diagonal line in* $H^{-1}(x)$, $l = 1, 2, \cdots, n$.

In the proof of the theorem 1.1, we need the following lemma, see [10].

Lemma 1.1. *For quasiregular mapping* $f(x) \in W^{1,n}_{loc}(\Omega, R^n)$ *and arbitrary constant vector* $a \in \mathbb{R}^n$, *in distribution meaning, we have*

$$\text{div}\{J_f(x) D^{-1} f(x) a\} = 0. \tag{1.10}$$

2. Proof of Theorem 1.1

Proof. From the Beltrami system with three characteristic matrices (1.7), we have

$$J_f(x)D^{-1}f(x)=J_f^{\frac{n-2}{n}}(x)G^{-1}(x)D^tf(x)H(x)-J_f^{\frac{n-2}{n}}(x)G^{-1}(x)K(x)D^tf(x). \tag{2.1}$$

From (1.12), for arbitrary constant vector $a\in\mathbb{R}^n$, we have

$$\mathrm{div}\{J_f^{\frac{n-2}{n}}(x)G^{-1}(x)D^tf(x)H(x)\cdot a-J_f^{\frac{n-2}{n}}(x)G^{-1}(x)K(x)D^tf(x)\cdot a\}=0. \tag{2.2}$$

Taking $x_0\in U$, assume $\{e^1,e^2,\cdots,e^n\}$ is a set of standard orthogonal basis in $\mathbb{R}^n$. Let $a=H^{-1}(x_0)e^l$, $l=1,2,\cdots,n$. from(2.2), we have

$$\begin{aligned}\mathrm{div}\{J_f^{\frac{n-2}{n}}(x)G^{-1}(x)[\nabla f^l(x)+D^tf(x)(H(x)H^{-1}(x_0)-Id)e^l]\\ -J_f^{\frac{n-2}{n}}(x)G^{-1}(x)K(x)D^tf(x)H^{-1}(x_0)e^l\}=0,\end{aligned} \tag{2.3}$$

From (1.7), we also get

$$J_f^{\frac{2}{n}}(x)H^{-1}(x)=Df(x)G^{-1}(x)D^tf(x)-Df(x)G^{-1}(x)K(x)D^tf(x)H^{-1}(x), \tag{2.4}$$

Consider the lth element of diagonal line in (2.4), we get

$$\begin{aligned}\langle J_f^{\frac{2}{n}}(x)H^{-1}(x)e^l,e^l\rangle&=\langle Df(x)G^{-1}(x)D^tf(x)e^l,e^l\rangle\\ &\langle -Df(x)G^{-1}(x)K(x)D^tf(x)H^{-1}(x)e^l,e^l\rangle,\end{aligned}$$

namely,

$$\begin{aligned}J_f^{\frac{2}{n}}(x)H^{ll}(x)&=\langle G^{-1}(x)D^tf(x)e^l,D^tf(x)e^l\rangle\\ &-H^{ll}(x)\langle G^{-1}(x)K(x)D^tf(x)e^l,D^tf(x)e^l\rangle\\ &=\langle G^{-1}(x)\nabla f^l,\nabla f^l\rangle-H^{ll}(x)\langle G^{-1}(x)K(x)\nabla f^l,\nabla f^l\rangle.\end{aligned} \tag{2.5}$$

Since $H(x)$ satisfies (ii), we have

$$\frac{1}{\beta_2}|\ \eta|^2\le\langle H^{-1}(x)\eta,\eta\rangle\le\frac{1}{\alpha_2}|\ \eta|^2,\ 0\ne\eta\in R^n.$$

Taking $\eta=e^l=(0,\cdots,0,\overbrace{1}^{lth},0,\cdots,0)^t$, then

$$\frac{1}{\beta_2}\le H^{ll}(x)\le\frac{1}{\alpha_2}, \tag{2.6}$$

and

$$J_f^{\frac{2}{n}}(x) = \frac{\langle G^{-1}(x)\nabla f^l, \nabla f^l\rangle}{H^{ll}(x)} - \langle G^{-1}(x)K(x)\nabla f^l, \nabla f^l\rangle. \tag{2.7}$$

From (2.1), (2.7) and (2.3), it follows

$$\begin{aligned}&\operatorname{div}\{(\frac{\langle G^{-1}(x)\nabla f^l, \nabla f^l\rangle}{H^{ll}(x)} - \langle G^{-1}(x)K(x)\nabla f^l, \nabla f^l\rangle)^{\frac{n-2}{2}} G^{-1}(x)\nabla f^l \\ &+(J_f(x)D^{-1}f(x)+J_f^{\frac{n-2}{n}}(x)G^{-1}(x)K(x)D^t f(x))(H^{-1}(x_0)-H^{-1}(x))e^l \\ &-J_f^{\frac{n-2}{n}}(x)G^{-1}(x)K(x)D^t f(x)H^{-1}(x_0)e^l\} = 0,\end{aligned}$$

namely,

$$\begin{aligned}&\operatorname{div}\{(\frac{\langle G^{-1}(x)\nabla f^l, \nabla f^l\rangle}{H^{ll}(x)} - \langle G^{-1}(x)K(x)\nabla f^l, \nabla f^l\rangle)^{\frac{n-2}{2}} G^{-1}(x)\nabla f^l \\ &+J_f(x)D^{-1}f(x)(H^{-1}(x_0) - H^{-1}(x))e^l \\ &-J_f^{\frac{n-2}{n}}(x)G^{-1}(x)K(x)D^t f(x)H^{-1}(x)e^l\} = 0.\end{aligned} \tag{2.8}$$

Taking $n = 2k, k = 1, 2, \cdots$, we have

$$\begin{aligned}&\operatorname{div}(\frac{\langle G^{-1}(x)\nabla f^l, \nabla f^l\rangle}{H^{ll}(x)})^{\frac{n-2}{2}} G^{-1}(x)\nabla f^l \\ &= \operatorname{div}(H^{ll}(x) - H^{ll}(x_0))J_f(x)D^{-1}f(x)e^l \\ &+ \operatorname{div}H^{ll}(x)J_f^{\frac{n-2}{n}}(x)G^{-1}(x)K(x)\nabla f^l vse \\ &+\operatorname{div}[\sum_{p=1}^{k-1}(-1)^{p+1}C_{k-1}^p(\frac{\langle G^{-1}(x)\nabla f^l, \nabla f^l\rangle}{H^{ll}(x)})^{k-1-p} \\ &\times\langle G^{-1}(x)K(x)\nabla f^l, \nabla f^l\rangle^p G^{-1}(x)\nabla f^l].\end{aligned} \tag{2.9}$$

Let $u = f^l$, and

$$\begin{aligned}A(x, \nabla u) &= (\frac{\langle G^{-1}(x)\nabla u, \nabla u\rangle}{H^{ll}(x)})^{\frac{n-2}{2}} G^{-1}(x)\nabla u, \\ B(x, Df) &= B_1(x, Df) + B_2(x, \nabla u) + B_3(x, \nabla u), \\ B_1(x, Df) &= (H^{ll}(x) - H^{ll}(x_0))J_f(x)D^{-1}f(x)e^l, \\ B_2(x, \nabla u) &= H^{ll}(x)J_f^{\frac{n-2}{n}}(x)G^{-1}(x)K(x)\nabla u(x), \\ B_3(x, \nabla u) &= \sum_{p=1}^{k-1}(-1)^{p+1}C_{k-1}^p(\frac{\langle G^{-1}(x)\nabla u, \nabla u\rangle}{H^{ll}(x)})^{k-1-p} \\ &\quad\times\langle G^{-1}(x)K(x)\nabla u, \nabla u\rangle^p G^{-1}(x)\nabla u,\end{aligned}$$

we complete the proof of Theorem 1.1.

3. The Properties of Operators A and B

Firstly, from (i) and (2.6), it is easy to get

$$\frac{1}{\beta_1}(\frac{\alpha_2}{\beta_1})^{\frac{n-2}{2}}|\xi|^n \le \langle A(x,\xi),\xi\rangle \le \frac{1}{\alpha_1}(\frac{\beta_2}{\alpha_1})^{\frac{n-2}{2}}|\xi|^n, \forall\xi\in R^n, \xi\neq 0. \quad (3.1)$$

Then, we prove

(1) Lipschitz condition

$$|A(x,h_1)-A(x,h_2)| \le c_1|h_1-h_2|(|h_1|+|h_2|)^{n-2},$$

where $c_1 = c_1(\alpha_1,\alpha_2,\alpha_3,\beta_1,\beta_2,\beta_3,n)$.

Proof. Since $G^{-1}(x)\in GL(n)$, there exist orthogonal matrix O_1 and diagonal matrix Γ_1, such that $G^{-1} = O_1\Gamma_1^2O_1^t = P_1^tP_1$, where $P_1 = (O_1\Gamma_1)^t$. So

$$A(x,h_i) = H^{ll}(x)^{\frac{2-n}{2}}|P_1h_i|^{n-2}P_1^tP_1h_i = H^{ll}(x)^{\frac{2-n}{2}}P_1^t|g_i|^{n-2}g_i,$$

where $g_i = P_1h_i,\ i=1,2$. Furthermore,

$$|A(x,h_1)-A(x,h_2)| = |H^{ll}(x)^{\frac{2-n}{2}}P_1^t(|g_1|^{n-2}g_1-|g_2|^{n-2}g_2)|. \quad (3.2)$$

In the following, we prove that

$$||g_1|^{n-2}g_1-|g_2|^{n-2}g_2| \le (n-1)|g_1-g_2|(|g_1|+|g_2|)^{n-2}. \quad (3.3)$$

In fact, with triangle inequality, we have

$$\begin{aligned}||g_1|^{n-2}g_1-|g_2|^{n-2}g_2| &\le |g_1|^{n-2}\cdot|g_1-g_2|+||g_1|^{n-2}-|g_2|^{n-2}|\cdot|g_2|\\ &\le (|g_1|+|g_2|)^{n-2}|g_1-g_2|+||g_1|^{n-2}-|g_2|^{n-2}||g_2|.\end{aligned} \quad (3.4)$$

From the symmetrical characteristic of g_1 and g_2, we have

$$||g_1|^{n-2}g_1-|g_2|^{n-2}g_2| \le (|g_1|+|g_2|)^{n-2}|g_1-g_2|+||g_1|^{n-2}-|g_2|^{n-2}|\cdot|g_1|. \quad (3.5)$$

Let $|g_1|\le|g_2|$. Then

$$\begin{aligned}||g_1|^{n-2}-|g_2|^{n-2}|\cdot|g_1| &\le (n-2)|g_2|^{n-3}||g_1|-|g_2||\cdot|g_1|\\ &\le (n-2)|g_2|^{n-3}|g_1-g_2|(|g_1|+|g_2|)\\ &\le (n-2)(|g_1|+|g_2|)^{n-2}|g_1-g_2|.\end{aligned} \quad (3.6)$$

From (3.5) and (3.6), we have (3.3). From (3.2) and (3.3), we have

$$|A(x,h_1)-A(x,h_2)| \le (n-1)|H^{ll}(x)^{\frac{2-n}{2}}|\cdot|P_1^t|\cdot|g_1-g_2|(|g_1|+|g_2|)^{n-2}. \quad (3.7)$$

From $G^{-1} = P_1^t P_1$ and $\frac{1}{\beta_1}|\xi|^2 \le \langle G^{-1}(x)\xi,\xi\rangle \le \frac{1}{\alpha_1}|\xi|^2$, we have

$$\frac{1}{\beta_1} \le |P_1|^2 \le \frac{1}{\alpha_1}. \quad (3.8)$$

From (3.7) and (3.8), we get the Lipschitz condition.

(2) Monotonous inequality

$$\langle A(x,h_1) - A(x,h_2), h_1 - h_2\rangle \ge c_2|h_1 - h_2|^2(|h_1| + |h_2|)^{n-2}. \quad (3.9)$$

Proof.

$$\begin{aligned}
&\langle A(x,h_1) - A(x,h_2), h_1 - h_2\rangle \\
&= \langle H^{ll}(x)^{\frac{2-n}{2}} P_1^t(|g_1|^{n-2}g_1 - |g_2|^{n-2}g_2),\ h_1 - h_2\rangle \\
&= \langle H^{ll}(x)^{\frac{2-n}{2}} (|g_1|^{n-2}g_1 - |g_2|^{n-2}g_2),\ g_1 - g_2\rangle \\
&= H^{ll}(x)^{\frac{2-n}{2}} \{\tfrac{1}{2}[|g_1 - g_2|^2(|g_1|^{n-2} + |g_2|^{n-2}) \\
&\quad +(|g_1|^2 - |g_2|^2)(|g_1|^{n-2} - |g_2|^{n-2})]\} \\
&\ge H^{ll}(x)^{\frac{2-n}{2}} \{\tfrac{1}{2}|g_1 - g_2|^2(|g_1|^{n-2} + |g_2|^{n-2})\} \\
&= H^{ll}(x)^{\frac{2-n}{2}} \{\tfrac{1}{2}\langle G^{-1}(x)(h_1 - h_2), h_1 - h_2\rangle \\
&\quad \times[\langle G^{-1}(x)h_1, h_1\rangle^{\frac{n-2}{2}} + \langle G^{-1}(x)h_2, h_2\rangle^{\frac{n-2}{2}}]\} \\
&\le C|h_1 - h_2|^2(|h_1|^{n-2} + |h_2|^{n-2}) \\
&\ge C_2|h_1 - h_2|^2(|h_1| + |h_2|)^{n-2}.
\end{aligned}$$

(3) Homogeneity condition

$$A(x,\lambda\xi) = |\lambda|^{n-2}\lambda A(x,\xi),\ \lambda \in \mathbb{R}. \quad (3.10)$$

Proof. It is easy to get from the definition of $A(x,\xi)$.

Finally, we give the condition which B is satisfied. From (1.7), we have

$$\begin{aligned}
J_f^{-\frac{2}{n}}(x)H(x) &= (D^{-1}f(x))^t G(x) D^{-1}f(x) \\
&\quad + J_f^{-\frac{2}{n}}(x)(D^{-1}f(x))^t K(x) D^t f(x),
\end{aligned}$$

then, for $\forall\xi \in \mathbb{R}^n$,

$$\begin{aligned}J_f^{-\frac{2}{n}}(x)\langle H(x)\xi,\xi\rangle &= \langle (D^{-1}f(x))^t G(x) D^{-1}f(x)\xi,\xi\rangle \\ &\quad + J_f^{-\frac{2}{n}}(x)\langle ((D^{-1}f(x))^t)^t K(x) D^t f(x)\xi,\xi\rangle \\ &= \langle G(x)D^{-1}f(x)\xi, D^{-1}f(x)\xi\rangle \\ &\quad + J_f^{-\frac{2}{n}}(x)\langle K(x)D^t f(x)\xi, D^{-1}f(x)\xi\rangle.\end{aligned}$$

So

$$\begin{aligned}&\langle G(x)D^{-1}f(x)\xi, D^{-1}f(x)\xi\rangle \\ &= J_f^{-\frac{2}{n}}(x)(\langle H(x)\xi,\xi\rangle - \langle K(x)D^t f(x)\xi, D^{-1}f(x)\xi\rangle).\end{aligned} \tag{3.11}$$

Consider (i), (ii) and (iii), we have

$$\begin{aligned}\alpha_1|D^{-1}f(x)\xi|^2 \le \langle G(x)D^{-1}f(x)\xi, D^{-1}f(x)\xi\rangle \le \beta_1|D^{-1}f(x)\xi|^2,\\ \alpha_2|\xi|^2 \le \langle H(x)\xi,\xi\rangle \le \beta_2|\xi|^2,\\ \alpha_3|\xi|^2 = \alpha_3|\langle D^t f(x)\xi, D^{-1}f(x)\xi\rangle| \le \langle K(x)D^t f(x)\xi, D^{-1}f(x)\xi\rangle\\ \le \beta_3|\langle D^t f(x)\xi, D^{-1}f(x)\xi\rangle| = \beta_3|\xi|^2.\end{aligned}$$

so

$$\begin{aligned}|D^{-1}f(x)\xi|^2 &\le \frac{1}{\alpha_1}J_f^{-\frac{2}{n}}(x)(\langle H(x)\xi,\xi\rangle - \langle K(x)D^t f(x)\xi, D^{-1}f(x)\xi\rangle)\\ &\le \frac{1}{\alpha_1}J_f^{-\frac{2}{n}}(x)(\beta_2-\alpha_3)|\xi|^2.\end{aligned}$$

Then,

$$|D^{-1}f(x)|^2 \le \frac{1}{\alpha_1}(\beta_2+\alpha_3)J_f^{-\frac{2}{n}}(x). \tag{3.12}$$

With (2.6),(2.7),(i) and (ii), we have

$$\begin{aligned}J_f^{\frac{2}{n}}(x) &= \frac{\langle G^{-1}(x)\nabla f^l, \nabla f^l\rangle}{H^{ll}(x)} - \langle G^{-1}(x)K(x)\nabla f^l, \nabla f^l\rangle\\ &= q\frac{\beta_2}{\alpha_1}|\nabla f^l|^2 - \langle K(x)\nabla f^l, (G^{-1}(x))^t\nabla f^l\rangle\\ &\le \frac{\beta_2}{\alpha_1}|\nabla f^l|^2 - \alpha_3|\langle G^{-1}(x)\nabla f^l, \nabla f^l\rangle|\\ &\le \frac{\beta_2}{\alpha_1}|\nabla f^l|^2 - \frac{\alpha_3}{\alpha_1}|\nabla f^l|^2 = (\frac{\beta_2-\alpha_3}{\alpha_1})|\nabla f^l|^2,\end{aligned} \tag{3.13}$$

so

$$\begin{aligned} |J_f(x)D^{-1}f(x)| &\le J_f(x)(\frac{\beta_2-\alpha_3}{\alpha_1})^{\frac{1}{2}}J_f^{-\frac{1}{n}}(x) \\ &= (\frac{\beta_2-\alpha_3}{\alpha_1})^{\frac{1}{2}}J_f^{\frac{n-1}{n}}(x) \le (\frac{\beta_2-\alpha_3}{\alpha_1})^{\frac{n}{2}}|\nabla f^l|^{n-1}, \end{aligned} \tag{3.14}$$

then

$$\begin{aligned} |B_1(x,Df)| &= |(H^{ll}(x)-H^{ll}(x_0))J_f(x)D^{-1}f(x)e^l| \\ &\le \frac{2}{\alpha_2}|J_f(x)D^{-1}f(x)| \le \frac{2}{\alpha_2}(\frac{\beta_2-\alpha_3}{\alpha_1})^{\frac{n}{2}}|\nabla f^l|^{n-1}, \end{aligned} \tag{3.15}$$

and

$$\begin{aligned} |B_2(x,\nabla f^l)| &= |H^{ll}(x)J_f^{\frac{n-2}{n}}(x)G^{-1}(x)K(x)\nabla f^l| \\ &\le \frac{1}{\alpha_2}J_f^{\frac{n-2}{n}}(x)|G^{-1}(x)K(x)\nabla f^l| \\ &\le \frac{1}{\alpha_2}(\frac{\beta_2-\alpha_3}{\alpha_1})^{\frac{n-2}{2}}|G^{-1}(x)|\cdot|K(x)|\cdot|\nabla f^l|^{n-1}. \end{aligned} \tag{3.16}$$

Since $K(x)\in GL(n)$, there exist orthogonal matrix O_2 and diagonal matrix Γ_2, such that $K=O_2\Gamma_2^2O_2^t=P_2^tP_2$, where $P_2=(O_2\Gamma_2)^t$, then

$$|\langle K(x)\xi,\xi\rangle| = \langle P_2^tP_2\xi,\xi\rangle = |P_2\xi|^2 \le \beta_3|\xi|^2,$$
$$|P_2|^2 = \sup_{|\xi|=1}|P_2\xi|^2 \le \beta_3.$$

Furthermore,

$$|K(x)| = |P_2^tP_2| \le |P_2|^2 \le \beta_3. \tag{3.17}$$

Similarly, we have

$$|G^{-1}(x)| \le \frac{1}{\alpha_1}. \tag{3.18}$$

Then

$$|B_2(x,\nabla f^l)| \le \frac{\beta_3}{\alpha_1\alpha_2}(\frac{\beta_2-\alpha_3}{\alpha_1})^{\frac{n-2}{2}}|\nabla f^l|^{n-1}. \tag{3.19}$$

When $n=2k,\ k=1,2,\cdots$, we have

$$\begin{aligned} |B_3(x,\nabla f^l)| &= |\sum_{p=1}^{k-1}(-1)^{p+1}C_{k-1}^p(\frac{\langle G^{-1}(x)\nabla f^l,\nabla f^l\rangle}{H^{ll}(x)})^{k-1-p} \\ &\quad\times\langle G^{-1}(x)K(x)\nabla f^l,\nabla f^l\rangle^pG^{-1}(x)\nabla f^l| \\ &\le \sum_{p=1}^{k-1}C_{k-1}^p|(\frac{\langle G^{-1}(x)\nabla f^l,\nabla f^l\rangle}{H^{ll}(x)})^{k-1-p}| \\ &\quad\times|\langle G^{-1}(x)K(x)\nabla f^l,\nabla f^l\rangle^p||G^{-1}(x)||\nabla f^l| \end{aligned}$$

$$\begin{aligned}
&\le \sum_{p=1}^{k-1} C_{k-1}^p (\frac{\beta_2}{\alpha_1}|\nabla f^l|^2)^{k-1-p} (\frac{\beta_3}{\alpha_1}|\nabla f^l|^2)^p \frac{1}{\alpha_1}|\nabla f^l| \\
&= \sum_{p=1}^{k-1} C_{k-1}^p (\frac{1}{\alpha_1})^k \beta_2^{k-1-p}\beta_3^p |\nabla f^l|^{2k-1} \\
&= \sum_{p=1}^{k-1} C_{k-1}^p (\frac{1}{\alpha_1})^k \beta_2^{k-1-p}\beta_3^p |\nabla f^l|^{n-1} \\
&= (\frac{1}{\alpha_1})^k \sum_{p=1}^{k-1} C_{k-1}^p \beta_2^{k-1-p}\beta_3^p |\nabla f^l|^{n-1}.
\end{aligned} \tag{3.20}$$

Then

$$\begin{aligned}
|B(x,Df)| &\le |B_1(x,Df)| + |B_2(x,\nabla f^l)| + |B_3(x,\nabla f^l)| \\
&\le (c_1+c_2+c_3)|\nabla f^l|^{n-1},
\end{aligned}$$

where

$$\begin{aligned}
c_1 &= \frac{2}{\alpha_2}(\frac{\beta_2-\alpha_3}{\alpha_1})^{\frac{n}{2}}, \\
c_2 &= \frac{\beta_3}{\alpha_1\alpha_2}(\frac{\beta_2-\alpha_3}{\alpha_1})^{\frac{n-2}{2}}, \\
c_3 &= (\frac{1}{\alpha_1})^k \sum_{p=1}^{k-1} C_{k-1}^p \beta_2^{k-1-p}\beta_3^p.
\end{aligned}$$

References

[1] T. Iwaniec aqnd G. Martin, Quasiregular mappings in even dimensions, *Acta Math.,* 170 (1993), 29-81.

[2] Hongya Gao, On weakly quasiregular mappings in high dimensions, *Acta Math. Sinica,* 45 (2002), 11:29-35.

[3] T. Iwaniec, p-Harmonic tensor and quasiregular mappings, *Ann. Math.,* 136 (1992), 651-685.

[4] Jinfa Cheng and Ainong Fang, Generalized Beltrami systems in even dimension domains and Beltrami systems in high-dimensional spaces, *Chinese Ann. of Math.*, 18A (1997), 789-798 (in Chinese).

[5] Jinfa Cheng and Ainong Fang, (G,H)-quasiregular mappings and B-harmonic equations, *Acta Math. Sinica,* 42 (1999), 883-888.

LOCAL BOUNDEDNESS AND LOCAL REGULARITY RESULTS IN DOUBLE OBSTACLE PROBLEMS[1]

YU-XIA TONG[a], JIAN-TAO GU[b] and XIU-LI HOU[c]

[a,b] *College of Science, Hebei United University, Tangshan 063009, China*

[a] *E-mail: tongyuxia@126.com*

[b] *E-mail: gujiantaolg@126.com*

[c] *Qianan College, Hebei United University, Hebei Qianan 064000, China*

E-mail: houxiuli000@163.com

This paper mainly concerns double obstacle problems for second order divergence type elliptic equation $\mathrm{div}A(x, u, \nabla u) = 0$. Firstly, we give local boundedness of solutions for double obstacle problems, then by using the similarly method, the local regularity of solutions for the above problems is proved.

Keywords: Double obstacle problems, local boundedness, local regularity.

AMS No: 35J60, 35B35.

1. Introduction

Let Ω be a bounded open set of $\mathbb{R}^n$ $(n \geq 2)$. We consider the second order divergence type elliptic equation (also called A-harmonic equation or Leray-Lions equation)

$$\mathrm{div}A(x, u(x), \nabla u(x)) = 0, \tag{1.1}$$

in which $A\colon \Omega \times \mathbb{R} \times \mathbb{R}^n \to \mathbb{R}^n$ is a Carathéodory function satisfying the coercivity and growth conditions: for almost all $x \in \Omega$, all $u \in \mathbb{R}$, and $\xi \in \mathbb{R}^n$,

(i) $\langle A(x, u, \xi), \xi\rangle \geq \alpha|\xi|^p$,

(ii) $|A(x, u, \xi)| \leq \beta_1|\xi|^{p-1} + \beta_2|u|^m + \varphi_1(x)$,

where $\alpha > 0$, β_1 and β_2 are some nonnegative constants, $1 < p < n$, $p - 1 \leq m \leq \frac{n(p-1)}{n-p}$ and $\varphi_1(x) \in L^{s/(p-1)}_{loc}(\Omega)$ for some $s > p$.

Suppose that ψ_1, ψ_2 are any functions in Ω with values in $\mathbb{R} \cup \{\pm\infty\}$, and that $\theta \in W^{1,p}(\Omega)$. Let

$$K^{\theta,p}_{\psi_1,\psi_2}(\Omega) = \{v \in W^{1,p}(\Omega) : \psi_1 \leq v \leq \psi_2,\ a.e. \text{ and } v - \theta \in W^{1,p}_0(\Omega)\}.$$

The function ψ_1, ψ_2 are two obstacles and θ determines the boundary values.

[1]The research is supported by Natural Science Foundation of Hebei Province (A2010000910) and Tangshan Science and Technology projects (09130206c).

Definition 1.1. A solution to the $K^{\theta,p}_{\psi_1,\psi_2}(\Omega)$-double obstacle problem for the A-harmonic equation (1.1) is a function $u \in K^{\theta,p}_{\psi_1,\psi_2}(\Omega)$, such that

$$\int_{\Omega} \langle A(x,u,\nabla u), \nabla(v-u)\rangle dx \geq 0, \tag{1.2}$$

whenever $v \in K^{\theta,p}_{\psi_1,\psi_2}(\Omega)$.

Remark 1.1. If there only one obstacle function ψ, we say it's the (single) obstacle problem. That is,

$$K_{\psi,\theta} = \{v \in W^{1,p}(\Omega) : v \geq \psi,\ a.e. \text{ and } v - \theta \in W_0^{1,p}(\Omega)\}.$$

For the details to see [2-3].

The obstacle problem has a strong background, and has many applications in physics and engineering. The local boundedness for solutions of obstacle problems plays a central role in many aspects. Based on the local boundedness, we can further study the regularity of the solutions. In [1], Gao Hongya et al. first considered the local boundedness for very weak solutions of obstacle problems to the A-harmonic equation in 2010. Precisely, the authors considered the local boundedness for very weak solutions of $K_{\psi,\theta}(\Omega)$-obstacle problems to the A-harmonic equation div $A(x, \nabla u(x)) = 0$ with the condition $\psi \geq 0$, where operator A satisfies conditions $\langle A(x,\xi),\xi\rangle \geq \alpha|\xi|^p$ and $|A(x,\xi)| \leq \beta|\xi|^{p-1}$ with $A(x,0) = 0$. For the local boundedness results of weak solutions of nonlinear elliptic equations, we refer the reader to [4].

In this article, we continue to consider the local boundedness property. Under some general conditions (i) and (ii) given above on the operator A, we obtain a local boundedness result for solutions of $K^{\theta,p}_{\psi_1,\psi_2}$-double obstacle problems to the A-harmonic equation (1.1).

Theorem A. *Let operator A satisfy conditions* (i) *and* (ii). *Suppose that $\psi_1 \in W^{1,\infty}_{loc}(\Omega)$. Then a solution u to the $K^{\theta,p}_{\psi_1,\psi_2}(\Omega)$-obstacle problem of* (1.1) *is locally bounded.*

Remark 1.2. Since we have assumed that operator A satisfies the growth condition (ii), in the proof of the theorem, we have to estimate the integral of some power of $|u|$ by means of $|\nabla u|$. To deal with this difficulty, we will make use of the Sobolev inequality that was used in [8].

Remark 1.3. Notice that we have restricted ourselves to the case $1 < p < n$ in the above theorem. In [1], Gao Hongya et al. consider the case $\max\{1, p-1\} < r \leq p$.

The second aim of this paper considers local regularity for solutions of $K^{\theta,p}_{\psi_1,\psi_2}(\Omega)$-double obstacle problems, here we will show that the condition $\psi_1 \geq 0$ in [8] is not necessary.

Theorem B. *Let operator A satisfies conditions* (i) *and* (ii). *Suppose that* $\psi_1, \psi_2 \in W^{1,s}_{loc}(\Omega)\,(1 < p < s < n)$. *Then, a solution u to the $K^{\theta,p}_{\psi_1,\psi_2}(\Omega)$-obstacle problem of* (1.1) *is belongs to* $L^{s^*}_{loc}(\Omega)\,(s^* = 1/(1/s - 1/n))$.

2. Preliminary Knowledge and Lemmas

We give some symbols and preliminary lemmas used in the proof. If $x_0 \in \Omega$ and $t > 0$, then B_t denotes the ball of radius t centered at x_0. For a function $u(x)$ and $k > 0$, let $A_k = \{x \in \Omega : |u(x)| > k\}$, $A_k^+ = \{x \in \Omega : u(x) > k\}$, $A_{k,t} = A_k \cap B_t$, $A^+_{k,t} = A^+_k \cap B_t$. Moreover, if $s < n$, s^* is always the real number satisfying $1/s^* = 1/s - 1/n$. Denote $t_k(u) = \min\{u, k\}$. Let $T_k(u)$ be the usual truncation of u at level $k > 0$, that is,

$$T_k(u) = \max\{-k, \min\{k, u\}\}. \tag{2.1}$$

Lemma 2.1[5]. *Let $u \in W^{1,r}_{loc}(\Omega)$, $\varphi_0 \in L^q_{loc}(\Omega)$, where $1 < r < n$ and q satisfies $1 < q < n/r$. Assume that the following integral estimate holds:*

$$\int_{A_{k,t}} |\nabla u|^r dx \le c_0\left[\int_{A_{k,t}} \varphi_0 dx + (t-\tau)^{-\alpha}\int_{A_{k,t}} |u|^r dx\right], \tag{2.2}$$

for every $k \in N$ and $R_0 \le \tau < t \le R_1$, where c_0 is a real positive constant that depends only on $N, q, r, R_0, R_1, |\Omega|$ and, α is a real positive constant. Then $u \in L^{(qr)^*}_{loc}(\Omega)$.

Lemma 2.2[6]. *Let $f(\tau)$ be a nonnegative bounded function defined for $0 \le R_0 \le t \le R_1$. Suppose that for $R_0 \le \tau < t \le R_1$, one has*

$$f(\tau) \le A(t-\tau)^{-\alpha} + B + \theta f(t),$$

where A, B, α, θ are nonnegative constants and $\theta < 1$. Then there exists a constant $c_2 = c_2(\alpha, \theta)$, depending only on α and θ, such that for every $\rho, R, R_0 \le \rho < R \le R_1$, one has

$$f(\rho) \le c_2[A(t-\tau)^{-\alpha} + B].$$

Definition 2.1[7]. A function $u \in W^{1,m}_{loc}(\Omega)$ belongs to the class $\mathbf{B}(\Omega, \gamma, m, k_0)$, if for all $k > k_0$, $k_0 > 0$ and all $B_\rho = B_\rho(x_0)$, $B_{\rho-\rho\sigma} = B_{\rho-\rho\sigma}(x_0)$, $B_R = B_R(x_0)$, one has

$$\int_{A^+_{k,\rho-\rho\sigma}} |\nabla u|^m dx \le \gamma\left\{\sigma^{-m}\rho^{-m}\int_{A^+_{k,\rho}} (u-k)^m dx + \left|A^+_{k,\rho}\right|\right\},$$

for $R/2 \le \rho - \rho\sigma < \rho < R$, $m < n$, where $|A^+_{k,\rho}|$ is the n-dimensional Lebesgue measure of the set $A^+_{k,\rho}$.

Lemma 2.3[7]. *Suppose that $u(x)$ is an arbitrary function belonging to the class $\boldsymbol{B}(\Omega, \gamma, m, k_0)$ and $B_R \subset\subset \Omega$. Then one has*

$$\max_{B_{R/2}} u(x) \leq c,$$

in which the constant c is determined only by $\gamma, m, k_0, R, \|\nabla u\|_m$.

3. Proof of Theorem A

Proof. Let u be a solution to the $K^{\theta,p}_{\psi_1,\psi_2}(\Omega)$-obstacle problem for the A-harmonic equation (1.1), and $B_{R_1} \subset\subset \Omega$ and $0 < R_1/2 \leq \tau < t \leq R_1$ be arbitrarily fixed. Fix a cutoff function $\phi \in C_0^\infty(B_{R_1})$, such that

$$\operatorname{supp} \phi \subset B_t, \ 0 \leq \phi \leq 1, \ \phi \equiv 1 \text{ in } B_\tau, \ |\nabla \phi| \leq 2(t-\tau)^{-1}. \tag{3.1}$$

Consider the function

$$v = u - \phi^p(u - \psi_k), \tag{3.2}$$

where $\psi_k = \min\{\max\{\psi_1, t_k(u)\}, \psi_2\}$, $t_k(u) = \min\{u, k\}$, $k \geq 0$. It is easy to see $\psi_1 \leq \psi_k \leq \psi_2$. Now, $v \in K^{\theta,p}_{\psi_1,\psi_2}(\Omega)$; indeed, since $u \in K^{\theta,p}_{\psi_1,\psi_2}(\Omega)$ and $\phi \in C_0^\infty(\Omega)$, then

$$v - \theta = u - \theta - \phi^p(u - \psi_k) \in W_0^{1,p}(\Omega), \tag{3.3}$$

$$v - \psi_1 = u - \psi_1 - \phi^p(u - \psi_k) \geq (1 - \phi^p)(u - \psi_1) \geq 0 \text{ a.e. in } \Omega, \tag{3.4}$$

$$v - \psi_2 = u - \psi_2 - \phi^p(u - \psi_k) \leq (1 - \phi^p)(u - \psi_2) \leq 0 \text{ a.e. in } \Omega. \tag{3.5}$$

For any fixed $k > 0$, let

$$v_0 = \begin{cases} u, \text{ if } u \leq k, \\ v, \text{ if } u > k. \end{cases}$$

It is easy to see that $v_0 \in K^{\theta,p}_{\psi_1,\psi_2}(\Omega)$. By Definition 1.1, we have

$$\begin{aligned} 0 &\leq \int_\Omega \langle A(x,u,\nabla u), \nabla(v_0 - u)\rangle dx \\ &= \Big(\int_{\Omega\cap\{u\leq k\}} + \int_{\Omega\cap\{u>k\}}\Big)\langle A(x,u,\nabla u), \nabla(v_0 - u)\rangle dx \\ &= \int_{\Omega\cap\{u>k\}} \langle A(x,u,\nabla u), \nabla(v_0 - u)\rangle dx \\ &= \int_{A^+_{k,t}} \langle A(x,u,\nabla u), \nabla(v - u)\rangle dx. \end{aligned} \tag{3.6}$$

That is

$$\int_{A_{k,t}^+} \langle A(x,u,\nabla u), \phi^p(\nabla u - \nabla\psi_k) + p\phi^{p-1}\nabla\phi(u-\psi_k)\rangle dx \le 0. \tag{3.7}$$

This implies

$$\begin{aligned}\int_{A_{k,t}^+} \langle A(x,u,\nabla u), \phi^p\nabla u\rangle dx &\le \int_{A_{k,t}^+} \langle A(x,u,\nabla u), \phi^p\nabla\psi_k\rangle dx \\ &\quad + \int_{A_{k,t}^+} \langle A(x,u,\nabla u), p\phi^{p-1}\nabla\phi(\psi_k - u)\rangle dx \\ &= I_1 + I_2.\end{aligned} \tag{3.8}$$

We now estimate the left-hand side and the right-hand side of (3.8), respectively. First,

$$\begin{aligned}\int_{A_{k,t}^+} \langle A(x,u,\nabla u), \phi^p\nabla u\rangle dx &\ge \int_{A_{k,\tau}^+} \langle A(x,u,\nabla u), \nabla u\rangle dx \\ &\ge \alpha\int_{A_{k,\tau}^+} |\nabla u|^p dx,\end{aligned} \tag{3.9}$$

here we have used condition (i). Secondly, $\psi_k = \max\{\psi_1, k\}$ in $A_{k,t}^+$, $|\nabla\psi_k| \le |\nabla\psi_1|$, by condition (ii),

$$\begin{aligned}|I_1| &= \left|\int_{A_{k,t}^+} \langle A(x,u,\nabla u), \phi^p\nabla\psi_k\rangle dx\right| \\ &\le \int_{A_{k,t}^+} [\beta_1|\nabla u|^{p-1} + \beta_2|u|^m + \varphi_1]|\nabla\psi_1| dx \\ &= I_{11} + I_{12} + I_{13}.\end{aligned} \tag{3.10}$$

By Young's inequality $ab \le \varepsilon a^{p'} + C(\varepsilon,p)b^p$ valid for $a, b \ge 0, \varepsilon > 0$ and $p > 1$, we have the estimates

$$|I_{11}| \le \beta_1\left[\varepsilon\int_{A_{k,t}^+} |\nabla u|^p dx + C(\varepsilon,p)\int_{A_{k,t}^+} |\nabla\psi_1|^p dx\right], \tag{3.11}$$

$$|I_{12}| \le \beta_2\left[\varepsilon\int_{A_{k,t}^+} |u|^{mp'} dx + C(\varepsilon,p)\int_{A_{k,t}^+} |\nabla\psi_1|^p dx\right]. \tag{3.12}$$

We observe now that, if $w \in W^{1,p}(B_t)$ and $|\operatorname{supp} w| \le 1/2|B_t|$, then we have the Sobolev inequality (see also [8])

$$\left(\int_{B_t} |w|^{p^*} dx\right)^{p/p^*} \le c_1(n,p)\int_{B_t} |\nabla w|^p dx. \tag{3.13}$$

Set

$$g_k(u) = \begin{cases} u, & \text{if } u \le k, \\ 0, & \text{if } u > k. \end{cases}$$

Since $p-1 \le m \le \frac{n(p-1)}{n-p}$ by assumption, then $p \le mp' \le p^*$. (3.13) implies

$$\begin{aligned} &\int_{A_{k,t}^+} |u|^{mp'} dx = \int_{B_t} |u - g_k(u)|^{mp'} dx \\ &\le \|u - g_k(u)\|_{p^*}^{mp'-p} |B_t|^{1-mp'/p^*} \left(\int_{B_t} |u - g_k(u)|^{p*} dx \right)^{p/p^*} \\ &\le c_1(n,p) \|u - g_k(u)\|_{p^*}^{mp'-p} |B_t|^{1-mp'/p^*} \int_{B_t} |\nabla(u - g_k(u))|^p dx \\ &= c_1(n,p) \|u - g_k(u)\|_{p^*}^{mp'-p} |B_t|^{1-mp'/p^*} \int_{A_{k,t}^+} |\nabla u|^p dx, \end{aligned} \tag{3.14}$$

provided that $|\text{supp}(u - g_k(u))|_{B_t}| \le 1/2|B_t|$. Since $\text{supp}(u - g_k(u))|_{B_t} \subset A_{k,t}^+$, then $|\text{supp}(u - g_k(u))|_{B_t}| \le |A_{k,t}^+|$. On the other hand, we have

$$\|u\|_{p^*,B_t}^{p^*} = \int_{B_t} u^{p^*} dx \ge \int_{A_{k,t}^+} |u|^{p^*} dx \ge k^{p^*} |A_{k,t}^+|.$$

Thus, there exists a constant $k_0 > 0$, such that for all $k \ge k_0$, we have $|A_{k,t}^+| \le 1/2|B_t|$. We can also suppose that k_0 such that

$$\int_{A_{k_0,t}} u^{p^*} dx \le 1.$$

For such values of k we then have inequality

$$\begin{aligned} \int_{A_{k,t}^+} |u|^{mp'} dx &\le c_1(n,p) \|u - g_k(u)\|_{p^*}^{mp'-p} |B_t|^{1-mp'/p^*} \int_{A_{k,t}^+} |\nabla u|^p dx \\ &\le c_1(n,p) \|u - g_k(u)\|_{p^*}^{mp'-p} |\Omega|^{1-mp'/p^*} \int_{A_{k,t}^+} |\nabla u|^p dx \\ &\le c_1(n,p) |\Omega|^{1-mp'/p^*} \int_{A_{k,t}^+} |\nabla u|^p dx \\ &= C \int_{A_{k,t}^+} |\nabla u|^p dx, \end{aligned} \tag{3.15}$$

where $C = C(n, m, p, k_0, |\Omega|)$.

We derive from (3.12) and (3.15) that

$$|I_{12}| \le \beta_2 C\varepsilon \int_{A^+_{k,t}} |\nabla u|^p dx + \beta_2 C(\varepsilon,p) \int_{A^+_{k,t}} |\nabla \psi_1|^p dx, \tag{3.16}$$

I_{13} can be estimated as follows:

$$|I_{13}| \le \varepsilon \int_{A^+_{k,t}} |\varphi_1|^{p'} dx + C(\varepsilon,p) \int_{A^+_{k,t}} |\nabla \psi_1|^p dx. \tag{3.17}$$

In conclusion, we derive from (3.10), (3.11), (3.16), and (3.17) that

$$\begin{aligned} I_1 \le (\beta_1+\beta_2 C)\varepsilon \int_{A^+_{k,t}} |\nabla u|^p dx + \varepsilon \int_{A^+_{k,t}} |\varphi_1|^{p'} dx \\ +(\beta_1+\beta_2+1)C(\varepsilon,p) \int_{A^+_{k,t}} |\nabla \psi_1|^p dx. \end{aligned} \tag{3.18}$$

We now estimate $|I_2|$. By condition (ii) and $|u-\psi_k| \le |u-k|$ a.e. in $A^+_{k,t}$,

$$\begin{aligned} |I_2| &= \left| \int_{A^+_{k,t}} \langle A(x,u,\nabla u), p\phi^{p-1}\nabla\phi(\psi_k - u)\rangle dx \right| \\ &\le p\int_{A^+_{k,t}} [\beta_1|\nabla u|^{p-1} + \beta_2|u|^m + \varphi_1]|\nabla\phi||u-\psi_k|dx \\ &\le \frac{2p\beta_1}{t-\tau}\int_{A^+_{k,t}} |\nabla u|^{p-1}|u-k|dx + \frac{2p\beta_2}{t-\tau}\int_{A_{k,t}} |u|^m|u-k|dx \\ &\quad + \frac{2p}{t-\tau}\int_{A^+_{k,t}} |\varphi_1||u-k|dx \\ &= I_{21}+I_{22}+I_{23}, \end{aligned} \tag{3.19}$$

I_{21} and I_{23} can be estimated as

$$I_{21} \le 2p\beta_1\Big[\varepsilon\int_{A^+_{k,t}} |\nabla u|^p dx + \frac{C(\varepsilon,p)}{(t-\tau)^p}\int_{A^+_{k,t}} |u-k|^p dx\Big], \tag{3.20}$$

$$I_{23} \le 2p\varepsilon \int_{A^+_{k,t}} |\varphi_1|^{p'} dx + \frac{2pC(\varepsilon,p)}{(t-\tau)^p}\int_{A^+_{k,t}} |u-k|^p dx. \tag{3.21}$$

By (3.15), we know that if $k \ge k_0$, then

$$\begin{aligned} I_{22} &\le 2p\beta_2\varepsilon \int_{A^+_{k,t}} |u|^{mp'} dx + \frac{2p\beta_2 C(\varepsilon,p)}{(t-\tau)^p}\int_{A^+_{k,t}} |u-k|^p dx \\ &\le 2p\beta_2\varepsilon C\int_{A^+_{k,t}} |\nabla u|^p dx + \frac{2p\beta_2 C(\varepsilon,p)}{(t-\tau)^p}\int_{A^+_{k,t}} |u-k|^p dx. \end{aligned} \tag{3.22}$$

Combining (3.19) with (3.20), (3.21), and (3.22), we obtain

$$\begin{aligned} I_2 \le 2p\varepsilon(\beta_1+\beta_2 C)\int_{A^+_{k,t}} |\nabla u|^p dx + 2p\varepsilon \int_{A^+_{k,t}} |\varphi_1|^{p'} dx \\ + \frac{2p(\beta_1+\beta_2+1)C(\varepsilon,p)}{(t-\tau)^p}\int_{A^+_{k,t}} |u-k|^p dx. \end{aligned} \tag{3.23}$$

Thus, the inequalities (3.8), (3.9), (3.18), and (3.23) imply that

$$\begin{aligned} \int_{A^+_{k,\tau}} |\nabla u|^p dx \le & \frac{(2p\beta_1+2p\beta_2 C+\beta_1+\beta_2 C)\varepsilon}{\alpha}\int_{A^+_{k,t}} |\nabla u|^p dx \\ & + \frac{(2p+1)\varepsilon}{\alpha}\int_{A^+_{k,t}} |\varphi_1|^{p'} dx \\ & + \frac{(\beta_1+\beta_2+1)C(\varepsilon,p)}{\alpha}\int_{A^+_{k,t}} |\nabla\psi_1|^p dx \\ & + \frac{2p(\beta_1+\beta_2+1)C(\varepsilon,p)}{\alpha(t-\tau)^p}\int_{A^+_{k,t}} |u-k|^p dx. \end{aligned} \tag{3.24}$$

Choosing ε small enough such that, the summation θ of the coefficients of the first term in the right-handside of (3.24) is smaller than 1. Let ρ, R be arbitrarily fixed with $R_1/2 \le \rho < R \le R_1$. Thus, from (3.24), we deduce that for every t and τ, such that $R_1/2 \le \tau < t \le R_1$, we have

$$\begin{aligned} \int_{A^+_{k,\tau}} |\nabla u|^p dx \le & \frac{c_3}{\alpha}\int_{A^+_{k,R}} \left[|\nabla\psi_1|^p + |\varphi_1|^{p'}\right] dx \\ & + \frac{c_4}{\alpha(t-\tau)^p}\int_{A^+_{k,R}} |u-k|^p dx + \theta\int_{A^+_{k,t}} |\nabla u|^p dx, \end{aligned} \tag{3.25}$$

in which c_3, c_4 are some constant depending only on α, β_1, β_2, and p. Applying Lemma 2.2, we conclude that

$$\begin{aligned} \int_{A^+_{k,\rho}} |\nabla u|^p dx \le & \frac{c_2 c_4}{\alpha(R-\rho)^p}\int_{A^+_{k,R}} |u-k|^p dx \\ & + \frac{c_2 c_3}{\alpha}\int_{A^+_{k,R}} \left[|\nabla\psi_1|^p + |\varphi_1|^{p'}\right] dx \\ \le & \frac{c_2 c_4}{\alpha(R-\rho)^p}\int_{A^+_{k,R}} |u-k|^p dx + \frac{c_2 c_3 c_5}{\alpha}|A^+_{k,R}|, \end{aligned} \tag{3.26}$$

where c_2 is the constant given by Lemma 2.2 and $c_5 = \|\nabla\psi_1\|^p_{L^\infty(\Omega)} + \|\varphi_1\|^{p'}_{L^{p'}(\Omega)}$. Thus u belongs to the class **B** with $\gamma = \max\{c_2c_4/\alpha, c_2c_3c_5/\alpha\}$

and $m = p$. Lemma 2.3 yields

$$\max_{B_{R/2}} u(x) \le c.$$

This result together with the assumptions $\psi_1 \le u \le \psi_2$ and $\psi_1 \in W^{1,\infty}_{loc}(\Omega)$ yields the desired result.

4. Proof of Theorem B

Proof. Let u be a solution to the $K^{\theta,p}_{\psi_1,\psi_2}(\Omega)$-obstacle problem for the A-harmonic equation (1.1). Owing to Lemma 2.1, it is sufficient to prove that u satisfies the integral estimate (2.2) with $\alpha = p$. Let $B_{R_1} \subset\subset \Omega$ and $0 < R_0 \le \tau < t \le R_1$ be arbitrarily fixed. Fix a cutoff function $\phi \in C_0^\infty(B_{R_1})$, such that

$$\text{supp}\phi \subset B_t,\ 0 \le \phi \le 1, \phi \equiv 1 \text{ in } B_\tau, |\nabla\phi| \le 2(t-\tau)^{-1}. \tag{4.1}$$

Consider the function

$$v = u - \phi^p(u - \tilde{\psi}_k), \tag{4.2}$$

where $\tilde{\psi}_k = \min\{\max\{\psi_1, T_k(u)\}, \psi_2\}$, $T_k(u)$ is the usual truncation of u at level $k \ge 0$ defined in (2.1). It is easy to see $\psi_1 \le \tilde{\psi}_k \le \psi_2$. Now, $v \in K^{\theta,p}_{\psi_1,\psi_2}(\Omega)$. Similarly as in the proof of Theorem A, let

$$v_0 = \begin{cases} u, \text{ if } |u| \le k, \\ v, \text{ if } |u| > k, \end{cases}$$

and to show that the condition $\psi_1 \ge 0$ in [8] is not necessary, we can using $|u - \tilde{\psi}_k| \le |u|$ a.e. in $A_{k,t}$, and obtain

$$\begin{aligned} \int_{A_{k,\tau}} |\nabla u|^p dx \le{}& \frac{(2p\beta_1 + 2p\beta_2 C + \beta_1 + \beta_2 C)\varepsilon}{\alpha} \int_{A_{k,t}} |\nabla u|^p dx \\ &+ \frac{(2p+1)\varepsilon}{\alpha} \int_{A_{k,t}} |\varphi_1|^{p'} dx \\ &+ \frac{(\beta_1 + \beta_2 + 1)C(\varepsilon, p)}{\alpha} \int_{A_{k,t}} |\nabla\tilde{\psi}_k|^p dx \\ &+ \frac{2p(\beta_1 + \beta_2 + 1)C(\varepsilon, p)}{\alpha(t-\tau)^p} \int_{A_{k,t}} |u|^p dx. \end{aligned} \tag{4.3}$$

Choosing ε small enough such that, the coefficient of the first term in the right-handside of (3.24) is smaller than 1. Let ρ, R be arbitrarily fixed with

$R_0 \le \rho < R \le R_1$. Thus, from (3.24), we deduce that for every t and τ, such that $\rho \le \tau < t \le R_1$, and get

$$\int_{A_{k,\tau}} |\nabla u|^p dx \le \frac{c_5}{\alpha} \int_{A_{k,R}} [|\nabla \psi_1|^p + |\nabla \psi_2|^p + |\varphi_1|^{p'}] dx + \frac{c_6}{\alpha(t-\tau)^p} \int_{A_{k,R}} |u|^p dx + \theta \int_{A_{k,t}} |\nabla u|^p dx, \tag{4.4}$$

where c_5, c_6 are some constant depending only on α, β_1, β_2, and p. Applying Lemma 2.2, we conclude that

$$\int_{A_{k,\rho}} |\nabla u|^p dx \le \frac{c_2 c_6}{\alpha(R-\rho)^p} \int_{A_{k,R}} |u|^p dx + \frac{c_2 c_5}{\alpha} \int_{A_{k,R}} [|\nabla \psi_1|^p + |\nabla \psi_2|^p + |\varphi_1|^{p'}] dx, \tag{4.5}$$

where c_2 is the constant given by Lemma 2.2. Thus, u satisfies the inequality (2.2) with $\phi_0 = |\nabla \psi_1|^p + |\nabla \psi_2|^p + |\varphi_1|^{p'}$ and $\alpha = p$. The theorem follows from Lemma 2.1.

References

[1] Hongya Gao, Jinjing Qiao and Yuming Chu, Local regularity and local boundedness results for very weak solutions of obstacle problems, *J. Inequal. Appl.*, 2010, Art. ID 878769, 12 pp.

[2] Gongbao Li and O. Martio, Local and global integrability of gradients in obstacle problems, *AAnn Acad Sci Fenn, Ser A I Math.*, 19 (1994), 25-34.

[3] Hongya Gao and Huiying Tian, Local regularity result for solutions of obstacle problems, *Acta Mathematica Scientia B,* 24 (2004), 1:71ÍC74.

[4] A. Bensoussan and J. Frehse, *Regularity results for nonlinear elliptic systems and applications,* vol.151 of Applied Mathematical Sciences, Springer, Berlin, 2002.

[5] D. Giachetti and M. M. Porzio, Local regularity results for minima of functionals of the calculus of variation, *Nonlinear Analysis: Theory, Methods & Applications*, 39 (2000), 4:463-482.

[6] M. Giaquinta, *Multiple integrals in the calculus of variations and nonlinear elliptic systems,* Annals of Mathematics Studies, Vol.105, Princeton University Press, Princeton, 1983.

[7] M. C. Hong, Some remarks on the minimizers of variational integrals with nonstandard growth conditions, *Bollettino dell'Unione Matematica Italiana*, 6 (1992), 1:91-101.

[8] Hongya Gao, Jing Guo et al, Local regularity result in obstacle problems, *Acta Mathematica Scientia B,* 30 (2010), 1:208-214.

[9] Gongbao Li and O. Martio, Stability and higher integrability of derivatives of solutions in double obstacle problems, *J. Math. Anal. Appl.*, 272 (2002), 19-29.

MULTIPLICITY OF SOLUTIONS FOR QUASILINEAR ELLIPTIC EQUATIONS WITH SINGULARITY[1]

JUAN LI

Department of Mathematics, Ningbo University, Ningbo 315211, China

E-mail: juanjuan-li@163.com

In this paper, we study the following quasilinear problem:

$$-\triangle_p u - \mu\frac{|u|^{p-2}u}{|x|^p} = \alpha\frac{u^{p^*(t)-2}}{|x|^t}u + \lambda|u|^{p-2}u + |u|^{q-2}u,\ x \in W_0^{1,p}(\Omega)$$

with Dirichlet boundary condition, where $N \geq 3$, $1 < p < N$, $0 \leq \mu < \bar{\mu} \equiv (\frac{N-p}{p})^p$, $0 \leq t < p$, $1 < q < p$, and $p^*(t) \equiv p(N-t)/(N-p)$ is the critical Sobolev-Hardy exponent. Via the variational method, we get the the multiplicity of solutions for the quasilinear problem.

Keywords: Quasilinear elliptic equations, singularity, critical growth.

AMS No: 35J60.

1. Introduction and Main Results.

In this paper, we consider the quasilinear elliptic equation

$$\begin{cases} -\triangle_p u - \mu\dfrac{|u|^{p-2}u}{|x|^p} = \alpha\dfrac{u^{p^*(t)-2}}{|x|^t}u + \lambda|u|^{p-2}u + |u|^{q-2}u, & x \in \Omega, \\ u(x) = 0, & x \in \partial\Omega, \end{cases} \tag{1.1}$$

where $\triangle_p u = \text{div}(|\nabla u|^{p-2}\nabla u)$ is the p-Laplacian of u and $\Omega \subset \Re^N (N \geq 3)$ is a smooth bounded domain containing the origin 0, and assume that $1 < p < N, 0 \leq \mu < \bar{\mu} \equiv (\frac{N-p}{p})^p$, α, λ are the real positive constant, $0 \leq t < p$, $p < q < p^*(t) \equiv p(N-t)/(N-p)$, $p^*(t)$ is the critical Sobolev-Hardy exponent, especially $p^*(0) = p^* \equiv pN/(N-p)$ is the critical Sobolev exponent.

The first result for nonlinear critical problems have been obtained in a celebrated paper by Brezis and Nirenberg[1]. This pioneering work has stimulated a vast amount of research on the class of problems, see [1–6]. It should be mentioned that the following quasilinear problems

$$\begin{cases} -\triangle_p u = |u|^{p^*-2}u + \lambda|u|^{p-2}u + \beta|u|^{q-2}u, & x \in \Omega, \\ u(x) = 0, & x \in \partial\Omega, \end{cases}$$

had been studied in [6], and existence results about k pairs of nontrivial solutions were obtained. On the other hand, in recent years, people have

[1] This research is supported by Ningbao Scientific Research Foundation (2009B21003) and K. C. Wong Magna Fund in Ningbo University

paid much attention to the existence of nontrivial solutions for the singular problems, see [7–12]. It should be mentioned that, for $\alpha, \beta > 0$, $p \le r \le p^*$ and $p \le q \le p^*(s)$, the following quasilinear problems

$$\begin{cases} -\mathrm{div}(|\nabla u|^{p-2}\nabla u) = \alpha|u|^{r-2}u + \beta\dfrac{|u|^{q-2}}{|x|^s}u, & x \in \Omega, \\ u(x) = 0, & x \in \partial\Omega, \end{cases}$$

had been studied in [8] by Ghoussoub and Yuan, and existence results about positive solutions and sign-changing solutions were obtained. In 2008, Dongsheng Kang in [7] had studied the quasilinear elliptic equation

$$\begin{cases} -\mathrm{div}(|\nabla u|^{p-2}\nabla u) - \mu\dfrac{|u|^{p-2}u}{|x|^p} = \dfrac{u^{p^*(s)-2}}{|x|^s}u + \lambda\dfrac{|u|^{q-2}}{|x|^t}u, & x \in \Omega, \\ u(x) = 0, & x \in \partial\Omega, \end{cases}$$

where $p \le q < p^*(t)$. Via variational methods, the authors establish the existence of one positive solution. Thus, it is natural for us to consider the multiplicity of solutions for the problem (1.1).

In the case $\mu = 0$, problem (1.1) is related to the well know Sobolev-Hardy inequalities, which is essentially due to Cafferelli, Kohn and Nirenberg (see [13]),

$$\left(\int_{\Re^N}\frac{|u|^q}{|x|^s}dx\right)^{p/q} \le C_{q,s,p}\int_{\Re^N}|\nabla u|^p dx, \ \forall u \in W_0^{1,p}(\Omega), \tag{1.2}$$

where $p \le q \le p^*$. As $q = s = p$, the above Sobolev inequality becomes the well known Hardy inequality (see [8,12,13]),

$$\int_\Omega \frac{|u|^p}{|x|^p}dx \le \frac{1}{\bar{\mu}}\int_\Omega |\nabla u|^p dx, \ \forall u \in W_0^{1,p}(\Omega). \tag{1.3}$$

In $W_0^{1,p}(\Omega)$, for $\mu \in [0, \bar{\mu}]$, we use the norm

$$\|u\| = \|u\|_{W_0^{1,p}(\Omega)} = \left(\int_\Omega\left(|\nabla u|^p - \mu\frac{|u|^p}{|x|^p}\right)dx\right)^{1/p},$$

by (1.3), this norm is equivalent to the usual norm $\left(\int_\Omega |\nabla u|^p dx\right)^{1/p}$.

By the Hardy inequality and the Sobolev-Hardy inequality, for $0 \le \mu < \bar{\mu}$, $0 \le t < p$, we can define the best Sobolev-Hardy constant:

$$A_{\mu,t,r}(\Omega) = \inf_{u \in W_0^{1,p}(\Omega)\setminus\{0\}} \frac{\int_\Omega\left(|\nabla u|^p - \mu\frac{|u|^p}{|x|^p}\right)dx}{\left(\int_\Omega \frac{|u|^r}{|x|^t}dx\right)^{p/r}}.$$

In the important case where $r = p^*(t)$, we shall simply denote $A_{\mu,t,p^*(t)}$ as $A_{\mu,t}$. Note that $A_{\mu,0}$ is the best constant in the Sobolev inequality, i.e.,

$$A_{\mu,0}(\Omega) = \inf_{u \in W_0^{1,p}(\Omega)\setminus\{0\}} \frac{\int_\Omega \left(|\nabla u|^p - \mu \frac{|u|^p}{|x|^p}\right) dx}{\left(\int_\Omega |u|^{p^*} dx\right)^{p/p^*}}.$$

Note that $A_{\mu,0,p}(\Omega)$ is nothing but the first eigenvalue of the positive operator L in $W_0^{1,p}(\Omega)$:

$$\lambda_1(\Omega) = \inf_{u \in W_0^{1,p}(\Omega)\setminus\{0\}} \frac{\int_\Omega \left(|\nabla u|^p - \mu \frac{|u|^p}{|x|^p}\right) dx}{\int_\Omega |u|^p dx}. \tag{1.4}$$

The energy functional corresponding to problem (1.1) is defined as follows,

$$\begin{aligned} I_\mu(u) = & \frac{1}{p}\int_\Omega \left(|\nabla u|^p - \mu \frac{|u|^p}{|x|^p}\right) dx - \frac{\alpha}{p^*(t)}\int_\Omega \frac{|u|^{p^*(t)}}{|x|^s} dx \\ & - \frac{\lambda}{p}\int_\Omega |u|^p dx - \frac{1}{q}\int_\Omega |u|^q dx, \end{aligned} \tag{1.5}$$

then $I_\mu(u)$ is well defined on $W_0^{1,p}(\Omega)$ and belongs to $C^1(W_0^{1,p}(\Omega), R)$. The solutions of problem (1.1) are precisely the critical points of the functional I_μ.

Now we may state our result.

Theorem 1.1. *Suppose $p < q < p^*$. Then, given $k \in \mathbf{N}$, there exists $\alpha_k \in (0, \infty]$ such that* (1.1) *possesses at least k pairs of nontrivial solutions for all $\alpha \in (0, \alpha_k)$.*

2. The Palais-Smale Condition

We first need to recall the following Lemma.

Lemma 2.1[7]**.** *Assume that $0 \le t \le p$, $p \le q \le p^*(t)$, and $0 \le \mu < \bar{\mu}$. Then* (1) *There exists a constant $C > 0$ such that*

$$\left(\int_\Omega \frac{|u|^q}{|x|^t} dx\right)^{1/q} \le C\|u\|, \ \forall\, u \in W_0^{1,p}(\Omega).$$

(2) *The map $u \mapsto u/|x|^{t/q}$ from $W_0^{1,p}(\Omega)$ into $L^q(\Omega)$ is compact if $p \le q < p^*(t)$.*

We recall that given E a real Banach space and $I \in C^1(E, \Re)$. We say that I satisfies the Palase-Smale condition on the level $c \in \Re$, denoted by $(PS)_c$, if every sequence $(u_n) \subset E$ such that $I(u_n) \longrightarrow c$ and $I'(u_n) \longrightarrow 0$, as $n \longrightarrow \infty$, possesses a convergent subsequence. In this paper, we shall take $I = I_\mu$ and $E = W_0^{1,p}(\Omega)$.

Lemma 2.2. *Given $M > 0$, there exists $\alpha_* > 0$ such that I_μ satisfies the $(PS)_c$ condition for all $c < M$, provided $0 < \alpha \leq \alpha_*$, where $\alpha_* = \left(\frac{p-t}{pM(N-t)}\right)^{\frac{p-t}{N-p}} A_{\mu,t}^{\frac{N-t}{N-p}}$.*

Proof. Given $c < M$, let $(u_m) \subset W_0^{1,p}(\Omega)$ be such that (i) $I_\mu(u_m) \longrightarrow c$, and (ii) $I'_\mu(u_m) \longrightarrow 0$ in $W^{-1,p'}(\Omega)$, as $m \longrightarrow \infty$. First, we claim that (u_m) is bounded in $W_0^{1,p}(\Omega)$. Indeed, by (i) and (ii), for n sufficiently large,

$$\begin{aligned} pc + 1 + o(1) &\geq pI_\mu(u_m) - \langle I'_\mu(u_m), u_m\rangle \\ = \alpha\left(1 - \frac{p}{p^*(t)}\right)&\int_\Omega \frac{|u_m|^{p^*(t)}}{|x|^t}dx + (1 - \frac{p}{q})\int_\Omega |u_m|^q dx. \end{aligned} \tag{2.1}$$

Also, Invoking (1.4),

$$\begin{aligned} \|u_m\|^p &= pI_\mu(u_m) + \frac{\alpha p}{p^*(t)}\int_\Omega \frac{|u_m|^{p^*(t)}}{|x|^t}dx + \lambda\int_\Omega |u_m|^p dx + \frac{p}{q}\int_\Omega |u_m|^q dx \\ &\leq pI_\mu(u_m) + \frac{\alpha p}{p^*(t)}\int_\Omega \frac{|u_m|^{p^*(t)}}{|x|^t}dx + \frac{p}{q}\int_\Omega |u_m|^q dx \\ &\quad + \frac{\lambda}{\lambda_1(\Omega)}\int_\Omega \left(|\nabla u|^p - \mu\frac{|u|^p}{|x|^p}\right)dx. \end{aligned} \tag{2.2}$$

Furthermore, by (2.1) and (2.2), we obtain

$$\begin{aligned} \left(1 - \frac{\lambda}{\lambda_1(\Omega)}\right)\|u_m\|^p &\leq pI_\mu(u_m) + \frac{\alpha p}{p^*(t)}\int_\Omega \frac{|u_m|^{p^*(t)}}{|x|^t}dx + \frac{p}{q}\int_\Omega |u_m|^q dx \\ &\leq C + o(1). \end{aligned} \tag{2.3}$$

This prove the claim. Hence, without loss of generality, we may assume that there is $u \in W_0^{1,p}(\Omega)$ such that (u_m) satisfies as $n \longrightarrow \infty$

$$\begin{cases} u_m \rightharpoonup u \text{ weakly in } W_0^{1,p}(\Omega), \\ u_m \rightharpoonup u \text{ weakly in } L^{p^*(t)}(\Omega, |x|^{-t}), \\ u_m \rightharpoonup u \text{ weakly in } L^p(\Omega, |x|^{-p}), \\ u_m \to u \text{ weakly in } L^q(\Omega), \\ u_m \to u \text{ weakly in } L^p(\Omega), \\ u_m \to u \text{ a.e. in } \Omega. \end{cases} \tag{2.4}$$

By the concentration compactness principle (see [14,15]), there exists a subsequence, still denoted by $\{u_m\}$, at most countable set $\jmath$, a set of different points $\{x_j\}_{j\in\jmath} \subset \Omega\setminus\{0\}$, sets of nonnegative real numbers $\{\tilde{\mu_j}\}_{j\in\cup\{0\}}$, $\{\tilde{\nu_j}\}_{j\in\jmath\cup\{0\}}$, and nonnegative real numbers $\tilde{\tau_0}$ and $\tilde{\gamma_0}$ such that

$$|\bigtriangledown u_m|^p \rightharpoonup d\tilde{\mu} \geq |\bigtriangledown u|^p + \sum_{j\in\jmath} \tilde{\mu_j}\delta_{x_j} + \tilde{\mu_0}\delta_0, \tag{2.5}$$

$$|u_m|^p \rightharpoonup d\tilde{\nu} = |u|^{p^*} + \sum_{j\in\jmath} \tilde{\nu_j}\delta_{x_j} + \tilde{\nu_0}\delta_0, \tag{2.6}$$

$$\frac{|u_m|^{p^*(t)}}{|x|^t} \rightharpoonup d\tilde{\tau} = \frac{|u|^{p^*(t)}}{|x|^t} + \tilde{\tau_0}\delta_0, 0<t<p, \tag{2.7}$$

$$\frac{|u_m|^p}{|x|^p} \rightharpoonup d\tilde{\gamma} = \frac{|u|^p}{|x|^p} + \tilde{\gamma_0}\delta_0, \tag{2.8}$$

where δ_x is the Dirac mass at x.

Case (1). $t=0$ and $p^*(t)=p^*$.

We claim that $\jmath$ is finite and for $j\in\jmath$, either

$$\tilde{\nu_j} = 0 \ \text{ or } \ \tilde{\nu_j} \geq (A_{\mu,0})^{\frac{N}{p}}.$$

In fact, let $\varepsilon>0$ be small enough such that $0 \overline{\in} B_\varepsilon(x_j)$ and $B_\varepsilon(x_i) \bigcap B_\varepsilon(x_j) = \emptyset$ for $i\neq j$, $i,j\in\jmath$. We consider $\varphi^j \in C_0^\infty(\Re^N)$, such that

$$\varphi^j \equiv 1 \text{ on } B(x_j, \frac{\varepsilon}{2}),\ \ \varphi^j \equiv 0 \text{ on } B(x_j,\varepsilon)^c,\ |\bigtriangledown \varphi^j| \leq \frac{2}{\varepsilon}.$$

It is clear that the sequence $\{\varphi^j u_m\}$ is bounded in $W_0^{1,p}(\Omega)$, then, by using (ii), we have that

$$\lim_{m\to\infty} \langle I_\mu'(u_m), \varphi^j u_m\rangle = 0,$$

that is

$$\lim_{m\to\infty}\Big(\int_\Omega |\nabla u_m|^p\varphi^j\,dx + \int_\Omega u_m|\nabla u_m|^{p-2}\nabla u_m\nabla\varphi^j\,dx - \mu\int_\Omega \frac{|u_m|^p}{|x|^p}\varphi^j\,dx$$
$$-\alpha\int_\Omega |u_m|^{p^*}\varphi^j\,dx - \lambda\int_\Omega |u_m|^p\varphi^j\,dx - \int_\Omega |u_m|^q\varphi^j\,dx\Big) = 0. \tag{2.9}$$

By (2.4)–(2.6), (2.8) and (2.9), we obtain

$$\alpha\int_\Omega \varphi^j\,d\tilde{\nu} + \int_\Omega |u|^q\varphi^j\,dx + \lim_{m\to\infty}\mu\int_\Omega \frac{|u_m|^p}{|x|^p}\varphi^j\,dx + \lambda\int_\Omega |u|^p\varphi^j\,dx - \int_\Omega \varphi^j\,d\tilde{\mu}$$
$$= \lim_{m\to\infty}\int_\Omega u_m|\nabla u_m|^{p-2}\nabla u_m\nabla\varphi^j\,dx. \tag{2.10}$$

By the Hölder inequality, we have

$$\lim_{\varepsilon\to 0}\lim_{m\to\infty}\left|\int_\Omega u_m|\nabla u_m|^{p-2}\nabla u_m\nabla\varphi^j\,dx\right| \le C\lim_{\varepsilon\to 0}\left(\int_{B(x_j,\varepsilon)}|u|^{p^*}dx\right)^{\frac{1}{p^*}}=0, \tag{2.11}$$

$$\lim_{\varepsilon\to 0}\lim_{m\to\infty}\left|\int_\Omega \frac{|u_m|^p}{|x|^p}\varphi^j\,dx\right| \le \lim_{\varepsilon\to 0}\lim_{m\to\infty}\left|\int_{B_\varepsilon(x_j)}\frac{|u_m|^p}{(|x_j|-\varepsilon)^p}\varphi^j\,dx\right| = 0. \tag{2.12}$$

Then, from (2.10)–(2.12), we get

$$\begin{aligned}0 &= \lim_{\varepsilon\to 0}(\alpha\int_\Omega \varphi^j\,d\widetilde{\nu} + \int_\Omega |u|^q\varphi^j\,dx + \lim_{m\to\infty}\mu\int_\Omega \frac{|u_m|^p}{|x|^p}\varphi^j\,dx \\ &\quad +\lambda\int_\Omega |u|^p\varphi^j\,dx - \int_\Omega \varphi^j\,d\widetilde{\mu}) \\ &\le \alpha\widetilde{\nu_j} - \widetilde{\mu_j}.\end{aligned}$$

By the Sobolev inequality, $A_{0,0}\widetilde{\nu_j}^{\frac{p}{p^*}} \le \widetilde{\mu_j}$, hence, we deduce that

$$\widetilde{\nu_j} = 0 \ \text{ or } \ \widetilde{\nu_j} \ge (\frac{A_{0,0}}{\alpha})^{\frac{N}{p}},$$

which implies that $\jmath$ is finite.

Now we consider the possibility of concentration at the origin. Let $\varepsilon > 0$ be small enough such that $x_j\overline{\in}B(0,\varepsilon)$, $\forall j\in\jmath$. Take $\varphi^0 \in C_0^\infty(\Re^N)$ such that

$$\varphi^0 \equiv 1 \text{ on } B(0,\frac{\varepsilon}{2}),\ \ \varphi^0\equiv 0 \text{ on } B(0,\varepsilon)^c,\ \ |\bigtriangledown\varphi^0| \le \frac{4}{\varepsilon}.$$

By (2.4)–(2.6), (2.8) and (2.9), we also get that

$$\begin{aligned}0 &= \lim_{\varepsilon\to 0}(\alpha\int_\Omega \varphi^0 d\widetilde{\nu} + \int_\Omega |u|^q\varphi^0 dx + \mu\int_\Omega \varphi^0 d\widetilde{\gamma} + \lambda\int_\Omega |u|^p\varphi^0 dx - \int_\Omega \varphi^0 d\widetilde{\mu}) \\ &\le \alpha\widetilde{\nu_0} + \mu\widetilde{\gamma_0} - \widetilde{\mu_0}.\end{aligned} \tag{2.13}$$

By the Sobolev inequalities, we infer that $A_{\mu,0}\widetilde{\nu_0}^{\frac{p}{p^*}} \le \widetilde{\mu_0} - \mu\widetilde{\gamma_0}$. From (2.12), we obtain $A_{\mu,0}\widetilde{\nu_0}^{\frac{p}{p^*}} \le \widetilde{\mu_0} - \mu\widetilde{\gamma_0} \le \alpha\widetilde{\nu_0}$, which implies that

$$\widetilde{\nu_0} = 0 \ \text{ or } \ \widetilde{\nu_0} \ge (\frac{A_{\mu,0}}{\alpha})^{\frac{N}{p}}.$$

Let us assume that there exists a $j\in\jmath\cup\{0\}$ such that $\widetilde{\nu_j}\ne 0$, then, by (i) and (ii), we infer that

$$c = \lim_{m\to\infty} I_\mu(u_m) = \lim_{m\to\infty}[I_\mu(u_m) - \frac{1}{p}\langle I'_\mu(u_m), u_m\rangle]$$

$$
\begin{aligned}
&= \lim_{m\to\infty}\{\alpha(\frac{1}{p}-\frac{1}{p^*})\int_\Omega |u_m|^{p^*}dx+(\frac{1}{p}-\frac{1}{q})\int_\Omega |u_m|^q dx\} \\
&\geq \frac{\alpha}{N}(\int_\Omega |u|^{p^*}dx + \sum_{j\in\jmath}\widetilde{\nu_j} + \widetilde{\nu_0}) \\
&\geq \frac{\alpha}{N}\min\{(\frac{A_{0,0}}{\alpha})^{\frac{N}{p}}, (\frac{A_{\mu,0}}{\alpha})^{\frac{N}{p}}\} \\
&= \frac{\alpha}{N}A_{\mu,0}^{\frac{N}{p}}\cdot\alpha^{-\frac{N}{p}} = \frac{\alpha^{1-\frac{N}{p}}}{N}A_{\mu,0}^{\frac{N}{p}} \geq M,
\end{aligned}
$$

and this inequality contradicts the hypothesis. Then,

$$\widetilde{\nu_j} = 0,\quad \forall j \in \jmath \cup \{0\}.$$

Hence, we obtain that $u_m \longrightarrow u$ strongly in $W_0^{1,p}(\Omega)$.

Case (2). $0 < t < p$, then $p < p^*(t) < p^*$.

We only need to consider the possibility of concentration at the origin. Let $\varepsilon > 0$ be small enough such that $B_\varepsilon(0) \subset \Omega$. Take $\varphi^0 \in C_0^\infty(\Re^N)$ such that

$$\varphi^0 \equiv 1 \text{ on } B(0,\frac{\varepsilon}{2}),\quad \varphi^0 \equiv 0 \text{ on } B(0,\varepsilon)^c,\quad |\bigtriangledown\varphi^0| \leq \frac{4}{\varepsilon}.$$

From (2.4),(2.5), (2.7)–(2.11), we get that

$$0 \leq \alpha\widetilde{\tau_0} + \mu\widetilde{\gamma_0} - \widetilde{\mu_0}. \tag{2.14}$$

By the definition of $A_{\mu,t}$, we deduce that

$$A_{\mu,t}\widetilde{\tau_0}^{\frac{p}{p^*(t)}} \leq \widetilde{\mu_0} - \mu\widetilde{\gamma_0}. \tag{2.15}$$

From (2.14) and (2.15) we have

$$A_{\mu,t}\widetilde{\tau_0}^{\frac{p}{p^*(t)}} \leq \alpha\widetilde{\tau_0},$$

which implies that

$$\widetilde{\tau_0} = 0,\quad \text{or}\quad \widetilde{\tau_0} \geq (\frac{A_{\mu,t}}{\alpha})^{\frac{N-t}{p-t}}.$$

If $\widetilde{\tau_0} \neq 0$, then we infer that

$$
\begin{aligned}
c &= \lim_{m\to\infty}[I_\mu(u_m) - \frac{1}{p}\langle I'_\mu(u_m), u_m\rangle] \\
&\geq \frac{(p-t)\alpha}{p(N-t)}(\int_\Omega \frac{|u|^{p^*(t)}}{|x|^t}dx + \widetilde{\tau_0}) \\
&\geq \frac{(p-t)\alpha}{p(N-t)}\widetilde{\tau_0} \geq \frac{(p-t)\alpha}{p(N-t)}(\frac{A_{\mu,t}}{\alpha})^{\frac{N-t}{p-t}} \geq M,
\end{aligned}
$$

which contradicts of the assumption that $c < M$. Hence, we have that $u_m \longrightarrow u$ strongly in $W_0^{1,p}(\Omega)$. Thus, the proof of the lemma is completed.

3. Proof of Theorem 1.1

In this article we shall be using the following version of the symmetric mountain pass theorem (see [16–18]).

Lemma 3.1. *Let $E = V \oplus X$, where E is a real Banach space and V is finite dimensional. Suppose $I \in C^1(E, \Re)$ is an even functional satisfying $I(0) = 0$ and*

(I_1) there is a constant $\rho > 0$ such that $I\,|_{\partial B_\rho \cap X} \geq 0$;

(I_2) there is a subspace W of E with $\dim V < \dim W < \infty$ *and there is $M > 0$ such that* $\max_{u \in W} I(u) < M$;

(I_3) considering $M > 0$ given by (I_2), I satisfies $(PS)_c$ for $0 \leq c \leq M$. Then I possesses at least $\dim W - \dim V$ *pairs of nontrivial critical points.*

First, we recall that each basis $(e_i)_{i \in \mathbf{N}}$ for a real Banach space E is a Schauder basis for E, i.e., given $n \in \mathbf{N}$, the functional $e_n^* : E \longrightarrow \Re$ defined by

$$e_n^*(v) = \alpha_n, \quad \text{for } v = \sum_{i=1}^{\infty} \alpha_i e_i \in E, \tag{3.1}$$

is a bounded linear functional [19, 20]. We observe that the existence of a Schauder basis for the space $W_0^{1,p}(\Omega)$ was proved by Fucik, John and Necas in [21].

Now, fixing a Schauder basis $(e_i)_{i \in \mathbf{N}}$ for $W_0^{1,p}(\Omega)$, for $j \in \mathbf{N}$ we set

$$\begin{aligned} V_j &= \{u \in W_0^{1,p}(\Omega) : e_i^*(u) = 0, i > j\}, \\ X_j &= \{u \in W_0^{1,p}(\Omega) : e_i^*(u) = 0, i \leq j\}. \end{aligned} \tag{3.2}$$

It follows by (3.1) that $W_0^{1,p}(\Omega) = V_j \oplus X_j$.

The following Lemma comes from [6].

Lemma 3.2. *Given $p \leq r < p^*$ and $\delta > 0$, there is $j \in \mathbf{N}$ such that, for all $u \in X_j$, $\|u\|_r^r \leq \delta \|u\|^r$.*

Lemma 3.3. *There exists $\widetilde{\alpha} > 0$ and $\rho_0, d > 0$ such that $I_\mu|_{\partial B_{\rho_0} \cap X_j} \geq d$ for all $0 < \alpha < \widetilde{\alpha}$.*

Proof. By Lemma 2.1, (1.4) and Lemma 3.2, we obtain

$$\begin{aligned} I_\mu(u) &\geq \frac{1}{p}\|u\|^p - \frac{c_1 \alpha}{p^*(t)}\|u\|^{p^*(t)} - \frac{\lambda}{\lambda_1(\Omega)p}\|u\|^p - \frac{c_2}{q}\|u\|^q \\ &= \frac{1}{p}\rho^p - \frac{c_1 \alpha}{p^*(t)}\rho^{p^*(t)} - \frac{\lambda}{\lambda_1(\Omega)p}\rho^p - \frac{c_2}{q}\rho^q, \end{aligned}$$

where $\rho = \|u\|$. Now, we take $\widetilde{\alpha} > 0$ so that

$$I_\mu(u) \geq \frac{1}{p}\rho^p - \frac{c_1 \widetilde{\alpha}}{p^*(t)}\rho^{p^*(t)} - \frac{\lambda}{\lambda_1(\Omega)p}\rho^p - \frac{c_2}{q}\rho^q > 0,$$

for every $u \in X_j$, $\|u\| = \rho$. The proof is completed.

Lemma 3.4. *Given $m \in \Re$, there exist a subspace W of $W_0^{1,p}(\Omega)$ and a constant $M_m > 0$, independent of α, such that* $\dim W = m$ *and* $\max_{u\in W} I_0(u) < M_m$.

Proof. Taking $\Omega_0 \subset \Omega$ with $|\Omega_0| > 0$ and $0 \in \Omega_0$. Let $x_0 \in \Omega_0$ and $r_0 > 0$ be such that $\overline{B(x_0, r_0)} \subset \Omega$ and $0 < |\overline{B(x_0, r_0)} \cap \Omega_0| < |\Omega_0|/2$. First, we take $\nu_1 \in C_0^\infty(\Omega)$ with $\text{supp}(v_1) = \overline{B(x_0, r_0)}$. Considering $\Omega_1 = \Omega_0 \setminus [\overline{B(x_0, r_0)} \cap \Omega_0] \subset \widehat{\Omega_0} = \Omega \setminus \overline{B(x_0, r_0)}$, we have $|\Omega_1| > |\Omega_0|/2 > 0$. Let $x_1 \in \Omega_1$ and $r_1 > 0$ be such that $\overline{B(x_1, r_1)} \subset \widehat{\Omega}$ and $0 < |\overline{B(x_1, r_1)} \cap \Omega_1| < |\Omega_1|/2$. Next, we take $v_2 \in C_0^\infty(\Omega)$ with $\text{supp}(v_2) = \overline{B(x_1, r_1)}$. After a finite number of steps, we get $v_1, \cdots, v_m$ such that $\text{supp}(v_i) \cap \text{supp}(v_j) = \emptyset$, $i \neq j$ and $|\text{supp}(v_j) \cap \Omega_0| > 0$, for all $i, j \in \{1, \cdots, m\}$. Let $W = \text{span}\{v_1, \cdots, v_m\}$. By construction, $\dim W = m$ and

$$\int_{\Omega_0} |v|^p dx > 0, \text{ for every } v \in W \setminus \{0\}.$$

Let $v = \frac{u}{\|u\|}$, by (1.5), we know

$$\max_{u\in W\setminus\{0\}} I_0(u) = \max_{t>0, v\in\partial B_1(0)\cap W} \{t^p[\frac{1}{p} - \frac{\lambda}{p}\int_\Omega |v|^p dx - \frac{1}{q}t^{q-p}\int_\Omega |v|^q dx]\}.$$

So, to prove the Lemma, it suffices to verify that

$$\lim_{t\longrightarrow\infty} (\frac{\lambda}{p}\int_\Omega |v|^p dx - \frac{1}{q}t^{q-p}\int_\Omega |v|^q dx) > \frac{1}{p} \tag{3.3}$$

uniformly for $v \in \partial B_1(0) \cap W$. Since $q > p$, the inequality (3.3) is right. The proof is completed.

Proof of Theorem 1.1. First, we recall that $W_0^{1,p}(\Omega) = V_j \oplus X_j$, where V_j and X_j are defined in (3.2). Invoking Lemma 3.3, we find $j \in \mathbf{N}$ and $\widetilde{\alpha} > 0$ such that I_μ satisfies (I_1) with $X = X_j$ for all $0 < \alpha < \widetilde{\alpha}$. Now, by Lemma 3.4, there is a subspace W of $W_0^{1,p}(\Omega)$ with $\dim W = k + j = k + \dim V_j$ and such that I_μ satisfies (I_2). By Lemma 2.2, taking $\widetilde{\alpha}$ smaller if necessary, we also have that I_μ satisfies (I_3) for $0 < \alpha < \widetilde{\alpha}$. Since $I_\mu(0) = 0$ and I_μ is even, we may apply Theorem 3.1 to conclude that I_μ possesses at least k pairs of nontrivial critical points for $\alpha > 0$ sufficiently small.

References

[1] H. Brézis, L. Nirenberg, Positive solutions of nonlinear elliptic equations involving critical Sobolev exponents, *Comm. Pure Appl. Math.* 36 (1983), 437-477.

[2] A. Capozzi, D. Fortunato and G. Palmieri, An existence result for nonlinear elliptic problems involving critical Sobolev exponent, *Ann. Inst. H. Poincaré, Analyse Non Linéaire* 2(6) (1985), 463-470.

[3] M. Guedda and L. Veron, Quasilinear elliptic equations involving critical Sobolev exponents, *Nonlinear Anal. TMA* 13(8) (1989), 879-902.

[4] Z. Wei and X. Wu, A multiplicity result for quasilinear elliptic equations involving critical Sobolev exponents, *Nonlinear Anal. TMA* 18(6) (1992), 559-567.

[5] D. G. Costa and E. A. B. Silva, A note on problems involving critical Sobolev exponents, *Differential and Integral Equations* 8(3) (1995), 673-679.

[6] E. A. B. Silva and M. S. Xavier, Multiplicity of solutions for quasilinear elliptic problems involving critical Sobolev Exponents, *Ann. I. H. Poincaré-AN* 20(2) (2003), 341-358.

[7] D. Kang, On the quasilinear elliptic problems with critical Sobolev-Hardy exponents and Hardy terms, *Nonlinear Anal.* 68 (2008), 1973-1985.

[8] N. Ghoussoub, C. Yuan, Multiple solutions for quasi-linear PDEs involving the critical Sobolev-Hardy exponents, *Trans. Amer. Math. Soc.* 352 (2000), 5703-5743.

[9] E. Janneli, The role played by space dimension in elliptic critical problems, *J. Differential Equations* 156 (1999), 407-426.

[10] A. Ferrero and F. Gazzola, Existence of solutions for singular critical growth semilinear elliptic equations, *J. Differential Equations* 177 (2001), 494-522.

[11] C. Cao, X. He and S. Peng, Positive solutions for some singular critical growth nonlinear elliptic equations, *Nonlinear Anal.* 60 (2005), no. 3, 589–609.

[12] J. Garcia Azorero and I. Peral, Hardy inequalities and some critical elliptic and parabolic problems, *J. Differential Equations* 144 (1998), 441-476.

[13] L. Caffarelli, R. Kohn and L. Nirenberg, First order interpolation inequality with weights, *Compos. Math.* 53 (1984), 259-275.

[14] P. L. Lions, The concentration compactness principle in the calculus of variations, the limit case (I), *Rev. Mat. Iberoamericana* 1(1) (1985), 145-201.

[15] P. L. Lions, The concentration compactness principle in the calculus of variations, the limit case (II), *Rev. Mat. Iberoamericana* 1(2) (1985), 45-121.

[16] A. Ambrosetti and P.H. Rabinowitz, Dual variational methods in critical point theory and applications, *J. Funct. Anal.* 14 (1973), 349-381.

[17] P. Bartolo, V. Benci and D. Fortunato, Abstract critical point theorems and applications to some nonlinear problems with "strong" resonance at infinity, *Nonlinear Anal. TMA* 7(9) (1983), 981-1012.

[18] E. A. B. Silva, *Critical point theorems and applications to differential equations,* Ph.D. Thesis, University of Wisconsin-Madison, 1988.

ON CERTAIN CLASSES OF P-HARMONIC MAPPINGS

JIN-JING QIAO
Hunan Normal University, Changsha, Hunan 410081, P. R. China
E-mail: qiaojinjing1982@yahoo.com.cn

A p times continuously differentiable complex-valued function F in a domain $D \subseteq C$ is *p-harmonic*, if F satisfies the p-harmonic equation $\underbrace{\Delta \cdots \Delta}_{p} F = 0$, where $p\ (\geq 1)$ is an integer and Δ represents the complex Laplacian operator. In this paper, the main aim is to introduce two classes $\mathcal{M}_{H_p}(\alpha)$ and $\mathcal{N}_{H_p}(\alpha)$ of p-harmonic mappings together with their subclasses $\mathcal{M}_{H_p}(\alpha) \cap T_1$ and $\mathcal{N}_{H_p}(\alpha) \cap T_2$, and investigate the properties of mappings in these classes. First, we obtain characterizations for mappings in $\mathcal{M}_{H_p}(\alpha) \cap T_1$ and $\mathcal{N}_{H_p}(\alpha) \cap T_2$ in terms of S-Inequality-I and S-Inequality-II, respectively. And then we prove that the image domains of the unit disk D under the mappings in $\mathcal{M}_{H_p}(\alpha)$ (resp. $\mathcal{N}_{H_p}(\alpha)$) satisfying Inequality-$I$ (resp. Inequality-II) are starlike (resp. convex) of certain order.

Keywords: P-harmonic mapping, (S-)inequality-I, (S-)inequality-II, characterization, starlikeness, convexity.
AMS No: 30C65, 30C45, 30C20.

1. Introduction

A p times continuously differentiable complex-valued function $F = u + iv$ in a domain $D \subseteq C$ is *p-harmonic*, if F satisfies the p-harmonic equation $\underbrace{\Delta \cdots \Delta}_{p} F = 0$, where $p\ (\geq 1)$ is an integer and Δ represents the complex Laplacian operator

$$\Delta = \frac{\partial^2}{\partial x^2} + \frac{\partial^2}{\partial y^2} = 4\frac{\partial^2}{\partial z \partial \bar{z}}.$$

A mapping F is p-harmonic in a simply connected domain D if and only if F has the following representation:

$$F(z) = \sum_{k=1}^{p} |z|^{2(k-1)} G_{p-k+1}(z),$$

where each G_{p-k+1} is harmonic, i.e., $\Delta G_{p-k+1}(z) = 0$ for $k \in \{1, \cdots, p\}$ (cf. Proposition 2.1, [5]).

Obviously, when $p = 1$ (resp. 2), F is harmonic (resp. biharmonic). The properties of harmonic mappings have been investigated by many authors, see [6, 7, 17] and the references therein.

Biharmonic mappings arise in a lot of physical situations, particularly in fluid dynamics and elasticity problems, and have many important applications in engineering and biology. See [8, 12, 13] for the details. Many references on biharmonic mappings have been in literature, see [1, 2, 3, 8, 12, 13].

In this paper, as a generalization of harmonic mappings and biharmonic mappings, we consider p-harmonic mappings of the unit disk D.

Let A denote the set of all analytic functions h of D with the normalization $h(0) = 0$ and $h^{'}(0) = 1$. Owa et al. introduced two classes $\mathcal{M}(\alpha)$ and $\mathcal{N}(\alpha)$, where $\mathcal{M}(\alpha)$ denotes the set of all functions $h \in A$ such that

$$\mathrm{Re}\frac{zh^{'}(z)}{h(z)} < \alpha,$$

for $\alpha > 1$, and $\mathcal{N}(\alpha)$ the set of all functions $h \in A$ such that

$$\mathrm{Re}(1 + \frac{zh^{''}(z)}{h^{'}(z)}) < \alpha,$$

for $\alpha > 1$. In [14, 15], the authors discussed the starlikeness, convexity and coefficient estimate of functions in $\mathcal{M}(\alpha)$ and $\mathcal{N}(\alpha)$. See [18, 20, 21] for related discussions when $\alpha \in (1, \frac{4}{3}]$.

In order to discuss the characterization, starlikeness, convexity of p-harmonic mappings, in Section 2, we will introduce the notations: $\mathcal{M}_{H_p}(\alpha)$ and $\mathcal{N}_{H_p}(\alpha)$, which are generalizations of $\mathcal{M}(\alpha)$ and $\mathcal{N}(\alpha)$ for p-harmonic mappings, respectively. Then we prove a sufficient condition for p-harmonic mappings to be in $\mathcal{M}_{H_p}(\alpha)$ (resp. $\mathcal{N}_{H_p}(\alpha)$) in terms of Inequality-I (resp. Inequality-II) (see Section 2 for the definitions). Some necessary notions are given and several elementary results are also proved in this section. In Section 3, we obtain characterizations for mappings in $\mathcal{M}_{H_p}(\alpha) \cap T_1$ and $\mathcal{N}_{H_p}(\alpha) \cap T_2$ by using S-Inequality-I and S-Inequality-II (see Sections 2 and 3 for the definitions). Our main results are Theorems 3.1 and 3.2. Finally, we discuss the starlikeness (resp. convexity) of mappings in $\mathcal{M}_{H_p}(\alpha)$ (resp. $\mathcal{N}_{H_p}(\alpha)$), when they satisfy Inequality-I (resp. Inequality-II). Our result is Theorem 4.1, which is a generalization of [Theorem 2.5, 14].

2. Preliminaries

Let H_p denote the class of mappings F of D with the form:

$$F(z) = \sum_{k=1}^{p} |z|^{2(k-1)} G_{p-k+1}(z), \tag{2.1}$$

where for each $k \in \{1, \cdots, p\}$, the harmonic mapping G_{p-k+1} has the expression:

$$G_{p-k+1} = h_{p-k+1} + \overline{g}_{p-k+1},$$

both h_{p-k+1}, g_{p-k+1} are analytic and satisfy the following conditions:

$$h_{p-k+1}(z) = \sum_{n=1}^{\infty} a_{n,p-k+1} z^n \text{ with } a_{1,p} = 1,$$

and

$$g_{p-k+1}(z) = \sum_{n=1}^{\infty} b_{n,p-k+1} z^n.$$

For $F \in H_p$, let $L(F) = zF_z - \overline{z}F_{\overline{z}}$. In [3], Abdulhadi et al. discussed the properties of this operator for biharmonic mappings. For p-harmonic mappings, we have

Proposition 2.1. (1) *If F is a p-harmonic mapping of D, then $L(F)$ is p-harmonic.*

(2) *Suppose F is a p-harmonic mapping of D. Then $L(F) = F$ if and only if $F(z) = \left(\sum_{k=1}^{p} a_{p-k+1}|z|^{2(k-1)}\right)z$.*

Proof. (1) Assume that

$$F(z) = \sum_{k=1}^{p} |z|^{2(k-1)} G_{p-k+1}(z),$$

where $G_{p-k+1} = h_{p-k+1} + \overline{g}_{p-k+1}$ for $k \in \{1, \cdots, p\}$. Since

$$F_z(z) = G_z(z) + \sum_{k=2}^{p} |z|^{2(k-1)} (G_{p-k+1})_z(z) + \sum_{k=2}^{p} (k-1) z^{k-2} \overline{z}^{k-1} G_{p-k+1}(z),$$

and

$$F_{\overline{z}}(z) = G_{\overline{z}}(z) + \sum_{k=2}^{p} |z|^{2(k-1)} (G_{p-k+1})_{\overline{z}}(z) + \sum_{k=2}^{p} (k-1) z^{k-1} \overline{z}^{k-2} G_{p-k+1}(z),$$

we see that

$$L(F) = \sum_{k=1}^{p} |z|^{2(k-1)} L(G_{p-k+1}).$$

By [Lemma 3, 3], we obtain that $L(F)$ is p-harmonic.

(2) Assume that $L(F) = F$. Then $L(G_{p-k+1}) = G_{p-k+1}$ for each $k \in \{1, \cdots, p\}$ which implies

$$zh'_{p-k+1}(z) = h_{p-k+1}(z) \text{ and } zg'_{p-k+1}(z) = -g_{p-k+1}(z).$$

It follows that $h_{p-k+1}(z) = a_{p-k+1}z$ and $g_{p-k+1}(z) \equiv 0$. The conversion is obvious.

By using the operator L, we introduce two classes of p-harmonic mappings. Let $\alpha > 1$ be a constant. We always use $\mathcal{M}_{H_p}(\alpha)$ to denote the set of all mappings $F \in H_p$ such that $F(z) \neq 0$, whenever $z \neq 0$ and for any $z \in D \setminus \{0\}$,

$$\operatorname{Re}\frac{L(F)(z)}{F(z)} < \alpha,$$

and $\mathcal{N}_{H_p}(\alpha)$ the set of all mappings $F \in H_p$ such that $L(F)(z) \neq 0$, whenever $z \neq 0$ and for any $z \in D \setminus \{0\}$,

$$\operatorname{Re}\frac{L(L(F))(z)}{L(F)(z)} < \alpha.$$

Obviously, we have

Proposition 2.2. *$F \in \mathcal{N}_{H_p}(\alpha)$ if and only if $L(F) \in \mathcal{M}_{H_p}(\alpha)$.*

For convenience, we introduce the following definitions.

Definition 2.1. For $k_0 \in [0, 1]$ and $\alpha > 1$, we say that $F \in H_p$ satisfies the inequality-I, if

$$\begin{aligned}
&\sum_{k=1}^{p}\sum_{n=2}^{\infty}(n - k_0 + |n - 2\alpha + k_0|)|a_{n,p-k+1}| \\
&+\sum_{k=1}^{p}\sum_{n=1}^{\infty} 2(n+\alpha)|b_{n,p-k+1}| \\
&+\sum_{k=2}^{p} 2(\alpha - k_0)|a_{1,p-k+1}| \leq 2(\alpha - 1);
\end{aligned} \tag{2.2}$$

and the inequality-II, if

$$\begin{aligned}
&\sum_{k=1}^{p}\sum_{n=2}^{\infty}n(n - k_0 + |n - 2\alpha + k_0|)|a_{n,p-k+1}| \\
&+\sum_{k=1}^{p}\sum_{n=1}^{\infty} 2n(n+\alpha)|b_{n,p-k+1}| \\
&+\sum_{k=2}^{p} 2(\alpha - k_0)|a_{1,p-k+1}| \leq 2(\alpha - 1).
\end{aligned} \tag{2.3}$$

In particular, when $k_0 = 1$ and $\alpha \in (1, \frac{3}{2}]$, (2.2) and (2.3) are changed into the following (2.4) and (2.5) respectively,

$$\sum_{k=1}^{p}\sum_{n=2}^{\infty}(n-\alpha)|a_{n,p-k+1}| + \sum_{k=1}^{p}\sum_{n=1}^{\infty}(n+\alpha)|b_{n,p-k+1}| + \sum_{k=2}^{p}(\alpha-1)|a_{1,p-k+1}| \leq \alpha - 1; \tag{2.4}$$

$$\sum_{k=1}^{p}\sum_{n=2}^{\infty}n(n-\alpha)|a_{n,p-k+1}| + \sum_{k=1}^{p}\sum_{n=1}^{\infty}n(n+\alpha)|b_{n,p-k+1}| + \sum_{k=2}^{p}(\alpha-1)|a_{1,p-k+1}| \leq \alpha - 1. \tag{2.5}$$

Definition 2.2. F is said to satisfy *S-Inequality-I* (resp. *S-Inequality-II*) if it satisfies (2.4) (resp. (2.5)).

Now we come to derive a sufficient condition for F to be in $\mathcal{M}_{H_p}(\alpha)$ (resp. $\mathcal{N}_{H_p}(\alpha)$) by using Inequalities-$I$ (resp. Inequality-II), which are generalizations of [14, Theorem 2.1].

Lemma 2.1. *If $F \in H_p$ satisfies Inequality-I for some $k_0 \in [0,1]$ and $\alpha > 1$, then $F \in \mathcal{M}_{H_p}(\alpha)$.*

Proof. Assume that $F \in H_p$ satisfies the inequality-I. Since

$$\begin{aligned}
&|z|\Big(2(\alpha-1) - \sum_{k=1}^{p}\sum_{n=1}^{\infty}2(n+\alpha)|b_{n,p-k+1} - \sum_{k=2}^{p}2(\alpha-k_0)|a_{1,p-k+1}| \\
&\quad - \sum_{k=1}^{p}\sum_{n=2}^{\infty}\big(n-k_0+|n-2\alpha+k_0|\big)|a_{n,p-k+1}||\Big) \\
&< |z|\Big(2(\alpha-1) - \sum_{k=2}^{p}2(\alpha-k_0)|a_{1,p-k+1}||z|^{2(k-1)} \\
&\quad - \sum_{k=1}^{p}\sum_{n=2}^{\infty}\big(n-k_0+|n-2\alpha+k_0|\big)|a_{n,p-k+1}||z|^{2(k-1)+n-1} \\
&\quad - \sum_{k=1}^{p}\sum_{n=1}^{\infty}2(n+\alpha)|b_{n,p-k+1}||z|^{2(k-1)+n-1}\Big) \\
&\leq 2(\alpha-1)\Big|z + \sum_{k=2}^{p}|z|^{2(k-1)}a_{1,p-k+1}z + \sum_{k=1}^{p}\sum_{n=2}^{\infty}|z|^{2(k-1)}a_{n,p-k+1}z^n \\
&\quad + \sum_{k=1}^{p}\sum_{n=1}^{\infty}|z|^{2(k-1)}\overline{b}_{n,p-k+1}\overline{z}^n\Big| = 2(\alpha-1)|F(z)|,
\end{aligned}$$

thus we know that $F(z) \neq 0$, whenever $z \neq 0$.

Next, we show that for any $z \neq 0$, F satisfies

$$|\frac{L(F)(z)}{F(z)} - k_0| < |\frac{L(F)(z)}{F(z)} - (2\alpha - k_0)|,$$

which is equivalent to $Re(\frac{L(F)(z)}{F(z)}) < \alpha$. Let

$$A(z) = |L(F)(z) - k_0 F(z)|$$

and

$$B(z) = |L(F)(z) - (2\alpha - k_0)F(z)|.$$

Then

$$\begin{aligned} A(z) = \Bigg|&(1-k_0)z + \sum_{k=2}^{p}(1-k_0)a_{1,p-k+1}|z|^{2(k-1)}z \\ &+ \sum_{k=1}^{p}\sum_{n=2}^{\infty}(n-k_0)a_{n,p-k+1}|z|^{2(k-1)}z^n \\ &- \sum_{k=1}^{p}\sum_{n=1}^{\infty}(n+k_0)\bar{b}_{n,p-k+1}|z|^{2(k-1)}\bar{z}^n\Bigg|, \end{aligned}$$

and

$$\begin{aligned} B(z) = \Bigg|&(1-2\alpha+k_0)z + \sum_{k=2}^{p}(1-2\alpha+k_0)a_{1,p-k+1}|z|^{2(k-1)}z \\ &+ \sum_{k=1}^{p}\sum_{n=2}^{\infty}(n-2\alpha+k_0)a_{n,p-k+1}|z|^{2(k-1)}z^n \\ &- \sum_{k=1}^{p}\sum_{n=1}^{\infty}(n+2\alpha-k_0)\bar{b}_{n,p-k+1}|z|^{2(k-1)}\bar{z}^n\Bigg|. \end{aligned}$$

Obviously,

$$A(z) < C = (1-k_0) + \sum_{k=2}^{p}(1-k_0)|a_{1,p-k+1}|$$
$$-\sum_{k=1}^{p}\sum_{n=2}^{\infty}(n-k_0)|a_{n,p-k+1}| + \sum_{k=1}^{p}\sum_{n=1}^{\infty}(n+k_0)|b_{n,p-k+1}|,$$

and

$$B(z) > D = (2\alpha-k_0-1) - \sum_{k=2}^{p}(2\alpha-k_0-1)|a_{1,p-k+1}|$$
$$-\sum_{k=1}^{p}\sum_{n=2}^{\infty}|n-2\alpha+k_0||a_{n,p-k+1}| - \sum_{k=1}^{p}\sum_{n=1}^{\infty}(n+2\alpha-k_0)|b_{n,p-k+1}|.$$

Since F satisfies Inequality-I, we see that $C \leq D$, which implies that

$$|\frac{L(F)(z)}{F(z)} - k_0| < |\frac{L(F)(z)}{F(z)} - (2\alpha - k_0)|,$$

for all $z \neq 0$. The proof of the lemma is finished.

The following is an easy consequence of Lemma 2.1.

Corollary 2.1. *If $F \in H_p$ satisfies S-Inequality-I for $\alpha \in (1, \frac{3}{2}]$, then $F \in \mathcal{M}_{H_p}(\alpha)$.*

It follows from Proposition 2.2 and Lemma 2.1 that

Lemma 2.2. *If $F \in H_p$ satisfies Inequality-II for $k_0 \in [0, 1]$ and $\alpha > 1$, then F belongs to $\mathcal{N}_{H_p}(\alpha)$.*

It follows from Lemma 2.2 that

Corollary 2.2. *If $F \in H_p$ satisfies S-Inequality-II for $\alpha \in (1, \frac{3}{2}]$, then $F \in \mathcal{N}_{H_p}(\alpha)$.*

3. Characterizations for Mappings in $\mathcal{M}_{H_p}(\alpha) \cap T_1$ and $\mathcal{M}_{H_p}(\alpha) \cap T_2$

The classes of analytic functions and harmonic mappings with nonnegative (or negative) coefficients possess many interesting properties, and many references have been in literature, see, for example, [10, 16, 19, 22]. In order to get some analogues for p-harmonic mappings, we introduce two notions:

$$\begin{aligned} T_1 = \Big\{F \in H_p : F(z) = z - \sum_{k=2}^{p} |z|^{2(k-1)} a_{1,p-k+1} z \\ + \sum_{k=1}^{p} |z|^{2(k-1)} \Big(\sum_{n=2}^{\infty} a_{n,p-k+1} z^n - \sum_{n=1}^{\infty} b_{n,p-k+1} \bar{z}^n \Big), \\ \text{where } a_{n,p-k+1} \geq 0,\ b_{n,p-k+1} \geq 0 \text{ for } k \in \{1, \cdots, p\} \text{ and } n \geq 1 \Big\}, \end{aligned}$$

and

$$\begin{aligned} T_2 = \Big\{F \in H_p : F(z) = z - \sum_{k=2}^{p} |z|^{2(k-1)} a_{1,p-k+1} z \\ + \sum_{k=1}^{p} |z|^{2(k-1)} \Big(\sum_{n=2}^{\infty} a_{n,p-k+1} z^n + \sum_{n=1}^{\infty} b_{n,p-k+1} \bar{z}^n \Big), \\ \text{where } a_{n,p-k+1} \geq 0,\ b_{n,p-k+1} \geq 0 \text{ for } k \in \{1, \cdots, p\} \text{ and } n \geq 1 \Big\}. \end{aligned}$$

Theorem 3.1. *Suppose $F \in T_1$. Then $F \in \mathcal{M}_{H_p}(\alpha)$ if and only if F satisfies S-Inequality-I for $\alpha \in (1, \frac{3}{2}]$.*

Proof. By Corollary 2.1, it suffices to prove the necessity. Assume $F \in \mathcal{M}_{H_p}(\alpha) \cap T_1$. Then for any $z \neq 0$,

$$\Big|\frac{L(F)(z)}{F(z)} - 1\Big| < \Big|\frac{L(F)(z)}{F(z)} - (2\alpha - 1)\Big|, \tag{3.1}$$

which implies that

$$2\big|L(F)(z)-\alpha F(z)\big| = \big|L(F)(z)-(2\alpha-1)F(z)+L(F)(z)-F(z)\big| > 0. \tag{3.2}$$

Since

$$L(F)(z) - \alpha F(z) = (1-\alpha)z + \sum_{k=2}^{p} |z|^{2(k-1)}(\alpha-1)a_{1,p-k+1}z$$
$$+\sum_{k=1}^{p} |z|^{2(k-1)}\Big(\sum_{n=2}^{\infty}(n-\alpha)a_{n,p-k+1}z^n + \sum_{n=1}^{\infty}(n+\alpha)b_{n,p-k+1}\bar{z}^n\Big),$$

it follows that for any $r \in (0,1)$,

$$L(F)(r) - \alpha F(r) = -rA(r),$$

where

$$A(r) = \alpha - 1 - \sum_{k=2}^{p}(\alpha-1)a_{1,p-k+1}r^{2(k-1)}$$
$$-\sum_{k=1}^{p}\sum_{n=2}^{\infty}(n-\alpha)a_{n,p-k+1}r^{2(k-1)+n-1}$$
$$-\sum_{k=1}^{p}\sum_{n=1}^{\infty}(n+\alpha)b_{n,p-k+1}r^{2(k-1)+n-1}.$$

Suppose F does not satisfy S-Inequality-I for $\alpha \in (1, \frac{3}{2}]$. Then $A(r)$ is negative when r sufficiently approaches to 1. Obviously, $A(r)$ is positive when r sufficiently approaches to 0. Hence there exists $r_0 \in (0,1)$ such that $A(r_0) = 0$, which yields $L(F)(r_0) - \alpha F(r_0) = 0$. By (3.2), this is a contradiction.

By Proposition 2.2 and similar arguments as in the proof of Theorem 3.1, we get

Theorem 3.2. *Suppose $F \in T_2$. Then $F \in \mathcal{N}_{H_p}(\alpha)$ if and only if S-Inequality-II holds for $\alpha \in (1, \frac{3}{2}]$.*

4. Starlikeness and Convexity of P-Harmonic Mappings

Before the statement of the main result in this section, we introduce the following concepts for p-harmonic mappings.

Definition 4.1. We say that a univalent p-harmonic mapping F with $F(0) = 0$ is *starlike of order* β with respect to the origin if $\frac{d}{d\theta}(argF(re^{i\theta})) > \beta$ for $z = re^{i\theta} \neq 0$, where $\beta \in [0, 1)$ is a constant.

Definition 4.2. A univalent p-harmonic mapping F with $F(0) = 0$ and $\frac{d}{d\theta}F(re^{i\theta}) \neq 0$ whenever $0 < r < 1$, is said to be *convex of order* β if $\frac{d}{d\theta}\big(arg\frac{d}{d\theta}F(re^{i\theta})\big) > \beta$ for $z = re^{i\theta} \neq 0$, where $\beta \in [0, 1)$ is a constant.

Let $S^*_{H_p}(\beta)$ and $K^*_{H_p}(\beta)$ denote the classes of all starlike p-harmonic mappings of order β and all convex p-harmonic mappings of order β, respectively, and set $S^*_H(\beta) = S^*_{H_1}(\beta)$ and $K^*_H(\beta) = K^*_{H_1}(\beta)$ which are corresponding classes for harmonic mappings. In [4, 9, 10, 11, 17], the authors discussed the properties of mappings in $S^*_H(\beta)$ and $K^*_H(\beta)$ such as the coefficient estimate, distortion theorem and covering theorem. Now we come to prove the starlikeness (resp. convexity) of mappings in $\mathcal{M}_{H_p}(\alpha)$ (resp. $\mathcal{N}_{H_p}(\alpha)$).

Theorem 4.1. *$F \in H_p$ satisfies Inequality-I for some $k_0 \in (0, 1]$ and $\alpha \in (1, \min\{\frac{k_0+2}{2}, \frac{4}{3}\})$, $a_{1,p-k+1} = 0$ $(k \in \{2, \cdots, p\})$ and $F(z_1) \neq F(z_2)$ whenever $|z_1| \neq |z_2|$. Then $F \in S^*_{H_p}(\frac{4-3\alpha}{3-2\alpha})$.*

If $F \in H_p$ subjects to Inequality-II for some $k_0 \in (0, 1]$ and $\alpha \in (1, \min\{\frac{k_0+2}{2}, \frac{4}{3}\})$, $a_{1,p-k+1} = 0$ $(k \in \{2, \cdots, p\})$ and $F(z_1) \neq F(z_2)$ whenever $|z_1| \neq |z_2|$, then $F \in K^*_{H_p}(\frac{4-3\alpha}{3-2\alpha})$.

Proof. Assume

$$F(z) = z + \sum_{k=1}^{p} |z|^{2(k-1)}\Big(\sum_{n=2}^{\infty} a_{n,p-k+1}z^n + \sum_{n=1}^{\infty} \overline{b}_{n,p-k+1}\bar{z}^n\Big)$$

satisfies Inequality-I for some $k_0 \in (0, 1]$ and $\alpha \in (1, \min\{\frac{k_0+2}{2}, \frac{4}{3}\})$. Then

$$\sum_{k=1}^{p}\sum_{n=2}^{\infty}(n-\alpha)|a_{n,p-k+1}| + \sum_{n=1}^{p}\sum_{n=1}^{\infty}(n+\alpha)|b_{n,p-k+1}| \leq \alpha - 1.$$

For $r \in (0, 1)$, let

$$F_r(z) = z + \sum_{k=1}^{p} r^{2(k-1)}\Big(\sum_{n=2}^{\infty} a_{n,p-k+1}z^n + \sum_{n=1}^{\infty} \overline{b}_{n,p-k+1}\bar{z}^n\Big),$$

for $z \in D$. Obviously, F_r is harmonic.

It is known that if a harmonic mapping f with the normalization $f_z(0) = 1$ satisfies

$$\sum_{n=2}^{\infty} \frac{n-\beta}{1-\beta}|a_n| + \sum_{n=1}^{\infty} \frac{n+\beta}{1-\beta}|b_n| \leq 1,$$

then $f \in S_H^*(\beta) = S_{H_1}^*(\beta)$, where $\beta \in [0,1)$ (cf. [9, Theorem 1]).

Let $\beta_0 = (4-3\alpha)/(3-2\alpha)$. Then

$$\beta_0 \leq \frac{(2-\alpha)n-\alpha}{n-2\alpha+1}, \ \beta_0 \leq \frac{(2-\alpha)n+\alpha}{n+2\alpha-1},$$

for $n \geq 2$ and

$$\frac{1+\beta_0}{1-\beta_0} \leq \frac{\alpha+1}{\alpha-1}.$$

Hence

$$\sum_{n=2}^{\infty} \frac{n-\beta_0}{1-\beta_0}\Big|\sum_{k=1}^{p} r^{2(k-1)} a_{n,p-k+1}\Big| + \sum_{n=1}^{\infty} \frac{n+\beta_0}{1-\beta_0}\Big|\sum_{k=1}^{p} r^{2(k-1)} b_{n,p-k+1}\Big|$$
$$\leq \sum_{k=1}^{p}\sum_{n=2}^{\infty} \frac{n-\alpha}{\alpha-1}|a_{n,p-k+1}| + \sum_{k=1}^{p}\sum_{n=1}^{\infty} \frac{n+\alpha}{\alpha-1}|b_{n,p-k+1}| \leq 1,$$

which shows $F_r \in S_H^*(\frac{4-3\alpha}{3-2\alpha}) = S_{H_1}^*(\frac{4-3\alpha}{3-2\alpha})$, that is, $\frac{d}{d\theta}(\arg F_r(r_1 e^{i\theta})) > \frac{4-3\alpha}{3-2\alpha}$ for $r_1 \in (0,1)$. Let $r_1 = r$. Then we have $\frac{d}{d\theta}(\arg F(re^{i\theta})) > \frac{4-3\alpha}{3-2\alpha}$. From $F_r \in S_H^*(\frac{4-3\alpha}{3-2\alpha})$, we know that F_r is univalent. By the univalence of F_r and the assumption $F(z_1) \neq F(z_2)$ whenever $|z_1| \neq |z_2|$, we deduce that F is univalent. Therefore $F \in S_{H_p}^*(\frac{4-3\alpha}{3-2\alpha})$.

Similarly, we can show that if $F \in H_p$ satisfies Inequality-II for some $k_0 \in (0,1]$ and α $(1 < \alpha \leq \min\{\frac{k_0+2}{2}, \frac{4}{3}\})$ with $a_{1,p-k+1} = 0$ $(k \in \{2,\cdots,p\})$ and $F(z_1) \neq F(z_2)$ whenever $|z_1| \neq |z_2|$, then $F \in K_{H_p}^*(\frac{4-3\alpha}{3-2\alpha})$.

References

[1] Z. Abdulhadi and Y. Abu-Muhanna, Landau's theorem for biharmonic mappings, *J. Math. Anal. Appl.* 338 (2008), 705-709.

[2] Z. Abdulhadi, Y. Abu-Muhanna and S. Khuri, On univalent solutions of the biharmonic equations, *J. Inequal. Appl.* 5 (2005), 469-478.

[3] Z. Abdulhadi, Y. Abu-Muhanna and S. Khuri, On some properties of solutions of the biharmonic equation, *Appl. Math. Comput.* 177 (2006), 346-351.

[4] Y. Avci and E. Zlotkiewicz, On harmonic univalent mappings, *Ann. Univ. Mariae Curie-Sklodowska, Sect. A* 44 (1990), 1-7.

[5] Sh. Chen, S. Ponnusamy and X. Wang, Some properties of planar p-harmonic mappings, *to appear.*

[6] J. G. Clunie and T. Sheil-Small, Harmonic univalent functions, *Ann. Acad. Sci. Fenn. Ser. A. I.* 9 (1984), 3-25.

[7] P. Duren, *Harmonic mappings in the plane,* Cambridge Univ. Press, New York, 2004.

[8] J. Happel and H. Brenner, *Low reynolds number hydrodynamics,* Prentice-Hall, Englewood Cliffs, 1965.

[9] J. M. Jahangiri, Harmonic functions starlike in the unit disk, *J. Math. Anal. Appl.* 235 (1999), 470-477.

[10] J. M. Jahangiri, Coefficient bounds and univalence criteria for harmonic functions with negative coefficients, *Ann. Univ. Marie Curie-Sklodowska Sect. A* 52 (1998), 57-66.

[11] A. Janteng, Properties of harmonic functions which are convex of order β with respect to conjugate points, *Int. Journal of Math. Analysis* 1 (2007), 1031-1039.

[12] S. A. Khuri, Biorthogonal series solution of Stokes flow problems in sectorial regions, *SIAM J. Appl. Math.* 56 (1996), 19-39.

[13] W. E. Langlois, *Slow viscous flow,* Macmillan Co., New York, 1964.

[14] S. Owa and J. Nishiwaki, Coefficient estimates for certain classes of analytic functions, *J. Inequal. Pure Appl. Math.* 3 (2002), Article 72, 1-5 (electronic).

[15] S. Owa and H. M. Srivastava, Some generalized convolution properties associated with certain subclasses of analytic functions, *J. Inequal. Pure Appl. Math.* 3 (2002), Article 42, 1-13 (electronic).

[16] H. Silverman, Integral means for univalent functions with negative coefficients, *Houston J. Math.* 23 (1997), 169-174.

[17] H. Silverman and E. M. Silvia, Subclasses of harmonic univalent functions, *N. Z. J. Math.* 28 (1999), 275-284.

[18] H. M. Srivastava and S. Owa, *Current topics in analytic function theory,* World Scientific Publishing Co., Singapore, 1992.

[19] B. A. Stephen, P. Nirmaladevi, T. V. Sudharsan and K. G. Subramanian, A class of harmonic meromorphic functions with negative coefficients, *Chamchuri J. Math.* 1 (2009), 87-94.

[20] B. A. Uralegaddi, M. D. Ganigi and S. M. Sarangi, Univalent functions with positive coefficients, *Tamkang J. Math.* 25 (1994), 225-230.

[21] B. A. Uralegaddi and A. R. Desal, Convolutions of univalent functions with positive coefficients, *Tamkang J.Math.* 25 (1998), 279-285.

[22] S. Yalçin and M. Öztürk, On a subclass of certain convex harmonic functions, *J. Korean Math. Soc.* 43 (2006), 803-813.

HOLOMORPHIC FUNCTION SPACES AND THEIR PROPERTIES IN CLIFFORD ANALYSIS[1]

YU-YING QIAO[a], HE-JU YANG[b] HAI-YING QIAO[c] and JING HU[c]

[a] *College of Mathematics and Information Science, Hebei Normal University, Shijiazhuang 050016, China*

E-mail: qiaoyuying@hebtu.edu.cn

[b] *College of Science, Hebei University of Science and Technology, Shijiazhuang 050018, China*

E-mail: earnestqin7384@sina.com

[c] *Hebei Radio and TV university, Shijiazhuang 050071, China*

E-mail : qhaiying@hebnetu.edu.cn, hujing@hebnetu.edu.cn

In this paper, we list some holomorphic function spaces in Clifford analysis and give their properties. Firstly, we give some properties of regular function space. Next we study hypermonogenic function space and its properties. Finally we study k-holomorphic function and give some of its properties in unbounded domain.

Keywords: Holomorphic function space, regular function, hypermonogenic function, k-holomorphic function, Clifford analysis.

AMS No: 30G30, 30G35.

1. Clifford Algebra Cl_n and Regular Function

Let Cl_n be a real Clifford algebra over a n-dimensional real vector space $\mathbb{R}^n$ with orthogonal basis $e := \{e_1, \cdots, e_n\}$. Then Cl_n has its basis

$$1,\ e_1, \cdots, e_n;\ e_1e_2, \cdots, e_{n-1}e_n;\ \cdots; e_1 \cdots e_n.$$

Hence an arbitrary element of the basis may be written as $e_A = e_{\alpha_1} \cdots e_{\alpha_h}$, where $A = \{\alpha_1, \cdots, \alpha_h\} \subseteq \{1, \cdots, n\}$, $1 \leq \alpha_1 < \alpha_2 < \cdots < \alpha_h \leq n$ and $e_\emptyset = 1$,

$$e_ie_j = \begin{cases} -e_je_i, & i \neq j, \\ -1, & i = j, \end{cases} \quad i, j = 1, \cdots, n.$$

Clifford number is

$$\begin{aligned} a &= \sum_A x_A e_A = \sum_{A, n \notin A} x_A e_A + \Big(\sum_{A, n \in A} x_A e_{A/n} \Big) e_n \\ &= Pa + (Qa)e_n, \end{aligned}$$

[1] This research is supported by NSFC (No.10771049)

where $x_A(\in \mathbb{R})$ are real numbers. We call $Pa,\ Qa \in Cl_{n-1}$ the P part and Q part of a.

And we define some involutions:

$$' : Cl_n \to Cl_n,\ (e_0)' = e_0,\ (e_i)' = -e_i,\ i=1,\cdots,n,$$

$$\widehat{\ }: Cl_n \to Cl_n,\ \widehat{e_n} = -e_n,\ \widehat{e_i} = e_i,\ i = 0,\cdots,n-1.$$

Clifford analysis deals with the function $f : \Omega \subset \mathbb{R}^{n+1} \longrightarrow Cl_n,\ x = x_0 + x_1e_1 + \cdots + x_ne_n,\ f(x) = \sum\limits_A f_A(x)e_A$. Dirac operator is defined as

$$Df(x) = \sum_{i=0}^{n} e_i \frac{\partial f(x)}{\partial x_i} = \sum_{i=0}^{n} e_i \sum_A \frac{\partial f_A(x)}{\partial x_i} e_A.$$

If $Df(x) = 0$ in domain Ω, we call $f(x)$ is a left regular (monogenic or holomorphic) function in Ω. Similarly we have defined right Dirac operation and right regular (monogenic or holomorphic) functions. In Clifford analysis this function space was firstly discussed by F. Brack, R. Delanghe and F. Sommen[1]. Huang Sha, Qiao Yuying et al. have done some work about the properties of this function Space. They also studied bimonogenic function in Clifford analysis and published a book basic on these results[2].

Remark. The regular function in Clifford analysis has many properties of monogenic function in complex, such as Cauchy integral formula, Taylor series, extension theorem and uniqueness theorem. But x and x^n are not regular (monogenic or holomorphic) functions.

2. Hypermonogenic Function Space

2.1 Hypermonogenic function

Definition 2.1. Let

$$Mf(x) = Df(x) + (n-1)\frac{Q'f}{x_n}.$$

If $Mf = 0$, then we call f is a hypermonogenic function.

In Clifford analysis, the hypermonogenic function have been discussed since 1992 by S. L. Eriksson, H. Leutwiler and J. Ryan[3]–[4]. It is the generalization of holomorphic function with the hyperbolic metric. S. L. Eriksson studied this function firstly and gave the Cauchy formula[3]. J. Ryan, Qiao Yuying and S. Bernstein have done some work about this function[4],[5].

2.2 Bihypermonogenic function

Definition 2.2. Suppose that Ω_1 and Ω_2 are open subsets of $\mathbb{R}^{m+1}\backslash\{x_m \leq 0\}$ and $\mathbb{R}^{k+1}\backslash\{y_k \leq 0\}$ respectively. If

$$\begin{cases} M_x^l f(x,y) = 0, \\ M_y^r f(x,y) = 0, \end{cases} \quad \text{in } \Omega_1 \times \Omega_2,$$

then the function $f(x,y)$ is called a bihypermonogenic function in $\Omega_1 \times \Omega_2$.

2.3 A equivalent condition of hypermonogenic function

Theorem 2.1. *Let $f(z) : \Omega \subset \mathbb{R}^{n+1} \to Cl_n$ be a function, $f \in C^1(\Omega)$. Then f is a hypermonogenic function if and only if*

$$\begin{cases} \dfrac{\partial f_A}{\partial x_0} - \displaystyle\sum_{j=1}^{n} \delta_{\overline{jA}} \dfrac{\partial f_{\overline{jA}}}{\partial x_j} = 0, \; n \in A, \\ \dfrac{\partial f_A}{\partial x_0} - \displaystyle\sum_{j=1}^{n} \delta_{\overline{jA}} \dfrac{\partial f_{\overline{jA}}}{\partial x_j} + \dfrac{n-1}{x_n}(-1)^{|A|} f_{\overline{nA}} = 0, \; n \notin A, \end{cases}$$

where $\delta_{\overline{jA}}$ is the symbol as stated in reference [2].

2.4 Cauchy-Riemann form condition

Theorem 2.2. *Let Ω be an open subset of $\mathbb{R}_+^{n+1}$ and $f : \Omega \to Cl_n$, $f \in C^2(\Omega)$. Then f is a k-hypermonogenic function if and only if*

$$\begin{cases} \dfrac{\partial P' f(x)}{\partial x_n} + \displaystyle\sum_{i=0}^{n-1} e_i \dfrac{\partial Q f(x)}{\partial x_i} = 0, \\ \displaystyle\sum_{i=0}^{n-1} e_i \dfrac{\partial P f(x)}{\partial x_i} - \dfrac{\partial Q' f(x)}{\partial x_n} + k \dfrac{Q' f(x)}{x_n} = 0. \end{cases}$$

2.5 The extension theorem of hypermonogenic function

Theorem 2.3. (Extension theorem) *Let Ω be a set as stated above and $\Omega' = \Omega \bigcap \{(x_0, x_1, \ldots, x_{n-1}, a_n)\}$, where $a_n > 0$ be a constant. If $f \in C^1(\Omega, Cl_n)$ and f is hypermonogenic in $\Omega\backslash\Omega'$, then f is a hypermonogenic function in Ω.*

2.6 The uniqueness theorem of hypermonogenic function

Theorem 2.4. (Uniqueness theorem) *Let Ω be the set as stated above, $f(x)$ be a hypermonogenic function in Ω and $f(x) \in C^1(\Omega, Cl_n)$. Then if $f(x) = 0$ in $\widetilde{\bigwedge} \subset \Omega \bigcap \{x_n = a_n\} \neq \phi (a_n > 0)$, we have $f(x) = 0$ in Ω.*

2.7 Cauchy integral formula and Plemelj formula for bihypermonogenic functions

The integral

$$\begin{aligned}\phi(t_1,t_2)=\lambda\int_{\partial\Omega_1\times\partial\Omega_2} & E_m(\mu,t_1)d\sigma_0(\mu)\varphi(\mu,v)d\sigma_0(v)E_k(v,t_2)\\ -\lambda\int_{\partial\Omega_1\times\partial\Omega_2} & E_m(\mu,t_1)d\sigma_0(\mu)\underbrace{\varphi(\mu,v)}\,\underbrace{d\sigma_0(v)}\,M_k(v,t_2)\\ -\lambda\int_{\partial\Omega_1\times\partial\Omega_2} & M_m(\mu,t_1)\widehat{d\sigma_0(\mu)}\widehat{\varphi(\mu,v)}d\sigma_0(v)E_k(v,t_2)\\ +\lambda\int_{\partial\Omega_1\times\partial\Omega_2} & M_m(\mu,t_1)\widehat{d\sigma_0(\mu)}\,\underbrace{\widehat{\varphi(\mu,v)}}\,\underbrace{d\sigma_0(v)}\,M_k(v,t_2)\end{aligned}$$

is called a singular integral on $\partial\Omega_1\times\partial\Omega_2$, where

$$\begin{aligned}\lambda &= \frac{2^{m-1}x_m^{m-1}2^{k-1}y_k^{k-1}}{\omega_{m+1}\omega_{k+1}},\\ E_l(\mu,x) &= \frac{(\mu-x)^{-1}}{|\mu-x|^{l-1}|\mu-\widehat{x}|^{l-1}},\quad (l=m,\,k),\\ M_l(\mu,x) &= \frac{(\widehat{\mu}-x)^{-1}}{|\mu-x|^{l-1}|\mu-\widehat{x}|^{l-1}},\quad (l=m,\,k),\end{aligned}$$

in which $\hat{e_i}=e_i$, $i=0,1,2,\cdots,m-1$, $\widehat{e_m}=-e_m$, $\underbrace{e_i}=e_i, i=0,1,2,\cdots,k-1$, $\underbrace{e_k}=-e_k$.

If $\lim\limits_{\delta\to 0}\phi_\delta(t_1,t_2)=I$, where

$$\begin{aligned}\phi_\delta(t_1,t_2)=\lambda\int_{\partial\Omega_1\times\partial\Omega_2-\lambda_\delta} & E_m(\mu,t_1)d\sigma_0(\mu)\varphi(\mu,v)d\sigma_0(v)E_k(v,t_2)\\ -\lambda\int_{\partial\Omega_1\times\partial\Omega_2-\lambda_\delta} & E_m(\mu,t_1)d\sigma_0(\mu)\underbrace{\varphi(\mu,v)}\,\underbrace{d\sigma_0(v)}\,M_k(v,t_2)\\ -\lambda\int_{\partial\Omega_1\times\partial\Omega_2-\lambda_\delta} & M_m(\mu,t_1)\widehat{d\sigma_0(\mu)}\widehat{\varphi(\mu,v)}d\sigma_0(v)E_k(v,t_2)\\ +\lambda\int_{\partial\Omega_1\times\partial\Omega_2-\lambda_\delta} & M_m(\mu,t_1)\widehat{d\sigma_0(\mu)}\,\underbrace{\widehat{\varphi(\mu,v)}}\,\underbrace{d\sigma_0(v)}\,M_k(v,t_2),\end{aligned}$$

then I is called the Cauchy principal value of the singular integral and written as $I=\phi(t_1,t_2)$.

Theorem 2.5. (Cauchy type formula) *Suppose $\Omega^{'}$ and $\Omega^{''}$ be open subsets of $\mathbb{R}^{m+1}\backslash\{x_m\le 0\}$ and $\mathbb{R}^{k+1}\backslash\{y_k\le 0\}$ respectively. Let Ω_1,Ω_2*

satisfy $\overline{\Omega_1} \subset \Omega', \overline{\Omega_2} \subset \Omega''$ *respectively. If* $\varphi(x,y)$ *is a bihypermonogenic function in* $\Omega' \times \Omega''$, $x \in \Omega_1$ *and* $y \in \Omega_2$, *then*

$$\begin{aligned}\varphi(x,y) = \lambda \int_{\partial\Omega_1\times\partial\Omega_2} & E_m(\mu,x)d\sigma_0(\mu)\varphi(\mu,v)d\sigma_0(v)E_k(v,y) \\ -\lambda \int_{\partial\Omega_1\times\partial\Omega_2} & E_m(\mu,x)d\sigma_0(\mu)\underbrace{\varphi(\mu,v)}\,\underbrace{d\sigma_0(v)}\,M_k(v,y) \\ -\lambda \int_{\partial\Omega_1\times\partial\Omega_2} & M_m(\mu,x)\widehat{d\sigma_0(\mu)}\widehat{\varphi(\mu,v)}d\sigma_0(v)E_k(v,y) \\ +\lambda \int_{\partial\Omega_1\times\partial\Omega_2} & M_m(\mu,x)\widehat{d\sigma_0(\mu)}\,\underbrace{\widehat{\varphi(\mu,v)}}\,\underbrace{d\sigma_0(v)}\,M_k(v,y).\end{aligned}$$

3. K-holomorphic Function

3.1 Definition of K-holomorphic function

Definition 3.1. Let $\Omega \subset \mathbb{R}^n$ be a nonempty and open set, $f \in C^{(r)}(\Omega, Cl_n)$ $(r \geq k,\ k < n)$, $D_* f(x) = \sum\limits_{i=1}^{n} e_i \frac{\partial f}{\partial x_i}$. If $D_*^k f(x) = 0$ $(f(x)D_*^k = 0)$ for any $x \in \Omega$, then f is called a left $k-$regular (right $k-$regular) function on Ω. Usually left $k-$regular function is called a $k-$regular function for short.

Remark. Let $H_l^k(\Omega)$ $(H_r^k(\Omega))$ denote all the $k-$regular functions (right $k-$regular functions) on Ω. If $k_1 < k_2$, then $H_l^{k_1}(\Omega) \subset H_l^{k_2}(\Omega)$ $(H_r^{k_1}(\Omega) \subset H_r^{k_2}(\Omega))$.

For any natural number k, $H_l^k(\Omega)$ is not empty. The higher order kernel functions introduced later in this paper are examples.

3.2 Higher order kernel function

$$H_j^*(x) = \frac{A_j}{\omega_n} \frac{\overline{x}^j}{|\, x\,|^n}, \quad j < n, \tag{1}$$

$$A_j = \begin{cases} \dfrac{1}{2^{i-1}(i-1)!\prod_{r=1}^{i}(2r-n)}, & j=2i,\ j<n,\ i=1,2,\cdots, \\ \dfrac{1}{2^i i!\prod_{r=1}^{i}(2r-n)}, & j=2i+1,\ j<n,\ i=0,1,\cdots, \end{cases} \tag{2}$$

where $x \in \mathbb{R}_0^n$, $\omega_n = \dfrac{2\pi^{\frac{n}{2}}}{\Gamma(\frac{n}{2})}$ is the area of the unit sphere in $\mathbb{R}^n$.

3.3 Cauchy integral formula of k-holomorphic function

Lemma 3.1[6]. *Let* Ω *be a bounded, connected and open set in* $\mathbb{R}^n$, $\partial\Omega$ *be its boundary,* $f \in C^{(r)}(\Omega \bigcup \partial\Omega, Cl_n)$, $k \leq r$, $k < n$, *and* $H_j^*(x)$ *be defined* *in* (1). *Then for any* $z \in \Omega$, *we have*

$$f(z) = \sum_{j=0}^{k-1}(-1)^j \int_{\partial\Omega} H_{j+1}^*(x-z)d\sigma_x D_*^j f(x) + (-1)^k \int_{\Omega} H_k^*(x-z) \cdot D_*^k f(x)dx^n.$$

Let U be an unbounded, connected and open set, ∂U be its boundary and $R^n \setminus \overline{U}$ be not an empty set. Then we can obtain some properties of k-holomorphic function in U.

Theorem 3.1. *Let* U *and* ∂U *be defined as above,* $f \in C^{(r)}(U \bigcup \partial U, Cl_n)$ $(r \geq k,\ k < n)$, $\mid D_*^j f(x) \mid \leq C \mid x \mid^{-j+s}$ $(s \in [0,1),\ j = 0,1,\cdots,k-1)$, *and* C *be a positive real constant. Then for any* $z \in U$, *we have*

$$f(z) = \sum_{j=0}^{k-1}(-1)^j \int_{\partial U} G_{j+1}^*(x,z)d\sigma_x D_*^j f(x) + (-1)^k \int_{U} G_k^*(x,z)(D_*^k f(x))dx^n, \tag{3}$$

or

$$f(z) = \sum_{j=0}^{k-1}(-1)^j \int_{\partial U} f(x) D_*^j d\sigma_x G_{j+1}^*(x,z) + (-1)^k \int_{U} (f(x)D_*^k) G_k^*(x,z)dx^n, \tag{4}$$

where $G_{j+1}^*(x,z) = H_{j+1}^*(x-z) - H_{j+1}^*(x-z_0)$ *and* $z_0 \in R^n \setminus \overline{U}$ *is a fixed point.*

Theorem 3.2. (Higher order Cauchy integral formula) *Let* U *be defined as above,* $f \in H^k(U)$ *and* $\mid D_*^j f(x) \mid \leq C \mid x \mid^{-j+s}$, $s \in [0,1)$, $j = 0,1,\cdots,k-1$. *Then*

$$\sum_{j=0}^{k-1}(-1)^j \int_{\partial U} G_{j+1}^*(x,z)d\sigma_x D_*^j f(x) = \begin{cases} f(z), & z \in U, \\ 0, & z \in \mathbb{R}^n \setminus \overline{U}, \end{cases} \tag{5}$$

or

$$\sum_{j=0}^{k-1}(-1)^j \int_{\partial U} (f(x)D_*^j)d\sigma_x G_{j+1}^*(x,z) = \begin{cases} f(z), & z \in U, \\ 0, & z \in \mathbb{R}^n \setminus \overline{U}, \end{cases} \tag{6}$$

where $G_{j+1}^*(x,z)$ *is the function in Theorem* 3.1.

Theorem 3.3. (Cauchy inequality) *Let* U *be defined as above,* $f \in H^k(U)$ $(r \geq k,\ k < n)$, *and* $|D_*^j f(x)| \leq C|x|^{-j+s}$ $(s \in [0,1), j = 0,1,\cdots,k-1)$. *Then for any* $a \in U$, $\overline{B(a,r)} \subset U$, *we have*

$$\left| \frac{\partial^p f(z)}{\partial z_{k_1} \cdots \partial z_{k_p}} \right|_{z=a} \leq \frac{C_p(n)}{r^p},$$

where $C_p(n)$ is a constant depending on $D_*^j f (j = 0, \cdots, k-1)$, p, n and r.

3.4 Boundary behavior of higher order Cauchy-type integrals over unbounded domain

The integral

$$\Phi(z) = \sum_{j=0}^{k-1}(-1)^j \int_{\partial U} G_{j+1}^*(x,z)d\sigma_x D_*^j f(x) \tag{7}$$

is called a higher order Cauchy-type integral, where U and ∂U are defined as above, $G_{j+1}^*(x,z) = H_{j+1}^*(x-z) - H_{j+1}^*(x-z_0)$, $f \in C^{(r)}(\partial U, Cl_n)\ (r \geq k)$, $z_0 \in \mathbb{R}^n \backslash \overline{U}$ is a fixed point, $x \in \partial U$, $\mid D_*^j f(x) \mid \leq C \mid x \mid^{-j+s}$ $(s \in [0,1),\ j = 0, 1, \cdots, k-1)$, $z \in \mathbb{R}^n \setminus U$ and C is a real constant.

If $z \in \partial U$, we construct a sphere E with the center at z and radius $\delta > 0$, where ∂U is divided into two parts by E and the part of ∂U lying in the interior of E is denoted by λ_δ. If $\lim\limits_{\delta \to 0} \Phi_\delta(z) = I$, in which

$$\Phi_\delta(z) = \sum_{j=0}^{k-1}(-1)^j \int_{\partial U - \lambda_\delta} G_{j+1}^*(x,z)d\sigma_x D_*^j f(x),$$

then I is called the Cauchy principal value of the singular integral and denoted by $I = \Phi(z)$.

Lemma 3.2. *Let U and ∂U be defined as above. Suppose that $\psi : \partial U \to Cl_n$ is bounded and Hölder continuous with exponent $s \in (0,1)$. Then the integral*

$$\int_{\partial U} G_1^*(x,z)d\sigma_x \psi(x)$$

is well defined for each $x \in \partial U$ in the context of Cauchy principal value.

Theorem 3.4. *Let U and ∂U be defined as above, $f \in C^{(r)}(\partial U, Cl_n)\ (r \geq k,\ k < n)$, and $\mid D_*^j f(x) \mid \leq C \mid x \mid^{-j+s}$ $(s \in [0,1),\ j = 0, 1, \cdots, k-1)$. Then for any $z \in \partial U$, $\Phi(z)$ is well defined in the context of Cauchy principal value.*

Lemma 3.3. *For any $x, z, z_0 \in \mathbb{R}^n$, when $j \geq 0$, we have*

$$\left| \frac{(x-z)^{j+1}}{\mid x-z \mid^n} - \frac{(x-z_0)^{j+1}}{\mid x-z_0 \mid^n} \right|$$

$$\leq \left[\sum_{k=1}^{n-1} \frac{(\mid x \mid + \mid z_0 \mid)^j}{\mid x-z \mid^k \mid x-z_0 \mid^{n-k}} + \sum_{k=1}^{j} \frac{(\mid x \mid + \mid z_0 \mid)^{j-k}}{\mid x-z \mid^{n-k}} \right].$$

Lemma 3.4. *Let U and ∂U be as above. Suppose that $\psi : \partial U \to Cl_n$ is Hölder continuous with exponent $s \in (0,1)$, $t \in \partial U$, and $\Psi(z) =$*

$\int_{\partial U} G_1^*(x,z)d\sigma_x\psi(x)$. *Then*

$$\Psi^{\pm}(t) = \pm\frac{1}{2}\psi(t) + \int_{\partial U} G_1^*(x,t)d\sigma_x\psi(x).$$

Theorem 3.5. (Plemelj formula) *Let U and ∂U be as above, $f \in C^{(r)}(\partial U, Cl_n)\,(r \geq k,\, k < n)$, and $\mid D_*^j f(x) \mid \leq C \mid x \mid^{-j+s}$ $(s \in [0,1),\, j = 0,1,\cdots,k-1)$. Then*

$$\begin{cases} \Phi^+(t) = \dfrac{1}{2}f(t) + \Phi(t), \\ \Phi^-(t) = -\dfrac{1}{2}f(t) + \Phi(t). \end{cases} \tag{8}$$

Theorem 3.6. *Let U, ∂U and $G_{j+1}^*(x,z)$ be as above, $z \in \mathbb{R}^n \setminus \partial U$ and $f \in C^{(r)}(\partial U, Cl_n)\,(k \leq r,\, k < n)$. Then*

$$\Phi(z) = \sum_{j=0}^{k-1}(-1)^j \int_{\partial U} G_{j+1}^*(x,z)d\sigma_x D_*^j f(x)$$

is a $k-$regular function.

Theorem 3.7. (Extension Theorem) *Let U and ∂U be as above. If $D_*^k f(z) = 0$ for any point $z \in \mathbb{R}^n \setminus \partial U$ and or $z \in \partial U$, $f^+(z) = f^-(z)\,(f^+(z), f^-(z) \in H(\partial U, \beta), 0 < \beta < 1)$. Then $D_*^k f(z) = 0$ for $z \in \mathbb{R}^n$.*

References

[1] F. Brack, R. Delanghe and F. Sommen. *Clifford analysis,* Research Notes in Mathematics 76, Pitman Books Ltd, London, 1982.

[2] Sha Huang, Yuying Qiao and Guochun Wen, *Real and complex Clifford analysis,* Springer-Verag, Berlin, 2006.

[3] S. L. Eriksson and H. Leutwiler, Hypermonogenic functions and Möbius transformations, *Advance in Applied Clifford Algebra,* 11(2) (2001), 67-76.

[4] R. S. Kruasshar, Y. Y. Qiao and J. Ryan. Harmonic, monogenic and hypermonogenic functions on some conformably flat manifolds in rn arising from special arithmetic groups of the vahlen group, in *The p-harmonic equation and recent advances in analysis,* Contemp. Math., 370, Amer. Math. Soc., Providence, RI, 2005, 159-173.

[5] Y. Y. Qiao, S. Bernstein and S. L. Eriksson, Function theory for Laplace and Dirac-Hodge operators on hyperbolic space, *Journal D'Analyse Mathematique,* 98 (2006), 43-63.

[6] H. Begehr, J. Y. Du and Z. X. Zhang, On higher order Cauchy-Pompeiu formula in Clifford analysis and its application, *Gen. Math.,* 11 (2003), 5-26.

BOUNDARY BEHAVIOR OF SOME INTEGRAL OPERATORS IN CLIFFORD ANALYSIS[1]

LE JIANG[a] and JIN-YUAN DU[b]

[a,b] *School of Mathematics and Statistics, Wuhan University, Wuhan 430072, China*

[a] *E-mail: jlmath@126.com*

[b] *E-mail: jydu@whu.edu.cn*

In this paper, we discussed the boundary behavior of some integral operators in Clifford analysis, which are very important for representing harmonic and bi-harmonic functions. By using the standard techniques, we obtained some Privalov theorems and Plemelj formulae for these integrals.

Keywords: Integral operator, Privalov theorem, Plemelj formula.

AMS No: 30G35, 31B05, 35C10.

1. Introduction

The method of Clifford analysis theory is a powerful tool for generalizing the classical one variable complex function theory to higher dimensions. In [1–3], some kinds of jumping boundary value problems for regular functions in Clifford analysis were studied, which generalized the classical one in [4]. The Privalov theorems and Plemelj formulae play crucial roles in solving these problems. In [5,6], they studied the boundary value problems for harmonic and bi-harmonic functions. With the help of high order cauchy formula in [7,8], we can represent k-regular functions by some integrals, furthermore represent harmonic and bi-harmonic functions. Therefore, the boundary properties of these integrals are very important for solving the boundary value problems for harmonic and bi-harmonic functions. In this article, we will study these integral operators' boundary behavior with the standard method in [9,10], then obtain a series of Privalov theorems and Plemelj formulae.

2. Integral Representation for K-Regular Functions

Let $V_{n,n}$ be an n-dimensional real linear space with basis $\{e_1, e_2, \cdots e_n\}$, $C(V_{n,n})$ be the 2^n-dimensional real linear space with the basis

$$\{e_{\mathcal{A}} : \mathcal{A} = \{h_1, \cdots, h_r\} \in \mathcal{P}N,\ 1 \le h_1 < \cdots < h_r \le n\},$$

where N stands for the set $\{1, \cdots, n\}$ and $\mathcal{P}N$ denotes the family of all order-preserving subsets of N in the above way. We denote $e_\emptyset$ as e_0 and $e_{\mathcal{A}}$

[1]This research is supported by NSFC (No.10471107) and RFDP of Higher Eduction of China(No.20060486001)

as $e_{h_1\cdots h_r}$ for $\mathcal{A}=\{h_1,\cdots,h_r\}\in\mathcal{PN}$. The product on $C(V_{n,n})$ is defined by

$$\begin{cases} e_{\mathcal{A}}e_{\mathcal{B}}=(-1)^{P(\mathcal{A},\mathcal{B})}e_{\mathcal{A}\Delta\mathcal{B}}, & \text{if } \mathcal{A},\mathcal{B}\in\mathcal{PN}, \\ \lambda\mu=\sum\limits_{\mathcal{A},\mathcal{B}\in\mathcal{PN}}\lambda_{\mathcal{A}}\mu_{\mathcal{B}}e_{\mathcal{A}}e_{\mathcal{B}}, & \text{if } \lambda=\sum\limits_{\mathcal{A}\in\mathcal{PN}}\lambda_{\mathcal{A}}e_{\mathcal{A}},\ \mu=\sum\limits_{\mathcal{B}\in\mathcal{PN}}\mu_{\mathcal{B}}e_{\mathcal{B}}, \end{cases}$$

where the number $P(\mathcal{A},\mathcal{B})=\sum_{j\in\mathcal{B}}P(\mathcal{A},j),\ P(\mathcal{A},j)=\#\{i:i\in\mathcal{A},i>j\}$. e_0 is the identity element written as 1.

Thus $C(V_{n,n})$ is real, linear, associative, but noncommutative algebra and it is called a universal Clifford algebra over $V_{n,n}$ (see [11,12]).

In the sequel, we constantly use the following conjugate

$$\begin{cases} \bar{e}_i=e_i, & \text{if } i=1,2,\cdots,n, \\ \bar{\lambda}=\sum\limits_{\mathcal{A}\in\mathcal{PN}}\lambda_{\mathcal{A}}\bar{e}_{\mathcal{A}}, & \text{if } \lambda\in C(V_{n,n}). \end{cases}$$

Hence we introduce the norm on $C(V_{n,n})$ as follows

$$|\lambda|=(\sum_{\mathcal{A}\in\mathcal{PN}}\lambda_{\mathcal{A}}^2)^{\frac{1}{2}},\ \text{and}\ |\lambda+\mu|\le|\lambda|+|\mu|,\ |\lambda\mu|\le 2^n|\lambda||\mu|.$$

Let Ω be an open nonempty subset of $\mathbb{R}^n$. Functions f defined in Ω and with values in $C(V_{n,n})$ will be considered, i.e., $f:\Omega\to C(V_{n,n})$, more clearly, $f(x)=\sum_{\mathcal{A}\in\mathcal{PN}}f_{\mathcal{A}}(x)e_{\mathcal{A}}, x\in\Omega$, where $f_{\mathcal{A}}$ is the $e_{\mathcal{A}}$-component of $f(x)$. Obviously, $f_{\mathcal{A}}$ are real-valued functions in Ω. Whenever a property such as continuity and differentiability is ascribed to f, it is clear that in fact all the component functions $f_{\mathcal{A}}$ possess the cited property. So $f\in C^r(\Omega,C(V_{n,n}))$ is very clear.

In this paper, we shall consider the following operator $\mathcal{D}$:

$$\mathcal{D}=\sum_{i=1}^n e_i\frac{\partial}{\partial x_i}:C^r(\Omega,\ C(V_{n,n}))\to C^{r-1}(\Omega,\ C(V_{n,n})),$$

its action on functions from the left being governed by the rules

$$\mathcal{D}[f]=\sum_{i=1}^n\sum_{\mathcal{A}}e_ie_{\mathcal{A}}\frac{\partial f_{\mathcal{A}}}{\partial x_i}.$$

Definition 2.1. A function $f(x)\in C^r(\Omega,C(V_{n,n}))$ $(r\ge1)$ is called regular in Ω, if $\mathcal{D}[f]=0$ in Ω; A function $f(x)\in C^r(\Omega\ ,C(V_{n,n}))$ $(r\ge1)$ is called k-regular in Ω if $\mathcal{D}^k[f]=0$ in Ω.

Thus the Laplace operator satisfies $\Delta = \mathcal{D}^2$, $\Delta^2 = \mathcal{D}^4$.

Denote the fundamental solution of $\mathcal{D}^k$(k=1,2,3,4) respectively as follows, for $n > 4$

$$\begin{cases} H_1(x) = \dfrac{x}{w_n \rho^n(x)}, \\ H_2(x) = \dfrac{-1}{(n-2)w_n \rho^{n-2}(x)}, \\ H_3(x) = \dfrac{-x}{2(n-2)w_n \rho^{n-2}(x)}, \\ H_4(x) = \dfrac{1}{2(n-2)(n-4)w_n \rho^{n-4}(x)}, \end{cases}$$

where $\rho(x) = (\sum\limits_{k=1}^{n} x_k^2)^{\frac{1}{2}}$, w_n denotes the area of the unit sphere in $\mathbb{R}^n$.

Lemma 2.1 ([7, 8]). *Let Ω be an open bounded nonempty subset of $\mathbb{R}^n$ with a Liapunov boundary $\partial\Omega$, $f \in C^2(\Omega, C(V_{n,n})) \cap C^1(\overline{\Omega}, C(V_{n,n}))$, $\Delta[f] = 0$ in Ω. Then for $x \in \Omega$,*

$$f(x) = \int\limits_{\partial\Omega} H_1(y-x)d\sigma_y f(y) - \int\limits_{\partial\Omega} H_2(y-x)d\sigma_y \mathcal{D}[f](y).$$

Lemma 2.2 ([7, 8]). *Let Ω be an open bounded nonempty subset of $\mathbb{R}^n$ with a Liapunov boundary $\partial\Omega$, $f \in C^4(\Omega, C(V_{n,n})) \cap C^3(\overline{\Omega}, C(V_{n,n}))$, $\Delta^2[f] = 0$ in Ω. Then for $x \in \Omega$,*

$$\begin{aligned} f(x) = &\int\limits_{\partial\Omega} H_1(y-x)d\sigma_y f(y) - \int\limits_{\partial\Omega} H_2(y-x)d\sigma_y \mathcal{D}[f](y) \\ &+ \int\limits_{\partial\Omega} H_3(y-x)d\sigma_y \Delta[f](y) - \int\limits_{\partial\Omega} H_4(y-x)d\sigma_y \mathcal{D}^3[f](y). \end{aligned}$$

In what follows, we suppose Ω be an open bounded nonempty subset of $\mathbb{R}^n$ with a Liapunov boundary $\partial\Omega$, denote $\Omega^+ = \Omega$ and $\Omega^- = \mathbb{R}^n \setminus \overline{\Omega}$.

For $f \in C(\partial\Omega, C(V_{n,n}))$, denote

$$(S_1 f)(x) \triangleq \int\limits_{\partial\Omega} H_1(y-x)d\sigma_y f(y), \quad x \in \mathbb{R}^n,$$

$$(S_2 f)(x) \triangleq \int\limits_{\partial\Omega} H_2(y-x)d\sigma_y f(y), \quad x \in \mathbb{R}^n,$$

$$(S_3f)(x) \triangleq \int\limits_{\partial\Omega} H_3(y-x)d\sigma_y f(y), \quad x \in \mathbb{R}^n,$$

$$(S_4f)(x) \triangleq \int\limits_{\partial\Omega} H_4(y-x)d\sigma_y f(y), \quad x \in \mathbb{R}^n,$$

where for $x \in \partial\Omega$, $(S_if)(x)\,(i = 1,2,3,4)$ are considered as principal value integrals.

3. Privalov Theorem and Plemelj Formula

In this section, we discuss the boundary properties of operators $S_i\,(i = 1,2,3,4)$.

Theorem 3.1. *Let $f \in H^\alpha(\partial\Omega,\, C(V_{n,n}))\,(0<\alpha\leq 1)$, $g \in C(\partial\Omega, C(V_{n,n}))$. Then the following principal value integrals are convergent and satisfy*

i) $(S_1f)(t) \in H^{\widetilde{\alpha}}(\partial\Omega, C(V_{n,n}))$, $0<\widetilde{\alpha}\leq\alpha$,

ii) $(S_2g)(t) \in H^{1-\varepsilon}(\partial\Omega, C(V_{n,n}))$, $\forall\, 0<\varepsilon\leq 1$,

iii) $(S_3g)(t) \in H^1(\partial\Omega, C(V_{n,n}))$,

iv) $(S_4g)(t) \in H^1(\partial\Omega, C(V_{n,n}))$.

Proof. i) has been proved in [9,10].

ii) For $t_1, t_2 \in \partial\Omega$, it is enough to consider the case of $\rho(t_1-t_2)=\delta>0$ being sufficiently small,

$$\begin{aligned}
&(n-2)\Big|\,(S_2g)(t_1)-(S_2g)(t_2)\,\Big| \\
&= \Big|\frac{1}{w_n}\int\limits_{\partial\Omega}\frac{1}{\rho^{n-2}(y-t_1)}d\sigma_y g(y) - \frac{1}{w_n}\int\limits_{\partial\Omega}\frac{1}{\rho^{n-2}(y-t_2)}d\sigma_y g(y)\Big| \\
&\leq \Big|\frac{1}{w_n}\int\limits_{\partial\Omega\setminus B(t_1,2\delta)}\Big(\frac{1}{\rho^{n-2}(y-t_1)}-\frac{1}{\rho^{n-2}(y-t_2)}\Big|\Big)d\sigma_y g(y)\Big| \\
&\quad +\Big|\,\frac{1}{w_n}\int\limits_{\partial\Omega\cap B(t_1,2\delta)}\frac{1}{\rho^{n-2}(y-t_1)}d\sigma_y g(y)\,\Big| \\
&\quad +\Big|\frac{1}{w_n}\int\limits_{\partial\Omega\cap B(t_2,4\delta)}\frac{1}{\rho^{n-2}(y-t_2)}d\sigma_y g(y)\Big| \\
&\triangleq I_1+I_2+I_3.
\end{aligned}$$

For $y \in \partial\Omega\setminus B(t_1,2\delta)$, we have $\rho(y-t_1)\leq 2\rho(y-t_2)$. For any $t_1,t_2,y\in\mathbb{R}^n$, $t_1,t_2\neq y$,

$$
\begin{aligned}
&\left|\frac{1}{\rho^{n-2}(y-t_1)}-\frac{1}{\rho^{n-2}(y-t_2)}\right| \\
\le &\left|\frac{1}{\rho^{n-2}(y-t_1)}-\frac{1}{\rho^{n-3}(y-t_1)\rho(y-t_2)}+\frac{1}{\rho^{n-3}(y-t_1)\rho(y-t_2)}\right. \\
&\left.-\cdots+\frac{1}{\rho(y-t_1)\rho^{n-3}(y-t_2)}-\frac{1}{\rho^{n-2}(y-t_2)}\right| \\
\le &\frac{\rho(t_1-t_2)}{\rho^{n-2}(y-t_1)\rho(y-t_2)}+\frac{\rho(t_1-t_2)}{\rho^{n-3}(y-t_1)\rho^2(y-t_2)}+\cdots+ \\
&\frac{\rho(t_1-t_2)}{\rho(y-t_1)\rho^{n-2}(y-t_2)}=\delta\cdot\sum_{i=1}^{n-2}\frac{1}{\rho^{n-1-i}(y-t_1)\rho^i(y-t_2)}.
\end{aligned}
\tag{3.1}
$$

So

$$
I_1 \le \frac{2^{2n-1}\cdot\delta \sup\limits_{y\in\partial\Omega}|g(y)|}{w_n}\int\limits_{\partial\Omega\setminus B(t_1,2\delta)}\frac{1}{\rho^{n-1}(y-t_1)}dS,
$$

$$
\begin{aligned}
I_2 &\le \frac{2^n \sup\limits_{y\in\partial\Omega}|g(y)|}{w_n}\int\limits_{\partial\Omega\bigcap B(t_1,2\delta)}\frac{1}{\rho^{n-2}(y-t_1)}dS \\
&\le M(n)\int_0^{2\delta}dr\int_0^{\pi}\sin^{n-3}\varphi_1 d\varphi_1\cdots\int_0^{\pi}\sin\varphi_{n-3}d\varphi_{n-3}\int_0^{2\pi}d\varphi_{n-2} \\
&\le M(n)\int_0^{2\delta}dr\le M(n)\delta,
\end{aligned}
$$

$$
I_3 \le \frac{2^n \sup\limits_{y\in\partial\Omega}|g(y)|}{w_n}\int\limits_{\partial\Omega\bigcap B(t_2,4\delta)}\frac{1}{\rho^{n-2}(y-t_2)}dS\le M(n)\int_0^{4\delta}dr\le M(n)\delta,
$$

where $M(n)$ denotes a nonnegative constant, depending on the quantities in the parentheses, and $M(n)$ always changes in different inequality.

Then there exists a constant δ_0 independent of t_1 and t_2, such that

$$
I_1\le M(n)\delta\int_{2\delta}^{\delta_0}\frac{dr}{r}\le M(n)\delta\,[|\log\delta_0|+|\log(2\delta)|]\le M(n)\delta^{1-\varepsilon},
$$

$$
\forall\, 0<\varepsilon<1.
$$

Thus the result ii) holds.

iii)
$$2(n-2)\Big|\ (S_3g)(t_1)-(S_3g)(t_2)\ \Big|$$
$$=\Big|\frac{1}{w_n}\int\limits_{\partial\Omega}\frac{y-t_1}{\rho^{n-2}(y-t_1)}d\sigma_y g(y)-\frac{1}{w_n}\int\limits_{\partial\Omega}\frac{y-t_2}{\rho^{n-2}(y-t_2)}d\sigma_y g(y)\Big|$$
$$\leq\Big|\frac{1}{w_n}\int\limits_{\partial\Omega\setminus B(t_1,2\delta)}\Big(\frac{y-t_1}{\rho^{n-2}(y-t_1)}-\frac{y-t_2}{\rho^{n-2}(y-t_2)}\Big)d\sigma_y g(y)\Big|$$
$$+\Big|\frac{1}{w_n}\int\limits_{\partial\Omega\cap B(t_1,2\delta)}\frac{y-t_1}{\rho^{n-2}(y-t_1)}d\sigma_y g(y)\Big|$$
$$+\Big|\frac{1}{w_n}\int\limits_{\partial\Omega\cap B(t_2,4\delta)}\frac{y-t_2}{\rho^{n-2}(y-t_2)}d\sigma_y g(y)\Big|$$
$$\triangleq I_1+I_2+I_3.$$

For any $t_1,t_2,y\in\mathbb{R}^n$, $t_1,t_2\neq y$,

$$\Big|\ \frac{y-t_1}{\rho^{n-2}(y-t_1)}-\frac{y-t_2}{\rho^{n-2}(y-t_2)}\ \Big|$$
$$\leq\Big|\ \frac{y-t_1}{\rho^{n-2}(y-t_1)}-\frac{y-t_2}{\rho^{n-2}(y-t_1)}\ \Big|+\Big|\ \frac{y-t_2}{\rho^{n-2}(y-t_1)}-\frac{y-t_2}{\rho^{n-2}(y-t_2)}\ \Big|$$
$$\leq\frac{\rho(t_1-t_2)}{\rho^{n-2}(y-t_1)}+\rho(t_1-t_2)\sum_{i=1}^{n-2}\frac{1}{\rho^{n-1-i}(y-t_1)\rho^{i-1}(y-t_2)}\ \text{(By (3.1))}.$$

So

$$I_1\leq M(n)\delta\int\limits_{\partial\Omega\setminus B(t_1,2\delta)}\frac{1}{\rho^{n-2}(y-t_1)}dS,$$

$$I_2\leq M(n)\int\limits_{\partial\Omega\bigcap B(t_1,2\delta)}\Big|\ \frac{y-t_1}{\rho^{n-2}(y-t_1)}\ \Big|dS\leq M(n)\int_0^{2\delta}rdr\leq M(n)\delta^2,$$

$$I_3\leq M(n)\int\limits_{\partial\Omega\bigcap B(t_2,4\delta)}\Big|\ \frac{y-t_2}{\rho^{n-2}(y-t_2)}\ \Big|dS\leq M(n)\int_0^{4\delta}rdr\leq M(n)\delta^2.$$

Then there exists a constant δ_0 independent of t_1 and t_2, such that

$$I_1\leq M(n)\delta\int_{2\delta}^{\delta_0}dr\leq M(n)\delta(\delta_0-2\delta).$$

Thus iii) holds. By using the same method, iv) follows immediately.

Theorem 3.2. (Privalov theorem) *Suppose* $f \in H^{\alpha}(\partial\Omega, C(V_{n,n}))$ $(0 < \alpha \leq 1)$, $g \in C(\partial\Omega, C(V_{n,n}))$. *Then*

i) $(S_1^{\pm} f)(x) \in H^{\widetilde{\alpha}}(\overline{\Omega^{\pm}}, C(V_{n,n}))$, $0 < \widetilde{\alpha} \leq \alpha$,

ii) $(S_2^{\pm} g)(x) \in H^{1-\varepsilon}(\overline{\Omega^{\pm}}, C(V_{n,n}))$, $\forall\, 0 < \varepsilon < 1$,

iii) $(S_3^{\pm} g)(x) \in H^{1}(\overline{\Omega^{\pm}}, C(V_{n,n}))$,

iv) $(S_4^{\pm} g)(x) \in H^{1}(\overline{\Omega^{\pm}}, C(V_{n,n}))$.

Proof. In the following, we only prove the results for $x \in \Omega^+$. For $x \in \Omega^-$, it can be proved similarly. Since i) has been proved in [9, 10], now we prove ii).

Case 1: For all $x_1, x_2 \in \partial\Omega$, by Theorem 3.1, we have

$$|(S_2^+ g)(x_1) - (S_2^+ g)(x_2)| \leq M(n)|x_1 - x_2|^{1-\varepsilon}, \forall\, 0 < \varepsilon < 1. \tag{3.2}$$

Case 2: For $x_1 \in \Omega, x_2 \in \partial\Omega$, denote $\rho(x_1 - x_2) = \delta > 0$ being sufficiently small. Since $\partial\Omega$ is compact, for all $x_1 \in \Omega$, there exists a point $x_{1,\partial\Omega} \in \partial\Omega$ such that $\rho(x_{1,\partial\Omega} - x_1) = \inf\limits_{y \in \partial\Omega} \rho(y - x_1)$. Obviously, $x_{1,\partial\Omega}$ exists, but is not unique necessarily. It satisfies

$$\rho(x_1 - x_{1,\partial\Omega}) \leq \delta\ ,\ \rho(x_{1,\partial\Omega} - x_2) \leq 2\rho(x_1 - x_2) = 2\delta,$$

and then

$$\begin{aligned}
&\Big| \int\limits_{\partial\Omega} \frac{1}{\rho^{n-2}(y - x_1)} d\sigma_y g(y) - \int\limits_{\partial\Omega} \frac{1}{\rho^{n-2}(y - x_2)} d\sigma_y g(y) \Big| \\
\leq &\Big| \int\limits_{\partial\Omega} \frac{1}{\rho^{n-2}(y - x_1)} d\sigma_y g(y) - \int\limits_{\partial\Omega} \frac{1}{\rho^{n-2}(y - x_{1,\partial\Omega})} d\sigma_y g(y) \Big| \\
&+ \Big| \int\limits_{\partial\Omega} \frac{1}{\rho^{n-2}(y - x_{1,\partial\Omega})} d\sigma_y g(y) \int\limits_{\partial\Omega} \frac{1}{\rho^{n-2}(y - x_2)} d\sigma_y g(y) \Big|.
\end{aligned}$$

By Theorem 3.1,

$$\Big|\int\limits_{\partial\Omega} \frac{1}{\rho^{n-2}(y - x_{1,\partial\Omega})} d\sigma_y g(y) - \int\limits_{\partial\Omega} \frac{1}{\rho^{n-2}(y - x_2)} d\sigma_y g(y)\Big| \leq M(n)\delta^{1-\varepsilon}, \tag{3.3}$$

and

$$\begin{aligned}
&\Big|\int\limits_{\partial\Omega}\frac{1}{\rho^{n-2}(y-x_1)}d\sigma_y g(y)-\int\limits_{\partial\Omega}\frac{1}{\rho^{n-2}(y-x_{1,\partial\Omega})}d\sigma_y g(y)\Big|\\
\leq\ & 2^n\sup_{y\in\partial\Omega}|g(y)|\int\limits_{\partial\Omega\backslash B(x_{1,\partial\Omega},2\delta)}\left|\frac{1}{\rho^{n-2}(y-x_1)}-\frac{1}{\rho^{n-2}(y-x_{1,\partial\Omega})}\right|dS\\
&+2^n\sup_{y\in\partial\Omega}|g(y)|\int\limits_{\partial\Omega\cap B(x_{1,\partial\Omega},2\delta)}\frac{1}{\rho^{n-2}(y-x_1)}dS\\
&+2^n\sup_{y\in\partial\Omega}|g(y)|\int\limits_{\partial\Omega\cap B(x_{1,\partial\Omega},2\delta)}\frac{1}{\rho^{n-2}(y-x_{1,\partial\Omega})}dS\\
\triangleq\ & I_1+I_2+I_3.
\end{aligned}$$

For $y\in\partial\Omega$, since $x_1,\partial\Omega$ is a nearest distance point between x_1 and $\partial\Omega$, $\rho(y-x_{1,\partial\Omega})-\rho(y-x_1)\leq\rho(x_{1,\partial\Omega}-x_1)\leq\rho(y-x_1)$, then

$$\rho(y-x_{1,\partial\Omega})\leq 2\rho(y-x_1).$$

So there exists a a constant δ_0 independent of x_1 and $x_1,\partial\Omega$, such that

$$\begin{aligned}
I_1 &\leq 2^{2n-1}\cdot\delta\sup_{y\in\partial\Omega}|g(y)|\int\limits_{\partial\Omega\backslash B(x_{1,\partial\Omega},2\delta)}\frac{1}{\rho^{n-1}(y-x_{1,\partial\Omega})}dS\\
&\leq M(n)\delta\int_{2\delta}^{\delta_0}\frac{dr}{r}\leq M(n)\delta[|\log(\delta_0)|+|\log(2\delta)|]\leq M(n)\delta^{1-\varepsilon}, \quad (3.4)\\
I_2 &\leq 2^{2n-2}\sup_{y\in\partial\Omega}|g(y)|\int\limits_{\partial\Omega\bigcap B(x_{1,\partial\Omega},2\delta)}\frac{1}{\rho^{n-2}(y-x_{1,\partial\Omega})}dS\leq M(n)\delta, \quad (3.5)\\
I_3 &\leq 2^n\sup_{y\in\partial\Omega}|g(y)|\int\limits_{\partial\Omega\bigcap B(x_{1,\partial\Omega},2\delta)}\frac{1}{\rho^{n-2}(y-x_{1,\partial\Omega})}dS\leq M(n)\delta. \quad (3.6)
\end{aligned}$$

Combining (3.3),(3.4),(3.5) and (3.6), (3.2)holds.

Case 3: For all $x_1,x_2\in\Omega$, denote $|x_1-x_2|=\delta$. Since segment $\overline{x_1x_2}$ and $\partial\Omega$ are compact, there exist $\widetilde{x}\in\overline{x_1x_2},\widetilde{x_{\partial\Omega}}\in\partial\Omega$, such that

$$\rho(\widetilde{x}-\widetilde{x_{\partial\Omega}})=\inf_{x\in\overline{x_1x_2},t\in\partial\Omega}\rho(x-t).$$

Denote $\rho(\widetilde{x}-\widetilde{x_{\partial\Omega}})=d$.

1) If d=0, then $\widetilde{x} \in \partial\Omega$. By Case 2, we have

$$\begin{aligned}
&|(S_2^+ g)(x_1) - (S_2^+ g)(x_2)| \\
&\leq |(S_2^+ g)(x_1) - (S_2^+ g)(\widetilde{x})| + |(S_2^+ g)(\widetilde{x}) - (S_2^+ g)(x_2)| \\
&\leq M(n)\delta^{1-\varepsilon}.
\end{aligned}$$

2) If $d > 0, d \leq 2\delta$, then $\rho(x_1 - \widetilde{x_{\partial\Omega}}) \leq 3\delta, \rho(x_2 - \widetilde{x_{\partial\Omega}}) \leq 3\delta$, thus by Case 2, we have

$$\begin{aligned}
&|(S_2^+ g)(x_1) - (S_2^+ g)(x_2)| \\
&\leq |(S_2^+ g)(x_1) - (S_2^+ g)(\widetilde{x_{\partial\Omega}})| + |(S_2^+ g)(\widetilde{x_{\partial\Omega}}) - (S_2^+ g)(x_2)| \\
&\leq M(n)\delta^{1-\varepsilon}.
\end{aligned}$$

3) If $d \geq 2\delta > 0$, then for all $y \in \partial\Omega$,

$$\begin{aligned}
&\rho(y - x_1) \leq 2\rho(y - x_2), \quad \rho(y - x_2) \leq 2\rho(y - x_1), \\
&\rho(y - \widetilde{x_{\partial\Omega}}) \leq 3\rho(y - x_1), \quad \rho(y - \widetilde{x_{\partial\Omega}}) \leq 3\rho(y - x_2).
\end{aligned}$$

Thus from the above inequalities, by the same technique, we have

$$|(S_2^+ g)(x_1) - (S_2^+ g)(x_2)| \leq M(n)\delta^{1-\varepsilon}.$$

So the result ii) holds, and iii), iv) follow in the same way.

Corollary 3.1. (Plemelj formula) *Suppose $f \in H^\alpha(\partial\Omega, C(V_{n,n}))$, $0 < \alpha \leq 1$, $g \in C(\partial\Omega, C(V_{n,n}))$, then for $x \in \mathbb{R}^n \setminus \partial\Omega$, we have*

$$\text{i)} \quad \lim_{\substack{x \to t \\ x \in \Omega^\pm \\ t \in \partial\Omega}} \frac{1}{w_n} \int_{\partial\Omega} \frac{y - x}{\rho^n(y - x)} d\sigma_y f(y) = \pm\frac{1}{2} f(t) + \frac{1}{w_n} \int_{\partial\Omega} \frac{y - t}{\rho^n(y - t)} d\sigma_y f(y),$$

$$\text{ii)} \quad \lim_{\substack{x \to t \\ x \in \Omega^\pm \\ t \in \partial\Omega}} \frac{1}{w_n} \int_{\partial\Omega} \frac{1}{\rho^{n-2}(y - x)} d\sigma_y g(y) = \frac{1}{w_n} \int_{\partial\Omega} \frac{1}{\rho^{n-2}(y - t)} d\sigma_y g(y),$$

$$\text{iii)} \quad \lim_{\substack{x \to t \\ x \in \Omega^\pm \\ t \in \partial\Omega}} \frac{1}{w_n} \int_{\partial\Omega} \frac{y - x}{\rho^{n-2}(y - x)} d\sigma_y g(y) = \frac{1}{w_n} \int_{\partial\Omega} \frac{y - t}{\rho^{n-2}(y - t)} d\sigma_y g(y),$$

$$\text{iv)} \quad \lim_{\substack{x \to t \\ x \in \Omega^\pm \\ t \in \partial\Omega}} \frac{1}{w_n} \int_{\partial\Omega} \frac{1}{\rho^{n-4}(y - x)} d\sigma_y g(y) = \frac{1}{w_n} \int_{\partial\Omega} \frac{1}{\rho^{n-4}(y - t)} d\sigma_y g(y).$$

Proof. It can be directly proved by Theorems 3.1 and 3.2.

References

[1] Z. Y. Xu , Riemann boundary value problems for regular functions on Clifford algebra, *Chinese Science Bulltin* 32 (23), 1987, 476–477.

[2] Z. X. Zhang and J. Y. Du, On certain Riemann boundary value problems and singular integral equations in Clifford analysis, *Chinese Math. Ann.* 22A, 4 (2001), 421–426.

[3] Y. F. Gong and J. Y. Du, A kind of Riemann and Hilbert boundary value problem for left monogenic functions in $\mathbb{R}^m (m \geq 2)$, *Complex Variables* 49 (2004), 303–318.

[4] J. K. Lu, *Boundary value problems for analytic functions,* World Sci. Publ., Singapore, 1993.

[5] K. Gürlebeck and Z. X. Zhang, Some Riemann boundary value problems in Clifford analysis, *Math. Meth. Appl. Sci.* 33 (2010), 287–302.

[6] K. Gürlebeck and U. Kähler, On a boundary value problem of biharmonic equation, *Math. Meth. Appl. Sci.* 20 (1997), 867–883.

[7] H. Begehr, J. Y. Du and Z. X. Zhang, On higher order Cauchy-pompeiu formula in Clifford analysis and its applications, *General Mathematics* 11 (2003), 5–26.

[8] Z. X. Zhang, A revised higher order Cauchy-pompeiu formula in Clifford analysis and its application, *J. Appl. Funct. Anal.* 2(3), 2004, 269-278.

[9] J. Y. Du, N. Xu and Z. X. Zhang, Boundary behavior of Cauchy-type integrals in Clifford analysis, *Acta Mathematica Scientia* 29B(1), 2009, 210–224.

[10] Z. X. Zhang, Some properties of operators in Clifford analysis, *Complex Variables and Elliptic Equations* 52 (2007), 455-473.

[11] F. Brack, R. Delanghe and F. Sommen, *Clifford analysis,* Pitman books Ltd, London, 1982.

[12] R. Delanghe, F. Sommen and V. Soućek, *Clifford algebra and Spinor-valued functions,* Kluwer Academic publishers, London, 1992.

A NOTE ON THE FUNDAMENTAL SOLUTIONS OF ITERATED DUNKL-DIRAC OPERATORS[1]

SHAN-SHAN LI[a] and MING-GANG FEI[b,c]

[a] *School of Computer Science and Technology, Southwest University for Nationalities, Chengdu 610041, China*
E-mail: shanshanli@swun.cn

[b] *Department of Mathematics, University of Aveiro, Aveiro, P-3810-193, Portugal*
E-mail: fei@ua.pt

[c] *School of Mathematical Sciences, University of Electronic Science and Technology of China, Chengdu 610054, China*
E-mail: fei@uestc.edu.cn

In this paper we give a new proof of certain identities for the fundamental solutions of iterated Dunkl-Dirac operators by using Dunkl transform.

Keywords: Reflection group, Clifford algebra, Dunkl transform, Dunkl-Dirac operator.

AMS No: 30G35, 33C80.

1. Introduction

We consider the differential-difference operators D_i, $i = 1, \cdots, d$, on $\mathbb{R}^d$, associated with a positive root system R_+, a nonnegative multiplicity function κ and index $\gamma_\kappa > 0$, introduced by C. F. Dunkl in [6] and called Dunkl operators in the literature which are invariant under finite reflection groups. These operators are very important in pure mathematics and in physics. Dunkl operators not only provide a useful tool in the study of special functions with root systems ([7,11,21]), but they are closely related to certain representations of degenerated affine Hecke algebras ([3,13]). Moreover the commutative algebra generated by these operators has been used in the study of certain exactly solvable models of quantum mechanics, namely the Calogero-Moser-Sutherland models, which describe quantum mechanical system of N identical particles on a circle or line which interact pairwise through long range potentials of inverse square type ([12,17]).

Based on Dunkl operators the authors in [2,14] were able to introduce a Dirac operator, called Dunkl-Dirac operator, which is invariant under finite reflection groups and also factorizes the Dunkl Laplacian. Starting from the latter operator the authors in [10] recently obtained the fundamental solutions of the iterated Dunkl-Dirac operators $D_h^l = (\sum_{i=1}^d \mathbf{e}_i D_i)^l$, $l \in \mathbf{Z}_+$ but for $l < \mu$, where $\mathbf{Z}_+$ denotes the set of all positive integers and $\mu = 2\gamma_\kappa + d$ is the so-called Dunkl-dimension(see Section 2). So, in this paper

[1]This research is supported by *Fundação para Ciência e a Tecnologia, Portugal, No. SFRH/BPD/41730/2007.*

we will further study the fundamental solutions of the iterated Dunkl-Dirac operators with a different approach, the so-called Dunkl transform ([4,8]), which generalizes the results in [10] to any positive integer power of Dunkl-Dirac operator.

Throughout this paper we use the convention that $c_{\alpha,\beta}$ and $c'_{\alpha,\beta}$ denote constants, depending on the parameters α and β, their values may change from line to line. Then the main result of this paper reads:

Theorem 1.1. *For any $l \in \mathbf{Z}_+$, the following function $E_{\mu,l}$ is a fundamental solution of the iterated Dunkl-Dirac operators D_h^l in $\mathbb{R}^d$:*
for μ odd,

$$E_{\mu,l}(x) = \begin{cases} c_{\mu,l}\dfrac{x}{|x|^{\mu-l+1}}, & l \text{ odd}, \\ c_{\mu,l}\dfrac{1}{|x|^{\mu-l}}, & l \text{ even}, \end{cases}$$

and for μ even,

$$E_{\mu,l}(x) = \begin{cases} c_{\mu,l}\dfrac{x}{|x|^{\mu-l+1}}, & l \text{ odd and } 1<\mu-1, \\ c_{\mu,l}\dfrac{1}{|x|^{\mu-l}}, & l \text{ even and } 1<\mu, \\ (c_{\mu,l}\log|x| + c'_{\mu,l})\dfrac{x}{|x|^{\mu-l+1}}, & l \text{ odd and } l \geq \mu-1, \\ (c_{\mu,l}\log|x| + c'_{\mu,l})\dfrac{1}{|x|^{\mu-l}}, & l \text{ even and } l \geq \mu. \end{cases}$$

2. Preliminaries and Dunkl Analysis

Let $\mathbf{e}_1, \cdots, \mathbf{e}_d$ be an orthonormal basis of $\mathbb{R}^d$ satisfying the anti-commutation relationship $\mathbf{e}_i\mathbf{e}_j + \mathbf{e}_j\mathbf{e}_i = -2\delta_{ij}$, where δ_{ij} is the Kronecker symbol. We define the universal real valued Clifford algebra $\mathbb{R}_{0,d}$ as the 2^d-dimensional associative algebra with basis given by $\mathbf{e}_0 = 1$ and $\mathbf{e}_A = \mathbf{e}_{h_1} \cdots \mathbf{e}_{h_n}$, where $A = \{h_1, h_2, \cdots, h_n\} \subset \{1, 2, \cdots, d\}$, for $1 \leq h_1 < h_2 < \cdots < h_n \leq d$. Hence, each element $x \in \mathbb{R}_{0,d}$ will be represented by $x = \sum_A x_A\mathbf{e}_A$, $x_A \in \mathbb{R}$. We now introduce the Dirac operator $\partial = \sum_{i=1}^d \mathbf{e}_i \frac{\partial}{\partial x_i}$. In particular we have that $\partial^2 = -\Delta$, where Δ is the d-dimensional Laplacian. For all what follows let Ω be an open set of $\mathbb{R}^d$. Then a function $f : \Omega \to \mathbb{R}_{0,d}$ is said to be left-monogenic (resp. right-monogenic) if it satisfies the equation $\partial f = 0$(resp. $f\partial = 0$) for each $x \in \Omega$. Basic properties of Dirac operator and left-monogenic functions can be found in [1,5].

For $\alpha \in \mathbb{R}^d\backslash\{0\}$, the reflection $\sigma_\alpha x$ of a given vector $x \in \mathbb{R}^d$ on the hyperplane $H_\alpha \subset \mathbb{R}^d$ orthogonal to α is given, in Clifford notation, by

$$\sigma_\alpha x := -\alpha x \alpha^{-1}.$$

A finite set $R \subset \mathbb{R}^d\backslash\{0\}$ is called a root system if $R\bigcap\mathbb{R}^d \cdot \alpha = \{\alpha, -\alpha\}$ and $\sigma_\alpha R = R$ for all $\alpha \in R$. For a given root system R the reflections σ_α, $\alpha \in R$, generate a finite group $W \subset O(d)$, called Coxeter group or reflection group associated to R. All reflections in W correspond to suitable pairs of roots. For a given $\beta \in \mathbb{R}^d\backslash\bigcup_{\alpha\in R} H_\alpha$, we fix the positive subsystem $R_+ = \{\alpha \in R | \langle\alpha, \beta\rangle > 0\}$, then for each $\alpha \in R$ either $\alpha \in R_+$ or $-\alpha \in R_+$.

A function $\kappa : R \to \mathbb{C}$ on a root system R is called a multiplicity function if it is invariant under the action of the associated reflection group W. If one regards κ as a function on the corresponding reflections, this means that κ is constant on the conjugacy classes of reflections in W. For abbreviation, we introduce the index $\gamma_\kappa = \sum_{\alpha\in R_+} \kappa(\alpha)$ and the weight function $h_\kappa(x) = \prod_{\alpha\in R_+} |\langle\alpha, x\rangle|^{\kappa_\alpha}$, which is homogeneous of degree γ_κ. In addition, we let $\mu = 2\gamma_\kappa + d$ which is the so-called Dunkl-dimension.

We now denote by $C^k(\mathbb{R}^d)$ the space of k-times continuously differentiable functions on $\mathbb{R}^d$ and by $\mathcal{S}(\mathbb{R}^d)$ the space of C^∞-functions on $\mathbb{R}^d$ which are rapidly decreasing as their derivatives. The Dunkl operators D_j, $j = 1, \cdots, d$, on $\mathbb{R}^d$ associated to the finite reflection group W and multiplicity function κ are given by

$$D_j f(x) = \frac{\partial}{\partial x_j} f(x) + \sum_{\alpha\in R_+} \kappa_\alpha \frac{f(x) - f(\sigma_\alpha(x))}{\langle\alpha, x\rangle} \alpha_j, \quad f \in C^1(\mathbb{R}^d).$$

In the case $\kappa = 0$, the D_j, $j = 1, \cdots, d$, reduce to the corresponding partial derivatives. In this paper, we will assume throughout that $k \geq 0$ and $\gamma > 0$. More importantly, these operators mutually commute; that is, $D_i D_j = D_j D_i$. This property allows us to define a Dunkl-Dirac operator for the reflection group W via

$$D_h f = \sum_{j=1}^{d} e_j D_j f, \ or\ f D_h = \sum_{j=1}^{d} (D_j f) e_j,$$

respectively. A function which is annihilated by the Dunkl-Dirac operator from the left or right is called left Dunkl-monogenic function or right Dunkl-monogenic function. In addition, the corresponding Dunkl Laplacian Δ_h on $\mathbb{R}^d$ is defined through $\Delta_h = -D_h^2 = \sum_{j=1}^{d} D_j^2$.

The Dunkl intertwining operator V_κ is a linear positive operator determined uniquely by([16])

$$V_\kappa \mathcal{P}_n^d \subset \mathcal{P}_n^d, \ V_\kappa \mathbf{1} = \mathbf{1}, \ D_j V_\kappa = V_\kappa \partial_j, \quad 1 \leq j \leq d,$$

where $\mathcal{P}_n^d$ is the space of homogeneous polynomials of degree n in d variables. This operator has been extended by K. Trimèche in [19] to the space $C^\infty(\mathbb{R}^d)$.

The Dunkl kernel $K(x,y)$ on $\mathbb{R}^d \times \mathbb{R}^d$ is now defined as

$$K(x,y) = \sum_{n=0}^{\infty} V_\kappa(\frac{\langle \cdot, y\rangle^n}{n!})(x) = V_\kappa(\exp(\langle \cdot, y\rangle))(x),$$

which generalizes the usual exponential function $\exp(\langle x,y\rangle)$. For $y \in \mathbb{R}^d$, the function $x \to K(x,y)$ may be also characterized as the unique analytic solution on $\mathbb{R}^d$ of the system

$$\begin{cases} D_j u(x,y) = y_j u(x,y), & j = 1, \cdots, d, \\ u(0,y) = 1, \quad \text{for all} \quad y \in \mathbb{R}^d. \end{cases} \tag{2.1}$$

This kernel has a unique holomorphic extension to $\mathbb{C}^d \times \mathbb{C}^d$. The detailed study of Dunkl kernel can be found in [7].

Now we need to define the equivalent of the Fourier transform in the Dunkl setting.

Definition 2.1. The Dunkl transform is given for any $f \in \mathcal{S}(\mathbb{R}^d)$ by

$$\hat{f}(y) = c_h \int_{\mathbb{R}^d} K(x, -iy) f(x) h_\kappa^2(x) dx,$$

where c_h is the constant defined by $c_h^{-1} = \int_{\mathbb{R}^d} e^{-|x|^2/2} h_\kappa^2(x) dx$.

Obviously, the extension of Dunkl transform to $\mathcal{S}'(\mathbb{R}^d)$ is by duality, as usual. The detailed study about Dunkl kernel and Dunkl transform can be found in [4,7,20].

Since the Dunkl transform is not invariant under the classical translation operator, we need to introduce the following translation operator (see [20]).

Definition 2.2. The Dunkl translation operators τ_y, $y \in \mathbb{R}^d$, are defined on $C^\infty(\mathbb{R}^d)$ by

$$\tau_y f(x) = V_\kappa^{(y)} V_\kappa^{(x)} [(V_\kappa)^{-1} f(x+y)], \quad \forall x \in \mathbb{R}^d,$$

where the superscript (y) in $V_\kappa^{(y)}$ means that V_κ is applied to the y-variable.

The properties of Dunkl translation operator can be found in [18,20]. Using this translation operator we have the Dunkl-convolution defined by

$$f * g(y) = \int_{\mathbb{R}^d} \tau_y f(-x) g(x) h_\kappa^2(x) dx,$$

for which the following relation holds

$$(f * g)\hat{} = \hat{f}\hat{g}. \tag{2.2}$$

In the following we will use the notation $\check{f}$ to denote the inverse Dunkl transform of the function f.

3. Proof of the Main Theorem

We will follow the approach in [15] for the unweighted case. To this end, we start with the following Lemma from [18], which we only state the special case that we will use in this paper.

Lemma 3.1. *For $0 < \beta < \mu$, the identity*

$$\left(\frac{1}{|x|^{\mu-\beta}}\right)^{\wedge} = c_{\mu,\beta}\frac{1}{|x|^{\beta}}, \quad c_{\mu,\beta} = 2^{-\mu/2+\beta}\frac{\Gamma(\beta/2)}{\Gamma(\mu/2-\beta/2)},$$

holds in the sense that

$$\int_{\mathbb{R}^d}\frac{1}{|x|^{\mu-\beta}}\hat{\phi}(x)h_\kappa^2(x)dx = c_{\mu,\beta}\int_{\mathbb{R}^d}\frac{1}{|x|^{\beta}}\phi(x)h_\kappa^2(x)dx$$

for every ϕ which is sufficiently rapidly decreasing at ∞, and whose Dunkl transform has the same property.

We are now ready to present our proof of Theorem 1.1.

Proof. Define $D_h^{-\beta}$, $\beta > 0$, by

$$D_h^{-\beta}f(x) = c_\mu\int_{\mathbb{R}^d}K(x,i\xi)(i\xi)^{-\beta}\hat{f}(\xi)h_\kappa^2(\xi)d\xi,$$

where $(i\xi)^{-\beta}$ is defined by

$$(i\xi)^{-\beta} = |\xi|^{-\beta}\chi_+(\xi) + (-|\xi|)^{-\beta}\chi_-(\xi) \text{ and } \chi_\pm(\xi) = \frac{1}{2}\left(1 \pm \frac{i\xi}{|\xi|}\right).$$

Thus, if $\beta = l$ is a positive integer, we have

$$(i\xi)^{-l} = \begin{cases} \dfrac{1}{|\xi|^l}, & l \text{ even;} \\ \dfrac{i\xi}{|\xi|^{l+1}}, & l \text{ odd.} \end{cases}$$

Therefore, using (2.1) there has

$$D_h^{-l}f(x) = \frac{c_\mu}{2}\left[\int_{\mathbb{R}^d}K(x,i\xi)|\xi|^{-l}\hat{f}(\xi)h_\kappa^2(\xi)d\xi\right.$$

$$+D_h \int_{\mathbb{R}^d} K(x, i\xi)|\xi|^{-l-1} \hat{f}(\xi) h_\kappa^2(\xi) d\xi$$
$$+ \int_{\mathbb{R}^d} K(x, i\xi)(-|\xi|)^{-l} \hat{f}(\xi) h_\kappa^2(\xi) d\xi$$
$$\left. +D_h \int_{\mathbb{R}^d} K(x, i\xi)(-|\xi|)^{-l-1} \hat{f}(\xi) h_\kappa^2(\xi) d\xi \right].$$

If $0 < l, l+1 < \mu$, by invoking relation (2.2) and Lemma 3.1 for $0 < m < \mu$, $(1/|x|^{\mu-m})^\vee = c_{\mu,m}/|x|^m$, we conclude that

$$D_h^{-l} f(x) = E_{\mu,l} * f(x), \tag{3.1}$$

where

$$E_{\mu,l}(x) = c_{\mu,l}(1 + e^{-il\pi}) \frac{1}{|x|^{\mu-l}} + c'_{\mu,l}(1 - e^{-il\pi}) D_h \frac{1}{|x|^{\mu-l-1}}.$$

Furthermore, for general $\beta > 0$, the same criterion gives

$$E_{\mu,\beta}(x) = c_{\mu,\beta}(1 + e^{-i\beta\pi}) G_{\mu,\beta}(x) + c'_{\mu,\beta}(1 - e^{-i\beta\pi}) D_h G_{\mu,\beta}(x),$$

where $G_{\mu,\beta}$ is the fundamental solution of $|D_h|^\beta$ that is the operator associated with symbol $|\xi|^\beta$.

Accordingly, from (3.1) and the spherical decomposition of D_h([9]), we can easily check that the function $E_{\mu,l}$ given by, for μ odd,

$$E_{\mu,l}(x) = \begin{cases} c_{\mu,l} \dfrac{x}{|x|^{\mu-l+1}}, & l \text{ odd}, \\ c_{\mu,l} \dfrac{1}{|x|^{\mu-l}}, & l \text{ even}, \end{cases}$$

and for μ even,

$$E_{\mu,l}(x) = \begin{cases} c_{\mu,l} \dfrac{x}{|x|^{\mu-l+1}}, & l \text{ odd and } l < \mu-1, \\ c_{\mu,l} \dfrac{1}{|x|^{\mu-l}}, & l \text{ even and } l < \mu, \\ (c_{\mu,l} \log|x| + c'_{\mu,l}) \dfrac{x}{|x|^{\mu-l+1}}, & l \text{ odd and } l \geq \mu-1, \\ (c_{\mu,l} \log|x| + c'_{\mu,l}) \dfrac{1}{|x|^{\mu-l}}, & l \text{ even and } l \geq \mu, \end{cases}$$

is a fundamental solution of D_h^l in $\mathbb{R}^d$.

Remark 3.1. We call "a" fundamental solution of D_h^l but not "the" fundamental solution of D_h^l in Theorem 1.1 since the fundamental solution of D_h^l is not unique.

Remark 3.2. Comparing with the classical case(see [15]), the crucial part in our treatment of Theorem 1.1 is the replacement of the classical Euclidean dimension d by the Dunkl-dimension μ.

References

[1] F. Brackx, R. Delanghe and F. Sommen, *Clifford analysis,* Research Notes in Mathematics, Vol. 76, Pitman Advanced Publishing Company, Boston, London, Melbourne, 1982.

[2] P. Cerejeiras, U. Kähler and G. Ren, Clifford analysis for finite reflection groups, *Complex Var. Elliptic Equ.* 51 (5-6) (2006), 487-495.

[3] L. Cherednik, A unification of the Knizhnik-Zamolodchikov equation and Dunkl operators via affine Hecke Algebras, *Invent. Math.* 106 (1991), 411-432.

[4] M. F. E. de Jeu, The Dunkl transform, *Invent. Math.* 113 (1993), 147-162.

[5] R. Delanghe, F. Sommen and V. Souček, *Clifford algebra and spinor valued functions,* A function theory for Dirac operator, Kluwer, Dordrecht, 1992.

[6] C. F. Dunkl, Differential-difference operators associated to reflection groups, *Trans. Amer. Math. Soc.* 311 (1989), 167-183.

[7] C. F. Dunkl, Integral kernels with reflection group invariance, *Canad. J. Math.* 43 (1991), 1213-1227.

[8] C. F. Dunkl, Hankel transforms associated to finite reflection groups, in *Hypergeometric functions on domains of positivity,* Jack polynomials, and applications (Tampa, FL, 1991), 123–138, Contemp. Math., 138, Amer. Math. Soc., Providence, RI, 1992.

[9] M. Fei, P. Cerejeiras and U. Kähler, Fueter's Theorem and its Generalizations in Dunkl-Clifford Analysis, *J. Phys. A: Math. Theor.* 42(2009), No. 39, 395209, 15pp.

[10] M. Fei, P. Cerejeiras and U. Kähler, Fundamental solutions of iterated Dunkl-Dirac operators and applications, submitted for publication.

[11] G. J. Heckman, An elementary approach to the hypergeometric shift operators of Opdam, *Invent. Math.* 103 (1991), 341-350.

[12] K. Hikami, Dunkl operator formalism for quantum many-body problems associated with classical root systems, *J. Phys. Soc. Japan* 65 (1996), 394-401.

[13] E. M. Opdam, Harmonic analysis for certain representations of graded Hecke algebras, *Acta Math.* 175 (1995), 75-121.

[14] B. Ørsted, P. Somberg and V. Souček, The Howe duality for Dunkl version of the Dirac operator, *Adv. Appl. Clifford Algebras* 19(2) (2009), 403-415.

[15] J. Peetre and T. Qian, Möbius covariance of iterated Dirac operators, *J. Austral. Math. Soc.* Ser. A 56(1994), No.3, 403-414.

[16] M. Rösler, Positivity of Dunkl's intertwining operator, *Duke Math. J.* 98 (1999), 445-463.

[17] M. Rösler, Dunkl Operators: Theory and Applications, in *Orthogonal polynomials and special functions* (Leuven, 2002), 93-135, Lecture Notes in Math. 1817, Springer, Berlin, 2003.

[18] S. Thangavelu and Y. Xu, Riesz transform and Riesz potentials for Dunkl transform, *J. Comput. Appl. Math.* 199 (2007), 181-195.

[19] K. Trimèche, The Dunkl intertwining operator on spaces of functions and distributions and integral representation of its dual, *Integral Transforms and Special Functions* 12(4) (2001), 349-374.

[20] K. Trimèche, Paley-Wiener theorems for the Dunkl transform and Dunkl translation operators, *Integral Transforms and Special Functions* 13 (2002), 17-38.

[21] J. F. van Diejen, Confluent hyperge2oetric orthogonal polynomials repated to the rational quantum Calogero system with harmonic confinement, *Comm. Math. Phys.* 188 (1997), 467-497.

ON THE LEFT LINEAR HILBERT PROBLEM IN CLIFFORD ANALYSIS

ZHONG-WEI SI[a] and JIN-YUAN DU[b]

[a,b]*School of Mathematics and Statistics, Wuhan University, Wuhan 430072, China*

[a]*E-mail: szhw8882005@163.com*

[b]*E-mail: jydu@whu.edu.cn*

In this paper, we consider a Hilbert boundary value problem in $\mathbb{R}_{0,m}$ Clifford analysis. Find a $\mathbb{R}_{0,m}$-valued left monogenic function $u(x)$ in $\mathbb{R}_+^m$, which can extend continuously to $\mathbb{R}_0^m$, whose positive boundary value u^+ satisfies

$$X^{(m)}\left(a(t)u^+(t)\right) = c(t),\ t \in \mathbb{R}_0^m,$$

where $c(t)$ is a $\mathbb{R}_{0,m-1}$-valued function, $a(x)$ is a given $\mathbb{R}_{0,m}$-valued function, whose inverse $a^{-1}(x)$ exsists. We may transform the Hilbert problem into a Riemann boundary value problem, in [1] the Riemann problem may be solved by the successive approximation, if $a(x)$ satisfies some conditions.

Keywords and Phrases: Riemann problem, Hilbert problem, monogenic function.

AMS No: 30G35.

1 Introduction

Recently many mathematicians have discussed the Riemann boundary value problem and the Hilbert boundary value problem in Clifford analysis. In this paper, we mainly discuss the Hilbert boundary value problem in Clifford analysis. Find an $\mathbb{R}_{0,m}$-valued left monogenic function $u(x)$ in $\mathbb{R}_+^m$, which can extend continuously to $\mathbb{R}_0^m$, whose positive boundary value u^+ satisfies

$$X^{(m)}\left(a(t)u^+(t)\right) = c(t),\ t \in \mathbb{R}_0^m,$$

where $c(t)$ is an $\mathbb{R}_{0,m-1}$-valued function, $a(x)$ is a given $\mathbb{R}_{0,m}$-valued function, whose inverse $a^{-1}(x)$ exsists. If a is a constant, the solution has been discussed in [14], but when $a(x)$ is an $\mathbb{R}_{0,m}$-valued function, the method may fail, we could not use the way again. Successive approximation for the singular integral equation in space of Hölder continuous functions was considered by Xu in [15] and Bernstein have discussed the Riemann boundary value problem in Clifford analysis when $a(x)$ is an $\mathbb{R}_{0,m}$-valued function. Here we mainly discuss the Hilbert boundary value problem by the successive approximation.

2 Preliminaries

Let $\mathbb{R}_{0,m}$ be the real Clifford algebra with generating vectors e_i, $i = 1, \cdots, n$, where, $e_i^2 = -1$ and $e_ie_j + e_je_i = 0$ if $i \neq j$ and $i, j = 1, 2, \ldots, n$.

Besides, let e_0 be the unit element. Then an arbitrary $b \in \mathbb{R}_{0,m}$ is given by $b = \sum_\beta b_\beta e_\beta$, where $b_\beta \in \mathbb{R}$ and $e_\beta = e_{\beta_1} e_{\beta_2} \cdots e_{\beta_h}$, and $\beta_1 < \beta_2 < \cdots < \beta_h$. A conjugation is defined by $\overline{b} = \sum_\beta b_\beta \overline{e_\beta}$, $\overline{e_\beta} = \overline{e_{\beta_h}} ... \overline{e_{\beta_2}} \cdot \overline{e_{\beta_1}}$ and $\overline{e_0} = e_0$, $\overline{e_j} = -e_j$, $j = 1, 2, \cdots n$. By $[b]_0 = b_0 e_0$ we denote the scalar part of b, whereas $\operatorname{Im} b = \sum_{\beta \neq 0} b_\beta e_\beta$ denotes the imaginary or multivector part.

In the sequel $x \in \mathbb{R}^m$ is represented as $x = \sum_{j=1}^m e_j x_j$ or $x = Z_x + e_m x_m$ or $x = (Z_x, x_m)$ equivalently, where $Z_x = \sum_{j=1}^{m-1} e_j x_j$. Define $\mathbb{R}^m_+ = \{x \mid x_m > 0\}$, $\mathbb{R}^m_- = \{x \mid x_m < 0\}$, $\mathbb{R}^m_0 = \{x \mid x_m = 0\}$, $\overline{\mathbb{R}^m_+} = \mathbb{R}^m_+ \cup \mathbb{R}^m_0$, $\overline{\mathbb{R}^m_-} = \mathbb{R}^m_- \cup \mathbb{R}^m_0$. Then $\mathbb{R}^m_0$ may be identified with $\mathbb{R}^{m-1}$, $\mathbb{R}^m_+, \mathbb{R}^m_-$ are the halt space locating above and below the hyperplane $\mathbb{R}^m_0$ respectively. For any $x \in \mathbb{R}^m$, since $x^2 = -|x|^2$, then $x^{-1} = -x|x|^{-2}$ is the inverse of $x \neq 0$, i.e., $x^{-1}x = xx^{-1} = 1$.

Supposed that Ω is a domain in $\mathbb{R}^m$, $D_x = \sum_{j=1}^m e_j(\partial/\partial x_j)$, $f(x) = \sum_A f_A(x) e_A$ is a $\mathbb{R}_{0,m}$-valued C^1-function, then we say $f \in C^1(\Omega; \mathbb{R}_{0,m})$ and f is left monogenic in Ω if $D_x f(x) = \sum_{j=1}^m e_j e_A (\partial f_A / \partial x_j) = 0$ as well as $f \in C^1(\Omega; \mathbb{R}_{0,m})$, and f is right monogenic in Ω if $f D_x = \sum_{j=1}^m e_A e_j (\partial f_A / \partial x_j) = 0$. Let $\mathbb{R}_{0,m-1}$ be the subalgebra of $\mathbb{R}_{0,m}$ constructed from $\{e_1, \cdots, e_{m-1}\}$, then $\mathbb{R}_{0,m}$ has the decomposition see [2]

$$\mathbb{R}_{0,m} = \mathbb{R}_{0,m-1} + e_m \mathbb{R}_{0,m-1}.$$

Thus any $\lambda \in \mathbb{R}_{0,m}$ can be decomposed as $\lambda = \lambda_1 + e_m \lambda_2$, $\lambda_1, \lambda_2 \in \mathbb{R}_{0,m-1}$, we may define $X^{(m)}(\lambda) = \lambda_1$, $Y^{(m)}(\lambda) = \lambda_2$. Particularly, $X^{(m)}(x) = Z_x$, $Y^{(m)}(x) = x_m$.

It is clear that the decomposition is the generalization of classical representation of complex number, i.e., operators $X^{(m)}$ and $Y^{(m)}$ acting on $\mathbb{R}_{0,m}$ are the generalizations of operators Re and Im acting on $\mathbb{C}$. For any $\mathbb{R}_{0,m}$-valued function $q(x) = \left(X^{(m)} q\right)(x) + e_m \left(Y^{(m)} q\right)(x)$, define the operator "$*$" as

$$q^*(x) = \left(X^{(m)} q\right)(x) - e_m \left(Y^{(m)} q\right)(x), \tag{2.1}$$

and for any $x = Z_x + x_m e_m \in \mathbb{R}^m_\pm$, $x^* = Z_x - x_m e_m \in \mathbb{R}^m_\mp$.

Theorem 2.1. *For any $\mathbb{R}_{0,m}$-valued function $a(x)$ and $b(x)$, then*

$$(a^*)^*(x) = a(x), (a(x)b(x))^* = a^*(x) b^*(x).$$

Proof.

$$\begin{aligned}(a^*)^* &= (X^{(m)}a - e_m(Y^{(m)}a))^* = X^{(m)}a + e_m(Y^{(m)}a),\\ (ab)^* &= ((X^{(m)}a + e_m(Y^{(m)}a))(X^{(m)}b + e_m(Y^{(m)}b)))^*\\ &= (X^{(m)}a)(X^{(m)}b) - (X^{(m)}a)e_m(Y^{(m)}b)\\ &\quad -e_m(Y^{(m)}a)(X^{(m)}b) + e_m(Y^{(m)}a)e_m(Y^{(m)}b),\end{aligned}$$

while

$$\begin{aligned}a^*b^* &= \left(X^{(m)}a - e_m\left(Y^{(m)}a\right)\right)\left(X^{(m)}b - e_m\left(Y^{(m)}b\right)\right)\\ &= \left(X^{(m)}a\right)\left(X^{(m)}b\right) - \left(X^{(m)}a\right)e_m\left(Y^{(m)}b\right)\\ &\quad -e_m\left(Y^{(m)}a\right)\left(X^{(m)}b\right) + e_m\left(Y^{(m)}a\right)e_m\left(Y^{(m)}b\right),\end{aligned}$$

so

$$(a^*)^*(x) = a(x), (a(x)b(x))^* = a^*(x)b^*(x).$$

Theorem 2.2. *Let $g(x)$ be a $\mathbb{R}_{0,m}$-valued left monogenic function defined in $\mathbb{R}^m$, and $g_*(x)$ be defined as*

$$g_*(x) = g^*(x^*) = \left(X^{(m)}g\right)(x^*) - e_m\left(Y^{(m)}g\right)(x^*), \tag{2.2}$$

which is the conjugate extension of g with respect to the hyperplane $x_m = 0$. Then $g_*(x)$ is a left monogenic function too.

Proof. $g(x)$ is left monogenic in $\mathbb{R}^m$ if and only if $X^{(m)}g$ and $Y^{(m)}g$ satisfy the equations

$$D_{Z_x}\left(X^{(m)}g\right)(Z_x, x_m) - \partial/\partial_{x_m}\left(Y^{(m)}g\right)(Z_x, x_m)=0, \tag{2.3}$$

$$\partial/\partial_{x_m}\left(X^{(m)}g\right)(Z_x, x_m) - D_{Z_x}\left(Y^{(m)}g\right)(Z_x, x_m)=0. \tag{2.4}$$

Since $D_{Z_x}e_m = -e_m D_{Z_x}$, we have

$$\begin{aligned}Dg_*(x) &= (D_{Z_x} + e_m\partial/\partial_{x_m})((X^{(m)}g)(Z_x, -x_m) - e_m(Y^{(m)}g)(Z_x, -x_m))\\ &= D_{Z_x}(X^{(m)}g)(Z_x, -x_m) + \partial/\partial_{x_m}(Y^{(m)}g)(Z_x, -x_m)\\ &\quad +e_m(\partial/\partial_{x_m}(X^{(m)}g)(Z_x, x_m) + D_{Z_x}(Y^{(m)}g)(Z_x, -x_m))\\ &= 0 + e_m 0 = 0.\end{aligned}$$

Theorem 2.3. *Let G be the lower half space $\mathbb{R}^m_+$ and assume that*

(1) $H(x) = \sum_\beta H_\beta(x)e_\beta$, *and all H_β are real-valued,*

(2) $(1+H(x))\overline{(1+H(x))} \in \mathbb{R}$, *and* $H(x)\overline{H(x)} \in \mathbb{R}$ *for all* $x \in \mathbb{R}^{m-1}$, *and*

(3) *There exists an* $\varepsilon > 0$ *such that* $H_0(x) > \varepsilon$ *for all* $x \in \mathbb{R}^{m-1}$. *Then the Riemann problem*

$$Du = 0 \text{ in } \mathbb{R}^m \setminus \mathbb{R}^m_0,$$
$$u^+ = H(x)u^- + h(x) \text{ on } \mathbb{R}^m_0,$$
$$|u| = \mathcal{O}(|x|^{\frac{m-1}{2}}) \text{ as } |x| \to \infty,$$

is uniquely solvable in $L_{2,\mathbb{R}}(\mathbb{R}^{m-1})$ *and the successive approximation*

$$u_n(x) := 2(1+H)^{-1}h - (1+H)^{-1}(1-H)Su_{n-1},\ n = 1, 2, \cdots$$

with arbitrary $u_0 \in L_{2,\mathbb{R}}(\mathbb{R}^{m-1})$ *converge to the unique solution* u *of*

$$\frac{1}{2}(I+S)u + \frac{1}{2}H(I-S)u = \frac{1}{2}(1+H)u + \frac{1}{2}(1-H)Su = h,$$

where

$$(Su)(x) = 2\int_{\mathbb{R}^m_0} E(x-y)n(y)u(y)dx,\ x \in \mathbb{R}^m_0,$$

$$E(x) = \frac{1}{A_m}\frac{x}{|x|^m},\quad A_m = \frac{2\pi^{\frac{m}{2}}}{\Gamma\left(\frac{m}{2}\right)}.$$

3 The Left Linear Hilbert Problem

Hilbert Boundary Value problem. Find a $\mathbb{R}_{0,m}$-valued left monogenic function $u(x)$ in $\mathbb{R}^m_+$, which can extend continuously to $\mathbb{R}^m_0$, whose positive boundary value u^+ satisfies

$$X^{(m)}\left(a(t)u^+(t)\right) = c(t),\ t \in \mathbb{R}^m_0, \tag{3.1}$$

where $c(t)$ is a $\mathbb{R}_{0,m-1}$-valued function, $a(x)$ is some given $\mathbb{R}_{0,m}$-valued function whose inverse $a^{-1}(x)$ exists. Since $u(x)$ is left monogenic in $\mathbb{R}^m_+$ and can extend continuously to the boundary $\mathbb{R}^m_0$, we may define a new function $\Omega(x)$ as

$$\Omega(x) = \begin{cases} u(x), & x \in \mathbb{R}^m_+, \\ u_*(x) = u^*(x^*), & x \in \mathbb{R}^m_-, \end{cases} \tag{3.2}$$

then by the Theorem 2.2 that $\Omega(x)$ is left monogenic in $\mathbb{R}^m_-$ and can also extend continuously to $\mathbb{R}^m_0$ in $\mathbb{R}^m_-$ too. Moreover

$$\Omega^+(t) = \left(X^{(m)}u\right)(Z_x, 0) + e_m\left(Y^{(m)}u\right)(Z_x, 0)),\quad t = Z_x \in \mathbb{R}^m_0, \tag{3.3}$$

$$\Omega^-(t) = \left(X^{(m)}u\right)(Z_x, 0) - e_m\left(Y^{(m)}u\right)(Z_x, 0)),\quad t = Z_x \in \mathbb{R}^m_0. \tag{3.4}$$

The relationships between $\Omega^+(t)$ and $\Omega^-(t)$:

$$\Omega^-(t) = u_*^-(t) = \left(u^+(t)\right)^* = (\Omega^+(t))^*,\ t = Z_x \in \mathbb{R}_0^m, \tag{3.5}$$

$$\left(\Omega^+(t)\right)^* = \Omega_*^-(t),\ \left(\Omega^-(t)\right)^* = \Omega_*^+(t),\ t = Z_x \in \mathbb{R}_0^m. \tag{3.6}$$

Thus, **Hilbert BVP** (3.1) is equivalent to

$$a(t)u^+(t) + \left(a(t)u^+(t)\right)^* = 2c(t),\ t \in \mathbb{R}_0^m, \tag{3.7}$$

since

$$\left(a(t)u^+(t)\right)^* = a^*(t)\left(u^+(t)\right)^*,$$

then (3.7) is equivalent to

$$a(t)u^+(t) + a^*(t)\left(u^+(t)\right)^* = 2c(t),\ \ t \in \mathbb{R}_0^m, \tag{3.8}$$

as $a(t)$ is required to have inverse $a^{-1}(t)$ in $\mathbb{R}_{0,m}$, and it can also be written as

$$a(t)\Omega^+(t) + a^*(t)\Omega^-(t) = 2c(t),\ \ t = Z_x \in \mathbb{R}_0^m, \tag{3.9}$$

$$\Omega^+(t) + a^{-1}(t)a^*(t)\Omega^-(t) = 2a^{-1}(t)c(t),\ t = Z_x \in \mathbb{R}_0^m. \tag{3.10}$$

Note that $c(t)$ is a $\mathbb{R}_{0,m-1}$-valued function defined on $\mathbb{R}_0^m$, then by the facts $c^*(t) = c(t)$, we use "*" operator to both sides in (3.9), we may get

$$\left(a(t)\Omega^+(t)\right)^* + \left((a(t))^*\,\Omega^-(t)\right)^* = 2c(t),\ t = Z_x \in \mathbb{R}_0^m, \tag{3.11}$$

thus, we have

$$a(t)\Omega_*^+(t) + a^*(t)\Omega_*^-(t) = 2c(t),\ t = Z_x \in \mathbb{R}_0^m, \tag{3.12}$$

it can be written

$$\Omega_*^+(t) + a^{-1}(t)a^*(t)\Omega_*^-(t) = 2a^{-1}(t)c(t),\ t = Z_x \in \mathbb{R}_0^m, \tag{3.13}$$

which shows that if $\Omega(x)$ is a solution of Riemann BVP (3.7), then $\Omega_*(x)$ is the solution of it too. But from the theorem 2.3, we know that the Riemann boundary value problem (3.10) have the unique solution if $H = -a^{-1}a^*$ satisfies the conditions in the Theorem 2.3, so we get that $\Omega(x) = \Omega_*(x)$.

Remark 3.1. In the formula (3.7), if $a(x) \in R_{0,m-1}$, we cannot use this method.

In fact, if $a(x) \in \mathbb{R}_{0,m-1}$, Hilbert BVP (3.7) is equivalent to

$$\Omega^+(t) + \Omega^-(t) = 2a^{-1}(t)c(t), \quad t = Z_x \in \mathbb{R}_0^m. \tag{3.14}$$

In the **Riemann Boundary Value problem** (3.14), we have

$$H \equiv -1 < 0.$$

So we cannot use the Theorem 2.3.

In the next section, we may see that not all the cases could be solved by this method.

4 Some Examples

In **Hilbert boundary value problem** (3.1), $a(x)$ should satisfy some conditions, or this method may fail. So here, we introduce the following examples.

Example 3.1. If $a(t) = x_1e_1 + \cdots + x_{m-1}e_{m-1} + x_1e_m,\ t \in \mathbb{R}_0^m$, we have $H(t) = -a^{-1}(t)a^*(t)$, while

$$\begin{aligned} a^{-1}(t) = & -\frac{x_1}{x_1^2 + x_2^2 + \cdots + x_{m-1}^2 + x_1^2}e_1 \\ & - \cdots - \frac{x_{m-1}}{x_1^2 + x_2^2 + \cdots + x_{m-1}^2 + x_1^2}e_{m-1} \\ & - \frac{x_1}{x_1^2 + x_2^2 + \cdots + x_{m-1}^2 + x_1^2}e_m,\ t \in \mathbb{R}_0^m, \end{aligned} \tag{4.1}$$

we get

(1) $(1 + H(t))\overline{(1 + H(t))} = (1 - a^{-1}a^*)\overline{(1 - a^{-1}a^*)} \in \mathbb{R}$, and
(2) $H\overline{H} = a^{-1}a^*\overline{a^{-1}a^*} \in \mathbb{R}$, but
(3)

$$H_0(x) = -\left(a^{-1}(t)a^*(t)\right)_0 = -\frac{x_2^2 + \cdots + x_{m-1}^2}{x_1^2 + x_2^2 + \cdots + x_{m-1}^2 + x_1^2} \le 0, \tag{4.2}$$

so $H(x)$ do not satisfies the condition (3) in the Theorem 1.3.

Example 3.2. If $a(t) = x_1e_1 + \cdots + x_{m-1}e_{m-1} + (x_1^2 + \cdots + x_{m-1}^2 + 1)e_m,\ t \in \mathbb{R}_0^m$, we have $H(t) = -a^{-1}(t)a^*(t)$, while

$$\begin{aligned} a^{-1}(t) = & -\frac{x_1}{x_1^2 + x_2^2 + \cdots + x_{m-1}^2 + (x_1^2 + \cdots + x_{m-1}^2 + 1)^2}e_1 \\ & - \cdots - \frac{x_{m-1}}{x_1^2 + x_2^2 + \cdots + x_{m-1}^2 + (x_1^2 + \cdots + x_{m-1}^2 + 1)^2}e_{m-1} \\ & - \frac{x_1^2 + \cdots + x_{m-1}^2 + 1}{x_1^2 + x_2^2 + \cdots + x_{m-1}^2 + (x_1^2 + \cdots + x_{m-1}^2 + 1)^2}e_m,\ t \in \mathbb{R}_0^m, \end{aligned}$$

we get

(1) $(1+H(t))(\overline{1+H(t)}) = (1-a^{-1}a^*)(\overline{1-a^{-1}a^*}) \in \mathbb{R}$,

(2) $H\overline{H} = a^{-1}a^*\overline{a^{-1}a^*} \in \mathbb{R}$, and

(3)

$$\begin{aligned} H_0(t) &= -\left(a^{-1}(t)a^*(t)\right)_0 \\ &= \frac{-x_1^2 - \cdots - x_{m-1}^2 + (x_1^2 + \cdots + x_{m-1}^2 + 1)^2}{x_1^2 + x_2^2 + \cdots + x_{m-1}^2 + (x_1^2 + \cdots + x_{m-1}^2 + 1)^2} \geq 1, \end{aligned} \tag{4.3}$$

so H do satisfies the conditions (1),(2) and (3) in the Theorem 2.3.

Remark 4.1. Let $\mathbb{R}^{0,m+1}$ be the real vector space $\mathbb{R}^{m+1}$ provided with a quadratic form of signature $(0, m+1)$ and let $e_0, e_1, \cdots, e_m$ be an orthonormal basis for $\mathbb{R}^{m+1}$, then $\mathbb{R}^{m+1}$ generates the universal Clifford algebra $\mathbb{R}_{0,m+1}$, which is a real linear associative algebra with identity 1, $e_i^2 = -1$, $e_ie_j + e_je_i = 0\,(i \neq j,\, 0 \leq i, j \leq m)$. Define $\mathbb{R}_+^{m+1} = \{x \mid x_0 > 0\}$, $\mathbb{R}_-^{m+1} = \{x \mid x_0 < 0\}$, $\mathbb{R}_0^{m+1} = \{x \mid x_0 = 0\}$, $\overline{\mathbb{R}_+^{m+1}} = \mathbb{R}_+^{m+1} \cup \mathbb{R}_0^{m+1}$, $\overline{\mathbb{R}_-^{m+1}} = \mathbb{R}_-^{m+1} \cup \mathbb{R}_0^{m+1}$. The space of vector $\mathbb{R}^{m+1}$ is identified with the subspace $\mathbb{R}^{0,m+1}$ by the correspondence

$$(x_0, x_1, \cdots, x_m) \longrightarrow x = \sum_{i=0}^{m} x_i e_i,$$

or with the subspace $\mathbb{R} + \overline{e_0}\mathbb{R}^{0,m}$ by the correspondence

$$(x_0, x_1, \cdots, x_m) = (x_0, \underline{x}) \longrightarrow x_0 + \overline{e_0}\underline{x} = x_0 + \overline{e_0}\sum_{i=1}^{m} x_i e_i.$$

Hereby $\mathbb{R}^{0,m} = span_{\mathbb{R}}(e_1, \cdots, e_m)$ generates inside $\mathbb{R}_{0,m+1}$ the Clifford algebra $\mathbb{R}_{0,m}$. By means of the identifications made, we have that for $x \in \mathbb{R}^{m+1}$ and $\underline{x} \in \mathbb{R}^m$, $x^2 = -|x|^2$, and $\underline{x}^2 = -|\underline{x}|^2$. Further $\mathbb{R}_{0,m+1}$ admits the decomposition

$$\mathbb{R}_{0,m+1} = \mathbb{R}_{0,m} + \overline{e_0}\mathbb{R}_{0,m}. \tag{4.4}$$

The Dirac operator ∂_x and the Cauchy-Riemann Operator D_x in $\mathbb{R}^{m+1}$ are defined, respectively by

$$\partial_x = \sum_{i=0}^{m} e_i \partial_{x_i}, \quad D_x = \partial_{x_0} + \overline{e_0}\partial_{\underline{x}},$$

where $\partial_{\underline{x}} = \sum_{j=1}^{m} e_j \partial_{x_j}$ is the Dirac operator in $\mathbb{R}^m$.

A function $f \in C_1(\Omega, \mathbb{R}_{0,m+1})$ satisfying $\partial_x f = 0$ in Ω (or equivalently $D_x f = 0$ in Ω) is called left monogenic in Ω.

By using the decomposition (4.4), a function $f \in C_1(\Omega, \mathbb{R}_{0,m+1})$ is written as

$$f = u + \overline{e_0}v,$$

where $u, v \in C_1(\Omega, \mathbb{R}_{0,m})$, we let $\mathrm{Re}f = u$, $\mathrm{Im}f = v$. Particularly, if $x = -x_0e_0 + x_1e_1 + \cdots + x_me_m = x_0\overline{e_0} + x_1e_1 + \cdots + x_me_m$, where $u = x_1e_1 + \cdots + x_me_m, v = x_0$, $x^* = u - \overline{e_0}v$, $\mathrm{Re}x^* = u$, $\mathrm{Im}x^* = -v$.

Lemma 4.1. *If f is left monogenic in Ω, we have*

$$D_x f = 0 \Leftrightarrow \begin{cases} \partial_{x_0}u + \partial_{\underline{x}}v = 0, \\ \partial_{\underline{x}}u + \partial_{x_0}v = 0. \end{cases}$$

For any $\mathbb{R}_{0,m+1}$-valued function $f(x) = u + \overline{e_0}v$, we may also define "*" operator as

$$f^*(x) = u - \overline{e_0}v.$$

Lemma 4.2. *Let $f(x)$ be a $\mathbb{R}_{0,m+1}$-valued left monogenic function defined in $\mathbb{R}^{m+1}$, and $f_*(x)$ be defined as*

$$f_*(x) = f^*(x^*),$$

which is the conjugate extension of f with respect to the hyperplane $x_0 = 0$, then $f_*(x)$ is a left monogenic function too.

Proof. $f(x)$ is left monogenic in $\mathbb{R}_{0,m}$ if and only if u and v satisfying the equations

$$\begin{cases} \partial_{x_0}u + \partial_{\underline{x}}v = 0, \\ \partial_{\underline{x}}u + \partial_{x_0}v = 0, \end{cases}$$

we have

$$\begin{aligned} D_x f_*(x) &= (\partial_{x_0} + \overline{e_0}\partial_{\underline{x}})\,(u(-x_0,\underline{x}) - \overline{e_0}v(-x_0,\underline{x})) \\ &= \partial_{x_0}u(-x_0,\underline{x}) - \partial_{\underline{x}}v(-x_0,\underline{x}) \\ &\quad + \overline{e_0}\left(\partial_{\underline{x}}u(-x_0,\underline{x}) - \partial_{x_0}v(-x_0,\underline{x})\right)(Z_x, -x_m) \\ &= 0. \end{aligned}$$

Because of Lemma 4.2, we may also solve the following **Hilbert BVP**: Find a $\mathbb{R}_{0,m+1}$-valued left monogenic function $u(x)$ in $\mathbb{R}^{m+1}_{+}$, which can extend continuously to $\mathbb{R}^{m+1}_{0}$, whose positive boundary value u^+ satisfies

$$\mathrm{Re}\left(a(t)u^+(t)\right) = c(t),\ t \in \mathbb{R}^{m+1}_{0}, \tag{4.5}$$

where $c(t)$ is a $\mathbb{R}_{0,m}$-valued function, $a(t)$ is some given $\mathbb{R}_{0,m+1}$-valued function whose inverse $a^{-1}(t)$ exists.

References

[1] S. Bernstein, On the left linear Riemann problem in Clifford analysis, *Bull. Belg. Math. Soc.* 3 (1996), 557-576.

[2] Zhenyuan Xu and Chiping Zhou, On boundary value problems of Riemann-Hilbert type for monogenic functions in a half space of $\mathbb{R}^m (m \geq 2)$, *Complex Variables* 22 (1993), 181-193.

[3] A. Erdèlyi et al., *Higher transcendental functions, Vols. I-II,* McGraw-Hill Book Company, Inc., New York, 1953.

[4] S. Bernstein, *Fundermental solutions for Dirac-type operators,* Banach Center Symposium on Generalizations of Complex Analysis and Their Applications in Physics, Banach Center Publ, 1994.

[5] S. Bernstein, On the index of Clifford algebra valued singular integral operators and the left linear Riemann problem, *Complex Variable* 35 (1998), 33-64.

[6] G. E. Gilbert and M. A. M. Murray, *Clifford algebra and Dirac operators in harmonic analysis,* Cambridge studies in advanced mathematics 26, Cambridge University Press, Cambridge, 1991.

[7] F. Brackx, R. Delanghe and F. Sommen, *Clifford analysis,* Pitman Advanced Publishing Program, Boston-London-Melbourne, 1982.

[8] R. Delanghe, F. Sommen and V. Souček, *Clifford algebra and Spinor-valued functions,* Mathematics and Its Applications 53, Kluwer Academic publishers, Dordrecht, 1992.

[9] Zhenyuan Xu, A function theory for the operator $D - \lambda$, *Complex Variables and Elliptic Equations* 16 (1991), 27-42.

[10] R. T. Seeley, The index of elliptic systerms of singular integral operators, *J. Math. Anal. Appl.* 7 (1963), 289-309.

[11] S. G. Mikhlin and S. Prößdorf, *Singular integral operators,* Akademie-Verlag, Berlin, 1986.

[12] M. V. Shapiro, On some boundary value problems for functions with values in a clifford algebra, *MaT. Vesnik* 40 (1988), 321-326.

[13] E. M. Stein and G. Weiss, Generalization of the Cauchy-Riemann equations and representations of the rotation group, *Amer. J. Math.* 90 (1992), 163-196.

[14] Yafang Gong, A kind of Riemann and Hilbert boundary value problem for left monogenic functions in $\mathbb{R}^m (m \geq 2)$, *Complex Variables* 49 (2004), 303-318.

[15] Zhenyuan Xu, Helmholtz equations and boundary value problems, in *Partial differential equations with complex analysis,* Pitman Res. Notes Math. Ser. 262, 1992, 204-214.

MODIFIED HELMHOLTZ EQUATION AND H_λ-REGULAR VECTOR FUNCTION[1]

PI-WEN YANG

Department of Mathematics, Sichuan Normal University, Chengdu 610066, P. R. China

E-mail: ypwen1@sina.com

Let $D = \begin{pmatrix} \lambda + \frac{\partial}{\partial x_1} & \frac{\partial}{\partial x_2} + i\frac{\partial}{\partial x_3} \\ \frac{\partial}{\partial x_2} - i\frac{\partial}{\partial x_3} & \lambda - \frac{\partial}{\partial x_1} \end{pmatrix}$, where λ is a positive real constant. In this paper, by using the methods from quaternion calculus, we investigate the complex vector solution $u(x) = \begin{pmatrix} u_1(x) \\ u_2(x) \end{pmatrix}$ of the equation $Du = 0$, and work out a systematic theory analogous to the quaternionic regular function. Differing from that the component functions of a quaternionic regular function are harmonic, the component functions of the solution satisfy the Modified Helmholtz equation, that is, $(\lambda^2 - \Delta)u_i = f$, $i = 1, 2$. In addition, we give out a distribution solution of the inhomogeneous equation $Du = f$ and study some properties of the solution.

Keywords: Quaternion calculus, modified Helmholtz equation, Pompeiu formula, Cauchy integral formula.

AMS No: 30G35, 35J05.

It is well known that the theories of the holomorphic function of one complex variable and the regular functions of quaternion as well as Clifford calculus are closely connected with the theory of harmonic functions, i.e. their component functions are all harmonic. But side by side with the Laplace operator is the Helmholtz operator and modified Helmholtz operator

$$\triangle_\lambda = \lambda^2 \pm \triangle, \lambda(\in \mathbb{R}) \neq 0,$$

which play an important role and are often met in application. In recent years, it was considered that by replacing the harmonic function with the solutions of Helmholtz equation and modified Helmholtz equation, the theory of the regular functions is naturally generalized in quaternion calculus and Clifford calculus. The theory has been well developed and has been applied to research of some partial differential equations such as Helmholtz equation, Klein-Cordon equation, Schroding equation. The corresponding results can be found in [1-9].

Let $\mathbb{H}(\mathbb{R})$ and $\mathbb{H}(\mathbb{C})$ denote the real and complex quaternion space respectively. Their basis elements $1, i, j, k$ satisfy relations: $i^2 = j^2 = k^2 = -1$, $ij = -ji = k$, $jk = -kj = i$.

In [1], the authors introduced a differential operator of first order $D_\lambda = \lambda + i\frac{\partial}{\partial x_1} + j\frac{\partial}{\partial x_2} + k\frac{\partial}{\partial x_3}$, where λ is a positive real constant. It is easy to see that

$$-D_{-\lambda}D_\lambda = \Delta + \lambda^2,$$

where $\Delta + \lambda^2$ is namely the 3-dimensional Helmholtz operator. A quaternion function theory associated with the operator was established, which involved the Pompeiu formula corresponding to D_λ, the Cauchy integral formula for the solution of equation $D_\lambda u = 0$, the Plemelj formula of Cauchy type integral and the theory of operator T_λ. By using these results, the Dirichlet boundary problems for Helmholtz equation

$$(\Delta + \lambda^2)u = f,$$

and the equation of linear elasticity

$$(\Delta + \frac{m}{m-2}\,\mathrm{grad}\,\mathrm{div})\, u = f,$$

as well as some related problems for the time-independent Maxwell equations and Stokes equations were investigated.

Since the operator $\lambda^2 - \Delta$ can not be decomposed into the product of two differential operators of first order, the quaternion function theory about the modified Helmholtz equation was developed in complex quaternion space $\mathbb{H}(\mathbb{C})$, namely the operator $\Delta + \lambda$, $\lambda \in \mathbb{C}$ and some related equations were directly investigated by $\mathbb{H}(\mathbb{C})$. However, different from the operator D_λ, the Dirac operators of first order corresponding to $\Delta + \lambda$, $\lambda \in \mathbb{C}$ are not elliptic in general. Some properties analogous to the regular function such as the Pompeiu formula, the Cauchy integral formula do not hold. There exists an essential difference between the two theories.

In this article, we shall use the quasi-quaternion space introduced in [10,11], transform the problem into matric form. By using the quaternion technique, we obtain a systematic theory about the modified Helmholtz operator analogous to the quaternion regular function.

1. Some Notations and Definitions

Denote

$$e_0 = \begin{pmatrix} 1 & 0 \\ 0 & 1 \end{pmatrix}, e_1 = \begin{pmatrix} 1 & 0 \\ 0 & -1 \end{pmatrix}, e_2 = \begin{pmatrix} 0 & 1 \\ 1 & 0 \end{pmatrix}, e_3 = \begin{pmatrix} 0 & i \\ -i & 0 \end{pmatrix}.$$

It is easy to see that

$$e_1^2 = e_2^2 = e_3^2 = e_0, e_1e_2 = -e_2e_1 = -ie_3,$$

$$e_2e_3 = -e_3e_2 = -ie_1, e_3e_1 = -e_1e_3 = -ie_2.$$

In the following we shall abbreviate e_0 to 1.

Introduce the 3-dimensional modified Helmholtz operator $D=\lambda+\nabla$, where $\nabla=\frac{\partial}{\partial x_1}e_1+\frac{\partial}{\partial x_2}e_2+\frac{\partial}{\partial x_3}e_3$, that is, the 3-dimensional gradient operator and λ is a positive real constant.Define $D'=\lambda-\nabla$, then $DD'=D'D=(\lambda^2-\triangle)e_0$, where $\triangle$ is the 3-dimensional Laplace operator. The matrix forms of D, D' are

$$D=\begin{pmatrix}\lambda+\frac{\partial}{\partial x_1} & \frac{\partial}{\partial x_2}+i\frac{\partial}{\partial x_3}\\ \frac{\partial}{\partial x_2}-i\frac{\partial}{\partial x_3} & \lambda-\frac{\partial}{\partial x_1}\end{pmatrix},\ D'=\begin{pmatrix}\lambda-\frac{\partial}{\partial x_1} & -\frac{\partial}{\partial x_2}-i\frac{\partial}{\partial x_3}\\ -\frac{\partial}{\partial x_2}+i\frac{\partial}{\partial x_3} & \lambda+\frac{\partial}{\partial x_1}\end{pmatrix},$$

and then

$$DD'=D'D=\begin{pmatrix}\lambda^2-\Delta & 0\\ 0 & \lambda^2-\Delta\end{pmatrix}.$$

Let Ω be a bounded domain in $\mathbb{R}^3$ with a piecewise smooth boundary. $u(x)=\begin{pmatrix}u_1(x)\\ u_2(x)\end{pmatrix}$ is a complex vector function defined in Ω. If $u(x)\in C^1(\Omega)$ and satisfies the equation

$$Du=0, \tag{1}$$

then $u(x)$ will be called H_λ-regular vector function in Ω.

2. Pompeiu Formula and Cauchy Integral Formula of H_λ-Regular Vector Function

Let Ω be a bounded domain in $\mathbb{R}^3$ with a piecewise smooth boundary S. $U(x), V(x)$ are 2-dimensional complex vector functions defined in Ω and $U(x), V(x)\in C^1(\Omega)\cap C(\overline{\Omega})$. By the divergence theorem

$$\int_\Omega [(U\nabla)V+U(\nabla V)]\,d\sigma$$

$$=\int_\Omega\left[\frac{\partial}{\partial x_1}(Ue_1V)+\frac{\partial}{\partial x_2}(Ue_2V)+\frac{\partial}{\partial x_3}(Ue_3V)\right]d\sigma=\int_S U\Im V\,dS,$$

where $\Im=e_1\cos\alpha_1+e_2\cos\alpha_2+e_3\cos\alpha_3$, $(\cos\alpha_1,\cos\alpha_2,\cos\alpha_3)$ denotes the unit outward normal to the surface S. From the equality (2), we have

$$\int_\Omega [U(DV)-(UD')V]\,d\sigma=\int_S U\Im V\,dS. \tag{2}$$

It is easy to show that $\frac{1}{4\pi r}e^{\lambda r}$, $r=|x|=(x_1^2+x_2^2+x_3^2)^{\frac{1}{2}}$, is a fundamental solution of the Helmholtz operator $\lambda^2-\Delta$. When $r\neq 0$, $(\lambda^2-\Delta)(\frac{1}{4\pi r}e^{\lambda r})=$

0. We write

$$E(x) = D\left(\frac{1}{4\pi r}e^{\lambda r}\right) = \frac{1}{4\pi}\left[\frac{\lambda}{r} + \left(\frac{\lambda}{r^2} - \frac{1}{r^3}\right)(x_1e_1 + x_2e_2 + x_3e_3)\right]e^{\lambda r}.$$

Suppose $u(x)$ is a complex vector function defined in Ω and $u(x) \in C^1(\Omega) \cap C(\overline{\Omega})$. Let $x = (x_1, x_2, x_3)$ be a fixed point in Ω and $B_\varepsilon(x)$ be an open ball with the center x, and the radius ε be so small such that $\overline{B_\varepsilon(x)} \subset \Omega$. Write $\Omega_\varepsilon = \Omega \setminus \overline{B_\varepsilon(x)}$. Using the formula (3) in Ω_ε and replacing U, V by $E(y - x)$, $u(y)$ respectively, we have

$$\int_{\Omega_\varepsilon} E(y-x)D_y u d\sigma_y$$
$$= \int_S E(y-x)\Im u(y)dS_y - \int_{\partial B_\varepsilon(x)} E(y-x)\Im u(y)dS_y. \tag{3}$$

Where

$$\int_{\partial B_\varepsilon(x)} E(y-x)\Im u(y)dS_y = I_1 + I_2 + I_3,$$
$$I_1 = \frac{\lambda e^{\lambda\varepsilon}}{4\pi\varepsilon^2}\int_{\partial B_\varepsilon(x)} [(y_1 - x_1)e_1 + (y_2 - x_2)e_2 + (y_3 - x_3)e_3]\, u(y)dS_y,$$
$$I_2 = -\frac{\lambda e^{\lambda\varepsilon}}{4\pi\varepsilon}\int_{\partial B_\varepsilon(x)} u(y)dS_y, \quad I_3 = \frac{\lambda e^{\lambda\varepsilon}}{4\pi\varepsilon^2}\int_{\partial B_\varepsilon(x)} u(y)dS_y.$$

It is easy to show

$$\lim_{\varepsilon\to 0} I_1 = 0,\ \lim_{\varepsilon\to 0} I_2 = 0,\ \lim_{\varepsilon\to 0} I_3 = u(x).$$

Then letting ε tend to zero in (4), we obtain the following Pompeiu formula corresponding to the operator D_λ.

Theorem 1. *Let Ω be a bounded domain in $\mathbb{R}^3$ with a piecewise smooth boundary S. If $u(x)$ is a complex vector function defined in Ω and $u(x) \in C^1(\Omega) \cap C(\overline{\Omega})$, then*

$$u(x) = \int_S E(y-x)\Im u(y)dS_y - \int_\Omega E(y-x)D_y u d\sigma_y, x \in \Omega. \tag{4}$$

By applying Theorem 1 we can deduce the following Cauchy integral formula of the H_λ-regular vector function.

Theorem 2. *If a complex vector function $u(x) \in C^1(\Omega) \cap C(\overline{\Omega})$ and satisfies the equation $Du = 0$ in Ω, then*

$$u(x) = \int_S E(y-x)\Im u(y)dS_y,\ x \in \Omega, \tag{5}$$

and if $x\overline{\in}\Omega$, *then*

$$\int_S E(y-x)\Im u(y)dS_y = 0. \tag{6}$$

Proof. The formula (6) follows directly from the Pompeiu formula (5), and the equality (7) can easily be derived from (3).

3. Cauchy Type Integral and Plemelj Formula

Let $\varphi(x)$ be a complex vector function defined on a smooth closed surface S in $\mathbb{R}^3$, $\varphi(x) \in C_\alpha(S)$, $0 < \alpha < 1$. Denote

$$\Phi(x) = \int_S E(y-x)\Im u(y)dS_y, \tag{7}$$

and call $\Phi(x)$ the Cauchy type integral with respect to the operator D, in the following we shall simply call it the Cauchy type integral. In addition, $\varphi(x)$ is called the density function of $\Phi(x)$.

For arbitrary $x\overline{\in}S$, there exists a neighborhood $B_\rho(x)$ of x, which does not intersect with S.In $B_\rho(x)$,

$$\begin{aligned} D\Phi(x) &= (\lambda+\nabla_x)\int_S(\lambda+\nabla_y)(\frac{1}{4\pi|y-x|}e^{\lambda|y-x|})\Im u(y)dS_y \\ &= (\lambda+\nabla_x)\int_S(\lambda-\nabla_x)(\frac{1}{4\pi|y-x|}e^{\lambda|y-x|})\Im u(y)dS_y \\ &= \int_S(\lambda^2-\triangle_x)(\frac{1}{4\pi|y-x|}e^{\lambda|y-x|})\Im u(y)dS_y = 0. \end{aligned}$$

Consequently $\Phi(x)$ is H_λ-regular in the exterior of S.

In addition, the Cauchy type integral $\Phi(x) = o(e^{\lambda|x|})$, $x \to \infty$. In fact, since $\varphi(x) = \begin{pmatrix} \varphi_1(x) \\ \varphi_2(x) \end{pmatrix} \in C_\alpha(S)$, hence $\varphi(x)$ is bounded on S, that is, there exists a positive real constant M, such that $|\phi(x)| = (\varphi_1(x)^2 + \varphi_2(x)^2)^{\frac{1}{2}}$ $\leq M$. From

$$\begin{aligned} |\Phi(x)e^{-\lambda|x|}| &\leq \frac{1}{4\pi}\int_S(\frac{2\lambda}{r}+\frac{1}{r^2})e^{\lambda(|y-x|-|x|)}|\varphi(y)|dS_y \\ &\leq \frac{M}{4\pi}\int_S(\frac{2\lambda}{r}+\frac{1}{r^2})e^{\lambda|y|}dS_y, \end{aligned}$$

letting $x \to \infty$, the integral on the right-hand side of this inequality tends to zero, therefore the desired result is achieved.

When $x \in S$, we provide that the integral on the right side of (8) represents Cauchy's principal value.

Lemma 1. *Let Ω be a bounded domain in $\mathbb{R}^3$ with a smooth boundary S.If $x \in S$, in the sense of Cauchy's principal value, we have*

$$\int_S E(y-x)\Im dS_y = \frac{1}{2}e_0 + \int_\Omega E(y-x)d\sigma_y. \tag{8}$$

Proof. Let $B_\varepsilon(x)$ be an open ball with the radius ε and the center x, write the component of $\partial B_\varepsilon(x)$ lying in the exterior of Ω as Γ. Then x is an interior point of the region inclosed by the closed surface $S' = (S \setminus (S \cap B_\varepsilon(x)) \cup \Gamma$. By the Pompeiu formula (5), we have

$$e_0 = (\int_{S\setminus(S\cap B_\varepsilon(x))} + \int_\Gamma)E(y-x)\Im dS_y - \lambda\int_{\Omega\cup B_\varepsilon(x)} E(y-x)d\sigma_y. \tag{9}$$

Similar to the proof of Theorem 1, we can derive

$$\lim_{\varepsilon\to 0}\int_\Gamma E(y-x)\Im dS_y = \frac{1}{2}e_0.$$

Letting $\varepsilon \to 0$ in (10), it follows that (9) holds.

By using Lemma 1, we can obtain the following Plemelj formula of the Cauchy type integral (8).

Theorem 3. *Write the domain Ω as Ω^+ and the complementary domain of $\overline{\Omega}$ as Ω^-. When x tends to $x_0\,(\in S)$ from Ω^+ and Ω^- respectively, the limits of the Cauchy type integral (8) exist, which will be written as $\Phi^+(x_0)$ and $\Phi^-(x_0)$ respectively, and*

$$\begin{cases} \Phi^+(x_0) = \displaystyle\int_S E(y-x_0)\Im\varphi(y)dS_y + \frac{1}{2}\varphi(x_0), \\ \Phi^-(x_0) = \displaystyle\int_S E(y-x_0)\Im\varphi(y)dS_y - \frac{1}{2}\varphi(x_0). \end{cases} \tag{10}$$

The above formula can be rewritten as

$$\begin{cases} \Phi^+(x_0) - \Phi^-(x_0) = \varphi(x_0), \\ \Phi^+(x_0) + \Phi^-(x_0) = 2\displaystyle\int_S E(y-x_0)\Im\varphi(y)dS_y. \end{cases} \tag{11}$$

Proof. Since $\varphi(x) \in C_\alpha(S)$, $0 < \alpha < 1$, therefore the improper integral $\int_S E(y-x_0)\Im(\varphi(y)-\varphi(x_0))dS_y$ is convergent. By Lemma 1, we have

$$\int_S E(y-x_0)\Im\varphi(y)dS_y$$

$$= \int_S E(y-x_0)\Im(\varphi(y)-\varphi(x_0))dS_y + \frac{1}{2}\varphi(x_0) + \lambda\int_\Omega E(y-x)d\sigma_y\cdot\varphi(x_0).$$

The Cauchy type integral (8) can be written in the following form

$$\int_S E(y-x)\Im\varphi(y)dS_y$$
$$=\int_S E(y-x)\Im(\varphi(y)-\varphi(x_0))dS_y+\int_S E(y-x)\Im dS_y\cdot\varphi(x_0). \quad (12)$$

By the Pompeiu formula, we obtain

$$\int_S E(y-x)\Im dS_y = \begin{cases} e_0+\lambda\int_\Omega E(y-x)d\sigma_y, \ x\in\Omega, \\ \lambda\int_\Omega E(y-x)d\sigma_y, \ x\overline{\in}\Omega. \end{cases}$$

When $x\,(\overline{\in}S)\to x_0\,(\in S)$, by using the method similar to one complex variable[10], we can show that

$$\lim_{x\to x_0}\int_S E(y-x)\Im(\varphi(y)-\varphi(x_0))dS_y=\int_S E(y-x_0)\Im(\varphi(y)-\varphi(x_0))dS_y.$$

Moreover, by the using Hölder inequality, it is easy to show that

$$\lim_{x\to x_0}\int_\Omega E(y-x)d\sigma_y=\int_\Omega E(y-x_0)d\sigma_y.$$

Thus letting x tend to $x_0(\in S)$ from Ω^+ and Ω^- respectively in (13), we get

$$\begin{aligned}\Phi^+(x_0)&=\int_S E(y-x_0)\Im(\varphi(y)-\varphi(x_0))dS_y+\varphi(x_0)+\lambda\int_\Omega(y-x_0)d\sigma_y\cdot\varphi(x_0)\\ &=\int_S E(y-x_0)\Im\varphi(y)dS_y+\frac{1}{2}\varphi(x_0),\\ \Phi^-(x_0)&=\int_S E(y-x_0)\Im(\varphi(y)-\varphi(x_0))dS_y+\lambda\int_\Omega E(y-x_0)d\sigma_y\cdot\varphi(x_0)\\ &=\int_S E(y-x_0)\Im\varphi(y)dS_y-\frac{1}{2}\varphi(x_0).\end{aligned}$$

This is (11), and (12) is easily deduced from (11).

The following result follows directly from Theorem 3.

Corollary 1. *Let Ω be a bounded domain in $\mathbb{R}^3$, whose boundary is a smooth closed surface S. $\varphi(x)$ is a complex vector function defined on the surface S,and $\varphi(x)\in C_\alpha(S)$, $0<\alpha<1$. Then the Cauchy type integral* (8) *whose density function is $\varphi(x)$ is a Cauchy integral if and only if $x_0\in S$,*

$$\Phi^-(x_0)=0.$$

4. Operator $T_\Omega f$

Let $f(x)$ be a complex vector function defined in a bounded domain Ω of $\mathbb{R}^3$. Denote

$$T_\Omega f = -\int_\Omega E(y-x)f(y)d\sigma_y, \tag{13}$$

where

$$E(y-x)=\frac{1}{4\pi}\left[\frac{\lambda}{|y-x|}+\left(\frac{\lambda}{|y-x|^2}-\frac{1}{|y-x|^3}\right)((y_1-x_1)e_1 + (y_2-x_2)e_2+(y_3-x_3)e_3)\right]e^{\lambda|y-x|}.$$

In this section, we shall get that if $f(x)\in L_1(\overline{\Omega})$, then $T_\Omega f$ is a distribution solution of the inhomogeneous equation

$$Du = f, \tag{14}$$

and shall discuss some properties of the operator $T_\Omega f$.

Similar to the quaternion calculus, we can obtain the following results.

Theorem 4. *If $f(x)\in L_1(\overline{\Omega})$, then $T_\Omega f$ exists for all x in the exterior of Ω. Beside $T_\Omega f$ is H_λ-regular in the exterior of Ω and*

$$T_\Omega f = o(e^{\lambda|x|}), x\to\infty.$$

Theorem 5. *Let $f(x)\in L_1(\overline{\Omega})$, then $T_\Omega f$ exists almost everywhere on $\mathbb{R}^3$ and belongs to $L_p(\overline{\Omega}_*), 1\le p<\frac{3}{2}$, where Ω_* denotes any bounded domain in $\mathbb{R}^3$.*

For complex vector functions $f(x)=\begin{pmatrix} f_1(x)\\ f_2(x)\end{pmatrix}$, $\varphi(x)=\begin{pmatrix}\varphi_1(x)\\ \varphi_2(x)\end{pmatrix}$ given on Ω, define

$$(f,\varphi)=\int_\Omega(\overline{f_1}\varphi_1+\overline{f_2}\varphi_2)d\sigma.$$

When $f(x)\in L_1(\overline{\Omega})$, $\varphi(x)\in C_0^\infty(\Omega)$, it is easy to show that $f(\varphi)=(f,\varphi)$ is a distribution on $C_0^\infty(\Omega)$.

Theorem 6. *Let $f(x)\in L_1(\overline{\Omega})$. Then for any $\varphi(x)=\begin{pmatrix}\varphi_1(x)\\ \varphi_2(x)\end{pmatrix}\in C_0^\infty(\Omega)$,*

$$(T_\Omega f, D'\varphi)=(f,\varphi)$$

holds.

Proof. By using the method analogous to the proof of the Pompeiu formula (5), we can derive the Pompeiu formula corresponding to the operator D', i.e. if $u(x) \in C^1(\Omega) \cap C(\overline{\Omega})$, we have

$$u(x) = \int_S E'(y-x)\Im u(y)dS_y + \int_\Omega E'(y-x)D'_y u d\sigma_y, \; x \in \Omega, \qquad (15)$$

where $E'(x) = D'\left(\frac{1}{4\pi r}e^{\lambda r}\right)$. Thus for any $\varphi(x) \in C_0^\infty(\Omega)$,

$$\varphi(x) = T'_\Omega \varphi = \int_\Omega E'(y-x)D'_y \varphi d\sigma_y$$

holds.

In accordance with Theorem 5, $T_\Omega f \in L_1(\overline{\Omega})$. Thereby by the Fubini theorem

$$(T_\Omega f, D'\varphi) = (f, T'_\Omega(D'\varphi)) = (f, \varphi),$$

the desired result follows.

Let complex vector functions f, $g \in L_1(\overline{\Omega})$. If for any $\varphi(x) \in C_0^\infty(\Omega)$,

$$(g, D'\varphi) = (f, \varphi),$$

then f is called generalized derivative corresponding to the operator D of g. The derivative is denoted by $f = (g)_D$. From Theorem 6 and the definition, $(T_\Omega f)_D = f$.

Theorem 7. *If a complex vector function $g \in C^1(\Omega)$ and satisfies the equation $Dg = f$, then*

$$(g)_D = f.$$

This shows that if the complex vector function g is a classical solution of the equation (15)*, then it is also a distributional solution of the equation.*

Proof. Since $g \in C^1(\Omega)$, using the divergence theorem we can obtain

$$(\nabla g, \varphi) = -(g, \nabla\varphi),$$

so that

$$(Dg, \varphi) = ((\lambda + \nabla)g, \varphi) = \lambda(g, \varphi) + (\nabla g, \varphi) = \lambda(g, \varphi) - (g, \nabla\varphi) = (g, D'\varphi).$$

And since $Dg = f$, thus the above equality implies namely

$$(g)_D = f.$$

References

[1] K. Cürlebeck and W. Sprößig, *Quaternionic analysis and elliptic boundary value problem,* Birkhausser, Boston, 1990.

[2] Z. Y. Xu, A function theory for the operator $(D - \lambda)$, *Complex Variables,* 16 (1991), 27-42.

[3] Z. Y. Xu, Helmholtz equations and boundary value problems, In *Partial differential equations with complex analysis,* Pitman Res, Notes Math. Ser.262, Longman Sci. Tech., Harlow, 1992, 204-214.

[4] V. V. Kravchenko and M. V. Sharpirom, Hypercomplex factorization of the multidimensional Helmholtz operator and some its applications, *Dokl. Sem. Inst. I. N. Vekua,* Tbilisi, 5(1) (1990), 106-109.

[5] V. V. Kravchenko and M. V. Sharpiro, Helmholtz operator with a quaternionic wave number and associated function theory. II: Integral representations, *Acta Applicandae Mathematicae,* 32 (1993), 243-265.

[6] Thi Ngoc Ha Vu, Integral representation in quaternionic analysis related to the Helmholt operator, *Complex Variables Theory and Appl.,* 48 (2003), 1005-1021.

[7] I. Constales and R. S. Kraußhar, A unified approach for the treatment of some higher dimesiosal dirac type equations on spheres, in *Proceedings of the 18th International Conference on the Application of Computer Science and Mathematics in Architecture and Civil Engineering,* ?, 2009.

[8] M. V. Sharpirom, On the properties of a class of singular integral equations connected with the three-dimensional Helmholtz equatio, *Abstracts of lectures at the 14 school on Operator Theory in Functional Spaces,* U.S.S.R., Novgorod, 1989, pp.88.

[9] H. Begehr, Z. X. Zhang and Thi Ngoc Ha Vu, Generalized integral representations in Clifford analysis, *Complex variables and elliptic equations,* 8-11 (2006), 745-762.

[10] P. W. Yang, M. L. Li and Y. Chen, Initial-boundary value problems of some hyperbolic systems of first order equations, *Acta Math. Appl. Sinnica,* 31 (2008), 61-70.

[11] P. W. Yang, S. Yang and M. L. Li, An initial-boundary value problem for the Maxwell equations, *J. Diff. Eqns.,* to appear.

[12] P. W. Yang, *Quaternion calculus and partial differential equations,* Science Press, Beijing, 2009.

[13] I. N. Vekya, *Generalized analytic functions,* Pergamon, Oxford, 1962.

[14] H. Liede, The modified Helmholtz equations and their boundary value problems, *Selected papers of Tongji University,* 1989, 37-54.

THE INTEGRAL EQUATION METHOD ON THE FRACTURE OF FGCMs[1]

XING LI, SHENG-HU DING and HUI-LI HAN
School of Mathematics and Computer Science, Ningxia University, Yinchuan 750021, China
E-mail:li_x@nxu.edu.cn

The changes that have occurred and advances that have been achieved in the behavior of fracture for functionally graded materials (FGMs) subjected to a mechanical and/or temperature change. This paper mainly reviews the research of functionally graded composites material involving the static and dynamic crack problem, thermal elastic fracture analysis, the elastic wave propagation and contact problem. Development of analytical methods to obtain the solution of the transient thermal/mechanical fields in FGMs are introduced.
Keywords: Functionally graded material, crack, integral equation method.
AMS No: 35J65, 35J55, 35J45.

1. Introduction

The name of functionally graded materials was first coined by Japanese materials scientists in the Sendai area in 1984 as a means of manufacturing thermal barrier coating materials. The advantages of FGMs are that which can reduce the magnitude of the residual and thermal stresses and increase the bonding strength and fracture toughness, so they have been introduced and applied in the development of structural components in extremely high temperature environment [1]. Due to their intrinsic coupling between mechanical deformation and electric fields, piezoelectric materials (PMs) are widely used as sensors and actuators to monitor and control the dynamic behavior of intelligent structural systems [2]. Functionally graded piezoelectric materials (FGPMs) is a kind of piezoelectric material with material composition and properties varying continuously along certain direction [3]. FGPMs is the composite material intentionally designed so that they possess desirable properties for some specific applications. The advantage of this new kind of materials can improve the reliability of life span of piezoelectric devices. Magnetoelectric coupling is a new product property of composites, since it is absent in all constituents [4]. In some cases, the coupling effect of piezoelectric/piezomagnetic composites can even be obtained a hundred times greater than that in a single-phase magnetoelectric material. Consequently, they are extensively used in magnetic field probes, electric packaging, acoustic, hydrophones, medical ultrasonic imaging, sensors, and actuators for magneto-electro-mechanical energy conversion. Sim-

[1]This research project was supported by NSFC (10962008) and (51061015)

ilar to FGPMs, functionally graded piezoelectric/piezomagnetic materials (FGPPMs) was developed.

2. The Integral Equation Method on the Static and Dynamic Fracture of Functionally Graded Composite Materials (FGCMs)

(1) The investigations of static and dynamic crack problems in FGCMs (Non-periodic cases)

The static or dynamic crack problems on the complex model of FGMs were studied by our group using the integral equation method or hypersingular integral equation method, which overcome the complexity on mathematics and obtained the high degree of accuracy. In 2004, Ma Hailong and Li Xing [5] investigated the anti-plane moving Yoffe crack problem in a strip of functionally graded piezoelectric materials (FGPMs) using integral equation method.

And model III crack problem in two bonded functionally graded magneto-electro-elastic materials (FGPMMs) was studied by Li Xing and Guo Lifang [6]. It was assumed that the material constants of the magneto-electro-elastic varied continuously along the thickness of the strip. Integral transforms and dislocation density functions were employed to reduce the problem with the Cauchy singular integral equations, which could be solved numerically by Gauss-Chebyshev method. An anti-plane shear crack in bonded functionally graded piezoelectric materials under electromechanical loading was investigated by Ding and Li [7]. A moving mode III crack at interface between two different functionally graded piezoelectric piezomagnetic materials has been studied by Lu and Li [8].

The anti-plane problem of functionally graded magneto-electro-elastic strip sandwiched between two functionally graded strips was investigated by Guo, Li and Ding [9]. It is assumed that the material properties vary exponentially with the coordinate parallel to the crack as follows

$$c_{44} = c_{440}e^{\beta_1 x}, \quad \varepsilon_{11} = \varepsilon_{110}e^{\beta_1 x}, \quad e_{15} = e_{150}e^{\beta_1 x}, \tag{1}$$

$$f_{15} = f_{150}e^{\beta_1 x}, \quad g_{11} = g_{110}e^{\beta_1 x}, \quad \mu_{11} = \mu_{110}e^{\beta_1 x}. \tag{2}$$

The crack is assumed to be either magneto-electrically impermeable or permeable. Fourier transforms are used to reduce the crack problems to following system of singular integral equations for impermeable case

$$\tau_{zy} - m_{20}D_y - m_{30}B_y = m_{10}\frac{1}{\pi}e^{\beta_1 x}\int_a^b (\frac{1}{t-x} + \Lambda_1(x,t))g_1(t)dt, \tag{3}$$

$$D_y = \frac{1}{\pi}e^{\beta_1 x}\int_a^b [(\frac{1}{t-x} + \Lambda_2(x,t))(e_{150}g_1(t) - \varepsilon_{110}g_2(t) - g_{110}g_3(t))]dt, \tag{4}$$

$$B_y=\frac{1}{\pi}e^{\beta_1 x}\int_a^b[(\frac{1}{t-x}+\Lambda_2(x,t))(f_{150}g_1(t)-g_{110}g_2(t)-\mu_{110}g_3(t))]dt, \quad (5)$$

where

$$\Lambda_1(x,t)=h_1(x,t)+K_2(x,t)+K_3(x,t), \quad (6)$$

$$\Lambda_2(x,t)=h_1(x,t)+K_4(x,t)+K_5(x,t), \quad (7)$$

$$\begin{aligned} h_1(x,t)=&\int_0^\infty\left[(1+\frac{\beta^2}{\alpha^2})^{\frac{1}{4}}\cos(\frac{\theta}{2})-1\right]\sin[\alpha(t-x)]\,d\alpha\\ &+\int_0^A(1+\frac{\beta^2}{\alpha^2})^{\frac{1}{4}}\sin(\frac{\theta}{2})\cos[\alpha(t-x)]\,d\alpha+\int_0^{A(t-x)}\frac{\cos\alpha-1}{\alpha}d\alpha\\ &+\int_A^\infty\left[(1+\frac{\beta^2}{\alpha^2})^{\frac{1}{4}}\sin(\frac{\theta}{2})-\frac{\beta}{2\alpha}\right]\cos\left[\alpha(t-x)\right]d\alpha-\frac{\beta}{2}\gamma_0-\frac{\beta}{2}\log A, \end{aligned} \quad (8)$$

$$\begin{aligned} K_2(x,t)=&e^{\frac{\beta_1}{2}(t-x)}\{\frac{\delta_1}{t+x}\\ &+\int_0^\infty[\frac{(-\alpha^2)(m_2t_2t_5e^{-\alpha_1(t+x)}-m_3t_2t_4e^{\alpha_1(t-x)})}{\alpha_1m_2m_3z_1}-\delta_1e^{-\alpha(t+x)}]d\alpha\}, \end{aligned} \quad (9)$$

$$\begin{aligned} K_3(x,t)=&e^{\frac{\beta_1}{2}(t-x)}\{\frac{-\delta_3}{2h_1-t-x}\\ &+\int_0^\infty[\frac{(-\alpha^2)(t_1t_4m_3e^{\alpha_1(t+x)}-m_2t_2t_4e^{-\alpha_1(t-x)})}{\alpha_1m_2m_3z_1}+\delta_3e^{-\alpha(2h_1-t-x)}]d\alpha\}, \end{aligned} \quad (10)$$

$$K_4(x,t)=e^{\frac{\beta_1}{2}(t-x)}\{\frac{1}{t+x}+\int_0^\infty[\frac{\alpha^2e^{-\alpha_1x}\sinh(\alpha_1(t-h_1))}{\alpha_1m_2\sinh(\alpha_1h_1)}-e^{-\alpha(t+x)}])d\alpha\}, \quad (11)$$

$$\begin{aligned} K_5(x,t)=&e^{\frac{\beta_1}{2}(t-x)}\{-\frac{1}{2h_1-t-x}\\ &+\int_0^\infty[\frac{(-\alpha^2)e^{\alpha_1(x-h_1)}\sinh(\alpha_1t)}{\alpha_1m_3\sinh(\alpha_1h_1)}+e^{-\alpha(2h_1-t-x)}]d\alpha\}. \end{aligned} \quad (12)$$

For the magneto-electrically permeable case, the singular integral equation can be derived by a similar method as

$$\tau_{zy}-m_{20}D_y-m_{30}B_y=m_{10}\frac{1}{\pi}e^{\beta_1x}\int_a^b\left(\frac{1}{t-x}+\Lambda_1(x,t)\right)g_1(t)dt, \quad (13)$$

$$D_y=e_{150}\frac{1}{\pi}e^{\beta_1x}\int_a^b\left(\frac{1}{t-x}+\Lambda_2(x,t)\right)g_1(t)dt, \quad (14)$$

$$B_y=f_{150}\frac{1}{\pi}e^{\beta_1x}\int_a^b\left(\frac{1}{t-x}+\Lambda_2(x,t)\right)g_1(t)dt. \quad (15)$$

Recently, the crack-tip fields in bonded functionally graded finite strips were studied by Ding, Li and Zhou [10]. A bi-parameter exponential function was introduced to simulate the continuous variation of material properties as

$$\begin{aligned} \mu_1(x) &= \mu_0\alpha_1^{\beta_1 x}, \quad \rho_1(x) = \rho_0\alpha_1^{\beta_1 x}, \\ \mu_2(x) &= \mu_0\alpha_2^{\beta_2 x}, \quad \rho_2(x) = \rho_0\alpha_2^{\beta_2 x}. \end{aligned} \tag{16}$$

The problem was reduced as a system of Cauchy singular integral equations of the first kind by Laplace and Fourier integral transforms. Various internal cracks and edge crack and crack crossing the interface configurations were investigated, respectively. The asymptotic stress field near the tip of a crack crossing the interface was examined and it is shown that, unlike the corresponding stress field in piecewise homogeneous materials, in this case the "kink" in material property at the interface does not introduce any singularity.

(2) The investigations of static and dynamic crack problems in FGCMs (Periodic cases)

In 2002, Li Xing and Wu Yaojun [11] got the numerical solutions of the periodic crack problems for an anisotropic strip by employing Lobotto-Chebyshev quadrature formulas and Gauss quadrature formulas. On the basis, Li Xing and Wang Wenshuai extended the anisotropic materials to PMs and investigated an antiplane problem of periodic cracks in piezoelectric medium by means of Riemann-Schwarz's symmetry principle, complex conformal mapping and analytical continuation [12]. And Ding and Li [13] extended the anisotropic materials to FGMs and analyzed the interface cracking between a functionally graded material and an elastic substrate under antiplane shear loads. Two crack configurations were considered, namely a FGM bonded to an elastic substrate containing a single crack and a periodic array of interface cracks, respectively.

For the periodic cracks problem, application of finite Fourier transform techniques reduces the solution of the mixed boundary value problem for a typical strip to triple series equations, then to a singular integral equation with a Hilbert-type singular kernel as follows

$$\begin{aligned} &\int_{-1}^{1} [\cot(\frac{a\gamma(s-r)}{2}) + \cot(\frac{a\gamma(s+r)}{2})]\alpha_l^*(s)ds \\ &+ \int_{-1}^{1} Q^*(r,s)\alpha_l^*(s)ds = \frac{4\tau^*(r)}{\mu_1}, \quad |\, r \,| < 1. \end{aligned} \tag{17}$$

The resulting singular integral equation is solved numerically by employing the direct quadrature method of Li and Wu [11].

3. IEM for Scattering of SH Wave of Functionally Graded Piezoelectric/Piezomagnetic Materials

The study of elastic wave propagation through FGCMs has many important applications. Through analysis, we can predict the response of composite materials to various types of loading, and obtain the high strength and toughness of materials. In 2006, Liu Junqiao and Li Xing [14] studied the scattering of the SH wave on a crack in an infinite of orthotropic functionally grade materials plane by using dual integral equation method, the latter DIE was solved employing the Copson method [15].

The problem of scattering of SH wave propagation in laminated structure of functionally graded piezoelectric strip was studied by Yang Juan and Li Xing [16]. Due to the same time factor of scattering wave and incident wave, the scattering model of the crack tip can be constructed by making use of the displacement function of harmonic load on any point of the infinite plane. It is found from numerical calculation that the dynamic response of the system depends significantly on the crack configuration, the material combination and the propagating direction of the incident wave. It is expected that specifying an appropriate material combination may retard the growth of the crack for a certain crack configuration.

The scattering of the anti-plane incident time-harmonic wave with arbitrary degree by the interface crack between the functionally graded coating and the homogeneous substrate is investigated [17]. By using the principle of superposition and Fourier transform, the singular integral equations are give by

$$\begin{aligned} &\int_{-c}^{c}\left[\frac{1}{\xi-x}+Q(x,\xi)\right]f(\xi)d\xi=-2\pi\tau_0(x), \\ &Q\left(x,\,\xi\right)=\int_{0}^{\infty}Q_0\left(\eta\right)\,\sin\left[\,\eta\left(\xi-x\right)\right]d\eta. \end{aligned} \tag{18}$$

There are some pole points in the integral path, an integral path in the complex plane consisting of four straight lines is adopted. The effects of the frequency of the incident wave, the incident direction of the wave, material gradient parameter and the crack configuration on the dynamic stress intensity factors (DSIF) are examined. The scattering of the plane incident wave (P wave, SV wave) with arbitrary degree by the interface crack between the functionally graded coating and the homogeneous substrate is investigated. An integral path in the complex plane consisting of four straight lines is adopted to avoid singular points. Numerical results show the effects of the frequency of the incident wave, the type of incident wave, the incident direction of the wave, material gradient parameter and the crack configuration on the DSIF.

Problems of SH-wave scattering from the crack of functionally graded piezoelectric/piezomagnetic composite materials had been studied by Yang Juan and Li Xing [18]. Fourier transforms are used to reduce the problem to the solution of a pair of dual integral equation, which are then reduced to a Fredholm integral equation of the second kind by the Copson method. Numerical results shown the effect of loading combination parameter, the angle of wave upon the normalized stress intensity factors. Scattering of the SH wave from a crack in a piezoelectric substrate bonded to a half-space of functionally graded materials was investigated by Li Xing and Liu Junqiao [19].

4. Thermal Elastic Fracture Analysis of FGMs

The transient thermal fracture problem of a crack (perpendicular to the gradient direction) in a graded orthotropic strip was investigated by Zhou Yueting, Li Xing and Qin Junqing [20]. The transient two-dimensional temperature problem was analyzed by the methods of Laplace and Fourier transformations. A system of singular integral equations are obtained as

$$\begin{aligned}&\int_{-1}^{1}\left\{[\frac{\chi_1}{s-\overline{x}}+H_{11}(\overline{x},p,s)]\psi_1^*(s,p)+H_{12}(\overline{x},p,s)\psi_2^*(s,p)\right\}ds=2\pi w_1^T(\overline{x},p),\\&\int_{-1}^{1}\left\{[\frac{\chi_2}{s-\overline{x}}+H_{22}(\overline{x},p,s)]\psi_2^*(s,p)+H_{21}(\overline{x},p,s)\psi_1^*(s,p)\right\}ds=2\pi w_2^T(\overline{x}p).\end{aligned} \tag{19}$$

The transient response of an orthotropic functionally graded strip with a partially insulated crack under convective heat transfer supply was considered by Zhou Yueting, Li Xing and Yu Dehao[21]. The thermal boundary conditions were given by

$$\frac{\partial\overline{T}(\overline{x},-\overline{a},\tau)}{\partial\overline{y}}-H_a\cdot\overline{T}(\overline{x},-\overline{a},\tau)=-H_a\cdot\overline{T_a}(\tau)\cdot f_a(\overline{x}), \tag{20}$$

$$\frac{\partial\overline{T}(\overline{x},\overline{b},\tau)}{\partial\overline{y}}+H_b\cdot\overline{T}(\overline{x},\overline{b},\tau)=H_b\cdot\overline{T_b}(\tau)\cdot f_b(\overline{x}), \tag{21}$$

$$\frac{\partial\overline{T}(\overline{x},0^+,\tau)}{\partial\overline{y}}=-Bi\cdot(\overline{T}(\overline{x},0^+,\tau)-\overline{T}(\overline{x},0^-,\tau)), \tag{22}$$

$$\overline{T}(\overline{x},0^+,\tau)=\overline{T}(\overline{x},0^-,\tau), \tag{23}$$

$$\frac{\partial\overline{T}(\overline{x},0^+,\tau)}{\partial\overline{y}}=\frac{\partial\overline{T}(\overline{x},0^-,\tau)}{\partial\overline{y}}. \tag{24}$$

The mixed boundary value problems of the temperature field and displacement field were reduced to a system of singular integral equations in Laplace domain. The expressions with high order asymptotic terms for the singular integral kernel were considered to improve the accuracy and efficiency. The numerical results present the effect of the material nonhomogeneous parameters, the orthotropic parameters and dimensionless thermal resistant on the temperature distribution and the transient thermal stress intensity factors with different dimensionless time τ.

5. Contact Problem of FGMs

A problem for the bonded plane material with a set of curvilinear cracks, which is under the action of a rigid punch with the foundation of convex shape, has been considered by Zhou Yueting, Li Xing and Yu Dehao [22]. Kolosov-Muskhelishvili complex potentials are constructed as integral representations with the Cauchy kernels with respect to derivatives of displacement discontinuities along the crack contours and pressure under the punch. The considered problem has been transformed to a system of complex Cauchy type singular integral equations of first and second kind.

The receding contact problem between the functionally graded elastic strip and the rigid substrate is considered by An Zhenghai and Li Xing [23]. The receding contact problem are solved for various different stamp profiles including flat, semicircular, cylindrical, parabolical. Under the mixed boundary conditions, the Fourier transform technique and effective singular integral equation methods are employed to reduce the receding contact problem to a set of Cauchy kernel singular integral equations as follows

$$\begin{aligned}
&\int_{-a}^{a} \frac{p(t)}{t-x} dt + \int_{-b}^{b} k_{11}(x,t)p(t)dt + \int_{-a}^{a} k_{12}(x,t)q(t)dt = 2\pi\mu_0 e^{\beta h} f(x), \ |x|<a, \\
&\int_{-b}^{b} \frac{q(t)}{t-x} dt + \int_{-b}^{b} k_{21}(x,t)q(t)dt + \int_{-a}^{a} k_{22}(x,t)p(t)dt = 0, \ |x|<b,
\end{aligned} \tag{25}$$

where

$$\begin{aligned}
k_{11}(x,t) &= \int_0^{+\infty} \left[\alpha^2 \sum_{j=1}^{4} m_j(\alpha) J_j(\alpha) e^{\lambda_j h} - 1\right] \sin[\alpha(t-x)] d\alpha, \\
k_{12}(x,t) &= -e^{\beta h} \int_0^{+\infty} \left[\alpha^2 \sum_{j=1}^{4} m_j(\alpha) L_j(\alpha) e^{\lambda_j h}\right] \sin[\alpha(t-x)] d\alpha.
\end{aligned} \tag{26}$$

The singular integral equations can be solved by using the Gauss-Chebyshev formulas numerically.

Pang Mingjun and Li Xing [24] discussed the thermal contact problem of a functionally graded material with a crack bonded to a homogeneous elastic strip. The problem has been reduced to singular integral equations with the first Cauchy kernel by using of the superposition principle as follows

$$\begin{aligned}
&\int_{-a}^{a}\left[\left(\frac{1}{t-y}+g_{11}(t,y)\right)\varphi_1(y)+g_{12}(t,y)\varphi_2(y)\right]dt=\pi\frac{1+\kappa}{2}\omega_1(y),\\
&\int_{-a}^{a}\left[g_{21}(t,y)\varphi_1(y)+\left(\frac{1}{t-y}+g_{22}(t,y)\right)\varphi_2(y)\right]dt=\pi\frac{1+\kappa}{2}\omega_2(y).
\end{aligned} \tag{27}$$

6. Future Work

Doubly-periodic crack problems of FGCMs will shade a light in the future investigation. Great efforts have been made to study the doubly-periodic crack problems of homogenous elastic materials. However, little literature is available on the investigation of the doubly-periodic crack problems of FGCMs despite the practical significance of this case.

References

[1] M. Koizumi, FGM activities in Japan, *Composites, Part B* 28 (1997), 1–4.

[2] O. Tomio, Optical beam deflection using a piezoelectric bimorph actuator, *Sensor Actuat* 21 (1990), 726–728.

[3] C. M. Wu, M. Kahn and W. Moy, Piezoelectric ceramics with functional gradients, a new application in material design, *J Am Ceram Soc* 79 (1996), 809–812.

[4] V. Suchtelen, Product properties: a new application of composite materials, *Phillips. Res. Rep* 27 (1972), 28–37.

[5] H. L. Ma and X. Li, Anti-plane moving Yoffe-crack problem in a strip of functionally graded piezoelectric materials, *The Chinese Journal Of Nonferrous Materials* 14 (2004), 124–128.

[6] X. Li and L.F. Guo, Model III crack in two bonded functionally graded magneto-electro-elastic materials, *Chinese Journal of Theoretical and Applied Mechanics* 6 (2007), 760–766.

[7] S. H. Ding and X. Li, An anti-plane shear crack in bonded functionally graded piezoelectric materials under electromechanical loading, *Computational Materials Science* 43 (2008), 337–344.

[8] W. S. Lu and X. Li, Scattering of SH wave on crack in functionally graded/ piezoelectric layers, *Journal of Nanjing Normal University* (*Natural Sciences*) 32 (2009), 55–59.

[9] L. F. Guo, X. Li and S. H. Ding, Crack in a bonded functionally graded magnet-electro-elastic strip, *Computational Materials Science* 46 (2009), 452–458.

[10] S. H. Ding, Xing Li and Y. T. Zhou, Dynamic stress intensity factors of mode I crack problem for functionally graded layered structures, *Computer Modeling in Engineering and Sciences* 56 (2010), 43–84.

[11] X. Li and Y. J. Wu, The numerical solutions of the periodic crack problems of anisotropic strips, *International Journal of Fracture* 118 (2002), 41–46.

[12] X. Li and W. S. Wang, Anti-plane problem of periodic cracks in piezoelectric medium, *Complex Variables* 50 (2005) 611-620.

[13] S. H. Ding and X. Li. Anti-plane problem of problem of periodic interface cracks in a functionally graded coating-substrate structure, *International Journal of Fracture,* 153 (2009), 53–62.

[14] J. Q. Liu, and X. Li, Scattering of the SH wave on a crack in an infinite of orthotropic functionally grade materials plane, *Acta Mechanica Solida Sinica* 27 (2006), 369–373.

[15] X. Li, *Integral equations,* Science Press, Beijing, 2008

[16] J. Yang and X. Li, Scattering of SH wave on cracks in functionally graded piezoelectric strip, *Chinese Quarterly of Mechanics* 3 (2007), 471–478.

[17] Y. T. Zhou, *Problems of wave scattering by interface crack and thermal fracture in functionally graded materials,* Ph.D. Thesis, Department of Mathematics, Shanghai Jiao Tong University, 2007.

[18] J. Yang and X. Li, SH wave scattering on cracks in functionally graded piezoelectric/piezomagnetic materials, *Chinese Journal of Applied Mechanics* 25 (2008), 279–283.

[19] X. Li and J. Q. Liu, Scattering of the SH wave from a crack in a piezoelectric substrate bonded to a half-space of functionally graded materials, *Acta Mechanica* 208 (2009), 299–308.

[20] Y. T. Zhou, X. Li, and J. Q. Qin, Transient thermal stress analysis of orthotropic functionally graded materials with a crack, *Journal of Thermal Stresses* 30 (2007), 1211–1231.

[21] Y. T. Zhou, X. Li and D. H. Yu, Transient thermal response of a partially insulated crack in an orthotropic functionally graded strip under convective heat supply, *CMES Comput. Model. Eng. Sci.* 43 (2009), 191–221.

[22] Y. T. Zhou, X. Li and D.H. Yu, Integral method for contact problem of bonded plane material with arbitrary cracks, *CMES Comput. Model. Eng. Sci.* 36 (2008), 147–172.

[23] Z. H. An, *Singular integral equation method for receding contact problem of functionally graded materials,* Master's Thesis, School of Mathematics and Computer Science, Ningxia University, 2010.

[24] M. J. Pang, *Thermal contact problems of bonded functionally graded materials with crack,* Master's Thesis, School of Mathematics and Computer Science, Ningxia University, 2010.

MODE I CRACK PROBLEM FOR A FUNCTIONALLY GRADED ORTHOTROPIC COATING-SUBSTRATE STRUCTURE[1]

LI-FANG GUO[a,b], XING LI[c] and SHENG-HU DING[c]

[a] *Higher Health Vocational and Technical College, Ningxia Medical University, Yinchuan 750021, China.*

[b] *Department of Mathematics, Shanghai Jiaotong University, Shanghai 200240, China.*

E-mail: glf1998@163.com

[c] *School of Mathematics and Computer Science, Ningxia University, Yinchuan 750021, China*

In this paper, the Fourier integral transform-singular integral equation method is presented for the Mode I crack problem of the functionally graded orthotropic coating-substrate structure. The elastic property of the material is assumed vary continuously along the thickness direction. The principal directions of orthotropy are parallel and perpendicular to the boundaries of the strip. Numerical examples are presented to illustrate the effects of the crack length, the material nonhomogeneity and the thickness of coating on the stress intensity factors.

Keywords: Functionally graded orthotropic material, coating-substrate structure, mode I crack problem, singular integral equation.

AMS No: 35J65, 35J55, 35J45.

1. Introduction

The analysis of functionally graded materials has become a subject of increasing importance motivated by a number of potential benefits from the use of such novel materials in a wide range of modern technological practices. The major advantages of the graded material, especially in elevated temperature environments stem from the tailoring capability to produce a gradual variation of its thermomechanical properties in the spatial domain. A Great efforts have been made to study the fracture behavior of FGMs [1–5]. As it is reported in the literature [6–7], the graded materials are rarely isotropic because of the nature of techniques used in fabricating them. Thus, it is necessary to consider the anisotropic character when studying the failure behaviors of FGMs. Guo et al. [8] investigated the mode I surface crack problem for an orthotropic graded strip. H. M. Xu et al. [9] studied the problems of a power-law orthotropic and half-space

[1]This research is supported by NSFC (10962008) and (51061015) and NSF of Ningxia (NZ1001)

functionally graded material (FGM) subjected to a line load. In the paper [10], asymptotic analysis coupled with Westergaard stress function approach is used to develop quasi-static stress fields for a crack oriented along one of the principal axes of inhomogeneous orthotropic medium. Kim and Paulino [11–12] examined mixed-mode stress intensity factors for cracks arbitrarily oriented in orthotropic FGMs using modified crack closure method and the path-independent J_k^*-integral, respectively.

Though there are lots of papers related to the crack problem of orthotropic functionally graded materials, very few papers on the plane crack problem of a functionally graded strip with a crack perpendicular to the boundary are published. It is very significant to study this kind of crack problem, since the geometry can be used as an approximation to a number of structural components and laboratory specimens, at the same time in line with the result of the Kawasaki and Watanabe[13]'s experiments about the thermal fracture behavior of metal/ceramic functionally graded materials(PSZ/IN 100 FGMS and PSZ/Inco 718 FGMs), the sequence of spalling behavior was found to be: crack vertical to the sample surface formed during cooling, then transverse crack formed in graded layer during heating, and subsequent growth of transverse cracks and their coalescence led eventually the ceramic coating to spall. Therefore, the surface crack problem is a very important issue to be considered during the design of FGMs. In these papers, the Mode I crack problem for a functionally graded orthotropic coating-substrate structure will be presented and the results could be useful in the laboratory test and the design of the orthotropic functionally graded materials .

2. Formulation of the Problem

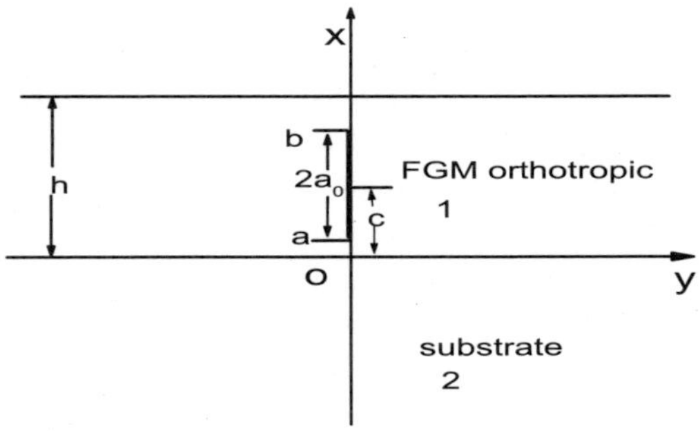

Figure 1: The geometry of the Mode I crack problem for the functionally graded orthotropic substrate-coating structure .

As shown in Fig.1: A functionally graded orthotropic strip with properties varying in the x-direction bonded to a half infinite orthotropic elastic substrate. The strip is infinite along the y-axis and has a thickness h along x-axis. The principal direction of orthotropy are parallel and perpendicular to the boundaries of the strip.

The material properties are defined as

$$c_{11}(x)=c_{110}e^{\delta x}, c_{12}(x)=c_{120}e^{\delta x}, c_{22}(x)=c_{220}e^{\delta x}, c_{66}(x)=c_{660}e^{\delta x}, \tag{1}$$

where c_{110}, c_{120}, c_{220}, c_{660} and δ are constants. c_{110}, c_{120}, c_{220} and c_{660} are the material parameters of $y = 0$, δ is the gradient parameters of functionally graded material.

The general constitutive relation can be written as

$$\begin{gathered}\sigma_{1xx}=c_{11}(x)\frac{\partial u}{\partial x}+c_{12}(x)\frac{\partial v}{\partial y}, \sigma_{1yy}=c_{12}(x)\frac{\partial u}{\partial x}+c_{22}(x)\frac{\partial v}{\partial y}, \\ \sigma_{1xy}=c_{66}(x)\left(\frac{\partial u}{\partial y}+\frac{\partial v}{\partial x}\right).\end{gathered} \tag{2}$$

The equilibrium equation in terms of the displacements can be given as:

$$\frac{\partial \sigma_{1xx}}{\partial x}+\frac{\partial \sigma_{xy}}{\partial y}=0, \quad \frac{\partial \sigma_{1yy}}{\partial y}+\frac{\partial \sigma_{xy}}{\partial x}=0, \tag{3}$$

make use of the equation Eqs. (1), (2) and (3), it can be obtained that

$$\begin{aligned} &c_{11}(x)\frac{\partial^2 u}{\partial x^2}+c_{66}(x)\frac{\partial^2 u}{\partial y^2}+[c_{12}(x)+c_{66}(x)]\frac{\partial^2 v}{\partial x\partial y} \\ &\qquad +c_{11}(x)\delta\frac{\partial u}{\partial x}+c_{12}(x)\delta\frac{\partial v}{\partial y}=0, \\ &c_{22}(x)\frac{\partial^2 v}{\partial y^2}+c_{66}(x)\frac{\partial^2 v}{\partial x^2}+[c_{12}(x)+c_{66}(x)]\frac{\partial^2 u}{\partial x\partial y} \\ &\qquad +c_{66}(x)\delta\frac{\partial u}{\partial y}+c_{66}(x)\delta\frac{\partial v}{\partial x}=0, \end{aligned} \tag{4}$$

if let the $\delta = 0$, then the equations of the elastic substrate can be obtained.

The mixed boundary conditions of the problem in Fig.1 can be written as

$$\begin{gathered}\sigma_{1xx}(h,y)=0, \ \sigma_{1xy}(h,y)=0, \ \sigma_{1xx}(0,y)=\sigma_{2xx}(0,y), \\ \sigma_{1xy}(0,y)=\sigma_{2xy}(0,y), \ u_1(0,y)=u_2(0,y), \ v_1(0,y)=v_2(0,y),\end{gathered} \tag{5}$$

$$\sigma_{2xx}(x,y)=0, \ \sigma_{2xy}(x,y)=0, \ x\to-\infty, \tag{6}$$

$$\begin{gathered}\sigma_{1xy}(x,0)=0, \ 0<x<h, \ \sigma_{1yy}(x,0)=-\sigma_0(x), \ a<x<b, \\ v_1(x,0)=0, \quad 0<x<a, \ b<x<h.\end{gathered} \tag{7}$$

By use of the Fourier transform method and thinking of the Eq. (6), the following displacement forms can be obtained, for functionally graded coating:

$$\begin{aligned} u_1 &= \frac{1}{2\pi}\int_{-\infty}^{\infty}\sum_{j=3}^{4} E_{1j}(s)\,A_{1j}e^{\lambda_{1j}(s)y-isx}ds \\ &+\frac{2}{\pi}\int_0^{\infty}\sum_{j=1}^{4} E_{2j}(\alpha)\,A_{2j}e^{\lambda_{2j}(\alpha)x}\cos\alpha y d\alpha, \\ v_1 &= \frac{1}{2\pi}\int_{-\infty}^{\infty}\sum_{j=3}^{4} A_{1j}e^{\lambda_{1j}(s)y-isx}ds \\ &+\frac{2}{\pi}\int_0^{\infty}\sum_{j=1}^{4} A_{2j}e^{\lambda_{2j}(\alpha)x}\sin\alpha y d\alpha, \end{aligned} \tag{8}$$

for elastic substrate:

$$\begin{aligned} u_2 &= \frac{2}{\pi}\int_0^{\infty}\sum_{j=1}^{2} A_{3j}(\alpha)\,E_{3j}e^{\lambda_{3j}(\alpha)x}\cos\alpha y d\alpha, \\ v_2 &= \frac{2}{\pi}\int_0^{\infty}\sum_{j=1}^{2} A_{3j}e^{\lambda_{3j}(\alpha)x}\sin\alpha y d\alpha, \end{aligned} \tag{9}$$

where the coefficient E_{ij}, $\lambda_{ij}\,(i = 1, 2, 3,\ j = 1, \cdots, 4)$ are shown in Appendix A, $A_{ij}\,(i=1, 2, 3, j=1, \cdots, 4)$ are unknown functions, which can be solved by the boundary conditions. To obtain the integral equations, let's introduce the following auxiliary function

$$g(x) = \frac{\partial}{\partial x}v_1(x, 0), \tag{10}$$

and $g(x)$ subjected to the following single-valuedness conditions

$$\int_a^b g(x)\,dt = 0. \tag{11}$$

By using Eqs. (7), (10) and applying the Fourier transform to Eq. (10), it can be obtained that

$$A_{13}=q_{13}\int_a^b g(u)\,e^{isu}du,\ A_{14}=q_{14}\int_a^b g(u)\,e^{isu}du, \tag{12}$$

where $q_{13} = \frac{-i\left(c_{120}s^2+c_{220}\lambda_{14}^2\right)}{c_{220}s\left(\lambda_{13}^2-\lambda_{14}^2\right)}$, $q_{14} = \frac{i\left(c_{120}s^2+c_{220}\lambda_{13}^2\right)}{c_{220}s\left(\lambda_{13}^2-\lambda_{14}^2\right)}$.

Then by residue theorem and the boundary conditions, we can get the following linear algebraic equations

$$\begin{vmatrix} B_{21}e^{\lambda_{21}h} & B_{22}e^{\lambda_{22}h} & B_{23}e^{\lambda_{23}h} & B_{24}e^{\lambda_{24}h} & 0 & 0 \\ F_{21}e^{\lambda_{21}h} & F_{22}e^{\lambda_{22}h} & F_{23}e^{\lambda_{23}h} & F_{24}e^{\lambda_{24}h} & 0 & 0 \\ B_{21} & B_{22} & B_{23} & B_{24} & -B_{31} & -B_{32} \\ F_{21} & F_{22} & F_{23} & F_{24} & -F_{31} & -F_{32} \\ E_{21} & E_{22} & E_{23} & E_{24} & -E_{31} & -E_{32} \\ 1 & 1 & 1 & 1 & 1 & 1 \end{vmatrix} \begin{vmatrix} A_{21} \\ A_{22} \\ A_{23} \\ A_{24} \\ A_{31} \\ A_{32} \end{vmatrix} = \begin{vmatrix} R_1 \\ R_2 \\ R_3 \\ R_4 \\ R_5 \\ R_6 \end{vmatrix}, \quad (13)$$

where

$$\begin{aligned} R_1(u) = {} & \frac{\left(-c_{120}^2 + c_{110}c_{220}\right)\alpha^2\lambda_{23}}{c_{110}\left(\lambda_{23}-\lambda_{21}\right)\left(\lambda_{23}-\lambda_{22}\right)\left(\lambda_{23}-\lambda_{24}\right)} e^{-\lambda_{23}(u-h)} \\ & + \frac{\left(-c_{120}^2 + c_{110}c_{220}\right)\alpha^2\lambda_{24}}{c_{110}\left(\lambda_{24}-\lambda_{21}\right)\left(\lambda_{24}-\lambda_{22}\right)\left(\lambda_{24}-\lambda_{23}\right)} e^{-\lambda_{24}(u-h)}, \\ R_2(u) = {} & -\frac{\left(-c_{120}^2 + c_{110}c_{220}\right)\alpha\lambda_{23}\left(\lambda_{23}+\delta\right)}{c_{110}\left(\lambda_{23}-\lambda_{21}\right)\left(\lambda_{23}-\lambda_{22}\right)\left(\lambda_{23}-\lambda_{24}\right)} e^{\lambda_{23}(h-u)} \\ & + c_{110}\left(\lambda_{24}-\lambda_{21}\right)\left(\lambda_{24}-\lambda_{22}\right)\left(\lambda_{24}-\lambda_{23}\right) e^{\lambda_{23}(h-u)}, \\ R_3(u) = {} & \frac{-\left(-c_{120}^2 + c_{110}c_{220}\right)\alpha^2\lambda_{21}}{c_{110}\left(\lambda_{21}-\lambda_{22}\right)\left(\lambda_{21}-\lambda_{23}\right)\left(\lambda_{21}-\lambda_{24}\right)} e^{-\lambda_{21}u} \\ & + \frac{-\left(-c_{120}^2 + c_{110}c_{220}\right)\alpha^2\lambda_{22}}{c_{110}\left(\lambda_{22}-\lambda_{21}\right)\left(\lambda_{22}-\lambda_{23}\right)\left(\lambda_{22}-\lambda_{24}\right)} e^{-\lambda_{22}u}, \\ R_4(u) = {} & \frac{\left(-c_{120}^2 + c_{110}c_{220}\right)\alpha\lambda_{21}\left(\lambda_{21}+\delta\right)}{c_{110}\left(\lambda_{21}-\lambda_{22}\right)\left(\lambda_{21}-\lambda_{23}\right)\left(\lambda_{21}-\lambda_{24}\right)} e^{-\lambda_{21}u} \\ & + \frac{\left(-c_{120}^2 + c_{110}c_{220}\right)\alpha\lambda_{22}\left(\lambda_{22}+\delta\right)}{c_{110}\left(\lambda_{22}-\lambda_{22}\right)\left(\lambda_{22}-\lambda_{23}\right)\left(\lambda_{22}-\lambda_{24}\right)} e^{-\lambda_{22}u}, \\ R_5(u) = {} & \frac{1}{c_{110}}\left[\frac{c_{120}\lambda_{21}^2 + 2\delta c_{120}\lambda_{21} + c_{120}\delta^2 + \alpha^2 c_{220}}{\left(\lambda_{21}-\lambda_{22}\right)\left(\lambda_{21}-\lambda_{23}\right)\left(\lambda_{21}-\lambda_{24}\right)} e^{-\lambda_{21}u}\right. \\ & \left. + \frac{c_{120}\lambda_{22}^2 + 2\delta c_{120}\lambda_{22} + c_{120}\delta^2 + \alpha^2 c_{220}}{\left(\lambda_{22}-\lambda_{21}\right)\left(\lambda_{22}-\lambda_{23}\right)\left(\lambda_{22}-\lambda_{24}\right)} e^{-\lambda_{22}u}\right], \\ R_6(u) = {} & \frac{-\alpha\left(-c_{120}^2 + c_{110}c_{220} - c_{120}c_{660}\right)\lambda_{21}^2}{c_{110}c_{660}\lambda_{21}\left(\lambda_{21}-\lambda_{22}\right)\left(\lambda_{21}-\lambda_{23}\right)\left(\lambda_{21}-\lambda_{24}\right)} e^{-\lambda_{21}u} \\ & + \frac{\alpha\delta\lambda_{21}\left(c_{120}^2 - c_{110}c_{220} + 2c_{120}c_{660}\right) - \alpha c_{660}\left(\delta^2 c_{120} + c_{220}\alpha^2\right)}{c_{110}c_{660}\lambda_{21}\left(\lambda_{21}-\lambda_{22}\right)\left(\lambda_{21}+\lambda_{23}\right) - \left(\lambda_{21}-\lambda_{24}\right)} e^{-\lambda_{21}u} \\ & + \frac{-\alpha\left(-c_{120}^2 + c_{110}c_{220} - c_{120}c_{660}\right)\lambda_{22}^2}{c_{110}c_{660}\lambda_{22}\left(\lambda_{22}-\lambda_{21}\right)\left(\lambda_{22}-\lambda_{23}\right)\left(\lambda_{22}-\lambda_{24}\right)} e^{-\lambda_{22}u} \end{aligned}$$

$$+\frac{\alpha\delta\lambda_{22}\left(c_{120}^2-c_{110}c_{220}+2c_{120}c_{660}\right)-\alpha c_{660}\left(\delta^2 c_{120}+c_{220}\alpha^2\right)}{c_{110}c_{660}\lambda_{22}\left(\lambda_{22}-\lambda_{21}\right)\left(\lambda_{22}-\lambda_{23}\right)\left(\lambda_{22}-\lambda_{24}\right)}e^{-\lambda_{22}u}$$
$$+\frac{\alpha\left(\delta^2 c_{120}+c_{220}\alpha^2\right)}{c_{110}\lambda_{21}\lambda_{22}\lambda_{23}\lambda_{24}}.$$

During the process of obtaining Eq. (13), the integral identities A6 [14] shown in Appendix are used. The B_j, C_j and F_j are some expressions of material consistents, E_{ij} and λ_{ij} $(i = 1, 2, 3,\ j = 1, \ldots, 4)$. Then the solution of the unknown functions can be obtained. Substitute the solution of the above equation and using the conditions of the crack surface, after considering the asymptotic when $s \to \infty$ and $\alpha \to 0$, the following equation can be obtained

$$\frac{1}{\pi}\int_a^b\left[-\frac{\mathrm{Im}(\omega_1)}{u-x}+h_1(u,x)+2K_{2s}+2h_2(u,x)\right]g(u)du=-\sigma_0(x)e^{-\delta x}, \quad (14)$$

the Eq. (14) is the first kind of Fredholm integral equation, which can be solved by the method of [15].

The stress intensity factors of the internal crack tips can be express as

$$\begin{aligned} K_I(a) &= \lim_{x\to a}\sqrt{2(a-x)\,\sigma_{yy}(x,0)} = -\mathrm{Im}(\omega_{11})\,e^{\delta a}\sum_{n=1}^{N} a_n, \\ g_I(b) &= \lim_{x\to b}\sqrt{2(x-b)\,\sigma_{yy}(x,0)} = -\mathrm{Im}(\omega_{11})\,e^{\delta b}\sum_{n=1}^{N}(-1)^n a_n. \end{aligned} \quad (15)$$

For convenience, the SIFs are normalized by $k_0 = \sigma_0\sqrt{a_0 h}$, where σ_0 is uniform crack surface pressure.

3. Numerical Results and Discussion

In the following analysis, we will study the influence of the length of the crack, the location of the crack, the functionally graded strip's width h and the gradient parameters of functionally graded material δ on the normalized stress intensity factors (SIFs) of the crack.

Firstly, the influence of the length of the crack and the gradient parameters of functionally graded material δ on the normalized stress intensity factors (SIFs) of the crack will be discussed. Here $h = 5.0$, $\delta h = -1, 0, 1, 2$, $c = 0.4h$. It can be found in the Fig.2–Fig.3 that the normalized intensity factors of both crack tips $K_I(a)$ and $K_I(b)$ increases with the increase of the normalized half crack length a_0 and the gradient parameters of functionally graded material δ. Therefore, to prevent the coating crack from growing toward the interface, the gradient parameters of functionally graded material δ should be chosen as $\delta < 0$.

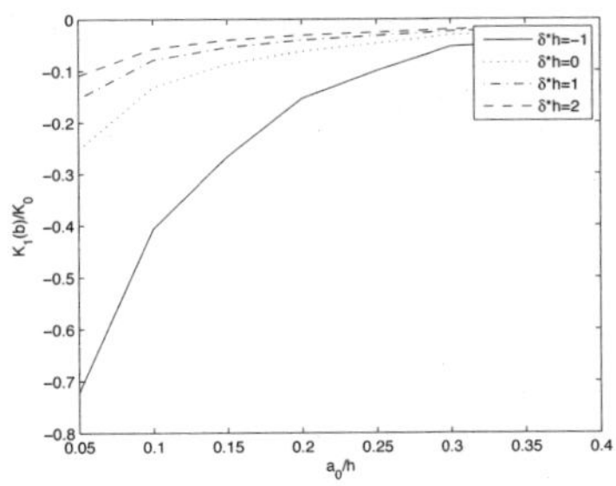

Figure 2: The normalized stress intensity factor $K_I(b)$ versus $\frac{a_0}{h}$

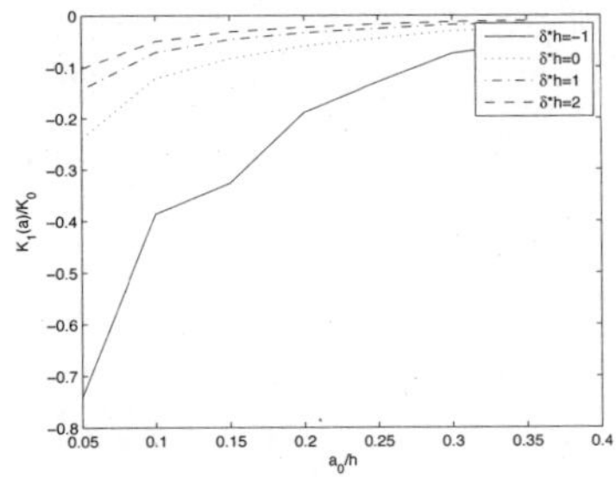

Figure 3: The normalized stress intensity factor $K_I(a)$versus $\frac{a_0}{h}$

Secondly, we will discuss the influence of the location of the crack on the normalized stress intensity factors (SIFs) of the crack. Here $\delta h = 1$, $c = 0.5, 1, 1.5, 2$. It can be found from Fig.4–Fig.5 that the normalized intensity factors of both crack tips $K_I(a)$ and $K_I(b)$ increases with the increase of the $c = \frac{b+a}{2}$.

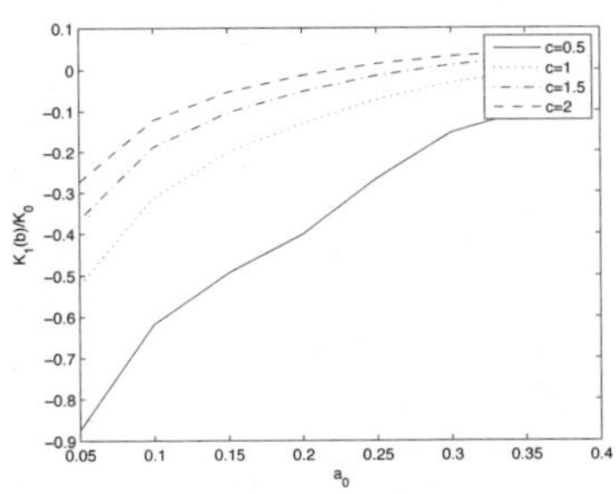

Figure 4: The normalized stress intensity factor $K_I(b)$versus a_0

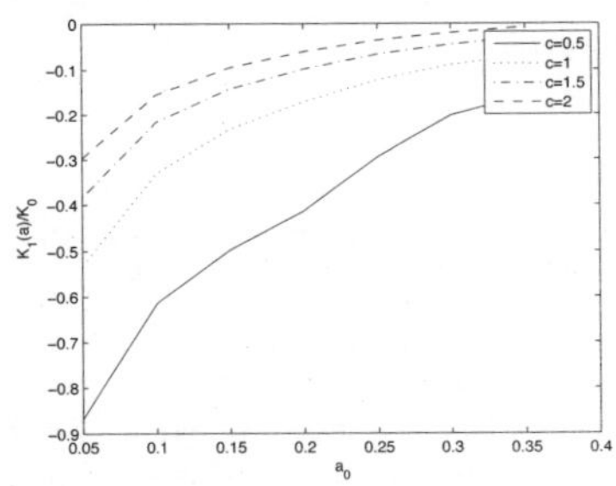

Figure 5: The normalized stress intensity factor $K_I(a)$versus a_0

Finally, we will discuss the influence of the width of the functionally graded orthotropic strip on the normalized stress intensity factors (SIFs) of the crack. Here $h = 1, 2, 3, 4$, $\delta h = 1$, $c = 0.5$. It can be found from the Fig.6–Fig.7 that the normalized intensity factors of both crack tips $K_I(a)$ and $K_I(b)$ decreases with the increase of the h, but as the increase of the strip width and the increase of the crack length, the effect is not obvious, so increasing the width of coating is not an effective way to restrain the expansion of the crack.

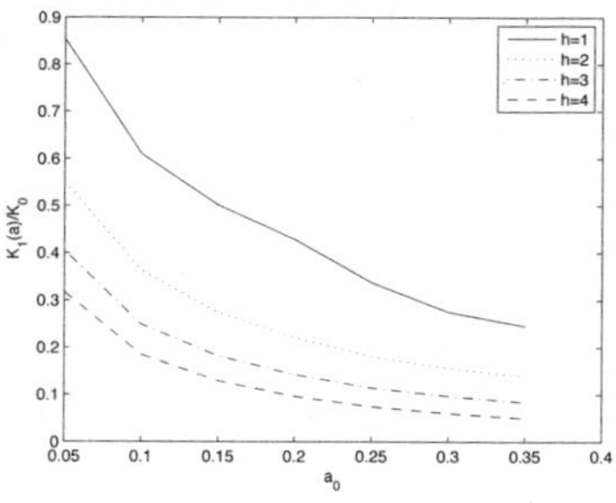

Figure 6: The normalized stress intensity factor $K_I(b)$versus a_0

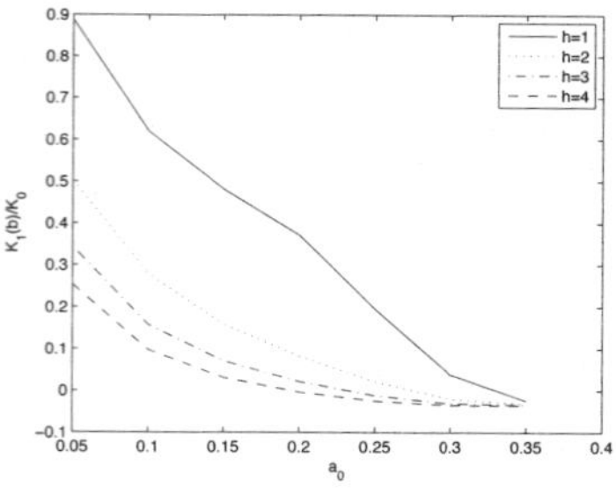

Figure 7: The normalized stress intensity factor $K_I(a)$versus a_0 for

4. Conclusion

In this paper, the mode I crack problem for a functionally graded orthotropic strip bonded to orthotropic substrate is studied analytically. The influences of the nonhomogeneity constants and geometric parameters on the stress intensity factors are investigated. The result may be help for the analysis and design of functionally graded orthotropic coating-substrate structures.

References

[1] F. Delale and F. Erdogan, On the mechanical modeling of the interfacial region in bonded half-planes, *International Journal of Applied Mechanic* 55 (1988), 317–324.

[2] S. Dag and F. Erdogan, A surface crack in a graded medium loaded by a sliding rigid stamp, *Engineering Fracture Mechanics* 69 (2002), 1729–1751.

[3] Y. D. Li, W. Tan and H. C. Zhang, Anti-plane transient fracture analysis of the functionally gradient elastic bi-material weak/infinitesimal-discontinuous interface, *International Journal of fracture* 142 (2006), 163–171.

[4] L. C. Guo, L. Z. Wu and L. Ma, The interface crack problem under a concentrated load for a functionally graded coating- substrate composite system, *Composite structure* 63 (2004), 397–406.

[5] L. C. Guo, L. Z. Wu, L. Ma and T. Zeng, Fracture analysis of a functionally graded coating-substrate structure with a crack perpendicular to the interface, Part I: static problem, *International Journal of fracture* 127 (2004), 21–38.

[6] M. Ozturk and F. Erdogan, Mode-I crack problem in an inhomogeneous orthotropic medium, *Journal of Engineering Science* 35 (1997), 869C-883.

[7] S. Dag, B. Yildirim and F. Erdogan, Interface crack problems in graded orthotropic media: Analytical and computational approaches. *International Journal of Fracture* 130 (2004), 471C-496.

[8] L. C. Guo, L. Z. Wu, T. Zeng and L. Ma, Mode I crack problem for a functionally graded orthotropic strip, *European Journal of Mechanics A/solids* 23 (2004), 219–234.

[9] H. M. Xu, X. F. Yao, X. Q. Feng and H. Y. Yeh, Fundamental solution of a power-law orthotropic and half-space functionally graded material under line loads, *Composites Science and Technology* 68 (2008), 27–34.

[10] Vijaya Bhaskar Chalivendra, Mode-I crack-tip stress fields for inhomogeneous orthotropic medium, *Mechanics of Materials* 40 (2008), 293–301.

[11] J. H. Kim and G. H. Paulino, Mixed-mode fracture of orthotropic functionally graded materials using finite elements and the modified crack closure method, *Engineer fracture mechanic* 69 (2002), 1557–1586.

[12] J. H. Kim and G. H. Paulino, Mixed-mode J-integral formulation and implementation using graded finite elements for fracture analysis of nonhomogeneous orthotropic materials, *mechanic material* 35 (2003), 107–128.

[13] A. Kawasaki and R. Watanabe, Thermal fracture behavior of metal/ceramic functionally graded materials, *Engineering Fracture Mechanics* 69 (2000), 1713–1728.

[14] I. S. Gradshteyn and I. M. Ryzhik, *Table of integrals, series, and products,* Elsevier/Academic Press, Amsterdam, 2007.

[15] X. Li, *Integral equations,* Science Press, Beijing, 2008.

Appendix

$$\left\{\begin{aligned} E_{1j} &= \frac{\left[c_{660}\left(s+i\delta\right)s - c_{220}\lambda_{1j}^2\right]i}{\left[c_{120}s + c_{660}\left(s+i\delta\right)\right]\lambda_{1j}}, \\ E_{2j} &= \frac{c_{660}\lambda_{2j}\left(\lambda_{2j}+\delta\right) - \alpha^2 c_{220}}{\alpha\left[c_{120}\lambda_{2j} + c_{660}\left(\lambda_{2j}+\delta\right)\right]}, \\ E_{3j} &= \frac{c_{660}\lambda_{3j}^2 - \alpha^2 c_{220}}{\alpha\left(c_{120}+c_{660}\right)\lambda_{3j}}, \end{aligned}\right\} j=1\ldots,4, \tag{A1}$$

$$\left\{\begin{aligned} \lambda_{1j} &= -\sqrt{-\frac{\Omega_{11}}{2} \pm \frac{1}{2}\sqrt{\Omega_{11}^2 - 4\Omega_{12}}}, \\ \lambda_{2j} &= \frac{-\delta \pm \sqrt{\delta^2 - 2\Omega_{21} \pm 2\sqrt{\Omega_{21}^2 - 4\Omega_{22}}}}{2}, \\ \lambda_{3j} &= \frac{\sqrt{2}}{2}\sqrt{-\Omega_{31} \pm \sqrt{\Omega_{31}^2 - 4\Omega_{32}}}, \end{aligned}\right\} j=1\ldots,4, \tag{A2}$$

$$\begin{gathered} \Omega_{11} = \frac{(c_{120}^2 - c_{120}c_{220})s(s+i\delta) + (2s^2 + 2is\delta - \delta^2)c_{120}c_{660}}{c_{220}c_{660}}, \\ \Omega_{12} = \frac{c_{110}s^2(s+i\delta)^2}{c_{220}}, \end{gathered} \tag{A3}$$

$$\Omega_{21} = \frac{\left[c_{120}^2 - c_{110}c_{220} + 2c_{120}c_{660}\right]\alpha^2}{c_{110}c_{660}}, \ \Omega_{22} = \left(\frac{c_{220}}{c_{110}}\alpha^2 + \frac{c_{120}}{c_{110}}\delta^2\right)\alpha^2, \tag{A4}$$

$$\Omega_{31} = \left[\frac{c_{120}^2}{c_{110}c_{660}} + 2\frac{c_{120}}{c_{110}} - \frac{c_{220}}{c_{660}}\right]\alpha^2, \ \Omega_{32} = \frac{c_{220}}{c_{110}}\alpha^4, \tag{A5}$$

$$\begin{aligned}\int_0^\infty e^{-px}\sin(qx+\lambda)\,dx &= \frac{1}{p^2+q^2}(q\cos\lambda + p\sin\lambda),\\ \int_0^\infty e^{-px}\cos(qx+\lambda)\,dx &= \frac{1}{p^2+q^2}(p\cos\lambda - q\sin\lambda),\end{aligned}\quad p>0, \tag{A6}$$

$$\begin{aligned}\omega_1 &= \frac{ic_{660}\left(c_{120}+c_{220}p_{11}^2\right)\left(c_{120}+c_{220}p_{12}^2\right)}{c_{220}\left(c_{120}+c_{660}\right)p_{11}p_{12}\left(p_{11}+p_{12}\right)},\\ h_1(u,x) &= \frac{1}{2}\int_{-\infty}^{\infty}\left(\sum_{j=3}^{4}C_{1j}q_{1j}-\omega_1\right)e^{is(u-x)}ds,\end{aligned} \tag{A7}$$

$$K_{2s} = \left[\frac{\Lambda_1}{p_{21}(x+u)} + \frac{\Lambda_2}{p_{21}x+p_{22}u} + \frac{\Lambda_3}{p_{22}x+p_{21}u} + \frac{\Lambda_4}{p_{22}(x+u)}\right], \tag{A8}$$

$$\begin{aligned}h_2(u,x) = \int_0^\infty\Bigg[\sum_{j=3}^{4}C_{2j}e^{\lambda_{2j}x}A_{2j} - \Lambda_1 e^{-p_{21}\alpha(x+u)}\\ -\Lambda_2 e^{-\alpha(p_{21}x+p_{22}u)} - \Lambda_3 e^{-\alpha(p_{22}x+p_{21}u)} - \Lambda_4 e^{-p_{22}\alpha(x+u)}\Bigg]d\alpha,\end{aligned} \tag{A9}$$

$$\begin{aligned}\Lambda_1 = &\frac{1}{4}\left(c_{120}p_{21}^2+c_{220}\right)\left(c_{120}c_{660}+c_{120}^2-c_{110}c_{220}+c_{110}c_{660}p_{22}^2\right)\\ &\times\left[-c_{110}c_{660}p_{21}^4+\left(c_{110}c_{220}-c_{120}^2-2c_{120}c_{660}\right)p_{21}^2-c_{220}c_{660}\right]\\ &\times\frac{1}{c_{660}\left(c_{660}+c_{120}\right)c_{110}^2p_{21}^2\left(p_{21}+p_{22}\right)^2\left(p_{21}-p_{22}\right)^2},\end{aligned} \tag{A10}$$

$$\begin{aligned}\Lambda_2 = &\frac{1}{4}(c_{120}p_{21}^2+c_{220})\\ &\times[-c_{110}c_{660}p_{22}^4+(c_{110}c_{220}-c_{120}^2-2c_{120}c_{660})p_{22}^2-c_{220}c_{660}]\\ &\times\left(c_{110}c_{220}-c_{110}c_{660}p_{21}p_{22}-c_{120}^2-c_{120}c_{660}\right)\\ &\times\frac{1}{c_{660}c_{110}^2\left(c_{660}+c_{120}\right)p_{22}^2\left(p_{21}+p_{22}\right)^2\left(p_{21}-p_{22}\right)^2},\end{aligned} \tag{A11}$$

$$\begin{aligned}\Lambda_3 = &\frac{1}{4}\left(c_{120}p_{22}^2+c_{220}\right)\\ &\times[-c_{110}c_{660}p_{21}^4+(c_{110}c_{220}-c_{120}^2-2c_{120}c_{660})p_{21}^2-c_{220}c_{660}]\\ &\times\left(c_{110}c_{220}-c_{110}c_{660}p_{21}p_{22}-c_{120}^2-c_{120}c_{660}\right)\\ &\times\frac{1}{c_{660}c_{110}^2\left(c_{660}+c_{120}\right)p_{21}^2\left(p_{21}+p_{22}\right)^2\left(p_{21}-p_{22}\right)^2},\end{aligned} \tag{A12}$$

$$\begin{aligned}
\Lambda_4 &= -\frac{1}{4}\left(c_{120}p_{22}^2 + c_{220}\right)\\
&\times\left[-c_{110}c_{660}p_{22}^4 + \left(c_{110}c_{220} - c_{120}^2 - 2c_{120}c_{660}\right)p_{22}^2 - c_{220}c_{660}\right]\\
&\times\left(-c_{120}c_{660} - c_{120}^2 + c_{110}c_{220} - c_{110}c_{660}p_{21}^2\right)\\
&\times\frac{1}{c_{660}c_{110}^2\left(c_{660}+c_{120}\right)p_{22}^2\left(p_{21}+p_{22}\right)^2\left(p_{21}-p_{22}\right)^2},
\end{aligned} \tag{A13}$$

$$\left.\begin{aligned}
p_{1j} &= -\frac{\sqrt{2}}{2}\xi_{11}\sqrt{aa_1 \pm bb_1},\ \xi_{1j} = \text{sign}\left[\text{Re}\sqrt{aa_1 \pm bb_1}\right],\\
p_{2j} &= \frac{\sqrt{2}}{2}\xi_{21}\sqrt{aa_2 \pm bb_2},\ \xi_{21} = \text{sign}\left[\text{Re}\sqrt{aa_2 \pm bb_2}\right],
\end{aligned}\right\} j = 1, 2, \tag{A14}$$

$$\begin{aligned}
aa_1 &= \frac{c_{110}c_{220} - c_{120}\left(c_{120} + 2c_{660}\right)}{c_{220}c_{660}},\\
bb_1 &= \frac{\sqrt{\left(c_{120}^2 - c_{110}c_{220}\right)\left[\left(c_{120}+2c_{660}\right)^2 - c_{110}c_{220}\right]}}{c_{220}c_{660}},\\
aa_2 &= \frac{c_{110}c_{220} - c_{120}^2 - 2c_{120}c_{660}}{c_{110}c_{660}},\\
bb_2 &= \frac{\sqrt{\left(c_{120}^2 - c_{110}c_{220}\right)\left[\left(c_{120}+2c_{660}\right)^2 - c_{110}c_{220}\right]}}{c_{110}c_{660}}.
\end{aligned} \tag{A15}$$

NÖETHER'S THEORY OF SOME TWO-DIMENSIONAL SINGULAR INTEGRAL OPERATORS WITH CONTINUOUS COEFFICIENTS AND APPLICATIONS[1]

SU-KUI LU, ZI-ZHI LI and SI-LE WANG

College of Mathematics and Computer Science

Hebei University, Baoding 071002, China

E-mail: lsk@hbu.edu.cn

Concerning Nöether's theory of some two-dimensional singular integral operator with continuous coefficients, K. Ch. Boimatov and G. Dzangibekov carried out a series of further effective research and then gave the complete effective Nöether's condition and index calculation formula. Besides, with the relevant results, we give the representation of solutions and the index formula of the non-homogeneous Dirichlet problem and non-homogeneous Neumann problem for general elliptic systems of second order equations.

Keywords: Singular integral operator, Nöether's condition, index, boundary value problem.

AMS No: 45P05, 47G10, 35J25.

1. Basic Knowledge and Notations

Suppose D is the bounded simply connected domain with the boundary $\partial D = \Gamma$ being a Lyapunov closed curve in the complex plane, and $z = 0 \in D$. When D is an arbitrary multiply connected bounded domain in the complex plane, the boundary $\partial D = \Gamma$ is composed of finite disjoint Lyapunov closed curves. It will not be announced below. Let $f(z)$ is a complex function in the domain D, while $L^p(D)(1 < p < \infty)$ and $L^p_{\beta-2/p}(D)(1 < p < \infty,\ 0 < \beta < 2)$ are the Banach spaces in real number field. Then

$$L^p_{\beta-2/p}(D)=\{f(z)\colon |z|^{\beta-2/p}f(z)=F(z)\in L^P(D),\ ||f||_{L^p_{\beta-2/p}}=||F||_{L^p}\}.$$

Through here the functions in $L^p(D)$ and $L^p_{\beta-2/p}(D)$ are complex-valued, the spaces themselves are regarded as real spaces, that is, they themselves can be as the linear combinations in the real number field, and also the singular operators' synthesis of the conjugate operation can be operated. The function

$$B(z,\zeta)=-\frac{2}{\pi}\frac{\partial^2 G(z,\zeta)}{\partial z\partial\bar{\zeta}}=\frac{1}{\pi(1-z\bar{\zeta})^2},\ G(z,\zeta)=\ln\left|\frac{1-z\bar{\zeta}}{z-\zeta}\right|,\ |z|<1, \quad (1)$$

[1]This research is supported by scientific and technological project of Hebei (No. 072135115)

is defined as Bergman kernel functions on the unit disk, while $G(z,\zeta)$ is the Green function of Dirichlet problem of Laplace equation in $|z|<1$. For the general single computational domain D, denote by $\sigma=\omega(z)$ a conformal mapping from D onto the circle $|\sigma|<1$ with the condition $\omega(0)=0$. Then

$$B(z,\zeta)=-\frac{2}{\pi}\frac{\partial^2 G(z,\zeta)}{\partial z\partial\bar{\zeta}}=\frac{\omega'(z)\overline{\omega'(\zeta)}}{\pi(1-\omega(z)\overline{\omega(\zeta)})^2},\ G(z,\zeta)=\ln\left|\frac{1-\omega(z)\overline{\omega(\zeta)}}{\omega(z)-\omega(\zeta)}\right| \quad (2)$$

is defined as the Bergman kernel functions in domain D, where ζ is analytic function of z and it possess the singularity when $z=\zeta\in\Gamma$.

Next, some simple operator notations will be introduced. For $z\in D$, I is identity operator, T is completely continuous operator in $L^p_{\beta-2/p}(D)$, and K is conjugate operator, that is, $(\mathrm{K}f)(z)=\overline{f(z)}$, and

$$\Pi=(\Pi f)(z)=-\frac{1}{\pi}\iint_D\frac{f(\zeta)}{(\zeta-z)^2}d\sigma_\zeta,\ (\bar{\Pi}f)(z)=-\frac{1}{\pi}\iint_D\frac{f(\zeta)}{(\bar{\zeta}-\bar{z})^2}d\sigma_\zeta; \quad (3)$$

$$\mathrm{B}=(\mathrm{B}f)(z)=\iint_D B(z,\zeta)f(\zeta)d\sigma_\zeta,\ (\bar{B}f)(z)=\iint_D\overline{B(z,\zeta)}f(\zeta)d\sigma_\zeta; \quad (4)$$

$$(S_v f)(z)=\frac{(-1)^{|v|}|v|}{\pi}\iint_D\frac{e^{-2iv\theta}}{|\zeta-z|^2}f(\zeta)d\sigma_\zeta,\ \theta=\arg(\zeta-z); \quad (5)$$

$$\bar{\Pi}=\Pi^{-1}=K\,\Pi\,K,\ \bar{B}=K\,B\,K. \quad (6)$$

There into, $d\sigma_\zeta$ is the plane measure element of Lebesgue, the integral in (3) is regarded as the integral with the Cauchy principal value, and v is an integer.

At last, the principal auxiliary function notations used in the following will be given. Let all $a(z)$, $b(z)$, $c(z)$, $d(z)$ etc. are continuous complex-valued functions on $\overline{D}$, marked as $a(z),b(z),c(z),d(z),\cdots\in C(\overline{D})$. Also as (Some leave out the function of variable z):

$$\begin{cases}\Delta=\Delta_1=|a|^2-|b|^2,\ \Delta_2=|c|^2-|d|^2,\ \Delta_3=|e|^2-|h|^2,\\ \lambda=\lambda_1=\bar{a}c-b\bar{d},\ \lambda_2=\bar{h}c-e\bar{d},\ \lambda_3=\bar{a}e-b\bar{h},\\ \mu=\mu_1=a\bar{d}-\bar{b}c,\ \mu_2=h\bar{d}-\bar{e}c,\ \mu_3=a\bar{h}-\bar{b}e;\end{cases} \quad (7)$$

$$\begin{cases}M(z)=\max\limits_{|t|=1}\mathrm{Re}\left[(\lambda_1(z)+\overline{\lambda_3}(z))t-\mu_2(z)t^2\right],\\ -m(z)=\min\limits_{|t|=1}\mathrm{Re}\left[(\lambda_1(z)+\overline{\lambda_3}(z))t-\mu_2(z)t^2\right];\end{cases} \quad (8)$$

$$\chi(z)=\begin{cases}M(z), & \text{if } \Delta_j(z)>0,\\ m(z), & \text{if } \Delta_j(z)<0,\end{cases}\quad j=1,2,3. \quad (9)$$

2. Nöether's Theory of Some Two-Dimensional Singular Integral Operator with Continuous Coefficients

In reference [1], the author gives the simplest two-dimensional singular integral operator, which plays an important role in the theory of quasi-conformal mappings and generalized analytic functions. In Chapter 2, [2], the author describes the theory about the general two-dimensional singular integral equation, which shows that there are close relations between two-dimensional singular integral operator and the Riemann boundary value problem of elliptic systems of first order equations. In [3] and [4], they make systematic researches on the elliptic systems of equations of different orders and the boundary value problem. However, little headway has been made in the deep and efficient research on Nöether's theory of operator for a long time. After 1970, the references from [5] to [8] make effective research on more general two-dimensional singular integral equation and present the effective Nöether's condition and the index formula. The main characteristics of the research is as follows: Firstly, the research is extended from $L^p(D)$ to $L^p_{\beta-2/p}(D)$. Secondly, operator's coefficients requires continuous functions with weak conditions or discontinuous points rather than very strong differentiability and even Hölder continuity. Finally, according to [9], the author constructed the relative operator matrix and the Bergman kernel function which play a key function in the proof. This method differs from that in [2], that is, research is carried out from operator itself without transforming the general boundary value problem related to the generalized analytic function, and Nöether's condition is set up and applied in the boundary value problem. Nöether's theory of two-dimensional singular integral operator with discontinuous coefficients has been discussed in [11], so this paper will present Nöether's theory of two-dimensional singular integral operator with continuous coefficients.

Next, we will provide an introduction about some two-dimensional singular integral operator with continuous coefficients discussed in the references from [5] to [8], the main Nöether theorems in the Banach space $L^p(D)(1 < p < \infty)$ and $L^p_{\beta-2/p}(D)$ $(1 < p < \infty,\ 0 < \beta < 2)$:

Theorem 1[5]. *Let D be a multi-connected bounded domain in the complex plane and an operator be*

$$A_1 \equiv (A_1 f)(z) \equiv aI + bK + c\Pi + dK\Pi,\ \forall z \in D, \tag{10}$$

with the conditions $a(z), b(z), c(z), d(z) \in C(\overline{D})$. Then in space $L^p(D)$ $(1 < p < \infty)$, A_1 is an Nöether operator, if and only if one of the following mutually exclusive conditions holds:

$$|\Delta_1| > |\lambda| + |\mu|,\ \ \forall z \in \overline{D}, \tag{11}$$

$$|\Delta_2| > |\lambda| + |\mu|, \ \forall z \in \overline{D}, \ \mu(t) \neq 0, \ \forall t \in \Gamma. \tag{12}$$

If the condition (11) is satisfied, then A_1 has bounded inverse operator. If the condition (12) is satisfied, the index κ of A_1 is

$$\kappa = 2\mathrm{Ind}_{\Gamma}\mu(t). \tag{13}$$

Nöether's condition and index calculation formula of another operator will be introduced in the following:

Let D be an arbitrary multiply connected bounded domain in the complex plane, and the operator

$$\mathrm{A}_2 \equiv (\mathrm{A}_2 f)(z) \equiv aI + b_0\mathrm{K} + \sum_{\nu=-m}^{n}{}^{*} \, b_\nu S_\nu K, \ z \in D, \tag{14}$$

where $\sum^*$ stands for all the items except ν=0, and K is conjugate operator and S_ν shown in (5) is marked as $\mathrm{P}_{n,m}(z,t) = \sum\limits_{\nu=-m}^{n} b_\nu(z)t^\nu$, $z \in \overline{D}$, $|t| = 1$.

First we introduce the following definitions:

Definition 1. If $\forall z \in \overline{D}$, $|t| = 1$, $|a(z)| > |P_{n,m}(z,t)|$, then $A_2 \in M$.

Definition 2. Suppose that $\forall z \in \overline{D}$, $|t| = 1$, $|a(z)| < |P_{n,m}(z,t)|$, and $\mathrm{Ind}_{|t|=1}P_{n,m}(z,t) = j$, then we say that $A_2 \in M_j$ (j is integer, $-m \le j \le n$).

On the basis of the above assumptions and definitions, we can prove following theorem:

Theorem 2[6]. *In space $L^p(D)$ $(1 < p < \infty)$, the operator A_2 is an Nöether operator if and only if one of the following mutually exclusive conditions holds:*

$$|a(z)| > |P_{n,m}(z,t)| \ ,\forall z \in \overline{D}, \ |t| = 1, \tag{15}$$

$$\begin{cases} |a(z)| < |P_{n,m}(z,t)|, \ \forall z \in \overline{D}, \ |t| = 1, \\ a(z) \neq 0, \ \text{for } \mathrm{Ind}_{|t|=1} \sum\limits_{\nu=-m}^{n} b_\nu(z)t^\nu \neq 0, \ \forall z \in \Gamma. \end{cases} \tag{16}$$

Moreover if $A_2 \in M$ or $A_2 \in M_0$, then index κ=0 of operator A_2. If $A_2 \in M_j (j \neq 0)$, then index κ of operator A_2:

$$\kappa = -2j\mathrm{Ind}_{\Gamma}a(t). \tag{17}$$

Theorem 3[7]. *Let D be the multiply onnected bounded domain, and $a(z)$, $b(z), c(z), d(z), e(z), h(z) \in C(\overline{D})$, and*

$$A_3 \equiv (A_3 f)(z) \equiv aI + bK + [cI + dK]\,\Pi + [eI + hK]\,\overline{\Pi}, \ \forall z \in D. \tag{18}$$

In space $L^p(D)\,(1 < p < \infty)$, the operator A_3 is an Nöether operator if and only if one of the following mutually exclusive conditions holds:

$$|\Delta_1| > \chi + [\chi^2 + |\mu_1|^2 - |\lambda_1|^2 + |\mu_3|^2 - |\lambda_3|^2]^{1/2},\ \forall z \in \overline{D}, \tag{19}$$

$$\begin{cases} |\Delta_2| > \chi + [\chi^2 + |\mu_1|^2 - |\lambda_1|^2 - |\mu_2|^2 + |\lambda_2|^2]^{1/2}, \forall z \in \overline{D}, \\ \lambda_1(t)\lambda_2(t) + \Delta_2(t)\mu_1(t) \neq 0, \forall t \in \Gamma; \end{cases} \tag{20}$$

$$\begin{cases} |\Delta_3| > \chi + [\chi^2(z) + |\mu_3|^2 - |\lambda_3|^2 - |\mu_2|^2 + |\lambda_2|^2]^{1/2},\ \forall z \in \overline{D}, \\ \lambda_2(t)\lambda_3(t) - \Delta_3(t)\mu_3(t) \neq 0,\ \forall t \in \Gamma, \end{cases} \tag{21}$$

where $\Delta_j, \mu_j, \lambda_j, \chi\,(\mathrm{j} = 1, 2, 3)$ are as stated in (7), (8) and (9). Under this case, the index κ of operator A_3 is:

$$\kappa = \begin{cases} 0, & \textit{if}\ \ (19)\ \ \textit{holds}, \\ 2\mathrm{Ind}_\Gamma[\lambda_1(t)\lambda_2(t) + \Delta_2(t)\mu_1(t)], & \textit{if}\ \ (20)\ \ \textit{holds}, \\ -2\mathrm{Ind}_\Gamma[\lambda_2(t)\lambda_3(t) - \Delta_3(t)\mu_3(t)], & \textit{if}\ \ (21)\ \ \textit{holds}. \end{cases} \tag{22}$$

Here we point out that the references [1] and [10] deal with the special circumstance of A_3 with $a(z) \equiv 1, b(z) \equiv 0$:

$$A_4 \equiv (A_4 f)(z) \equiv I + [cI + dK]\ \Pi + [eI + hK]\ \overline{\Pi},\ \forall z \in D, \tag{23}$$

and when the coefficient satisfies stronger conditions:

$$|a(z)| + |d(z)| + |e(z)| + |h(z)| < 1,\ z \in \overline{D},$$

applying the compressing reflection principle, we can prove that A_4 possesses a bounded inverse operator, when p is sufficiently close to 2.

Theorem 4[8]. *Let D be bounded simply connected domain and $a(z)$, $b(z), c(z), d(z), e(z), h(z), \alpha(z), \gamma(z), \delta(z), \nu(z) \in C(\overline{D})$, and operator A_5:*

$$\begin{aligned} A_5 \equiv (A_5 f)(z) &\equiv aI + bK + [cI + dK]\ \Pi + [eI + hK]\ \overline{\Pi} \\ &+ [\alpha I + \gamma K]B + [\delta I + \nu K]\overline{B},\ z \in D. \end{aligned} \tag{24}$$

In space $L^p_{\beta-2/p}(D)\,(1 < p < \infty,\ 0 < \beta < 2)$, the operator A_5 is an Nöether operator, if and only if one of the following mutually exclusive conditions holds:

$$\begin{cases} |\Delta_1| > \chi + [\chi^2 + |\mu_1|^2 - |\lambda_1|^2 + |\mu_3|^2 - |\lambda_3|^2]^{1/2},\ \forall z \in \overline{D}, \\ \omega_1(t) \neq 0,\ \forall t \in \Gamma; \end{cases} \tag{25}$$

$$\begin{cases} |\Delta_2| > \chi + [\chi^2 + |\mu_1|^2 - |\lambda_1|^2 - |\mu_2|^2 + |\lambda_2|^2]^{1/2}, \ \forall z \in \overline{D}, \\ \omega_2(t) \neq 0, \ \forall t \in \Gamma; \end{cases} \tag{26}$$

$$\begin{cases} |\Delta_3| > \chi + [\chi^2 + |\mu_3|^2 - |\lambda_3|^2 - |\mu_2|^2 + |\lambda_2|^2]^{1/2}, \ \forall z \in \overline{D}, \\ \omega_3(\mathrm{t}) \neq 0, \ \forall \mathrm{t} \in \Gamma, \end{cases} \tag{27}$$

in which $\Delta_j, \mu_j, \lambda_j, \chi\,(\mathrm{j} = 1, 2, 3)$ *as shown in* (7), (8) *and* (9), *and* $\omega_j(t)$ $(j = 1, 2, 3)$ *is determined by the following formulas:*

$$\alpha_{11} = \bar{a}\alpha - b\bar{\gamma}, \ \alpha_{21} = \bar{c}\alpha - d\bar{\gamma}, \ \alpha_{31} = \bar{e}\alpha - h\bar{\gamma},$$

$$\gamma_{11} = \bar{a}\gamma - b\bar{\alpha}, \ \gamma_{21} = \bar{c}\gamma - d\bar{\alpha}, \ \gamma_{31} = \bar{e}\gamma - h\bar{\alpha},$$

$$\delta_{11} = \bar{a}\delta - b\bar{\nu}, \ \delta_{21} = \bar{c}\delta - d\bar{\nu}, \ \delta_{31} = \bar{e}\delta - h\bar{\nu},$$

$$\nu_{11} = \bar{a}\nu - b\delta, \ \nu_{21} = \bar{c}\nu - d\bar{\delta}, \ \nu_{31} = \bar{e}\nu - h\bar{\delta};$$

$$\begin{cases} \alpha_{j2} = \nu_j\alpha_{j1} + \bar{w}_j\nu_{j1}, \gamma_{j2} = \bar{\nu}_j\gamma_{j1} + w_j\delta_{j1}, \\ \delta_{j2} = \nu_j\delta_{j1} + \bar{w}_j\gamma_{j1}, \nu_{j2} = \bar{\nu}_j\nu_{j1} + w_j\alpha_{j1}, \end{cases} \quad j = 1, 2, 3;$$

$$\beta_1 = (\bar{\lambda}_2\bar{\lambda}_3 - \Delta_3\bar{\mu}_3 - \mu_2\bar{\lambda}_2 q_1)[\Delta_3^2 - \bar{\lambda}_2 - (\lambda_2\bar{\mu}_3 - \Delta_3\bar{\lambda}_3)q_1 - \mu_2\Delta_3 q_1^2]^{-1},$$

$$\omega_3 = (\bar{\beta}_1 + \bar{\delta}_{32})(|\beta_1|^2 - 1 + \alpha_{32}\bar{\beta}_1 + q_1\bar{q}_2\,|\beta_1|^2) - (\bar{\gamma}_{32} - q_1)(\nu_{32} - \bar{q}_2)\beta_1,$$

$$\nu_3 = -\bar{\lambda}_2 + \Delta_3 q_1\beta_1, \quad w_3 = -\Delta_3 + \bar{\lambda}_2 q_1\beta_1;$$

$$\begin{cases} \beta_3 = [-\Delta_1\bar{\mu}_3 + (\bar{\lambda}_3\bar{\mu}_1 + \lambda_1\mu_3)q_1 \\ \qquad \times [|\lambda_3|^2 - |\mu_3|^2 + \Delta_1\lambda_3 q_1 + (\lambda_3\bar{\lambda}_1 - \mu_3\bar{\mu}_1)q_1^2]^{-1}, \\ \omega_1 = (1 + \bar{\delta}_{12}q_2)(1 - |\beta_3|^2 + \alpha_{12}) - \bar{\gamma}_{12}(\nu_{12} - q_1\beta_3), \\ \nu_1 = \bar{\lambda}_3 + q_1\beta_3\mu_3, \quad w_1 = \bar{\mu}_3 + q_1\beta_3\lambda_3, if\,|\lambda| \neq |\mu|; \end{cases}$$

$$\begin{cases} \beta_3 = \bar{\mu}_1/\Delta_1 + \bar{\lambda}_1 q_1), \\ \omega_1 = (1 + \bar{\delta}_{12}q_2)(1 - |\beta_3|^2 + \alpha_{12}) - \bar{\gamma}_{12}(\nu_{12} - q_1\beta_3), \\ \nu_1 = 1, \ w_1 = q_1\beta_3, \ if \ \ |\gamma| = |\mu|; \end{cases}$$

where $q_1(z), q_2(z)$ *are the root of trigonometric polynomial* $P_0 = \Delta_1 + \Delta_2 + \Delta_3 - 2Re\left[\mu_2 e^{4i\varphi} - (\lambda_1 + \bar{\lambda}_3)e^{2i\varphi}\right]$, *and* $\bar{\lambda}_3, \bar{\mu}_3, \Delta_3, \bar{\lambda}_2\alpha_{32}, \gamma_{32}, \delta_{32}, \nu_{32}$ *is substituted by* $\bar{\lambda}_3, \bar{\mu}_3, \Delta_3, \bar{\lambda}_2, \alpha_{32}, \gamma_{32}, \ \delta_{32}, \nu_{32}$ *in* $\omega_2 = \omega_3$. *If one of the above conditions is satisfied, then* κ *of operator* A_5*:*

$$\kappa = \mathrm{Ind}_\Gamma A_5 = \mathrm{Ind}_\Gamma \omega_j(t), \ j = 1, 2, 3. \tag{28}$$

3. Applications of Nöether's Theory in Second Order Elliptical Equations

We will discuss Nöether's theory of two-dimensional singular integral operator in [8] and its application to boundary value problems for the general second order elliptical systems of equations. Let D be a simply connected bounded domain in the complex plane, and the second order elliptical systems of equations in D be:

$$\begin{aligned} &\mathrm{a}W_{Z\overline{Z}} + b\overline{W}_{\overline{Z}Z} + cW_{ZZ} + d\overline{W}_{\overline{Z}\overline{Z}} + eW_{\overline{z}\overline{z}} + h\overline{W}_{ZZ} + \\ &+ a_1 W_{\overline{z}} + b_1\overline{W}_z + c_1 W_Z + d_1\overline{W}_{\overline{z}} + e_1 W + h_1\overline{W} = g(z), \end{aligned} \tag{29}$$

in which $z = x + iy$, $W = U(x,y) + iV(x,y)$, $\frac{\partial}{\partial\overline{z}} = \frac{1}{2}\left(\frac{\partial}{\partial x} + i\frac{\partial}{\partial y}\right)$, $\frac{\partial}{\partial z} = \frac{1}{2}(\frac{\partial}{\partial x} - i\frac{\partial}{\partial y})$, and all coefficients of (29) belong to $C(\overline{D})$ and the nonhomogeneous item $g(z) \in L^P(D)\,(2 < p < \infty)$.

On the basis of the conditions in (29), when each former inequality of (19), (20) and (21) are satisfied, then (29) can be transformed into:

$$\Delta_1 W_{Z\overline{Z}} + \lambda_1 W_{ZZ} + \overline{\mu_1}\overline{W}_{\overline{Z}\overline{Z}} + \lambda_3 W_{\overline{z}\overline{z}} + \overline{\mu_3}\overline{W}_{ZZ} + T_1(W) = g_1(z), \quad (30)_1$$

$$\lambda_1 W_{Z\overline{Z}} - \mu_1\overline{W}_{\overline{Z}Z} + \Delta_2 W_{ZZ} - \overline{\mu_2} W_{\overline{z}\overline{z}} + \overline{\lambda_2}\overline{W}_{ZZ} + T_2(W) = g_2(z), \quad (30)_2$$

$$\overline{\lambda_3} W_{Z\overline{Z}} - \overline{\mu_3}\overline{W}_{\overline{Z}Z} + \mu_2 W_{ZZ} - \overline{\lambda_2}\overline{W}_{\overline{Z}\overline{Z}} + \Delta_3 W_{\overline{z}\overline{z}} + T_3(W) = g_3(z), \quad (30)_3$$

respectively, where $g_1 = \overline{a}g - b\overline{g}$, $g_2 = \overline{c}g - d\overline{g}$, $g_3 = \overline{e}g - h\overline{g}$ (here the variable z is omitted); $T_j(W)\,(j = 1,2,3)$ is the lower order items of W. Then the results of two boundary value problems in book [8] will be discussed.

We introduce the Sobolev space $C(\overline{D}) \cap W_p^2(D)\,(2 < p < \infty)$, which can be seen as in [1,4,10].

Problem $\mathbf{D_0}$. Find a solution $W(z) \in C(\overline{D}) \cap W_p^2(D)$ of the elliptic systems (29) in simply connected domain D satisfying the boundary condition

$$W(t)\,|_\Gamma = 0. \tag{31}$$

The problem discussed above can be defined as homogeneous Dirichlet problem (Problem $\mathrm{D_0}$). Without loss of generality, we assume $D = \{Z : |z| < 1\}$. It is known from [1], [2] and [10], the solution of Poblem D_0 can be represented as:

$$W(z) = \iint_D G(z,\zeta) f(\zeta) d\sigma_\zeta, \tag{32}$$

in which $G(z,\zeta)$ as discussed in (1), $f(z)$ is the unknown function satisfying $L^p(D)\,(1 < p < \infty)$. Three equivalent singular integral equations are obtained from (29), (32) and $(30)_j$ $(j = 1, 2, 3)$. On the basis of Theorem 4, we can get

Theorem 5. *The necessary and sufficient condition of Problem D_0 for the systems* (29) *as the Nöether's problem is that one of the* (19), (20) *and* (21) *of Theorem* 3 *holds.*

Problem N_0. If the boundary condition (31) is replaced by $\frac{\partial W}{\partial n}|_\Gamma = 0$, where n is the outwards normal direction of Γ, then the boundary value problem is called the homogeneous Neumann problem (ProblemN_0). For this, the same result to theorem 5 can be obtained, provided that the Green function in the (32) is repaled by the Neumann function

$$N(z,\zeta) = -\frac{2}{\pi}\ln\left|\frac{z-\zeta}{1-z\overline{\zeta}}\right| - \frac{1}{\pi}(|z|^2 + |\zeta|^2) + \frac{3}{4}.$$

In [12], we deal with non-homogeneous Dirichlet problem (Problem D) and non-homogeneous Neumann problem (Problem N).

Problem D. Find a solution $W(z) \in C(\overline{D}) \cap W_p^2(\mathrm{D})$ of the elliptic systems (29) in simply connected domain D, such that $W(z)$ on Γ satisfies

$$W(t) = r(t),\ t \in \Gamma;\ r(t) \in C_\mu^1(\Gamma),\ 1/2 < \mu < 1. \tag{33}$$

Obviously when $r(t) \equiv 0$, Problem D is just Problem D_0. According to [8], the solution of problem D can be similarly represented as

$$W(z) = U(z) + (Hf)(z), \tag{34}$$

in which $f(z) \in L_p^2(D)\,(2 < p < \infty), U(z)$ is harmonic complex function, and $U(z) = \frac{1}{2\pi i}\int_\Gamma \mathrm{Re}\left(\frac{t+z}{t-z}\right)\frac{r(t)}{t}dt$, $(Hf)(z) = \iint_D G(z,\zeta)f(\zeta)d\sigma_\zeta$, $G(z,\zeta) = \ln\left|\frac{z-\zeta}{1-\overline{\zeta}z}\right|$. The conclusions of [12] is educed:

Theorem 6. *The necessary and sufficient condition of Problem D for the elliptic systems* (29) *as Nöether's problem is one of* (19), (20) *and* (21) *as in Theorem* 3 *to be satisfied, and we have the index formula* (22).

Problem N. Find a solution $W(z) \in C(\overline{D}) \cap W_p^2(\mathrm{D})$ of elliptic systems (29) in simply connected domain D, which satisfies the boundary condition

$$\left.\frac{\partial W}{\partial n}\right|_\Gamma = h(t),\ h(t) \in C_\mu^1(\Gamma),\ 1/2 < \mu < 1. \tag{35}$$

Provided that $W(0) = 0$, and $h(t) = h_1(t) + ih_2(t)$. Especially if $h(t) \equiv 0$, then Problem N is Problem N_0. We can assume that Domain D is the unit disk. And when $h_j(t)\,(j = 1, 2)$ satisfies:

$$\frac{1}{\pi}\int_\Gamma \frac{h_j(\tau)}{\tau}d\tau - \mathrm{Re}\left[\frac{1}{\pi i}\int_\Gamma \frac{h_j(\tau)}{\tau - t}d\tau\right] = 0,\ t \in \Gamma,\ j = 1, 2.$$

Then the solution of Problem N can be only represented as

$$\begin{cases} W(Z) = (H_1 f)(z) + V(Z), \\ (H_1 f)(z) = \iint_D N(z,\zeta) f(\zeta) d\sigma_\zeta + \frac{1}{\pi} \iint_D |\zeta|^2 f(\zeta) d\sigma_\zeta, \\ V(Z) = -\frac{2}{\pi} \int_\Gamma h(t) \ln|t-z|\, d\theta,\ \theta = \text{ang}(t-z). \end{cases} \tag{36}$$

If Problem N of elliptic systems (29) can be transformed into three similar singular integral equations with completely continuous operators, thus:

Theorem 7. *In Theorem* 6, *the conclusion is still valid, when Problem D is replaced by Problem N. And the necessary and sufficient condition of Problem N for elliptic systems* (29) *as Nöether's problem is that one of the* (19), (20) *and* (21) *of Theorem* 3 *is satisfied, and the index formula is as stated in* (22).

References

[1] I. N. Vekua, *Generalized analytic functions,* Pergamon, Oxford, 1962.

[2] A. D. Drulaev, *Method of singular integral equation,* Nauk, Moscow, 1987.

[3] G. C. Wen, *Linear and nonlinear elliptic complex equations*, Shanghai Science and Technology Publishing House, Shanghai, 1986.

[4] G. C. Wen and H. Begehr, *Boundary value problems for elliptic systems of equation and systems,* Longman Scientific and Technical, Harlow, 1990.

[5] K. Ch. Boimatov and G. Dzangibekov, *A singular integral operator,* Ucpehi. Mat. Nauk, 43 (1988), 3:171-172.

[6] G. Dzangibekov, *A class of singular integral operator,* Dokl. Akad. Nauk. SSSR, 314 (1990), 5:1055-1059.

[7] G. Dzangibekov, *Nöether's condition and index of the two-dimensional singular integral operators,* Dokl. Akad. Nauk. SSSR, 319 (1991), 4:811-815.

[8] G. Dzangibekov, *A class of singular integral operator and it is applied to boundary value problem for elliptic systems of equations in the plane,* Dokl. Akad. Nauk. SSSR, 330 (1993), 4:415-419.

[9] P. Duduchava, *On multidimensional singular integral operators, I. The half-space case; II. The case of compact manifolds,* J. Operator Theory, 11 (1984), 41-76, 199-214.

[10] B. V. Bojarski, *The Riemann-Hilbert problem for a holomorphic vector (Russian),* Dokl. Akad. Nauk SSSR, 126 (1959), 695–698.

[11] Su-kui Lu, Zi-zhi Li and Si-Le Wang, *Nöether's theory of some two-dimensional singular integral operators with interrupted coefficients,* Journal of Hebei University (Natural Science Edition), 28 (2008), 5:555-560.

[12] Da-Zeng Tian and Zi-Zhi Li, *The Nöether condition and index for basic boundary value problem to elliptic system of second order,* Journal of Ningxia University (Natural Science Edition), 27 (2006), 2:137-140.

SOME PROPERTIES OF CERTAIN NEW CLASSES OF ANALYTIC FUNCTIONS[1]

SHU-HAI LI[a] and HUO TANG[a,b]

[a]*School of Mathematics, Chifeng University, Chifeng 024000, China*

[b]*School of Mathematics and Science, Beijing Normal University, Beijing 100875, China*

E-mail: lishms66@sina.com, thth2009@tom.com

New classes of analytic functions defined by using the Salagean operator are introduced and studied. We provide coefficient inequalities, integral means inequalities and subordination relationships of these classes.

Keywords: Analytic functions, Salagean operator, coefficient inequalities, integral means inequalities, subordination relationships.

AMS No: 30C45, 30C50, 26D15.

1. Introduction, Definitions and Preliminaries

Let $\mathcal{H}$ denote the class of functions of the form

$$f(z) = z + \sum_{j=2}^{\infty} a_j z^j, \tag{1.1}$$

which are analytic in the open unit disk $U = \{z : |z| < 1\}$. Also let $\mathcal{H}^+$ denote the class of functions of the form

$$f(z) = z + \sum_{j=2}^{\infty} a_j z^j \quad (a_j \geq 0), \tag{1.2}$$

which are analytic in U. We denote by $\mathcal{S}^*(A, B)$ and $\mathcal{K}(A, B)$ $(-1 \leq B < A \leq 1)$ the subclasses of starlike functions and the subclasses of convex functions, respectively, where (see, for details, [1] and [2])

$$\mathcal{S}^*(A, B) = \left\{ f(z) \in \mathcal{H} : \frac{zf'(z)}{f(z)} \prec \frac{1 + Az}{1 + Bz},\ z \in U, 1 \leq B < A \leq 1 \right\},$$

and

$$\mathcal{K}(A, B) = \left\{ f(z) \in \mathcal{H} : 1 + \frac{zf''(z)}{f'(z)} \prec \frac{1 + Az}{1 + Bz},\ z \in U, -1 \leq B < A \leq 1 \right\}.$$

[1]This research is supported by the Natural Science Foundation of Inner Mongolia (No.2009MS0113) and Higher School Research Foundation of Inner Mongolia (No.NJzy08150, No.NJzc08160).

Clearly, we have

$$f(z) \in \mathcal{K}(A,B) \Longleftrightarrow zf'(z) \in \mathcal{S}^*(A,B).$$

A function $f(z) \in \mathcal{H}$ is said to be in the class of uniformly convex functions, denoted by $\mathcal{UK}$ (see [3-5]), if

$$\mathrm{Re}\left(1+\frac{zf''(z)}{f'(z)}\right) > \left|\frac{zf''(z)}{f'(z)}\right|, \tag{1.3}$$

and is said to be in a corresponding class denoted by $\mathcal{US}$, if

$$\mathrm{Re}\left(\frac{zf'(z)}{f(z)}\right) > \left|\frac{zf'(z)}{f(z)}-1\right|. \tag{1.4}$$

A function $f(z) \in \mathcal{H}$ is said to be in the class of α-uniformly convex functions of order β, denoted by $\mathcal{UK}(\alpha,\beta)$ (see [6]), if

$$\mathrm{Re}\left(1+\frac{zf''(z)}{f'(z)}\right) > \alpha\left|\frac{zf''(z)}{f'(z)}\right| + \beta,\ \alpha \geq 0,\ 0 \leq \beta < 1, \tag{1.5}$$

and is said to be in a corresponding class denoted by $\mathcal{US}(\alpha,\beta)$, if

$$\mathrm{Re}\left(\frac{zf'(z)}{f(z)}\right) > \alpha\left|\frac{zf'(z)}{f(z)}-1\right| + \beta,\ \alpha \geq 0,\ 0 \leq \beta < 1. \tag{1.6}$$

It is obvious that $f(z) \in \mathcal{UK}(\alpha,\beta)$ if and only if $zf'(z) \in \mathcal{US}(\alpha,\beta)$ (see [6]). The properties of various subclasses of functions $\mathcal{UK}(\alpha,\beta)$ and $\mathcal{US}(\alpha,\beta)$ were studied in [7]. For $f(z) \in \mathcal{H}$, Salagean [8] introduced the following operator, which is called the Salagean operator:

$$D^0 f(z) = f(z), D^1 f(z) = zf'(z), \cdots, D^n f(z) = D(D^{n-1} f(z)),$$
$$n \in N = \{1, 2, \cdots\}.$$

We note that

$$D^n f(z) = z + \sum_{j=2}^{\infty} j^n a_j z^j,\ n \in N_0 = N \cup \{0\}. \tag{1.7}$$

Let $\mathcal{U}_{m,n}(\alpha, A, B)$ denote the subclass of $\mathcal{H}$ consisting of functions $f(z)$, which satisfy the following inequality:

$$\frac{D^m f(z)}{D^n f(z)} - \alpha\left|\frac{D^m f(z)}{D^n f(z)} - 1\right| \prec \frac{1+Az}{1+Bz},$$
$$\alpha \geq 0,\ -1 \leq B < A \leq 1,\ m \in N,\ n \in N_0. \tag{1.8}$$

Also let $\mathcal{V}^s_{m,n}(\alpha, A, B)$ $(s \in N_0)$ be the subclass of $\mathcal{H}$ consisting of functions $f(z)$ which satisfy the following condition:

$$f(z) \in \mathcal{V}^s_{m,n}(\alpha, A, B) \Longleftrightarrow D^s f(z) \in \mathcal{U}_{m,n}(\alpha, A, B). \tag{1.9}$$

For $s = 0$, it is easily verified that

$$\mathcal{V}^0_{m,n}(\alpha, A, B) = \mathcal{U}_{m,n}(\alpha, A, B).$$

When $m = 1, n = 0$ and $m = 2, n = 1$ of inequality (1.8), respectively, we get two classes of functions

$$\mathcal{US}(\alpha, A, B)$$

$$= \left\{ f(z) \in \mathcal{H} : \frac{zf'(z)}{f(z)} - \alpha \left| \frac{zf'(z)}{f(z)} - 1 \right| \prec \frac{1+Az}{1+Bz}, \alpha \geq 0, -1 \leq B < A \leq 1 \right\},$$

and

$$\mathcal{UK}(\alpha, A, B)$$

$$= \left\{ f(z) \in \mathcal{H} : 1 + \frac{zf''(z)}{f'(z)} - \alpha \left| \frac{zf''(z)}{f'(z)} \right| \prec \frac{1+Az}{1+Bz}, \alpha \geq 0, -1 \leq B < A \leq 1 \right\}.$$

It is clear from two of the above definitions that

$$f(z) \in \mathcal{UK}(\alpha, A, B) \Longleftrightarrow zf'(z) \in \mathcal{US}(\alpha, A, B),$$

$$\mathcal{US}(1, 1, -1) = \mathcal{US}, \quad \mathcal{UK}(1, 1, -1) = \mathcal{UK}.$$

By specializing the parameters α, A, B, m and n involved in the class $\mathcal{U}_{m,n}(\alpha, A, B)$, we also obtain the following subclasses, which were studied in many earlier works:

(1) $\mathcal{U}_{1,0}(\alpha, 1-2\beta, -1) = \mathcal{US}(\alpha, \beta)$ and $\mathcal{U}_{2,1}(\alpha, 1-2\beta, -1) = \mathcal{UK}(\alpha, \beta)$ (see [6]).

(2) $\mathcal{U}_{n+1,n}(\alpha, 1-2\beta, -1) = \mathcal{US}_n(\alpha, \beta)$ (see [9],[10]).

(3) $\mathcal{U}_{m,n}(\alpha, 1-2\beta, -1) = \mathcal{U}_{m,n}(\alpha, \beta)$ and $\mathcal{V}^s_{m,n}(\alpha, 1-2\beta, -1) = \mathcal{V}^s_{m,n}(\alpha, \beta)$, $0 \leq \alpha, 0 \leq \beta < 1$ (see [11],[12]).

Let

$$\tilde{\mathcal{US}}(\alpha, A, B) = \mathcal{H}^+ \cap \mathcal{US}(\alpha, A, B), \tilde{\mathcal{UK}}(\alpha, A, B) = \mathcal{H}^+ \cap \mathcal{UK}(\alpha, A, B),$$

$$\tilde{\mathcal{U}}_{m,n}(\alpha, A, B) = \mathcal{H}^+ \cap \mathcal{U}_{m,n}(\alpha, A, B), \tilde{\mathcal{V}}^s_{m,n}(\alpha, A, B) = \mathcal{H}^+ \cap \mathcal{V}^{s,+}_{m,n}(\alpha, A, B).$$

Then we obtain contain relations and the close properties of integral operators. This paper mainly concerns the classes $\mathcal{U}_{m,n}(\alpha, A, B)$ and $\mathcal{V}^s_{m,n}(\alpha, A, B)$. We provide coefficient inequalities, integral means inequalities and subordination relationships of these classes.

2. Coefficient Inequalities for Classes $\mathcal{U}_{m,n}(\alpha, A, B)$ and $\mathcal{V}^s_{m,n}(\alpha, A, B)$

Theorem 2.1. If $f(z) \in \mathcal{H}$ satisfies

$$\sum_{j=2}^{\infty} \phi(m,n,j,\alpha,A,B)|a_j| \leq A - B, \tag{2.1}$$

where

$$\phi(m,n,j,\alpha,A,B) = (1+2\alpha)|j^m - j^n| + |Bj^m - Aj^n| \tag{2.2}$$

for some $\alpha \geq 0$, $-1 \leq B < A \leq 1$, $m \in N$, $n \in N_0 = N \cup \{0\}$, then $f(z) \in \mathcal{U}_{m,n}(\alpha, A, B)$.

Proof. Suppose that (2.1) is true for $\alpha \geq 0, -1 \leq B < A \leq 1, m \in N, n \in N_0$. For $f(z) \in \mathcal{H}$, let us define the function $p(z)$ by

$$p(z) = \frac{D^m f(z)}{D^n f(z)} - \alpha \left| \frac{D^m f(z)}{D^n f(z)} - 1 \right|.$$

It suffices to show that

$$\left| \frac{p(z) - 1}{A - Bp(z)} \right| < 1, \quad z \in U.$$

We note that

$$\left| \frac{p(z) - 1}{A - Bp(z)} \right| = \left| \frac{D^m f(z) - \alpha e^{i\theta}|D^m f(z) - D^n f(z)| - D^n f(z)}{AD^n f(z) - B(D^m f(z) - \alpha e^{i\theta}|D^m f(z) - D^n f(z)|)} \right|$$

$$= \left| \frac{(D^m f(z) - D^n f(z)) - \alpha e^{i\theta}|D^m f(z) - D^n f(z)|}{(A-B)D^n f(z) - B((D^m f(z) - D^n f(z)) - \alpha e^{i\theta}|D^m f(z) - D^n f(z)|)} \right|$$

$$= \left| \frac{\sum_{j=2}^{\infty}(j^m - j^n)a_j z^{j-1} - \alpha e^{i\theta}|\sum_{j=2}^{\infty}(j^m - j^n)a_j z^{j-1}|}{(A-B) - \sum_{j=2}^{\infty}(Bj^m - Aj^n)a_j z^{j-1} - \alpha e^{i\theta}|\sum_{j=2}^{\infty}(j^m - j^n)a_j z^{j-1}|)} \right|$$

$$\leq \frac{\sum_{j=2}^{\infty}|j^m - j^n||a_j||z|^{j-1} + \alpha |e|^{i\theta} \sum_{j=2}^{\infty}|j^m - j^n||a_j||z|^{j-1}}{(A-B) - \sum_{j=2}^{\infty}|Bj^m - Aj^n||a_j||z|^{j-1} - \alpha |e|^{i\theta}\sum_{j=2}^{\infty}|j^m - j^n||a_j||z|^{j-1}}$$

$$\leq \frac{\sum_{j=2}^{\infty}|j^m - j^n||a_j| + \alpha \sum_{j=2}^{\infty}|j^m - j^n||a_j|}{(A-B) - \sum_{j=2}^{\infty}|Bj^m - Aj^n||a_j| - \alpha \sum_{j=2}^{\infty}|j^m - j^n||a_j|}.$$

The last expression is bounded above by 1, if

$$\sum_{j=2}^{\infty}|j^m - j^n||a_j| + \alpha \sum_{j=2}^{\infty}|j^m - j^n||a_j|$$
$$\leq (A-B) - \sum_{j=2}^{\infty}|Bj^m - Aj^n||a_j| - \alpha \sum_{j=2}^{\infty}|j^m - j^n||a_j|,$$

which is equivalent to the condition (2.1). This completes the proof of Theorem 2.1. By using Theorem 2.1, we have

Theorem 2.2. *If $f(z) \in \mathcal{H}$ satisfies*

$$\sum_{j=2}^{\infty} j^s \phi(m,n,j,\alpha,A,B)|a_j| \leq A - B, \tag{2.3}$$

where $\phi(m,n,j,\alpha,A,B)$ is defined by (2.2) for some $\alpha \geq 0, -1 \leq B < A \leq 1, m \in N, n \in N_0$, then $f(z) \in \mathcal{V}^s_{m,n}(\alpha,A,B)$.

Proof. From (1.7), replacing a_j by $j^s a_j$ in Theorem 2.1, we have Theorem 2.2.

3. Integral Mean Inequalities

We need the following definitions and results.

Definition 3.1. Let us consider two functions $f(z)$ and $g(z)$, which are analytic in U. The function $f(z)$ is said to be subordinate to $g(z)$ in U, if there exists a function $\omega(z)$ analytic in U with

$$\omega(0) = 0 \text{ and } |\omega(z)| < 1, \; z \in U,$$

such that

$$f(z) = g(\omega(z)), \; z \in U.$$

We denote this subordination by

$$f(z) \prec g(z).$$

Definition 3.2. (Owa [13] and Srivastzva [14]) The fractional derivative of order λ is defined, for a function $f(z)$, by

$$D_z^{-\lambda} f(z) = \frac{1}{\Gamma(1-\lambda)} \frac{d}{dz} \int_0^z \frac{f(t)}{(z-t)^\lambda} dt, \; 0 \leq \lambda < 1, \tag{3.1}$$

where the function $f(z)$ is an analytic function in a simply connected region of the z-plane containing the origin, and the multiplicity of $(z-t)^\lambda$ is removed by requiring $\log(z-t)$ to be real when $(z-t) > 0$.

Definition 3.3. Under the hypotheses of Definition 3.2, the fractional derivative of order $(p+\lambda)$ is defined, for a function $f(z)$, by

$$D^{p+\lambda} f(z) = \frac{d^p}{dz^p} D_z^\lambda f(z),$$

where $0 \le \lambda < 1$ and $p \in N_0 = N \cup \{0\}$. It readily follows from (3.1) in Definition 3.2 that

$$D_z^{\lambda} z^k = \frac{\Gamma(k+1)}{\Gamma(k-\lambda+1)} z^{k-\lambda},\ 0 \le \lambda < 1. \tag{3.2}$$

Theorem 3.1. (Littlewood [15]) *If $f(z)$ and $g(z)$ are analytic in U with $f(z) \prec g(z)$, then for $\mu > 0$ and $z = re^{i\theta}\,(0 < r < 1)$,*

$$\int_0^{2\pi} |f(z)|^{\mu} d\theta \le \int_0^{2\pi} |g(z)|^{\mu} d\theta.$$

Theorem 3.2. *Let $f(z) \in \mathcal{H}$ given by (1.2) be in the class $\tilde{\mathcal{U}}_{m,n}(\alpha, A, B)$ and suppose that*

$$\sum_{j=2}^{\infty} (j-p)_{p+1} a_j \le \frac{(A-B)\Gamma(k+1)\Gamma(3-\lambda-p)}{\phi_k(m,n,\alpha,A,B)\Gamma(k+1-\lambda-p)\Gamma(2-p)}$$

for some $0 \le p \le 2, 0 \le \lambda < 1$, where $(j-p)_{p+1}$ denotes the Pochhammer symbol defined by $(j-p)_{p+1} = (j-p)(j-(p-1))\cdots j$. Also given the function $f_k(z)$ by

$$f_k(z) = z + \frac{(A-B)}{\phi_k(m,n,\alpha,A,B)} z^k,\ \ k \ge 2. \tag{3.3}$$

If there exists an analytic function $\omega(z)$ given by

$$\begin{aligned} \omega(z)^{k-1} &= \frac{\phi_k(m,n,\alpha,A,B)\Gamma(k+1-\lambda-p)}{(A-B)\Gamma(k+1)} \\ &\quad \times \sum_{j=2}^{\infty} (j-p)_{p+1} \frac{\Gamma(j-p)}{\Gamma(j+1-\lambda-p)} a_j z^{j-1}, \end{aligned}$$

then for $z = re^{i\theta}$ $(0 < r < 1)$ and $\mu > 0$,

$$\int_0^{2\pi} |D_z^{p+\lambda} f(z)|^{\mu} d\theta \le \int_0^{2\pi} |D_z^{p+\lambda} f_k(z)|^{\mu} d\theta.$$

Proof. By virtue of the fractional derivative formula (3.1) and Definition 3.3, we find from (1.1) that

$$\begin{aligned} D_z^{p+\lambda} f(z) &= \frac{z^{1-\lambda-p}}{\Gamma(2-\lambda-p)} \{1 + \sum_{j=2}^{\infty} \frac{\Gamma(2-\lambda-p)\Gamma(j+1)}{\Gamma(j+1-\lambda-p)} a_j z^{j-1}\} \\ &= \frac{z^{1-\lambda-p}}{\Gamma(2-\lambda-p)} \{1 + \sum_{j=2}^{\infty} \Gamma(2-\lambda-p)(j-p)_{p+1} \varphi(j) a_j z^{j-1}\}, \end{aligned}$$

where

$$\varphi(j) = \frac{\Gamma(j-p)}{\Gamma(j+1-\lambda-p)}.$$

Since $\varphi(j)$ is a decreasing function of j, we have

$$0 < \varphi(j) \leq \varphi(2) = \frac{\Gamma(2-p)}{\Gamma(3-\lambda-p)} \quad (0 \leq \lambda < 1; 0 \leq p \leq 2 \leq j).$$

Similarly, from (3.2), (3.3) and Definition 3.3, we obtain

$$D_z^{p+\lambda} f_k(z) = \frac{z^{1-\lambda-p}}{\Gamma(2-\lambda-p)}\{1 + \frac{(A-B)\Gamma(2-\lambda-p)\Gamma(k+1)}{\phi_k(m,n,\alpha,A,B)\Gamma(k+1-\lambda-p)} z^{k-1}\}.$$

For $z = re^{i\theta}$ $(0 < r < 1)$, we must show that

$$\int_0^{2\pi} |1 + \sum_{j=2}^{\infty} \Gamma(2-\lambda-p)(j-p)_{p+1}\varphi(j) a_j z^{j-1}|^{\mu} d\theta$$
$$\leq \int_0^{2\pi} |1 + \frac{(A-B)\Gamma(2-\lambda-p)\Gamma(k+1)}{\phi_k(m,n,\alpha,A,B)\Gamma(k+1-\lambda-p)} z^{k-1}|^{\mu} d\theta \quad (\mu > 0).$$

Thus by applying Littlewood's subordination theorem, it is sufficient to show that

$$\begin{aligned} &1 + \sum_{j=2}^{\infty} \Gamma(2-\lambda-p)(j-p)_{p+1}\varphi(j) a_j z^{j-1} \\ &\prec 1 + \frac{(A-B)\Gamma(2-\lambda-p)\Gamma(k+1)}{\phi_k(m,n,\alpha,A,B)\Gamma(k+1-\lambda-p)} z^{k-1}. \end{aligned} \tag{3.4}$$

By setting

$$\begin{aligned} &1 + \sum_{j=2}^{\infty} \Gamma(2-\lambda-p)(j-p)_{p+1}\varphi(j) a_j z^{j-1} \\ &= 1 + \frac{(A-B)\Gamma(2-\lambda-p)\Gamma(k+1)}{\phi_k(m,n,\alpha,A,B)\Gamma(k+1-\lambda-p)} \omega(z)^{k-1}, \end{aligned}$$

we find that

$$\omega(z)^{k-1} = \frac{\phi_k(m,n,\alpha,A,B)\Gamma(k+1-\lambda-p)}{(A-B)\Gamma(k+1)} \sum_{j=2}^{\infty} (j-p)_{p+1}\varphi(j) a_j z^{j-1},$$

which readily yields $\varphi(0) = 0$. Therefore, we have

$$|\omega(z)|^{k-1} = |\frac{\phi_k(m,n,\alpha,A,B)\Gamma(k+1-\lambda-p)}{(A-B)\Gamma(k+1)} \sum_{j=2}^{\infty} (j-p)_{p+1}\varphi(j) a_j z^{j-1}|$$

$$\leq \frac{\phi_k(m,n,\alpha,A,B)\Gamma(k+1-\lambda-p)}{(A-B)\Gamma(k+1)} \sum_{j=2}^{\infty}(j-p)_{p+1}\varphi(j)a_j|z|^{j-1}$$

$$\leq \frac{\phi_k(m,n,\alpha,A,B)\Gamma(k+1-\lambda-p)}{(A-B)\Gamma(k+1)}\varphi(2)|z|\sum_{j=2}^{\infty}(j-p)_{p+1}a_j$$

$$= \frac{\phi_k(m,n,\alpha,A,B)\Gamma(k+1-\lambda-p)}{(A-B)\Gamma(k+1)} \frac{\Gamma(2-p)}{\Gamma(2-\lambda-p)}|z|\sum_{j=2}^{\infty}(j-p)_{p+1}a_j, \ |z|<1,$$

by means of the hypothesis of Theorem 3.2.

Theorem 3.3. *Let $f(z) \in \mathcal{H}$ given by (1.2) be in the class $\tilde{\mathcal{U}}^s_{m,n}(\alpha, A, B)$ and suppose that*

$$\sum_{j=2}^{\infty}(j-p)_{p+1}a_j \leq \frac{(A-B)\Gamma(k+1)\Gamma(3-\lambda-p)}{k^s\phi_k(m,n,\alpha,A,B)\Gamma(k+1-\lambda-p)\Gamma(2-p)}$$

for some $0 \leq p \leq 2, 0 \leq \lambda < 1$, where $(j-p)_{p+1}$ denotes the Pochhammer symbol defined by $(j-p)_{p+1} = (j-p)(j-(p-1))\cdots j$. Also introduce the function $f_k(z)$ by

$$f_k(z) = z + \frac{(A-B)}{k^s\phi_k(m,n,\alpha,A,B)}z^k, \ k \geq 2.$$

If there exists an analytic function $\omega(z)$ given by

$$\omega(z)^{k-1} = \frac{k^s\phi_k(m,n,\alpha,A,B)\Gamma(k+1-\lambda-p)}{(A-B)\Gamma(k+1)} \times \sum_{j=2}^{\infty}(j-p)_{p+1}\frac{\Gamma(j-p)}{\Gamma(j+1-\lambda-p)}a_j z^{j-1},$$

then for $z = re^{i\theta}$ $(0 < r < 1)$ and $\mu > 0$,

$$\int_0^{2\pi} |D_z^{p+\lambda}f(z)|^\mu d\theta \leq \int_0^{2\pi} |D_z^{p+\lambda}f_k(z)|^\mu d\theta.$$

4. Subordination Result

Definition 4.1. (Hadamard Product or Convolution) Given two functions f and g in the class $\mathcal{H}$, where $f(z)$ is given by (1.1) and $g(z)$ is given by

$$g(z) = z + \sum_{j=2}^{\infty} b_j z^j, \tag{4.1}$$

the Hadamard product (or convolution) $f * g$ is defined (as usual) by

$$(f * g)(z) = z + \sum_{j=2}^{\infty} a_j b_j z^j = (g * f)(z), \ z \in U.$$

Definition 4.2. (Subordinating Factor Sequence) A sequence $\{b_j\}_{j=1}^{\infty}$ of complex numbers is said to be a Subordinating Factor Sequence if, whenever $f(z)$ of the form (1.1) is analytic, univalent and convex in U, we have the subordination given by

$$\sum_{j=1}^{\infty} a_j b_j z^j \prec f(z), \ z \in U, \ a_1 = 1. \tag{4.2}$$

Theorem 4.1. (Wilf [16]) *The sequence $\{b_j\}_{j=1}^{\infty}$ is a subordinating factor sequence if and only if*

$$\mathrm{Re}(1 + 2\sum_{j=1}^{\infty} b_j z^j) > 0, \ z \in U. \tag{4.3}$$

Theorem 4.2. *Let the function $f(z)$ defined by (1.1) be in the class $\tilde{\mathcal{U}}_{m,n}(\alpha, A, B)$, where $-1 \leq B < A \leq 1$. Suppose also that $\mathcal{K} = \mathcal{K}(0)$ denotes the familiar class of functions $f(z) \in \mathcal{H}$, which are univalent and convex in U. Then*

$$\frac{\phi_2(m,n,\alpha,A,B)}{2[(A-B)+\phi_2(m,n,\alpha,A,B)]}(g * f)(z) \prec g(z), \ z \in U, \ g(z) \in \mathcal{K}, \tag{4.4}$$

and

$$\mathrm{Re}(f(z)) > -\frac{(A-B)+\phi_2(m,n,\alpha,A,B)}{\phi_2(m,n,\alpha,A,B)}, \ z \in U. \tag{4.5}$$

The constant $\frac{\phi_2(m,n,\alpha,A,B)}{2[(A-B)+\phi_2(m,n,\alpha,A,B)]}$ is the best estimate.

Proof. Let $f(z) \in \tilde{\mathcal{U}}_{m,n}(\alpha, A, B)$ and suppose that

$$g(z) = z + \sum_{j=2}^{\infty} b_j z^j \in \mathcal{K}.$$

Then, for $f(z) \in \mathcal{H}$ given by (1.1), we have

$$\begin{aligned} &\frac{\phi_2(m,n,\alpha,A,B)}{2[(A-B)+\phi_2(m,n,\alpha,A,B)]}(g * f)(z) \\ &= \frac{\phi_2(m,n,\alpha,A,B)}{2[(A-B)+\phi_2(m,n,\alpha,A,B)]}(z + \sum_{j=2}^{\infty} a_j b_j z^j). \end{aligned} \tag{4.6}$$

Thus, by Definition 4.2, the subordination result (4.4) will hold true if the sequence

$$\left\{\frac{\phi_2(m,n,\alpha,A,B)}{2[(A-B)+\phi_2(m,n,\alpha,A,B)]}a_j\right\}_{j=1}^{\infty}$$

is a subordinating factor sequence, with (of course) $a_1 = 1$. In view of Theorem 4.1, this will be true if and only if

$$\mathrm{Re}\{1+\sum_{j=1}^{\infty}\frac{\phi_2(m,n,\alpha,A,B)}{(A-B)+\phi_2(m,n,\alpha,A,B)}a_jz^j\} > 0,\ z\in U. \tag{4.7}$$

Since

$$\begin{aligned}
&\mathrm{Re}\{1+\sum_{j=1}^{\infty}\frac{\phi_2(m,n,\alpha,A,B)}{(A-B)+\phi_2(m,n,\alpha,A,B)}a_jz^j\}\\
&=\mathrm{Re}\{1+\frac{\phi_2(m,n,\alpha,A,B)}{(A-B)+\phi_2(m,n,\alpha,A,B)}a_1z\\
&\quad+\frac{1}{(A-B)+\phi_2(m,n,\alpha,A,B)}\sum_{j=2}^{\infty}\phi_2(m,n,\alpha,A,B)a_jz^j\}\\
&\geq 1-\left\{\left|\frac{\phi_2(m,n,\alpha,A,B)}{(A-B)+\phi_2(m,n,\alpha,A,B)}\right|r\right.\\
&\quad\left.+\frac{1}{|(A-B)+\phi_2(m,n,\alpha,A,B)|}\sum_{j=2}^{\infty}\phi_2(m,n,\alpha,A,B)a_jz^j\right\}\\
&\geq 1-\left\{\frac{\phi_2(m,n,\alpha,A,B)}{(A-B)+\phi_2(m,n,\alpha,A,B)}r+\frac{A-B}{(A-B)+\phi_2(m,n,\alpha,A,B)}r\right\}\\
&=1-r>0,\ |z|=r<1,
\end{aligned}$$

thus (4.7) holds true in $\mathcal{U}$. This proves the inequality (4.4). The inequality (4.5) follows by taking the convex function $g(z)=\frac{z}{1-z}=z+\sum_{j=2}^{\infty}z^j$, $g(z)\in\mathcal{K}$ in (4.4). To prove the sharpness of the constant

$$\frac{\phi_2(m,n,\alpha,A,B)}{2[(A-B)+\phi_2(m,n,\alpha,A,B)]},$$

we consider $f_0(z)\in\tilde{\mathcal{U}}_{m,n}(\alpha,A,B)$ given by

$$f_0(z)=z-\frac{A-B}{\phi_2(m,n,\alpha,A,B)}z^2\ ,\ -1\leq B<A\leq 1.$$

Thus from (4.4), we have

$$\frac{\phi_2(m,n,\alpha,A,B)}{2[(A-B)+\phi_2(m,n,\alpha,A,B)]}f_0(z)\prec\frac{z}{1-z}. \tag{4.8}$$

It can easily be verified that

$$\min\{\mathrm{Re}(\frac{\phi_2(m,n,\alpha,A,B)}{2[(A-B)+\phi_2(m,n,\alpha,A,B)]}f_0(z))\}=-\frac{1}{2},\ z\in U.$$

This shows that the constant $\frac{\phi_2(m,n,\alpha,A,B)}{2[(A-B)+\phi_2(m,n,\alpha,A,B)]}$ is best possible.

Theorem 4.3. *Let the function* $f(z)$ *defined by* (1.1) *be in the class* $\tilde{\mathcal{V}}^s_{m,n}(\alpha,A,B)$, *where* $-1\le B<A\le 1$. *Suppose also that* $\mathcal{K}=\mathcal{K}(0)$ *denotes the familiar class of functions* $f(z)\in\mathcal{H}$, *which are univalent and convex in* U. *Then*

$$\frac{2^s\phi_2(m,n,\alpha,A,B)}{2[(A-B)+2^s\phi_2(m,n,\alpha,A,B)]}(g*f)(z)\prec g(z),\ z\in U,\ g(z)\in\mathcal{K},$$

and

$$\mathrm{Re}[f(z)]>-\frac{(A-B)+2^s\phi_2(m,n,\alpha,A,B)}{2^s\phi_2(m,n,\alpha,A,B)},\ z\in U.$$

The constant $\frac{2^s\phi_2(m,n,\alpha,A,B)}{2[(A-B)+2^s\phi_2(m,n,\alpha,A,B)]}$ *is the best estimate.*

We remark in conclusion that, by suitably specializing the parameters involved in the results presented in this paper, we can deduce numerous further corollaries and consequences of each of these results.

References

[1] W. Janowski, Some extremal problem for certain families of analytic functions, *Ann. Polon. Math.*, 28 (1973), 648-658.

[2] K. S. Padmanabhan, M. S. Ganesan, Convolutions of certain classes of univalent functions with negative coefficients, *Indian J. pure appl. Math.*, 19(9) (1988), 880-889.

[3] A. W. Goodman, On uniformly starlike functions, *J. Math. Anal. Appl.*, 155 (1991), 364-370.

[4] W. C. Ma, Uniformly convex functions, *Ann. Polon. Math.*, 57 (1992), 165-175.

[5] F. Ronning, Uniformly convex functions and a corresponding class of starlike functions, *Proc. Amer. Math. Soc.*, 118(1) (1993), 189-196.

[6] S. S. Hams, S. R. Kulkarni and J. M. Jahangiri, Classes of uniformly starlike and convex functions, *Internat. J. Math. Math. Sci.*, 2004 (2004), Issue 55, 2959-2961.

[7] S. Shams and S. R. Kulkarni, On a class of univalent functions defined by Ruscheweyh derivatives, *Kyungpook Math. J.*, 43 (2003), 579–585.

[8] G. S. Salagean, Subclasses of univalent functions, *Complex analysis-Pro. 5th Rom.-Finn Semin., Bucharest 1981, Part 1, Lect. Notes Math.*, 1013 (1983), 362-372.

[9] R. Bharati, R. Parvatham and A. Swaminathan, On subclasses of uniformly convex functions and corresponding class of starlike functions, *Tamkang J. Math.*, 28 (1997), 17-32.

[10] W. P. Kuang, Y. Sun and Z. G. Wang, On Quasi-Hadamard product of certain classes of analytic functions, *Bulletin of Mathematical Analysis and Applications*, 1(2) (2009), 36-46.

[11] S. S. Eker and S. Owa, Certain classes of analytic functions involving Salagean operator, *J. Inequal. Pure and Appl. Math.,* 10(1) (2009), 12-22.

[12] H. M. Srivastava and S. S. Eker, Some applications of a subordination theorem for a class of analytic functions, *Applied Mathematics Letters,* 21 (2008), 394-399.

[13] S. Owa, On the distortion theorems I, *Kyungpook Math. J.,* 18 (1978), 53-59.

[14] H. M. Srivastava and S. Owa (Eds.), *Univalent functions, fractional calculus, and their applications,* Halsted Press (Ellis Horwood Limited, Chichester) John Wiley and Sons, New York, Chichester, Brisbane and Toronto, 1989.

[15] J. E. Littlewood, On inequalities in the theory of functions, *Proc. London Math. Soc.,* 23 (1925), 481-519.

[16] H. S. Wilf, Subordinating factor sequences for convex maps of the unit circle, *Proc. Amer. Math. Soc.,* 12 (1961), 689-693.

STABILITY OF CAUCHY TYPE INTEGRAL APPLIED TO THE FUNDAMENTAL PROBLEMS IN PLANE ELASTICITY[1]

JUAN LIN[a,b]

[a] *Department of Mathematics, Wuhan University, Wuhan 430072, China*

[b] *Department of Foundation, Fujian Commercial College, Fuzhou 350012, China*

E-mail: lj7862124@163.com

Let the elastic domain be a disk, and its boundary the unit circle. By dint of the stability of Cauchy-type integral with respect to the perturbation of integral curve, the stability of the first fundamental problem and the second fundamental problem in plane elasticity will be discussed under the smooth perturbation for the boundary curve.

Keywords: Elastic domain, Cauchy-type integral, complex stress functions, normal stress, smooth perturbation

MR(2000) Subject Classification: 30E20, 45E99 .

1. The Fundamental Problems in Plane Elasticity

Under the assumption of lack of volume force (we always maintain this assumption afterwards), we have the equilibrium equations

$$\begin{cases} \dfrac{\partial \sigma_x}{\partial x} + \dfrac{\partial \tau_{xy}}{\partial y} = 0, \\ \dfrac{\partial \tau_{xy}}{\partial x} + \dfrac{\partial \sigma_y}{\partial y} = 0, \end{cases}$$

and the compatibility equation

$$\left[\frac{\partial^2}{\partial x^2} + \frac{\partial^2}{\partial y^2}\right](\sigma_x + \sigma_y) = 0,$$

where σ_x is the normal stress parallel to the x-axis, σ_y is that parallel to the y-axis, τ_{xy} is the shear stress.

Let the elastic region S be a disk with boundary L, a unit circle, oriented counter clockwise. We always assume in the sequel, appearing functions in S and on $L \in H^\mu\,(0 < \mu < 1)$, and so do their derivatives on L, whether they are known or not. We introduce holomorphic functions $\varphi(z)$ and $\psi(z)$, also called complex stress functions. Stress and displacement $u + iv$ at the

[1]This research is supported by Natural Science Foundation of Fujian Province (2008J0187) and the Science and Technology Foundation of Education Department of Fujian Province (JA08255), China.

point z in S could be expressed by $\varphi(z)$ and $\psi(z)$[1,2] as follows:

$$\begin{cases} \sigma_x + \sigma_y = 2\left[\varphi'(z) + \overline{\varphi'(z)}\right] = 4\mathrm{Re}\left[\varphi'(z)\right], \\ \sigma_y - \sigma_x + 2i\tau_{xy} = 2\left[\bar{z}\varphi''(z) + \psi'(z)\right], \end{cases} \tag{1.1}$$

and

$$2\mu(u + iv) = \kappa\varphi(z) - z\overline{\varphi'(z)} - \overline{\psi(z)}. \tag{1.2}$$

respectively, where κ, μ are the elastic constants and $\frac{5}{3} < \kappa < 3$ (We only consider the case of stress state in this paper).

According to [2], we know the fundamental problems can be reduced to the following boundary value problem for analytic functions:

(I) The first fundamental problem: find two holomorphic functions $\varphi(z)$, $\psi(z)$ such that the boundary condition

$$\varphi(t) + t\overline{\varphi'(t)} + \overline{\psi(t)} = f(t), \tag{1.3}$$

and

$$\varphi(0) = 0, \ \mathrm{Im}\varphi'(0) = 0, \tag{1.4}$$

where $f(t) = i\int_{t_0}^{t}(X_n(t) + iY_n(t))\mathrm{d}s$, $t \in L$, and $\mathrm{Re}\int_L f(t)\mathrm{d}\bar{t} = 0$, integral path is from t_0 to t clockwisely along L.

(II) The second fundamental problem: given the displacement function $g(t) = u(t) + iv(t)$, $t \in L$, find two holomorphic functions $\varphi(z)$ and $\psi(z)$ in S, satisfying the boundary value condition

$$\kappa\varphi(t) - t\overline{\varphi'(t)} - \overline{\psi(t)} = 2\mu g(t), \tag{1.5}$$

and

$$\varphi(0) = 0 \ (\text{or } \psi(0) = 0). \tag{1.6}$$

2. The Stability of the Perturbed First Fundamental Problem

Denote by $C^2(L)(C^1(L))$ the space of all functions with continuous second(first) derivatives and with norm

$$\|\delta\|_1 = \|\delta\|_L + \|\delta'\|_L, \|\delta\|_2 = \|\delta\|_1 + \|\delta''\|_L,$$

where $\|\cdot\| = \max\limits_{t\in L}|\delta(t)|$ and $\delta \in C^2(L)$. Denote

$$B(\rho) = \left\{\delta, \ \delta \in C^2(L), \ \|\delta\|_2 < \rho < \frac{2}{5\pi}\right\}.$$

When smooth perturbation occurs $\delta \in B(\rho)$ for L, we set $L_\delta = \{\xi, \xi = t + \delta(t), t \in L\}$, and L_δ is also a smooth curve[3].

After perturbation δ to L, the first fundamental problem about S_δ can be reduced to analytic function boundary problem: Find two holomorphic functions $\varphi_*(z)$ and $\psi_*(z)$ in S_δ, satisfying the boundary value condition

$$\varphi_*(\xi) + \xi\overline{\varphi'_*(\xi)} + \overline{\psi_*(\xi)} = f_*(\xi), \ \ \xi \in L_\delta, \tag{2.1}$$

where $f_*(\xi)$ such that $\int_{L_\delta} f_*(\xi)d\overline{\xi} = 0$. We can introduce a new function $\omega_*(\zeta)$, such that

$$\varphi_*(z) = \frac{1}{2\pi i}\int_{L_\delta} \frac{\omega_*(\zeta)}{\zeta - z} d\zeta, \ \ z \in S_\delta,$$

$$\psi_*(z) = \frac{1}{2\pi i}\int_{L_\delta} \frac{\overline{\omega_*(\zeta)}}{\zeta - z} \mathrm{d}\zeta - \frac{1}{2\pi i}\int_{L_\delta} \frac{\overline{\zeta}\omega'_*(\zeta)}{\zeta - z} \mathrm{d}\zeta, \ \ z \in S_\delta.$$

By Plemelj formula, we can obtain $\varphi_*(\xi), \varphi'_*(\xi), \psi_*(\xi)$ and substitute them into (2.1) and gain

$$\omega_*(\xi) + \frac{1}{2\pi i}\int_{L_\delta} \omega_*(\zeta)\mathrm{d}\ln\frac{\zeta - \xi}{\overline{\zeta} - \overline{\xi}} - \frac{1}{2\pi i}\int_{L_\delta} \overline{\omega_*(\zeta)}\mathrm{d}\frac{\zeta - \xi}{\overline{\zeta} - \overline{\xi}} = f_*(\xi), \ \xi \in L_\delta. \tag{2.2}$$

If denote $\omega_*(\xi) = \omega_\delta(t), g_*(\xi) = g_\delta(t)$, then (2.2) can be written in the form

$$\omega_\delta(t) = A_\delta + B_\delta t + f^*_\delta(t), \tag{2.3}$$

where

$$A_\delta = -\frac{1}{2\pi i}\int_L \frac{\omega_\delta(\tau)}{\tau} d\tau, \ \ B_\delta = -\frac{1}{2\pi i}\int_L \overline{\omega_\delta(\tau)} d\tau. \tag{2.4}$$

Substituting (2.3) into (2.4), we get

$$A_\delta = -\frac{1}{4\pi i}\int_L \frac{f^*_\delta(\tau)}{\tau}\mathrm{d}\tau, \ B_\delta + \overline{B_\delta} = -\frac{1}{2\pi i}\int_L \overline{f^*_\delta(\tau)}\mathrm{d}\tau, \tag{2.5}$$

$$\varphi'_*(z) = \frac{1}{2\pi i}\int_L \frac{f^{*'}_\delta(t)}{t + \delta(t) - z}(1 + \delta'(t))\mathrm{d}t + B_\delta. \tag{2.6}$$

By (1.1), (2.5) and (2.6), we know

$$\begin{aligned}
&(\sigma_x + \sigma_y)_\delta - (\sigma_x + \sigma_y) \\
&= 2[\varphi'_\delta(z) + \overline{\varphi'_\delta(z)}] - 2[\varphi'(z) + \overline{\varphi'(z)}] \\
&= \left(\frac{1}{2\pi i}\int_L \frac{f^{*'}_\delta(\tau)}{\tau + \delta(\tau) - z}(1 + \delta'(\tau))\mathrm{d}\tau - \frac{1}{2\pi i}\int_L \frac{f'(\tau)}{\tau - z}\mathrm{d}\tau\right) \\
&\quad + 2\overline{\left(\frac{1}{2\pi i}\int_L \frac{f^{*'}_\delta(\tau)}{\tau + \delta(\tau) - z}(1 + \delta'(\tau))\mathrm{d}\tau - \frac{1}{2\pi i}\int_L \frac{f'(\tau)}{\tau - z}\mathrm{d}\tau\right)} \\
&\quad - 2\left(\frac{1}{2\pi i}\int_L \overline{f^*_\delta(\tau)}\mathrm{d}\tau - \frac{1}{2\pi i}\int_L \overline{f(\tau)}\mathrm{d}\tau\right).
\end{aligned}$$

By dint of [3], we have[4]

Theorem 2.1. *If* $\delta \in B(\rho)$, $A_1^\nu < \frac{\nu(2-5\pi\rho)}{(\pi/2+1)^{1-\nu}2^{3+\nu}\rho^{1-\nu}}$, *then for a* $\nu \in (0,1)$ *chosen, we have* $\|f_\delta^*(t) - f(t)\|_L \le C(\rho,\nu)\|\delta\|_1^{\min\{\mu,\,1-\nu\}}$.

Theorem 2.2. *If* $\delta \in B(\rho)$, $A_1^\nu < \frac{\nu(2-5\pi\rho)}{(\pi/2+1)^{1-\nu}2^{3+\nu}\rho^{1-\nu}}$, *then for a* $\nu \in (0,1)$ *chosen, we have* $\left|f_\delta^{*\prime}(t) - f'(t)\right| \le C(\rho,\nu)\|\delta\|_2^{\min\{\mu,1-\nu\}}$.

Theorem 2.3. *If* $\delta \in B(\rho)$, $A_1^\nu < \frac{\nu(2-5\pi\rho)}{(\pi/2+1)^{1-\nu}2^{3+\nu}\rho^{1-\nu}}$, *for a* $\nu \in (0,1)$ *chosen and* $\varepsilon \in (0,1)$, *we have*

$$\left|(\sigma_x)_\delta - \sigma_x\right|_\Omega \le C(\rho,\varepsilon,\nu)\|\delta\|_2^{(1-\varepsilon)\min\{\mu,1-\nu\}},$$

$$\left|(\sigma_y)_\delta - \sigma_y\right|_\Omega \le C(\rho,\varepsilon,\nu)\|\delta\|_2^{(1-\varepsilon)\min\{\mu,1-\nu\}}.$$

3. The Stability of the Perturbed Second Fundamental Problem

Now we let

$$B(\rho) = \{\delta \in C^1(L) \mid \|\delta\|_1 < \rho\},\ \rho < \frac{1}{30\pi}.$$

After perturbation $\delta(t)$ to L, the second fundamental problem about S_δ can be reduced to analytic function boundary problem: Find two holomorphic functions $\varphi_*(z)$ and $\psi_*(z)$ in S_δ, satisfying the boundary value condition

$$\kappa\varphi_*(\xi) - \xi\overline{\varphi_*'(\xi)} - \overline{\psi_*(\xi)} = 2\mu g_*(\xi),\ \xi \in L_\delta. \tag{3.1}$$

where $g_*(\xi)$ is the perturbed displacement of $g(t)$, $\xi = t + \delta(t)$. Similarly, we can introduce a new function $\omega_*(\zeta)$, such that

$$\varphi_*(z) = \frac{1}{2\pi i}\int_{L_\delta}\frac{\omega_*(\zeta)}{\zeta - z}d\zeta,\ z \in S_\delta, \tag{3.2}$$

$$\psi_*(z) = -\frac{\kappa}{2\pi i}\int_{L_\delta}\frac{\overline{\omega_*(\zeta)}}{\zeta - z}d\zeta + \frac{1}{2\pi i}\int_{L_\delta}\frac{\omega_*(\zeta)}{\zeta - z}d\overline{\zeta} - \frac{1}{2\pi i}\int_{L_\delta}\frac{\overline{\zeta}\omega_*(\zeta)}{(\zeta - z)^2}d\zeta,\ z \in S_\delta, \tag{3.3}$$

namely,

$$\psi_*(z) = -\frac{\kappa}{2\pi i}\int_{L_\delta}\frac{\overline{\omega_*(\zeta)}}{\zeta - z}d\zeta - \frac{1}{2\pi i}\int_{L_\delta}\frac{\overline{\zeta}\omega_*'(\zeta)}{\zeta - z}d\zeta,\ z \in S_\delta.$$

Substituting (3.2) and (3.3) into (3.1), (3.1) can be changed to the integral equation on L_δ

$$\kappa\omega_*(\xi) + \frac{\kappa}{2\pi i}\int_{L_\delta}\omega_*(\zeta)d\ln\frac{\zeta - \xi}{\overline{\zeta} - \overline{\xi}} + \frac{1}{2\pi i}\int_{L_\delta}\overline{\omega_*(\zeta)}d\frac{\zeta - \xi}{\overline{\zeta} - \overline{\xi}} = 2\mu g_*(\xi),\ \xi \in L_\delta. \tag{3.4}$$

If denote $\omega_*(\xi) = \omega_\delta(t), g_*(\xi) = g_\delta(t)$, then (3.4) can be rewritten as

$$\omega_\delta(t) = a_\delta + b_\delta t + \frac{\widetilde{g_\delta}(t)}{\kappa}, \tag{3.5}$$

where

$$a_\delta = -\frac{1}{2\pi i}\int_L \frac{\omega_\delta(\tau)}{\tau} d\tau, b_\delta = \frac{1}{2\kappa\pi i}\int_L \overline{\omega_\delta(\tau)} d\tau. \tag{3.6}$$

The same as the above method, we can get

$$a_\delta = -\frac{1}{4\kappa\pi i}\int_L \frac{\widetilde{g_\delta}(\tau)}{\tau} d\tau, \tag{3.7}$$

$$b_\delta = \frac{1}{\kappa^2 - 1}\left[\frac{1}{2\pi i}\int_L \overline{\widetilde{g_\delta}(\tau)} d\tau - \frac{1}{2\kappa\pi i}\int_L \widetilde{g_\delta}(\tau) d\overline{\tau}\right]. \tag{3.8}$$

By (3.5), we have[5]

Theorem 3.1. *If* $\delta \in B(\rho)$, $\rho < \frac{1}{30\pi}$, $|\delta(\tau) - \delta(t) - \delta'(\tau)(\tau - t)| \le A|\tau - t|^{1+\varepsilon}$, $\varepsilon_0 < \varepsilon < 1\,(\varepsilon_0 > 0)$, *and*

$$A^\nu < \frac{59\kappa^2 - 62\kappa + 1}{30(3\kappa - 1)(\kappa + 1)} \frac{\varepsilon_0 \nu}{2^{\varepsilon_0 \nu}} \left(\frac{60\pi}{\pi + 2}\right)^{1-\nu}, \ \nu \in (0,1) \text{ fixed},$$

then

$$\|\omega_\delta(t) - \omega(t)\|_L \le C(\kappa, \rho, \mu, \nu, \varepsilon)\|\delta\|_1^{\min\{1-\nu,\varepsilon\}}, \tag{3.9}$$

where $C(\kappa, \mu, \nu, \varepsilon)$ *is a constant depending on* $\kappa, \mu, \nu, \varepsilon$.

According to (3.2) and (3.3), we obtain[5]

Theorem 3.2. *Assumptions as those in Theorem 3.1. Then the complex stress functions* $\varphi_*(z)$ *and* $\varphi(z)$ *satisfy*

$$\|\varphi_*(z) - \varphi(z)\|_\Omega \le C(\kappa, \mu, \nu, \varepsilon, \epsilon)\|\delta\|_1^{(1-\epsilon)\min\{\varepsilon, 1-\nu\}}, \ 0 < \epsilon < 1.$$

Theorem 3.3. *If* $\delta \in B(\rho)$, $\rho < \frac{1}{30\pi}$, $|\delta(\tau) - \delta(t) - \delta'(\tau)(\tau - t)| \le A|\tau - t|^{1+\varepsilon}$, $\varepsilon_0 < \varepsilon < 1\,(\varepsilon_0 > 0)$,

$$A^\nu < \frac{59\kappa^2 - 62\kappa + 1}{30(3\kappa - 1)(\kappa + 1)} \frac{\varepsilon_0 \nu}{2^{\varepsilon_0 \nu}} \left(\frac{60\pi}{\pi + 2}\right)^{1-\nu}, \ \nu \in (0,1) \text{ fixed},$$

and $S_1 \subset \Omega$ *is a closed subset,* $d > \rho$ *is the shortest distance between point* z *in* S_1 *and point* t *on* L. *For* $z \in S_1$, $\psi_*(z)$ *and* $\psi(z)$ *satisfy*

$$|\psi_*(z) - \psi(z)| \le C(\kappa, \mu, \nu, \varepsilon, \rho, d, \epsilon)\|\delta\|_1^{(1-\epsilon)\min\{\varepsilon, 1-\nu\}}, \ \epsilon \in (0,1),$$

$$|\varphi_*'(z) - \varphi'(z)| \le C(\kappa, \mu, \nu, \varepsilon, \rho, d)\|\delta\|_1^{\min\{\varepsilon, 1-\nu\}}.$$

Theorem 3.4. *Assumptions as those in Theorem 3.3. Then the displacement $u + iv$ at a $z \in \Omega$ and the perturbed $(u + iv)_\delta$ at the same point satisfy*

$$|(u+iv)_\delta - (u+iv)| \leq C(\kappa, \mu, \nu, \varepsilon, \rho, d, \epsilon)\|\delta\|_1^{(1-\epsilon)\min\{\varepsilon, 1-\nu\}}.$$

References

[1] N. I. Muskhelishvili, *Some basic problems of mathmetical theory of elasticity,* Groningen, Noordhoff, 1963.

[2] J. K. Lu, *Complex variable methods in plane elasticity,* World Scientific, Singapore, 1995.

[3] X. L. Wang and Y. F. Gong, On stability of a class of singular integral with respect to path of integration, *Acta Math. Sinica,* 42(2) (1999), 343–350 (in Chinese).

[4] J. Lin and J. Y. Du, Stability of the first fundamental problem under perturbation of the boundary curve in plane elasticity, to appear.

[5] J. Lin, C. R. Wang and K. M. Yan, Stability of the second fundamental problem under perturbation of the boundary curve in plane elasticity, in *Mathematical methods in the applied sciences,* 33(14) (2010), 1762-1770.

EFFECTS OF STRETCHING FUNCTIONS ON NON-UNIFORM FDM FOR POISSON-TYPE EQUATIONS ON A DISK WITH SINGULAR SOLUTIONS[1]

XIAO-YU ZHANG[a] and QING FANG[b]

[a,b] *Department of Mathematical Sciences, Faculty of Science*
Yamagata University, Yamagata 990-8560, Japan
[a] *E-mail: xyzhang@kdw.kj.yamagata-u.ac.jp*
[b] *E-mail: fang@sci.kj.yamagata-u.ac.jp*

In this paper, we consider the Dirichlet boundary value problem of Poisson-type equations on a disk. We assume that the exact solution performs singular properties that its derivatives go to infinity at the boundary of the disk. A stretching polynomial-like function with a parameter is used to construct local grid refinements and the Swartztrauber-Sweet scheme is considered over the non-uniform partition. The effects of the parameter are analyzed completely by numerical experiments, which show that there exists an optimal value for the parameter to have a best approximate solution. Moreover, we show that the discrete system can be considered as a stable one by exploring the concept of the effective condition number.

Keywords: Dirichlet boundary value problem, Poisson-type equations, singular solutions, finite difference methods, Swartztrauber-Sweet scheme, local grid refinements, effective condition number.
AMS No: 65N06, 65N15.

1. Introduction

The numerical schemes for solving ordinary and partial differential equations with singular exact solutions have been considered by many researchers in various computational fields as well as numerical analysis. In this paper, we are concerned with the Dirichlet boundary value problem of elliptic equations on a disk Ω. Matsunaga and Yamamoto [4] considered the same problem and proved that if the exact solution u is very smooth so that $u \in C^4(\overline{\Omega})$, then the approximate solution by Swartztrauber-Sweet scheme with uniform partition of Ω has an accuracy of $O(h^2 + k^2)$, where h and k are uniform mesh sizes in polar coordinate (r, θ). In this paper, we assume that the exact solution of the problem has singular properties that its derivatives go to infinity at the boundary of Ω. In this case, the numerical scheme over uniform partition could not yield satisfactory accuracy. Therefore, we use a stretching polynomial-like function with a parameter p to carry out local grid refinements and construct the Swartztrauber-Sweet

[1] This research is supported by Scientific Research Grant-in-Aid from JSPS (No. 22444 and No. 21540106).

scheme over the non-uniform partition. Since the scheme is inconsistent in the sense that truncation errors at the grid points near the boundary go to infinity, the usual error analysis can not be applied to our case. But by exploring the matrix analysis results to our discrete system, we can show that the approximate solution by our scheme is convergent and the convergency can be accelerated by adjusting the parameter p.

In the present paper, we first describe our problem and the non-uniform Swartztrauber-Sweet method in Section 2. Then in Section 3 we obtain the estimates of truncation errors at different grid points and the convergence result, which indicates that the parameter p can be chosen, so that our scheme yields its best convergent order. Numerical results are shown in Section 4, where the effects of the parameter are made clearly by specific example. Finally, in Section 5, we explore the concept of the effective condition number to show that our discrete system can be considered as a stable one, compared with the usual condition number.

2. Derivation of the Problem and the Non-uniform FDM

Let $\Omega_1 = \{(x,y) \mid x^2+y^2 < R^2\} \subset \mathbb{R}^2$ with $R>0$, and f_1, g_1, q_1 be given functions, $q_1(x,y) \geq 0$ in Ω_1. We consider the Dirichlet boundary value problem of elliptic equations

$$-\Delta u + q_1(x,y)u = f_1(x,y) \quad \text{in} \quad \Omega_1, \tag{2.1}$$

$$u = g_1(x,y) \quad \text{on} \quad \Gamma_1 = \partial\Omega_1. \tag{2.2}$$

The problem is usually solved by using polar coordinates, so that in fact in this paper we consider the following rewritten problem

$$-\left[\frac{1}{r}\frac{\partial}{\partial r}\left(r\frac{\partial u}{\partial r}\right)+\frac{1}{r^2}\frac{\partial^2 u}{\partial\theta^2}\right]+q(r,\theta)u = f(r,\theta) \quad \text{in} \quad \Omega, \tag{2.3}$$

$$u(R,\theta) = g(\theta) \qquad \text{on} \quad \Gamma, \tag{2.4}$$

where $\Omega = \{(r,\theta) \mid 0<r<R,\ 0\leq\theta<2\pi\}$ and $\Gamma = \partial\Omega = \{(R,\theta) \mid 0 \leq \theta < 2\pi\}$.

Let $\varphi(t) = R-(R-t)^{p+1}/R^p$ $(0 \leq t \leq R)$ for $p>0$, which satisfies $\varphi(0)=0$, $\varphi(R)=R$. We take the following partition of Ω and apply Swartztrauber-Sweet method to (2.3)–(2.4).

$$h = \frac{R}{m+1}, \quad t_i = ih, \quad r_i = \varphi(t_i), \quad i=0,1,2,\cdots,m+1, \tag{2.5}$$

$$r_{i+1/2} = (r_i+r_{i+1})/2, \quad i=0,1,2,\cdots,m, \tag{2.6}$$

$$h_i = r_i - r_{i-1}, \quad i=1,2,\cdots,m+1, \tag{2.7}$$

$$k = \tfrac{2\pi}{n}, \quad \theta_j = jk, \quad j=0,1,2,\cdots,n, \tag{2.8}$$

$$-\Big[\frac{1}{r_i}\Big\{\frac{r_{i+1/2}(U_{i+1,j}-U_{i,j})}{h_{i+1}}-\frac{r_{i-1/2}(U_{i,j}-U_{i-1,j})}{h_i}\Big\}$$
$$\Big/\Big(\frac{h_i+h_{i+1}}{2}\Big)+\frac{1}{r_i^2k^2}(U_{i,j+1}-2U_{i,j}+U_{i,j-1})\Big]$$
$$+q_{i,j}U_{i,j}=f_{i,j},\ i=1,2,\cdots,m,\ j=0,1,2,\cdots,n-1, \tag{2.9}$$
$$\frac{4}{h_1^2}\Big[U_{0,0}-\frac{1}{n}\sum_{j=0}^{n-1}U_{1,j}\Big]+q_{0,0}U_{0,0}=f_{0,0}, \tag{2.10}$$
$$U_{i,n}=U_{i,0},\ U_{i,-1}=U_{i,n-1},\ i=0,1,2,\cdots,m+1, \tag{2.11}$$
$$U_{0,j}=U_{0,0},\quad U_{m+1,j}=g_j,\ \ j=0,1,2,\cdots,n. \tag{2.12}$$

Following assumptions are supposed for the exact solution u of (2.3)–(2.4).

(H1) $u\in C(\overline{\Omega})\cap C^4(\overline{\Omega}\backslash\Gamma)$, $\partial^4u/\partial\theta^4$ is bounded over $\overline{\Omega}$ and there exist positive constants $\sigma\in(0,2)$ and K such that

$$\sup_{r\in(0,1)}\frac{(R-r)^j|(\partial^ju/\partial r^j)(r,\theta)|}{(R-r)^\sigma}\le K,\ j=1,2,3,4.$$

(H2) For small $\delta>0$, there exists a positive constant C_0 such that

$$\omega(d)\equiv\sup_{\mathrm{dist}(P,Q)\le d}|u(P)-u(Q)|\le C_0d^\sigma$$

for any $P,Q\in\Omega,\ \mathrm{dist}(P,\Gamma)\le\delta,\ \mathrm{dist}(Q,\Gamma)\le\delta.$

3. Convergence Result

Let $P=(r_i,\theta_j)\in\Omega_h$, and $\kappa<R/4$ be a small positive constant. We arrange the grid points in the following order

$$\Omega_h=\Big(\cup_{i=1}^I\Omega_h^{(i)}\Big)\cup\Omega_h^{(0)},$$

where $I=\lfloor\kappa/h\rfloor$, which denotes the largest integer not greater than κ/h,

$$\Omega_h^{(1)}=\{P\in\Omega_h\mid \text{at least one of the neighbors of } P\in\Gamma_h\},$$
$$\Omega_h^{(i)}=\{P\in\Omega_h\backslash\cup_{j=1}^{i-1}\Omega_h^{(j)}\mid \text{at least one of the neighbors of}$$
$$P\in\Omega_h^{(i-1)}\},\ \ 2\le i\le I,$$

and $\Omega_h^{(0)}=\Omega_h\backslash\cup_{j=1}^{I}\Omega_h^{(j)}$.

The number of points in $\Omega_h^{(i)}$ and $\Omega_h^{(0)}$ are denoted by m_i and m_0, respectively, in which $m_0+m_1+\cdots+m_I=mn+1$.

The truncation error of the discretization of $-\Delta$ at the grid point $P = (r_i, \theta_j) \in \Omega_h \backslash \{r = 0\}$ is defined by

$$\tau(P) := -\Delta_h u(P) + \Delta u(P),$$

where u is the exact solution of (2.3)–(2.4) and $-\Delta_h$ is the discretization of $-\Delta$ defined as the first term in (2.9).

In this paper, c, c_1 etc. denote positive constants independent of h.

When $P \in \Omega_h^{(1)}$, using the assumptions (H1) and (H2), we obtain

$$|\tau(P)| \leq c' 2^{p\sigma}(p+1)^{\sigma} h^{(p+1)(\sigma-2)} + ck^2. \tag{3.1}$$

When $P = P_{m+1-i,j} = (r_{m+1-i}, \theta_j) \in \Omega_h^{(i)}$, $i = 2, 3, \cdots, I$,

$$\begin{aligned}|\tau(P)| \leq\ & c_1 p(p+1) h^{(p+1)(\sigma-2)}(i+1)^{p-1} i^{(p+1)(\sigma-3)} + c_2(p+1)^2 \\ & \times h^{(p+1)(\sigma-2)}(i+1)^{2p}(i-1)^{(p+1)(\sigma-4)} + ck^2.\end{aligned} \tag{3.2}$$

When $P = (r_i, \theta_j) \in \Omega_h^{(0)} \backslash \{r = 0\}$,

$$|\tau(P)| \leq c\frac{1}{r_i}(L(p)h^2 + k^2), \tag{3.3}$$

where c is a positive constant dependent on κ, $\max\limits_{0 \leq r \leq R-\kappa} |\partial^3 u/\partial r^3|$ and $\max\limits_{0 \leq r \leq R-\kappa} |\partial^4 u/\partial r^4|$ but independent of h, $L(p)$ is a positive constant dependent on p but independent of h.

Finally, at origin $P_0 = (x, y) = (0, 0)$,

$$|\tau(P_0)| \leq O(h_1^2) + O\left(\frac{k^4}{h_1^2}\right) \leq c\left((p+1)^2 h^2 + \frac{k^4}{h_1^2}\right). \tag{3.4}$$

Since the truncation errors $|\tau(P)|$ for $P \in \Omega_h^{(i)}$, $i = 1, 2, 3, \cdots, I$, go to infinity when $\sigma < 2$ under the assumptions (H1) and (H2), it is known that the usual convergence analysis can not be applied to our case. But we note that if we denote the coefficients matrix of (2.9)–(2.19) rearranged in the order described by A, A is an irreducible and strictly diagonally dominant matrix. Therefore we can explore the properties of matrix analysis to show that A is an M-matrix and, especially, is invertible. Furthermore, we know that $A^{-1} \geq O$, which means that all the elements of A are nonnegative. We omit the detailed proof but refer to Varga [7] and Young [9].

Then by a detailed matrix analysis approach and using the truncation errors $|\tau(P)|$, we can get the following convergence result.

Theorem 1. *Suppose that the exact solution u of (2.1)–(2.2) satisfies the assumptions (H1) and (H2). Let $\boldsymbol{u}$ be the vector of exact values at grid points and $\boldsymbol{U}$ the vector of solution of (2.9)–(2.12). If $\mu = (p+1)\sigma < 2$ and $k^2 \le M_0 h$ for some positive constant M_0, then there exists a positive constant c such that*

$$\max |\boldsymbol{u} - \boldsymbol{U}| \le c\left(K(p)h^{\mu} + L(p)h^2 + k^2\right) \tag{3.5}$$

holds, where $K(p)$ and $L(p)$ are positive constants dependent on p but independent of h. Also, if $\mu = (p+1)\sigma = 2$ and $k^2 \le M_0 h$, then we have

$$\max |\boldsymbol{u} - \boldsymbol{U}| \le c\left(K(p)h^2|\log h| + L(p)h^2 + k^2\right). \tag{3.6}$$

Remark 1. If p is chosen larger, we get larger μ so that h^μ becomes smaller. But $K(p)$ becomes larger. On the other hand, if p is chosen smaller, we have larger h^μ but smaller $K(p)$. The situation is the same for the second term in (3.5). Therefore the analysis result suggests the existence of an optimal value for which the FDM solution approximates the exact solution best. This is illustrated by numerical results in next section.

4. Numerical Example

We show some numerical results to illustrate our error estimates.

Example. Consider the problem

$$\begin{cases} -\left[\dfrac{1}{r}\dfrac{\partial}{\partial r}\left(r\dfrac{\partial u}{\partial r}\right) + \dfrac{1}{r^2}\dfrac{\partial^2 u}{\partial \theta^2}\right] + q(r,\theta)u = f(r,\theta) & \text{in } \Omega, \\ u(1,\theta) = 0 & \text{on } \partial\Omega, \end{cases}$$

where $\Omega = \{(r,\theta) \mid 0 < r < 1, 0 \le \theta < 2\pi\}$ is the unit disk, $q(r,\theta) = r$ and

$$f(r,\theta) = (r^3(1-r)^\sigma + 5\sigma r(1-r)^{\sigma-1} - \sigma(\sigma-1)r^2(1-r)^{\sigma-2})\sin 2\theta.$$

The exact solution is $u = r^2(1-r)^\sigma \sin 2\theta$.

We take $h = 1/N$, $k = 2\pi/\lfloor 2\pi/h \rfloor$ in Section 2, when $\sigma = 0.5$, the convergence results as $h \to 0$ are shown in Table 4.1 for every different value of p.

We choose p as a parameter and plot the maximum error results in detail in Figure 4.1 for $\sigma = 0.5$ and in Figure 4.2 for $\sigma = 1.5$. They show that although we can choose p as large as possible to get the higher accuracy h^μ, where $\mu = \sigma(p+1)$, we get larger coefficients $K(p)$ and $L(p)$. Therefore we have an intermediate value for p around 1.5 for $\sigma = 0.5$ and around 0.5 for $\sigma = 1.5$ to obtain the best accuracy. We have similar results of maximum errors with respect to p for other values of σ, which denotes the strength of singularities of the exact solution.

The 3d-plot of the approximate solution with $h = 1/100$ for $p = 1$ and the error distribution are shown in Figure 4.3 and Figure 4.4, respectively. We note that if the exact solution is sufficiently smooth, the approximate solution shows superconvergence property in the meaning that the accuracy near the boundary is higher than inside the domain (refer to [5]). Figure 4.4 shows that this superconvergence property does not appear in general for the problems with singular solutions.

We note that when σ is chosen larger, the optimal value of p which has a minimal maximum error gets smaller and tends to zero. This implies that stretching functions used in non-uniform spatial partitions do not take effect for problems having much weak singularities.

Table 4.1: Maximum errors with $\sigma = 0.5$.

h	$p = 0$	$p = 1$	$p = 2$	$p = 3$
1/20	7.155e-2	1.032e-2	4.578e-3	8.032e-3
1/50	4.751e-2	4.541e-3	1.214e-3	1.613e-3
1/100	3.453e-2	2.359e-3	4.433e-4	4.667e-4
1/200	2.486e-2	1.206e-3	1.610e-4	1.328e-4

Figure 4.1: Plot of maximum errors with respect to p when $\sigma = 0.5$

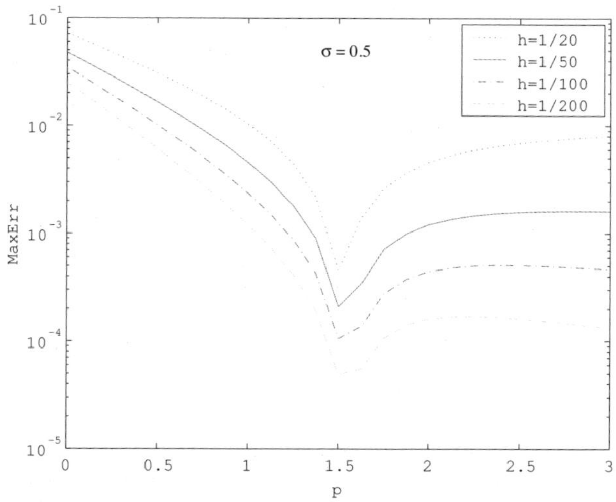

Figure 4.2: Plot of maximum errors with respect to p when $\sigma = 1.5$

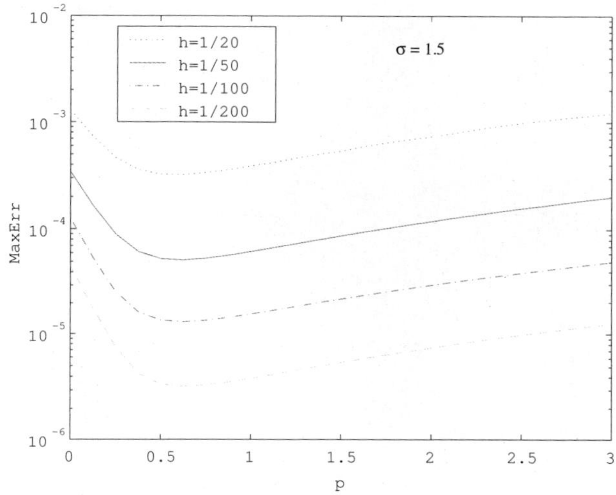

Figure 4.3: 3d-plot of the approximate solution with $h = 1/100, p = 1$ when $\sigma = 0.5$

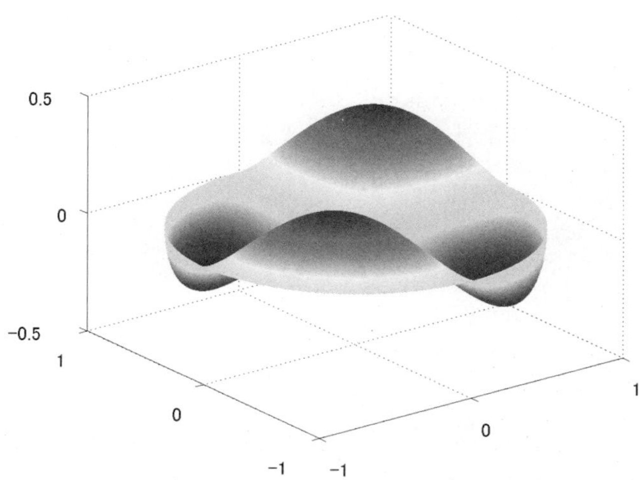

Figure 4.4: Plot of error distribution of the approximate solution with $h = 1/100, p = 1$ when $\sigma = 0.5$

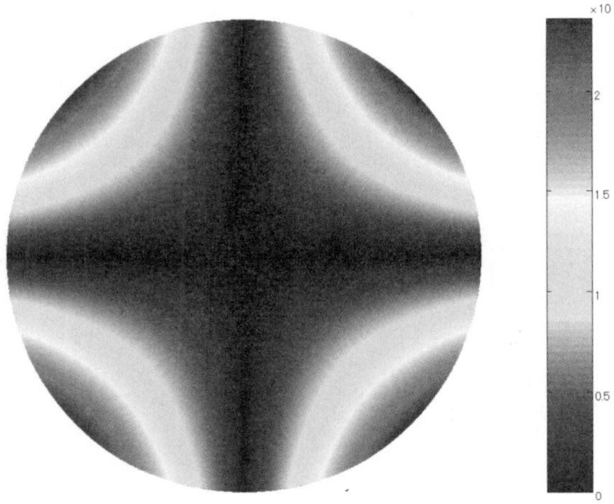

5. Stability Analysis

Since we only get approximate values when we carry out numerical computations by floating-point numbers, we should pay our attention to the changes of solutions when data are perturbed in the problem.

We consider the linear system obtained from our method in Section 2:

$$Ax = b, \tag{5.1}$$

where $A \in \mathbb{R}^{n\times n}$ has elements of discretization of the left hand of (2.3) and $b \in \mathbb{R}^n$ has values of f and g. Let $\|\cdot\|_2$ be the Euclid norm for a vector and the subordinate matrix norm for a matrix. When there exists a perturbation of b in (5.1),

$$A(x + \Delta x) = b + \Delta b, \tag{5.2}$$

the usual condition number $\mathrm{Cond}(A) = \|A\|_2 \, \|A^{-1}\|_2$ of the matrix A is often used to be as the upper bound of the relative errors of the perturbation solution (refer to Wilkinson [8]).

In Section 3, we have shown that the non-uniform local refined Swartztrauber-Sweet scheme constructed in Section 2 is inconsistent. That is, the truncation errors at those nodes near the boundary are disconvergent or unbounded as $h \to 0$. But in Section 3, we obtain the convergence result that the approximate solution of this inconsistent FDM is still convergent to the exact solution and even has accuracy of almost second order, if the parameter in the stretching function is chosen appropriately. It implies that our method can be used for singular problems. However, the inconsistence may result in stability problem since the matrix A in the algebraic system $Ax = b$ obtained from our FDM is ill-conditioned in the meaning of the usual condition number. That is, $\text{Cond}(A)$ is very large in our case.

As stated above, the usual condition number $\text{Cond}(A)$ is defined to measure the bound of relative errors without considering the nonhomogeneous term b. However, in practical applications, we deal with only a certain vector b and the true relative errors may be smaller, or even much smaller than the worst usual condition number. Such a case was first studied by Chan and Foulser [1] in 1988, and called the effective condition number. In 2006, Li, et al. [3] proved that the effective condition number is the appropriate criterion for stability analysis, when the huge condition number is misleading. Recently, Li, et al. [2] derived the new computational formulas for effective condition numbers and applied to Poisson's equation by FDM. In our case, we have singular solutions so that the usual condition number of the discretized linear systems by our scheme may be very large, which seems the scheme is bad. But in fact, error estimates show that the scheme has good and stable convergence. Therefore it is necessary to revisit the concept of condition numbers.

Let the singular value of A be $\lambda_1 \geq \lambda_2 \geq \cdots \geq \lambda_n > 0$ and the singular value decomposition of A be $UAV = \Lambda$, where U, V are orthogonal matrices and $\Lambda = \text{diag}(\lambda_1, \lambda_2, \cdots, \lambda_n)$. Denote $U = (u_1, \cdots, u_n)$, then it is known that we have the expansions

$$b = \sum_{i=1}^{n} \beta_i u_i, \quad \triangle b = \sum_{i=1}^{n} \alpha_i u_i, \tag{5.3}$$

where the expansion coefficients α_i and β_i are given by

$$\beta_i = u_i^T b, \quad \alpha_i = u_i^T \triangle b. \tag{5.4}$$

From (5.3) and (5.4), we have

$$\|b\|_2^2 = \sum_{i=1}^{n} \beta_i^2, \quad \|\triangle b\|_2^2 = \sum_{i=1}^{n} \alpha_i^2. \tag{5.5}$$

Since the inverse matrix A^{-1} exists, we have from (5.3)

$$x = A^{-1} b = \sum_{i=1}^{n} \beta_i A^{-1} u_i = \sum_{i=1}^{n} \frac{\beta_i}{\lambda_i} u_i, \quad \|x\|_2^2 = \sum_{i=1}^{n} \frac{\beta_i^2}{\lambda_i^2}.$$

Also from (5.1) and (5.2), $\triangle x = A^{-1}\triangle b = V\Lambda^{-1}U^T\triangle b$. Since U and V are orthogonal, we obtain

$$\|\triangle x\|_2^2 = \sum_{i=1}^{n} \frac{\alpha_i^2}{\lambda_i^2} \leq \frac{1}{\lambda_n^2} \sum_{i=1}^{n} \alpha_i^2 = \frac{1}{\lambda_n^2} \|\triangle b\|_2^2. \tag{5.6}$$

Hence we have

$$\frac{\|\triangle x\|_2}{\|x\|_2} \leq \frac{1}{\lambda_n} \frac{\|\triangle b\|_2}{\|x\|_2} = \text{Cond\_eff}(A) \frac{\|\triangle b\|_2}{\|b\|_2}, \tag{5.7}$$

where Cond_eff(A) is called the effective condition number, and defined by

$$\text{Cond\_eff}(A) = \frac{\|b\|_2}{\lambda_n \sqrt{\sum_{i=1}^{n} \beta_i^2 / \lambda_i^2}}. \tag{5.8}$$

Note that if the vector b (i.e. x) is just parallel to u_1, i.e.,

$$\beta_2 = \cdots = \beta_n = 0, \tag{5.9}$$

then Cond_eff(A) $= \lambda_1/\lambda_n$ which agrees with the usual condition number Cond(A). However, in practice, (5.9) may not happen for a given b. Hence, the effective condition number may provide a better estimation on the upper bound of relative errors of x. This implies that the analysis of the effective condition number is important for FDM on non-uniform grids for singularity problems.

Table 5.1: Maximal errors, usual condition numbers and effective condition numbers for $p = 1.5$ and $\sigma = 1.5$.

h	$\lambda_{max}(A)$	$\lambda_{min}(A)$	$\max_{i,j} \lvert\varepsilon_{i,j}\rvert$	$\|\varepsilon\|_2$	Cond(A)	Cond_eff(A)
1/70	7.36e+8	1.43e-2	4.46e-5	2.70e-3	5.13e+10	2.92e+4
1/90	2.58e+9	1.09e-2	2.70e-5	2.10e-3	2.37e+11	4.63e+4
1/110	7.05e+9	8.78e-3	1.81e-5	1.72e-3	8.02e+11	6.69e+4
1/130	1.63e+10	7.36e-3	1.29e-5	1.45e-3	2.21e+12	9.06e+4
1/150	3.32e+10	6.32e-3	9.71e-6	1.26e-3	5.26e+12	1.17e+5
1/170	6.21e+10	5.54e-3	7.56e-6	1.11e-3	1.12e+13	1.47e+5
1/190	1.08e+11	4.93e-3	6.05e-6	9.94e-4	2.20e+13	1.80e+5
1/210	1.79e+11	4.44e-3	4.95e-6	8.99e-4	4.03e+13	2.15e+5

In Table 5.1 and Table 5.2, we give numerical results of maximal errors, usual condition numbers and effective condition numbers, where $\varepsilon = (\varepsilon_{i,j})$ is the vector of errors $\varepsilon_{i,j}$, $\lambda_{max}(A)$ and $\lambda_{min}(A)$ denote the maximal and

Table 5.2: Maximal errors, usual condition numbers and effective condition numbers for $p = 0.8$ and $\sigma = 1.2$.

h	$\lambda_{max}(A)$	$\lambda_{min}(A)$	$\max_{i,j} \lvert\varepsilon_{i,j}\rvert$	$\lVert\varepsilon\rVert_2$	Cond(A)	Cond_eff(A)
1/70	2.99e+7	1.20e-2	3.42e-5	2.49e-3	2.48e+9	1.58e+4
1/90	8.16e+7	9.18e-3	2.07e-5	1.94e-3	8.88e+9	2.63e+4
1/110	1.82e+8	7.42e-3	1.39e-5	1.59e-3	2.45e+10	3.95e+4
1/130	3.54e+8	6.22e-3	9.95e-6	1.35e-3	5.69e+10	5.52e+4
1/150	6.28e+8	5.36e-3	7.48e-6	1.17e-3	1.17e+11	7.35e+4
1/170	1.04e+9	4.70e-3	5.83e-6	1.04e-3	2.20e+11	9.43e+4
1/190	1.61e+9	4.19e-3	4.67e-6	9.27e-4	3.85e+11	1.18e+5
1/210	2.41e+9	3.77e-3	3.82e-6	8.40e-4	6.38e+11	1.43e+5

minimal singular value of A, respectively. We know from these results that although Cond(A) is very large to show A is ill-conditioned, Cond_eff(A) is not so large, which implies the linear system obtained from our FDM can be considered as a stable system, since the relative errors by perturbations are much smaller than the usual condition number Cond(A).

References

[1] T. F. Chan and D. E. Foulser, Effectively well-conditioned linear systems, *SIAM J. Sci. and Stat. Comput.*, 9 (1988), 963-969.

[2] Z. C. Li, C. S. Chien and H. T. Huang, Effective condition number for finite difference method, *J. Comp. and Appl. Math.*, 198 (2007), 208-235.

[3] Z. C. Li, H. T. Huang and J. T. Chen, *Effective condition number for collocation Trefftz methods,* Technical report, Department of Applied Mathematics, National Sun Yat-sen University, Kaohsiung, Taiwan, 2006.

[4] N. Matsunaga and T. Yamamoto, Convergence of Swartztrauber-Sweet's approximation for the Poisson-type equation on a disk, *Numer. Funct. Anal. Optimz.*, 20 (1999), 917–928.

[4] Matsunaga-Yamamoto, N. Matsunaga and T. Yamamoto, Superconvergence of the Shortley-Weller approximation for Dirichlet problems, *Journal Comput. Appl. Math.*, 116 (2000), 263-273.

[5] P. N. Swartztrauber and R.A. Sweet, The direct solution of the discrete Poisson equation on a disk, *SIAM J. Numer. Anal.*, 10 (1973), 900-907.

[6] R. S. Varga, *Matrix iterative analysis, 2nd ed.,* Springer Series in Computational Mathematics, Vol. 27, Springer-Verlag, Berlin, 2000.

[7] J. H. Wilkinson, *The algebraic eigenvalue problem,* Clarendon Press, Oxford, 1965.

[8] D. M. Young, *Iterative solution of large linear systems,* Academic Press, New York, 1971.

A FAST MULTISCALE GALERKIN METHOD FOR SOLVING ILL-POSED INTEGRAL EQUATIONS VIA LAVRENTIEV REGULARIZATION[1]

XING-JUN LUO, FAN-CHUN LI and SU-HUA YANG
School of Mathematics and Computer Science,
Gannan Normal University, Ganzhou 341000, China
E-mail: lxj0298@126.com

We develop a fast multiscale Galerkin method to solve the linear ill-posed integral equations with approximately known right-hand sides and operators via Lavrentiev regularization. This method leads to fast solutions of discrete Lavrentiev regularization. The convergence rates of Lavrentiev regularization are achieved by using a modified discrepancy principle.
Keywords: Lavrentiev regularization, multiscale Galerkin method, a posteriori parameter choice, convergence rates.
AMS No: 65J20, 65J10.

1. The Fast Multiscale Galerkin Method for Lavrentiev Regularization

In this section, we describe the fast multiscale Galerkin method for solving ill-posed integral equations of the first kind via Lavrentiev regularization, and present the computational complexity for the truncation scheme.

Suppose that E is a bounded closed domain in $\mathbb{R}^d$ for $d \geq 1$ and $X := L^2(E)$ is the Hilbert space with norm $\|\cdot\|$. Let $\mathcal{A}$ be a linear bounded self-adjoint nonnegative operator from a Hilbert space X to itself, that is, $\mathcal{A} = \mathcal{A}^*$ and $(\mathcal{A}g, g) \geq 0, \forall g \in X$. We consider the Fredholm integral equation of the first kind

$$\mathcal{A}g = u, \tag{1.1}$$

where u is given element in $L^2(E)$ and g is the unknown element in $L^2(E)$, and the Fredholm integral operator $\mathcal{A}$ is defined by

$$\mathcal{A}g(t) := \int_E k(s,t)g(s)ds = u(t),\ s \in E,$$

where $k \in C(E \times E)$ is non-degenerate kernel. The operator $\mathcal{A}$ can be considered as a compact operator from $L^2(E)$ to $L^2(E)$. We assume that (1.1) has a solution, possibly non-unique, and denote by g_T the unique minimal-norm solution to (1.1). We assume that the range of $\mathcal{A}$, $R(\mathcal{A})$, is not closed, so problem (1.1) is ill-posed [1,2]. Suppose that instead of exact

[1]This research is supported by NSFC (No. 11061001), NSF (No. 2008GZS0025) and SF (No. GJJ10586).

input data $\mathcal{A}$ and u_T of equation (2.1), we have only some approximations $\mathcal{A}_h \in \mathcal{L}(X)$ and $u^\delta \in X$, such that

$$\|\mathcal{A} - \mathcal{A}_h\| \leq c_1 h, \ \|u^\delta - u_T\| \leq \delta, \tag{1.2}$$

and

$$\mathcal{A}_h = \mathcal{A}_h^*, \ (\mathcal{A}_h g, g) \geq 0, \ \forall g \in X,$$

where $c_1 > 0$, $h > 0$ and $\delta > 0$. For given $\{\delta, u^\delta, \mathcal{A}_h\}$, we want to construct efficient finite-dimensional approximations to the element g_T, under the assumption that g_T lies in the range of $\mathcal{A}^\nu$, i.e.

$$g_T \in M_{\nu,\rho} = \{x = \mathcal{A}^\nu \omega, \ \|\omega\| \leq \rho, \ \rho > 0, \ 0 < \nu \leq 1\}. \tag{1.3}$$

To get an approximation to the minimum norm solution g_T of (1.1), a widely used regularization method is Lavrentiev regularization. In this paper, we take the Lavrentiev's regularization to get a family of well-posed equations

$$(\alpha I + \mathcal{A}_h) g^{\alpha,\delta} = u^\delta. \tag{1.4}$$

We shall study the problem of efficient finite-dimensional approximation to the solution g_T. We next describe the multiscale Galerkin method to solve the equation (1.4), we use the same setting as the multiscale Galerkin method for solving the first kind integral equations as in [3]. We denote $N := \{1, 2, \cdots\}$, $N_0 := \{0, 1, 2, \cdots\}$ and $Z_n = \{0, 1, 2, \cdots, n-1\}$. Suppose that there is a multiscale partition of the set E, which consists of a family of partitions $\{E_i : i \in N_0\}$ of E such that for each $i \in N_0$ the partition E_i consists of a family of subsets$\{E_{i,j} : j \in Z_{e(i)}\}$ of E with the properties

$$\bigcup_{j \in Z_{e(i)}} E_{i,j} = E, \ \ \text{meas}\,(E_{i,j} \bigcap E_{i,j'}) = 0, \ j, j' \in Z_{e(i)}, \ j \neq j',$$

and

$$\max\{d(E_{i,j}) : j \in Z_{e(i)}\} \sim \mu^{-i/d},$$

where $e(i)$ denotes the cardinality of E_i, $d(A)$ denotes the diameter of the set A and $a_1 \sim b_1$ will always mean that a_1 and b_1 can be bounded by constant multiplies of each other. For $n \in N_0$ and $k \in N$, let X_n be the piecewise polynomial space associated with the partition E_n with total degree less than k. Then we have

$$\overline{\bigcup_{n \in N_0} X_n} = X, \ X_n \subset X_{n+1}, \ n \in N_0. \tag{1.5}$$

For each $i \in N$, let W_i be the orthogonal complement of X_{i-1} in X_i. This yields the multiscale space decomposition

$$X_n = W_0 \oplus^\perp W_1 \oplus^\perp \cdots \oplus^\perp W_n,$$

where $W_0 = X_0$. Denote $s(n) := \dim X_n$ and $w(i) := \dim W_i$, then we have $s(n) \sim \mu^n$ and $w(i) \sim \mu^i$.

We assume that W_i has a basis $\{w_{ij},\ j \in Z_{w(i)}\}$ satisfying

$$(w_{ij}, w_{i'j'}) = \delta_{ii'}\delta_{jj'},\ (i,j),(i',j') \in U_n.$$

This means that $X_n = \mathrm{span}\{w_{ij} :\ (i,j) \in U_n\}$, where $U_n := \{(i,j) :\ j \in Z_{w(i)},\ i \in Z_{n+1}\}$.

We now formulate the Galerkin method for solving equation (1.4). To this end, for each $n \in N_0$, denote by P_n the orthogonal projection from X onto X_n. The tradition Galerkin method for solving equation (1.4) is to find $g_{h,n}^{\alpha,\delta} \in X_n$ such that

$$(\alpha I + \mathcal{A}_{h,n}) g_h^{\alpha,\delta} = P_n u^\delta, \tag{1.6}$$

where $\mathcal{A}_{h,n} := P_n \mathcal{A}_h P_n$. The matrix representation of the operator $\mathcal{A}_{h,n} + \alpha I$ under the basis functions is a dense matrix [3]. To compress this matrix, we write it in the following form

$$\widetilde{\mathcal{A}}_{h,n} = \sum_{i \in Z_{n+1}} (P_i - P_{i-1}) \mathcal{A}_h P_{n-i}, \tag{1.7}$$

where $P_{-1} = 0$. In the Galerkin method (1.6), we replace $\mathcal{A}_{h,n}$ by $\widetilde{\mathcal{A}}_{h,n}$ and obtain a new approximation scheme for solving equation (1.6). That is, we find $\widetilde{g}_{h,n}^{\alpha,\delta} \in X_n$ such that

$$(\alpha I + \widetilde{\mathcal{A}}_{h,n}) \widetilde{g}_{h,n}^{\alpha,\delta} = P_n u^\delta, \tag{1.8}$$

We will show that this modified Galerkin method leads to a fast algorithm. To write equation (1.8) in its equivalent matrix form, we make use of the multiscale basis function. We write the solution $\widetilde{g}_{h,n}^{\alpha,\delta} \in X_n$ as $\widetilde{g}_{h,n}^{\alpha} = \sum_{(i,j) \in U_n} g_{ij}^h w_{ij} \in X_n$ and introduce solution vector $\mathbf{g}^h := [g_{ij}^h :\ (i,j) \in U_n]^T$. We introduce matrix

$$\mathbf{E}_n := [(w_{i'j'}, w_{ij}) :\ (i'j'),\ (i,j) \in U_n],$$

$$\widetilde{\mathbf{A}}_{\mathbf{h},\mathbf{n}} := [\widetilde{A}_{i'j',ij} :\ (i'j'),\ (i,j) \in U_n], \tag{1.9}$$

where

$$\widetilde{A}_{i'j',ij} = \begin{cases} (w_{i'j'}, \mathcal{A}_h w_{ij}), & i' + i \le n, \\ 0, & \text{otherwise}, \end{cases} \tag{1.10}$$

and vector $\mathbf{f}_n := [(w_{i'j'}, u^\delta)]$. Upon using this notations, equation (1.8) is written in the matrix form

$$(\alpha \mathbf{E}_n + \widetilde{\mathbf{A}}_{\mathbf{h},\mathbf{n}})\mathbf{g}^h = \mathbf{f}_n, \tag{1.11}$$

In the following, we analyse the computational complexity of the the truncation, which form a basis for fast multiscale algorithms.

Theorem 1.1. *If the truncated matrix* $\widetilde{\mathbf{A}}_{\mathbf{h},\mathbf{n}}$ *is obtained by the truncation strategy* (1.9) *and* (1.10), *then*

$$N \sim (n+1)\mu^n, \tag{1.12}$$

where N *denotes the inner products for solving equation* (1.11).

Proof. To find the number (1.12), it is required to estimate the amount N of the discrete information $(w_{i'j'}, \mathcal{A}_h w_{ij})$ and $(w_{i'j'}, u^\delta)$ used for solving equation (1.11). It follows from (1.10) that

$$N = \sum_{k=0}^{n} \text{dim} W_k \text{dim} W_{n-k} + \sum_{k=0}^{n} \text{dim} W_k \sim (n+1)\mu^n,$$

which completes the proof of the Theorem 1.1.

2. Error Estimate

We now turn to estimating the convergence rate of the modified Galerkin method (1.8). For $r \in (0, \infty)$, denote by $H^r(a,b)$ a linear subspace of X which is equipped with a norm $\|\varphi\|_{H^r} = \|\varphi\|_X + \|D_r\varphi\|_X$, where D_r is some linear (non-bounded) operator acting from H^r to X. We impose the following hypothesis

(H_1) There exists a positive constant c_r such that

$$\|(I - P_j)\|_{H^r \to X} \leq c_r \mu^{-rj/d}.$$

(H_2) There exists a positive constant $\gamma \geq 1$ such that $\mathcal{A}, \mathcal{A}_h \in \mathcal{H}^r_\gamma$, where

$$\mathcal{H}^r_\gamma := \{\mathcal{A} : \|\mathcal{A}\|_{X \to H^r} + \|\mathcal{A}^*\|_{H^r \to X} + \|(D_r\mathcal{A})^*\|_{H^r \to X} \leq \gamma\}.$$

It is well-known that ([3])

$$\|(\alpha I + \mathcal{A}_h)^{-1}\| \leq \frac{1}{\alpha}, \;\; \|(\alpha I + \mathcal{A})^{-1}\| \leq \frac{1}{\alpha}. \tag{2.1}$$

Theorem 2.1. *If hypothesis* (1.2) *and* H_1, H_2 *hold, then the following estimates hold*

$$\|\mathcal{A}_{h,n} - \widetilde{\mathcal{A}}_{h,n}\| \leq c_2 n \mu^{-rn/d}, \tag{2.2}$$

$$\|\mathcal{A}_h - \widetilde{\mathcal{A}}_{h,n}\| \le c_3(n+1)\mu^{-rn/d}, \tag{2.3}$$

where $c_2 := 2\gamma\mu^r c_r^2$, and $c_3 := \max\{2rc_r, 2\gamma\mu^r c_r^2\}$.

Proof. To proof (2.2), from (1.7) we have

$$\begin{aligned}\mathcal{A}_{h,n} - \widetilde{\mathcal{A}}_{h,n} &= P_n\mathcal{A}_hP_n - \sum_{i\in Z_{n+1}} (P_i - P_{i-1})\mathcal{A}_hP_{n-i} \\ &= \sum_{i=1}^{n}(P_i - P_{i-1})\mathcal{A}_h(p_n - P_{n-i}).\end{aligned} \tag{2.4}$$

From assumption H_1, H_2, it follows that

$$\begin{aligned}\|\mathcal{A}_{h,n} - \widetilde{\mathcal{A}}_{h,n}\| &\le \sum_{i=1}^{n} \|I - P_{i-1}\|_{H^r\to X}\|\mathcal{A}_h(I - P_{n-i})\|_{X\to H^r} \\ &\le \sum_{i=1}^{n} \|I - P_{i-1}\|_{H^r\to X}(\|\mathcal{A}_h(I - P_{n-i})\|_{X\to X} \\ &\quad + \|(I - P_{n-i})(D_r\mathcal{A}_h)^*\|_{X\to X}) \\ &\le c_2 n\mu^{-rn/d}.\end{aligned} \tag{2.5}$$

We write

$$\mathcal{A}_h - \widetilde{\mathcal{A}}_{h,n} = (I - P_n)\mathcal{A}_h + P_n\mathcal{A}_h(I - P_n) + (\mathcal{A}_{h,n} - \widetilde{\mathcal{A}}_n). \tag{2.6}$$

Recalling that $\|P_n\|$ is uniformly bounded by a constant, from hypothesis $(H_1), H_2$, (2.5) and (2.6), we get

$$\|\mathcal{A}_h - \widetilde{\mathcal{A}}_{h,n}\| \le c_3(n+1)\mu^{-rn/d}.$$

In order to analyze the convergence of the truncated multiscale Galerkin scheme (1.8), we need the following estimates.

Lemma 2.2. *Suppose that hypothesis $(H_1), (H_2)$ hold, and c_0 is a positive constant satisfying $0 < c_0 < 1$. If*

$$(n+1)\mu^{-rn/d} \le \frac{c_0\alpha}{c_3}, \tag{2.7}$$

where c_3 is the constant in Lemma 2.1, then $\alpha I + \widetilde{\mathcal{A}}_{h,n}$ is invertible such that

$$\|(\alpha I + \widetilde{\mathcal{A}}_{h,n})^{-1}\| \le \frac{1}{(1-c_0)\alpha}. \tag{2.8}$$

Proof. The proof is similar to Lemma 2.5 in [3].

We next proceed to estimate for $\|\widetilde{g}_{h,n}^{\alpha,\delta} - g_T\|$. To this end, we denote

$$g_h^\alpha = (\alpha I + \mathcal{A}_h)^{-1}u_T, \ \widetilde{g}_{h,n}^\alpha = (\alpha I + \widetilde{\mathcal{A}}_{h,n})^{-1}P_nu_T. \tag{2.9}$$

Theorem 2.3. *Let g_T satisfy condition* (1.3). *If the hypothesis* (1.2), (H_1) *and* (H_2) *hold, the integer n is chosen to satisfy inequality* (2.7), *then the following estimate holds*

$$\|\widetilde{g}_{h,n}^{\alpha,\delta} - g_T\| \leq \alpha^\nu \rho + \frac{c_4\delta + c_5 h}{\alpha} + \frac{c_6(n+2)\mu^{-rn/d}}{\alpha}, \tag{2.10}$$

where

$$c_4 = \frac{1}{1-c_0} > 1, c_5 = (2c_1 + c_1 c_4)c_{\gamma,\rho},$$

$$c_6 = \max\{c_r c_4 c_{\gamma,\rho},\ c_3 c_4 c_{\gamma,\rho}\}, c_{\gamma,\rho} := \gamma\rho.$$

Proof. We write

$$\widetilde{g}_{h,n}^{\alpha,\delta} - g_T = \widetilde{g}_{h,n}^{\alpha,\delta} - \widetilde{g}_{h,n}^{\alpha} + \widetilde{g}_{h,n}^{\alpha} - g_h^\alpha + g_h^\alpha - g_T.$$

Combing inequalities (1.2) and (2.8), we conclude that

$$\|\widetilde{g}_{h,n}^{\alpha,\delta} - \widetilde{g}_{h,n}^{\alpha}\| \leq \|(\alpha I + \widetilde{\mathcal{A}}_{h,n})^{-1}P_n\|\|u_T^\delta - u_T\| \leq \frac{\delta}{(1-c_0)\alpha}. \tag{2.11}$$

It follows from (2.9) that

$$\begin{aligned}\widetilde{g}_{h,n}^{\alpha} - g_h^\alpha &= (\alpha I + \widetilde{\mathcal{A}}_{h,n})^{-1}P_n u_T - (\alpha I + \mathcal{A}_h)^{-1}u_T \\ &= (\alpha I + \widetilde{\mathcal{A}}_{h,n})^{-1}(P_n - I)u_T \\ &\quad + \Big[(\alpha I + \widetilde{\mathcal{A}}_{h,n})^{-1} - (\alpha I + \mathcal{A}_h)^{-1}\Big]u_T.\end{aligned} \tag{2.12}$$

We write

$$\begin{aligned}J_1 :&= [(\alpha I + \widetilde{\mathcal{A}}_{h,n})^{-1} - (\alpha I + \mathcal{A}_h)^{-1}]u_T \\ &= (\alpha I + \widetilde{\mathcal{A}}_{h,n})^{-1}(\mathcal{A} - \widetilde{\mathcal{A}}_{h,n})(\alpha I + \mathcal{A})^{-1}\mathcal{A}g_T \\ &\quad + (\alpha I + \mathcal{A}_h)^{-1}(\mathcal{A}_h - \mathcal{A})(\alpha I + \mathcal{A})^{-1}\mathcal{A}g_T.\end{aligned} \tag{2.13}$$

Using hypothesis (H_2), (1.2), (1.3), (2.1), (2.3) and (2.8), we conclude that

$$\|J_1\| \leq \frac{\|\mathcal{A} - \widetilde{\mathcal{A}}_{h,n}\|}{(1-c_0)\alpha}\gamma^\nu\rho + \frac{\|\mathcal{A}_h - \mathcal{A}\|}{\alpha}\gamma^\nu\rho \leq \frac{c_1 h}{\alpha}\gamma\rho + \frac{c_1 h + c_3(n+1)\mu^{-rn/d}}{(1-c_0)\alpha}\gamma\rho. \tag{2.14}$$

It follows from (2.12) and (2.14) that

$$\|\widetilde{g}_{h,n}^{\alpha} - g_h^\alpha\| \leq \frac{c_r\mu^{-rn/d}}{(1-c_0)\alpha}\gamma\rho + \frac{c_1 h}{\alpha}\gamma\rho + \frac{c_1 h + c_3(n+1)\mu^{-rn/d}}{(1-c_0)\alpha}\gamma\rho. \tag{2.15}$$

On the other hand,

$$g_h^\alpha - g_T = (\alpha I + \mathcal{A}_h)^{-1}(\mathcal{A} - \mathcal{A}_h)(\alpha I + \mathcal{A})^{-1}\mathcal{A}g_T + [(\alpha I + \mathcal{A})^{-1}\mathcal{A} - I]g_T. \tag{2.16}$$

Using (1.3), (2.1) and (2.14), we have

$$\|g_h^\alpha - g_T\| \le \frac{c_1 h}{\alpha}\gamma\rho + \rho\alpha^\nu. \tag{2.17}$$

Combing estimates (2.11), (2.15) and (2.17), the desired estimate is derived.

3. Regularization Parameter Choice Strategies

In this section, we consider an a posterior parameter choice strategy for choosing the regularization parameter, which ensures the optimal convergence for the fast galerkin method. For more previous work on this kinds of discrepancy principle, reader can refer to [4–7] and the references therein. We follow a method investigated by Tautenhahn [8]. The discrepancy principle under our consideration is

$$d_{h,n}^\delta(\alpha) := \|\alpha(\alpha I + \widetilde{\mathcal{A}}_{h,n})^{-1}(\widetilde{\mathcal{A}}_{h,n}\widetilde{g}_{h,n}^{\alpha,\delta} - P_n u_T^\delta)\| = c'\delta + d'h, \tag{3.1}$$

where $c' := c_4^2 + 2$, $d' := c_1(c_4 + c_4^2)c_{\gamma,\rho} + 1$.

We will prove that there are two points in $(0, +\infty)$ at which the values of $d_{h,n}^\delta(\alpha)$ are nonpositive and nonnegatively. To this end, we define

$$\mathcal{R}(\alpha) := \alpha(\alpha I + \mathcal{A})^{-1},\ \mathcal{R}_{h,n}(\alpha) := \alpha(\alpha I + \widetilde{\mathcal{A}}_{h,n})^{-1}, \tag{3.2}$$

$$\Delta(\alpha) := \mathcal{R}^2(\alpha)u_T,\ \Delta_{h,n}^\delta(\alpha) := \mathcal{R}_{h,n}^2(\alpha)P_n u_T^\delta, \tag{3.3}$$

and

$$D(\delta, h) := c_7\delta + c_8 h, \tag{3.4}$$

where $c_7 := c_4^2 + 1$, $c_8 := c_1(c_4 + c_4^2)c_{\gamma,\rho}$.

Lemma 3.1. *Let g_T satisfy condition (1.3). If the hypothesis (1.2), (H_1) and (H_2) hold, the integer n is chosen to satisfy inequality (3.5),*

$$(n+2)\mu^{-rn/d} \le \min\left\{\frac{c_0}{c_3}\alpha, \frac{1}{d_*}\delta\right\}, \tag{3.5}$$

where $d_* := \max\{c_r c_4^2 c_{\gamma,\rho}, c_3(c_4 + c_4^2)c_{\gamma,\rho}\}$, then

$$\|\Delta_{h,n}^\delta(\alpha) - \Delta(\alpha)\| \le D(\delta, h), \tag{3.6}$$

and

$$\|\Delta(\alpha)\| \leq \rho\alpha^{1+\nu}. \tag{3.7}$$

Proof. It is easy to see

$$\begin{aligned}\Delta^{\delta}_{h,n}(\alpha)-\Delta(\alpha) &= \mathcal{R}^2_{h,n}(\alpha)P_n(u^{\delta}_T - u_T) + \mathcal{R}^2_{h,n}(\alpha)(P_n - I)u_T \\ &\quad +\mathcal{R}^2_{h,n}(\alpha)u_T - \mathcal{R}^2(\alpha)u_T.\end{aligned}$$

It follows from (3.2) that

$$\begin{aligned}J :&= \mathcal{R}^2_{h,n}(\alpha)u_T - \mathcal{R}^2(\alpha)u_T \\ &= \alpha^2(\alpha I + \widetilde{\mathcal{A}}_{h,n})^{-1}(\mathcal{A} - \widetilde{\mathcal{A}}_{h,n})(\alpha I + \mathcal{A})^{-2}u_T \\ &\quad +\alpha^2(\alpha I + \widetilde{\mathcal{A}}_{h,n})^{-2}(\mathcal{A} - \widetilde{\mathcal{A}}_{h,n})(\alpha I + \mathcal{A})^{-1}u_T.\end{aligned}$$

Thus, by (1.2), (1.3), (2.1), (2.2), (2.3) and (2.8), we get

$$\|J\| \leq (c_4 + c_4^2)(c_1 h + c_3(n+1)\mu^{-rn/d})c_{\gamma,\rho}.$$

Combining these inequalities we conclude that

$$\begin{aligned}\|\Delta^{\delta}_{h,n}(\alpha) - \Delta(\alpha)\| &\leq c_4^2\delta + c_r c_4^2 c_{\gamma,\rho}\mu^{-rn/d} \\ &\quad +(c_4 + c_4^2)(c_1 h + c_3(n+1)\mu^{-rn/d})c_{\gamma,\rho} \\ &\leq D(\delta, h).\end{aligned}$$

The inequalities (3.7) is easy to verify.

In the next lemma we show that there is an $\alpha \in (0, +\infty)$ such that

$$d^{\delta}_{h,n}(\alpha) \geq c'\delta + d'h,$$

where $c' := c_7 + 1$, $d' := c_8 + 1$. To this end, we require the following additional condition.

(H$_3$) $\|u^{\delta}_T\| \geq \max\{(1-c_0)^2((c_7+1)\delta+(c_8+1))h,\ 4[2(c_7+1)\delta+(2c_8+1)h]\}$.

Lemma 3.2. *Assume that hypothesis $(H_1), (H_2), (H_3)$ hold. If $\alpha_1 := \gamma$, then*

$$d^{\delta}_{h,n}(\alpha_1) \geq c'\delta + d'h. \tag{3.8}$$

Proof. From (3.3) and (3.6), it follows that

$$\begin{aligned}d^{\delta}_{h,n}(\alpha) &\geq \|\Delta(\alpha)\| - \|\|\Delta^{\delta}_{h,n}(\alpha) - \Delta(\alpha)\| \\ &\geq \|\mathcal{R}^2(\alpha)u^{\delta}_T\| - \|\mathcal{R}^2(\alpha)(u_T - u^{\delta}_T)\| - D(\delta, h) \\ &\geq \|\mathcal{R}^2(\alpha)u^{\delta}_T\| - \delta - D(\delta, h).\end{aligned}$$

Suppose $\{E_\lambda\}$ is the spectral family generated by the operator $\mathcal{A}$, then we have

$$\|\mathcal{R}^2(\alpha)u_T^\delta\|^2 = \int_0^{\|\mathcal{A}\|} (\frac{\alpha}{\alpha+\lambda})^4 d(E_\lambda u_T^\delta, u_T^\delta)$$
$$\geq (\frac{\alpha}{\alpha+\gamma})^4 \|u_T^\delta\|^2.$$

Consequently

$$d_{h,n}^\delta(\alpha_1) \geq \frac{1}{4}\|u_T^\delta\| - \delta - D(\delta, h). \tag{3.9}$$

Combing estimates (3.9) and (H_3) yields the desired estimate.

The following theorem shows the existence of the solution of equation (3.1). To do this, we set $\alpha_0 := \min\{\frac{1}{\rho}(\delta+h),\ 1\}$.

Lemma 3.3. *Assume that the hypothesis* $(H_1), (H_2), (H_3)$ *hold. Then the equation* (3.1) *has a solution* $\alpha \in [\alpha_0, \alpha_1]$.

Proof. By the definition of $d_{h,n}^\delta(\alpha)$,

$$d_{h,n}^\delta(\alpha) \leq \|\Delta(\alpha)\| + \|\Delta_{h,n}^\delta(\alpha) - \Delta(\alpha)\| \leq \rho\alpha^{1+\nu} + c_7\delta + c_8 h, \tag{3.10}$$

if $\alpha_0 := \min\{\frac{1}{\rho}(\delta+h),\ 1\}$, then

$$d_{h,n}^\delta(\alpha_0) \leq \delta + h + c_7\delta + c_8 h = c'\delta + d'h.$$

By the Lemma 3.2, we have

$$d_{h,n}^\delta(\alpha_1) \geq c'\delta + d'h.$$

Since $d_{h,n}^\delta(\alpha)$ is a continuous function on the interval $[\alpha_0, \alpha_1]$, hence, by intermediate mean value theorem, the equation (3.1) has a solution $\alpha \in [\alpha_0, \alpha_1]$.

4. Convergence Rate Analysis

In this section, we establish the optimal convergence rate for the approximation solution stabilized by the Lavrentiev regularization and obtained by the multiscale Galerkin method with the a posteriori parameter choice strategy given in the previous section.

Theorem 4.1. *Assume that hypothesis* $(H_1), (H_2), (H_3)$ *hold, and* n *is chosen according* (3.5). *Let* α_* *be the solution of equation* (3.1). *Then the following estimate holds*

$$\|\widetilde{g}_{h,n}^{\alpha_*,\delta} - g_T\| \leq c_{13}\delta^{\frac{\nu}{1+\nu}} + c_{14}h^{\frac{\nu}{1+\nu}}, \tag{4.1}$$

where $c_{13} := c_9 + c_{12}$, $c_{14} := c_{10} + c_{12}$.

Proof. It follows from (3.10) that

$$d_{h,n}^{\delta}(\alpha) \leq \rho\alpha^{1+\nu} + c_7\delta + c_8 h,$$

thus

$$\rho\alpha_*^{1+\nu} \geq (c' - c_7)\delta + (d' - c_8)h = \delta + h,$$

which gives

$$\alpha_* \geq \left(\frac{1}{\rho}\right)^{\frac{1}{1+\nu}} (\delta + h)^{\frac{1}{1+\nu}}.$$

Consequently,

$$\|\widetilde{g}_{h,n}^{\alpha_*,\delta} - g^{\alpha_*}\| \leq \frac{c_4\delta + c_5 h}{\alpha_*} + \frac{c_6(n+2)\mu^{-rn}}{\alpha_*} \leq c_9\delta^{\frac{\nu}{1+\nu}} + c_{10}h^{\frac{\nu}{1+\nu}}, \tag{4.2}$$

where $c_9 := (c_4 + \frac{c_6}{d_*})\rho^{\frac{1}{1+\nu}}$, $c_{10} := c_5\rho^{\frac{1}{1+\nu}}$.

In the next part we estimate the regularization error $\|g^{\alpha_*} - g_T\|$. We use moment inequality to get

$$\begin{aligned} \|g^{\alpha_*} - g_T\| &= \|(\mathcal{R}(\alpha_*)\mathcal{A})^{\nu}\mathcal{R}^{1-\nu}(\alpha_*)\omega\| \\ &\leq 2\|\mathcal{R}^2(\alpha_*)\mathcal{A}g_T\|^{\frac{\nu}{\nu+1}}\|\mathcal{R}^{1-\nu}(\alpha_*)\omega\|^{\frac{1}{1+\nu}} \\ &\leq 2\|\Delta(\alpha_*)\|^{\frac{\nu}{\nu+1}}\rho^{\frac{1}{1+\nu}}. \end{aligned} \tag{4.3}$$

By using (3.1) and (3.6), we obtain

$$\begin{aligned} &\|\Delta_{h,n}^{\delta}(\alpha_*)\| + \|\Delta_{h,n}^{\delta}(\alpha_*) - \Delta(\alpha_*)\| \\ &\leq (2c_7 + 1)\delta + (2c_8 + 1)h \leq c_{11}(\delta + h), \end{aligned} \tag{4.4}$$

where $c_{11} := \max\{2c_7 + 1, 2c_8 + 1\}$. Now from (4.3) and (4.4), it follows

$$\|g^{\alpha_*} - g_T\| \leq c_{12}(\delta^{\frac{\nu}{1+\nu}} + h_m^{\frac{\nu}{1+\nu}}), \tag{4.5}$$

in which $c_{12} := 2c_{11}^{\frac{\nu}{1+\nu}} 2^{\frac{\nu}{1+\nu}} \rho^{\frac{1}{1+\nu}}$.

Now it follows from (4.2) and (4.5) that

$$\|\widetilde{g}_{h,n}^{\alpha_*,\delta} - g_T\| \leq c_{13}\delta^{\frac{\nu}{1+\nu}} + c_{14}h^{\frac{\nu}{1+\nu}},$$

where $c_{13} := c_9 + c_{12}$, $c_{14} := c_{10} + c_{12}$, which with the above estimates of $\|\widetilde{g}_{h,n}^{\alpha_*,\delta} - g_T\|$ leads to the conclusion of this theorem.

References

[1] S. V. Pereverzev and S. G. Solodky, On one approach to the discretization of the Lavrentiev method, *Ukrainian Mathematical journal* 48 (1996), 239-247.

[2] H. W. Engl, Martin Hanke and Andreas Neubauer, *Regularization of inverse problems,* Kluwer Academic Publishers, Dordrecht, 2000.

[3] Z. Chen, S. Chenga, G. Nelakantib and H. Yang, A fast multiscale Galerkin method for the first kind ill-posed integral equations via Tikhonov regularization, *Inter. J. Comput. Math.* 27 (2008), 1029-0265.

[4] Z. Chen , Y. Jiang, L. Song and H. Yang, A parameter choice strategy for a multilevel augmentation method solving ill-posed operator equations, *J. Integral Equations Appl.* 20 (2008), 569-590.

[5] Xingjun Luo and Fanchun Li, An optimal regularized projection method for solving ill-posed problems via dynamical systems method, *J. Math. Anal. Appl.* 370 (2010), 379-391.

[6] M. P. Rajan, A posteriori parameter choice strategy and an efficient discretization scheme for solving ill-posed problems, *Applied Mathematics and Computation* 204 (2008), 891-904.

[7] P. Maass, S. V. Pereverzev, R. Ramlau and S.G. Solodky, An adaptive discretization for Tikhonov Phillips regularization with a posteriori parameter selection, *Numer. Math.* 87 (2001), 485-502.

[8] U. Tautenhahn, On the method of Lavrentiev regularization for nonlinear ill-posed problems, *Inverse Problems*, 18 (2002), 191-207.

APPROXIMATE METHODS FOR AN INVERSE BOUNDARY VALUE PROBLEM FOR ELLIPTIC COMPLEX EQUATIONS[1]

ZUO-LIANG XU, RAN SUN and LI YAN

School of Information, Renmin University of China, Beijing 100872, China

E-mail: xuzl@ruc.edu.cn, ranran011@sina.com

In this paper, approximate solutions of an inverse boundary value problem in gas dynamics are discussed. We mainly deal with a two dimensional compressible subsonic gas flow over an uneven bottom, which can be transformed into a mixed boundary value problem for nonlinear elliptic complex equations by means of the theory of generalized analytic functions and complex boundary value problems. By using Newton imbedding method, approximate solutions of the mixed boundary value problem are obtained. And under suitable conditions, we could also give error estimates of the approximate solutions.

Keywords: Inverse boundary value problem, elliptic complex equations, approximate solutions.

AMS No: 35J65, 35J55.

1. Introduction

Inverse boundary value problems or free boundary problems are very important in many scientific areas. Among these, inverse boundary value problems for elliptic complex equations in mechanics have recently received a great deal attention from many researchers. In [1] and [2], V. N. Monakhov studied some inverse boundary value problems for elliptic systems of equations by using function theoretic methods. In [3] and [4], R. P. Gilbert and G. C. Wen et al. discussed the solvability of some free boundary problems occurring in continuum mechanics. Z. L. Xu and G. Q. Zhang considered an inverse boundary value problem for cavity flows in gas dynamics ([5]), and proved the existence and uniqueness of the problem. It was seen that the inverse boundary value problem in planar fluid dynamics and gas dynamics might be transformed into a mixed boundary value problems for linear and nonlinear elliptic complex equations. In this paper, we investigate the approximate methods for the inverse boundary value problem for elliptic complex equations in gas dynamics.

Let us consider a subsonic gas flow over an uneven bottom. It is assumed that the subsonic gas flow is steady irrotational and compressible, and the problem is shown in Figure 1, where D_z is the flow region, the bottom of the curve AB(Γ_0),$\widetilde{R_1CR_2}(\Gamma_1)$ and the coordinates of the points R_1, R_2 are

[1]This research is supported by the NSFC(No.10971224, No.11010301015)

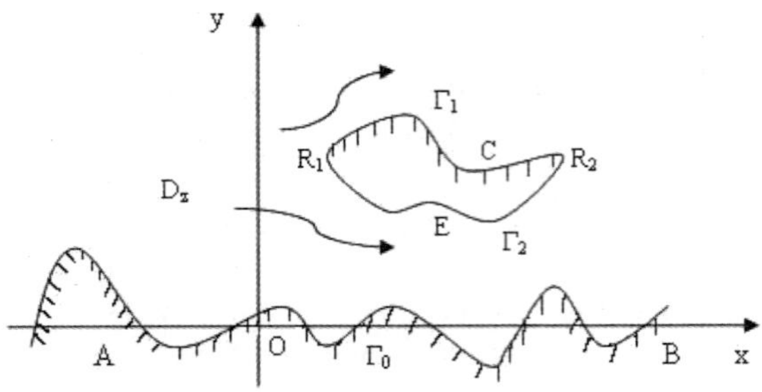

Figure 1: A steady irrotational and compressible subsonic gas flow

given, $\widetilde{R_1ER_2}$ (Γ_2) is an unknown arc, which is a free boundary, and the speed distribution $q = q(x)$ of the flow is given on the unknown boundary Γ_2, the flow is also supposed to be uniform far upstream.

Let ρ be the density of the fluid, u and v the projections of the velocity vector on x- and y-axis, the conditions of irrotational flow and continuity become

$$\begin{cases} \dfrac{\partial v}{\partial x} - \dfrac{\partial u}{\partial y} = 0, \\ \dfrac{\partial(\rho u)}{\partial x} + \dfrac{\partial(\rho v)}{\partial y} = 0. \end{cases} \tag{1.1}$$

We introduce a potential $\varphi(x,y)$ and a stream function $\psi(x,y)$ by setting $\varphi_x = u, \varphi_y = v$ and $\psi_x = -\rho v, \psi_y = \rho u$, then the following equations are obtained

$$\rho\frac{\partial\varphi}{\partial x} = \frac{\partial\psi}{\partial y}, \quad \rho\frac{\partial\varphi}{\partial y} = -\frac{\partial\psi}{\partial x}. \tag{1.2}$$

Since the flow is compressible, let $\rho = \rho(q)$, $q = (u^2 + v^2)^{\frac{1}{2}}$.

Obviously, $\Gamma_0, \Gamma_1, \Gamma_2$ are streamlines, we denote Γ_0 by $y = y_0(x)$, Γ_1 by $y = y_1(x)$, then the boundary condition can be written as follows

$$\begin{cases} \Gamma_0 : y = y_0(x), \quad \psi = \psi_0, \\ \Gamma_1 : y = y_1(x), \quad \psi = \psi_1, \\ \Gamma_2 : q = q(x), \quad \psi = \psi_1. \end{cases} \tag{1.3}$$

Without loss of generality, we may assume that $\psi_0 = 0$. Our Problem is to find a continuous solution of (1.2) in D_z satisfying the boundary condition (1.3), and determine the unknown boundary Γ_2. This problem

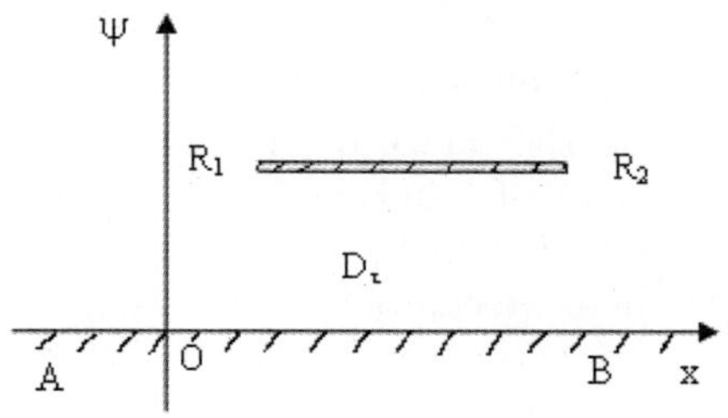

Figure 2: The domain D_τ corresponding to the flow region D_z

will be called Problem A which is an inverse boundary value problem, and is also called a free boundary problem.

2. Problem Transformation

We introduce a complex variable $\tau = x + \mathrm{i}\psi$, noting that the boundary condition (1.3), it is obvious to see that the domain D_τ in τ-plane corresponding to the flow region D_z is a known doubly connected domain (see Figure 2).

Letting $\theta = \tan^{-1} v/u$, the ficticious speed introduced as

$$q^* := \int_1^q \frac{\sqrt{1 - M^2(q)}}{q} dq,$$

where $M(q)$ is the Mach number. In terms of Joukowsky function $w = q^* - \mathrm{i}\theta$, we can obtain the nonlinear Beltrami type equation

$$w_{\overline{\tau}} = \tilde{\mu}(w) w_\tau. \tag{2.1}$$

We note the right side of (2.1) as $G(\tau, w, w_\tau)$, then, (2.1) can be written as follows

$$w_{\overline{\tau}} = G(\tau, w, w_\tau), \ \tau \in D_\tau, \tag{2.2}$$

and the uniformly elliptic condition can be written as

$$|G(\tau, w, V_1) - G(\tau, w, V_2)| \leq q_0 |V_1 - V_2|,$$

where, $\tau \in D_\tau$, V_1, V_2, $w \in \mathbf{C}$, q_0 $(0 \leq q < 1)$ is a constant.

If the flow is constantly subsonic $(M(q) \leq m_0 < 1)$ and q^*, θ satisfies the following condition in the flow domain

$$|q^*| \leq N < \infty, \quad |\theta| \leq \frac{\pi}{2} - \delta, \quad 0 < \delta < \frac{\pi}{2}, \tag{2.3}$$

then, (2.1) is uniformly elliptic, i.e. $|\tilde{\mu}(w)| \leq \mu_0 < 1$, where N, δ, μ_0, m_0 are constants (see [1]).

Assume that $\tau = \tau(\zeta)$ conformally map the annulus $D_\zeta = \{\zeta\colon \ell < |\zeta| < 1\}$ onto domain D_τ, and the image of unit circle $\Gamma_0^{'} : |\zeta| = 1$ is Γ_0, the image of the upper half-circle Γ_1^* and the lower half-circle Γ_1^{**} of $\Gamma_1^{'}\colon |\zeta| = \ell$ is Γ_1 and Γ_2, respectively. The image of the points ζ_1, ζ_2 on $\Gamma_1^{'} : |\zeta| = \ell$ is R_1, R_2, respectively. Then, we can obtain the following complex equation:

$$w_{\bar{\zeta}} = F(\zeta, w, w_\zeta), \tag{2.4}$$

where $F = \overline{\tau'(\zeta)}G(\tau(\zeta), w, w_\zeta/\tau'(\zeta))$, and (2.4) is also uniformly elliptic.

Next, we consider the boundary condition. On the unknown boundary Γ_2, we have $q = q(x)$, thus,

$$q^* = q^*(\mathrm{Re}\tau(\zeta))\colon = f(\zeta),\ \zeta \in \Gamma_1^{**}\colon |\zeta| = \ell,\ \mathrm{Im}(\zeta) < 0. \tag{2.5}$$

On the curve Γ_1, since the equation of the boundary is $y = y_1(x)$, then $\theta = \arctan y_1'(x)$, i.e.

$$\theta = \arctan y_1'[\mathrm{Re}(\tau(\zeta))]\colon = g_1(\zeta),\ \zeta \in \Gamma_1^*\colon |\zeta| = \ell,\ \mathrm{Im}(\zeta) > 0. \tag{2.6}$$

On $\Gamma_0^{'}$, it is obvious that $\theta = \arctan y_0'(x)$, this condition can be written as

$$\theta = \arctan y_0'[\mathrm{Re}(\tau(\zeta))]\colon = g_0(\zeta),\ \ \zeta \in \Gamma_0^{'}. \tag{2.7}$$

Thus, we come to the mixed boundary value problem for (2.4), i.e. to find a continuous solution $w(\zeta)$ of the complex equation (2.4) in D_ζ with the uniform ellipticity condition

$$|F(\zeta, w, U_1) - F(\zeta, w, U_2)| \leq q_0|U_1 - U_2|, \tag{2.8}$$

for any $\zeta \in D_\zeta$ and $w, U_1, U_2 \in \mathbf{C}$, and the boundary condition

$$\mathrm{Re}[\overline{\lambda(\zeta)}w(\zeta)] = r(\zeta), \tag{2.9}$$

where

$$\lambda(\zeta) = \begin{cases} 1, & \zeta \in \Gamma_1^{**}, \\ \mathrm{i}, & \zeta \in \Gamma_0^{'} \cup \Gamma_1^*, \end{cases} \tag{2.10}$$

$$r(\zeta) = \begin{cases} f(\zeta), & \zeta \in \Gamma_1^{**} = \widetilde{\zeta_2\zeta_1}, \\ g_1(\zeta), & \zeta \in \Gamma_1^* = \widetilde{\zeta_1\zeta_2}, \\ g_0(\zeta), & \zeta \in \Gamma_0^{'}. \end{cases} \tag{2.11}$$

This problem will be simply called Problem M. Therefore, Problem A is transformed into Problem M for the nonlinear elliptic complex equation (2.4).

In the following section, we will prove Problem M has bounded continuous solution $w(\zeta)$ in D_ζ. Furthermore, if $F(\zeta, w, w_\zeta)$ satisfies the following condition

$$|F(\zeta, w_1, V) - F(\zeta, w_2, V)| \leq R(\zeta, w_1, w_2, V)|w_1 - w_2|, \tag{2.12}$$

where $\zeta \in D_\zeta, w_1, w_2, V \in \mathbf{C}$, $R(\zeta, w_1, w_2, V) \in L_p(\overline{D_\zeta})$, $p > 2$, thus, the solution is unique (see [5,6]). Assume that $w(\zeta)$ is the solution of Problem M of complex equation (2.4), using $\zeta(\tau)$ represents the inverse function of $\tau(\zeta)$, thus, $w(\tau) = w(\zeta(\tau))$ is the solution of complex equation (2.1). Next, by using the Newton imbedding method we shill prove the existence of the solution for Problem M and discuss the error estimates of the approximate solutions.

3. The Approximate Solution of the Elliptic Problem

In order to seek the approximate solution of Problem M, we consider the complex equation with a parameter $t \in [0, 1]$:

$$w_{\overline{\zeta}} - tF(\zeta, w, w_\zeta) = B(\zeta),\ B(\zeta) = (1-t)F(\zeta, 0, 0), \tag{3.1}$$

and first suppose that $F(\zeta, w, w_\zeta) = 0$ in the $\varepsilon_m = 1/m$ (m is a positive integer) neighborhood U_m of points $\zeta_j\,(j = 1, 2)$, it sufficies to multiply $F(\zeta, w, w_\zeta)$ by the function

$$\eta_m(\zeta) = \begin{cases} 0, & \zeta \in U_m = \cup_{j=1}^2\{|\zeta - \zeta_j| < 1/m\}, \\ 1, & \zeta \in D_m = \overline{D_\zeta}\backslash U_m. \end{cases} \tag{3.2}$$

we know that when $t = 0$, Problem M for the complex equation (3.1) has a unique bounded solution $w(\zeta)$, which possesses the form

$$w(\zeta) = \Phi(\zeta) + \Psi(\zeta), \Psi(\zeta) = TB = -\frac{1}{\pi}\int_{D_\zeta}\frac{B(\tau)}{\tau - \zeta}d\sigma_\tau, \tag{3.2}$$

where $\Phi(\zeta)$ is a bounded analytic function in D_ζ satisfying the boundary condition

$$\mathrm{Re}[\overline{\lambda(\zeta)}\Phi(\zeta)] = r(\zeta) - \mathrm{Re}[\overline{\lambda(\zeta)}\Psi(\zeta)],\ \zeta \in \Gamma^* = \Gamma_1' \cup \Gamma_0' \backslash\{\zeta_1, \zeta_2\}. \tag{3.3}$$

Introduce a function $g(\zeta) = \prod_{j=1}^2 (\zeta - \zeta_j)^\eta$, it can be derived $w_*(\zeta) = g(\zeta)w(\zeta) \in C_\beta(\overline{D_\zeta}) \cap W_{p_0}^1(D_\zeta)$, herein $p_0\,(2 < p_0 < p)$, $\beta = 1 - \frac{2}{p_0}$, $0 < \eta < 1$.

Suppose that when $t = t_0\,(0 < t_0 < 1)$, Problem M for (2.4) has a bounded solution, we shall prove that there exists a neighborhood of t_0 : $E = \{|t - t_0| \le \delta, 0 \le t \le 1, \delta > 0\}$, such that for every $t \in E$ and any function $B(\zeta) \in L_{p_0}(\overline{D_\zeta})$, Problem M for (3.1) is solvable. Practically, the complex equation (3.1) can be written in the form

$$w_{\overline{\zeta}} - t_0 F(\zeta, w, w_\zeta) = (t - t_0)F(\zeta, w, w_\zeta) + B(\zeta). \tag{3.4}$$

We are free to choose a function $w_0(\zeta)$ so that $g(\zeta)w_0(\zeta) \in C_\beta(\overline{D_\zeta}) \cap W^1_{p_0}(D_\zeta)$, in particular $w_0(\zeta) = 0$ on $\overline{D_\zeta}$. Let $w_0(\zeta)$ in the right hand side of (3.4). From the condition mentioned in the Introduction, it is obvious that

$$B_0(\zeta) = (t - t_0)F(\zeta, w_0, w_0\zeta) + B(\zeta) \in L_{p_0}(\overline{D_m}). \tag{3.5}$$

Noting the assumption as before, Problem M for the complex equation (3.4) has a solution $w_1(\zeta)$, so that $g(\zeta)w_1(\zeta) \in C_\beta(\overline{D_\zeta}) \cap W^1_{p_0}(D_\zeta)$. By using the successive iteration, we can find out a sequence of functions: $w_n(\zeta)$, $g(\zeta)w_n(\zeta) \in C_\beta(\overline{D_\zeta}) \cap W^1_{p_0}(D_\zeta)$, which satisfies the complex equations

$$w_{n+1\overline{\zeta}} - t_0 F(\zeta, w_{n+1}, w_{n+1\zeta}) = (t - t_0)F(\zeta, w_n, w_{n\zeta}) + B(\zeta),\ n = 1, 2, \cdots. \tag{3.6}$$

The difference of the above equations for $n + 1$ and n is as follows

$$\begin{aligned}&(w_{n+1} - w_n)_{\overline{\zeta}} - t_0[F(\zeta, w_{n+1}, w_{n+1\zeta}) - F(\zeta, w_n, w_{n\zeta})]\\ &= (t - t_0)[F(\zeta, w_n, w_{n\zeta}) - F(\zeta, w_{n-1}, w_{n-1\zeta})],\ n = 1, 2, \cdots.\end{aligned} \tag{3.7}$$

From the condition (2.4) and (2.12), it can be seen that

$$\begin{aligned}&F(\zeta, w_{n+1}, w_{n+1\zeta}) - F(\zeta, w_n, w_{n\zeta}) = [F(\zeta, w_{n+1}, w_{n+1\zeta})\\ &- F(\zeta, w_{n+1}, w_{n\zeta})] + [F(\zeta, w_{n+1}, w_{n\zeta}) - F(\zeta, w_n, w_{n\zeta})]\\ &= \widetilde{Q}_{n+1}(\zeta)(w_{n+1} - w_n)_\zeta + \widetilde{B}_{n+1}(\zeta)(w_{n+1} - w_n),\\ &|\widetilde{Q}_{n+1}(\zeta)| \le Q_0 < 1,\ L_{p_0}[\widetilde{B}_{n+1}(\zeta), D_\zeta] \le K_0, n = 1, 2, \cdots,\end{aligned} \tag{3.8}$$

K_0 is a constant, and

$$\begin{aligned}&L_{p_0}[F(\zeta, w_n, w_{n\zeta}) - F(\zeta, w_{n-1}, w_{n-1\zeta}), D_m]\\ &\le Q_0 L_{p_0}[(w_n - w_{n-1})_\zeta, D_m] + K_0 C[w_n - w_{n-1}, D_m]\\ &\le (Q_0 + K_0)\{C_\beta[g(w_n - w_{n-1}), \overline{D_\zeta}] + L_{p_0}[|(g(w_n - w_{n-1}))_{\overline{\zeta}}|\\ &+ |(g(w_n - w_{n-1}))_\zeta|, D_\zeta]\} = (Q_0 + K_0)S_n,\end{aligned} \tag{3.9}$$

in which $S_n = C_\beta[g(w_n - w_{n-1}), \overline{D_\zeta}] + L_{p_0}[|(g(w_n - w_{n-1}))_{\overline{\zeta}}| + |(g(w_n - w_{n-1}))_\zeta|, D_\zeta]$. Moreover, $w_{n+1}(\zeta) - w_n(\zeta)$ satisfies the homogeneous boundary condition

$$\text{Re}[\overline{\lambda(\zeta)}(w_{n+1}(\zeta) - w_n(\zeta))] = 0,\ \zeta \in \Gamma^*. \tag{3.10}$$

On the basis of [6], we have

$$\begin{aligned} S_{n+1} = C_\beta[g(w_n - w_{n-1}), \overline{D_\zeta}] + L_{p_0}[|(g(w_n - w_{n-1}))_{\overline{\zeta}}| \\ + |(g(w_n - w_{n-1}))_\zeta|, \overline{D_\zeta}] \le M_1|t - t_0|(Q_0 + K_0)S_n, \end{aligned} \tag{3.11}$$

where M_1 is a positive integer. Provided that $\delta\,(>0)$ is small enough, so that $\eta = \delta M_1(Q_0 + K_0) < 1$, we can obtain that

$$S_{n+1} \le \eta^n S_1 = \eta^n [C_\beta[gw_1, \overline{D_\zeta}] + L_{p_0}[|(gw_1)_{\overline{\zeta}}| + |(gw_1)_\zeta|, \overline{D_\zeta}] \tag{3.12}$$

for every $t \in E$. Thus

$$\begin{aligned} S_{nl} = C_\beta[g(w_n - w_l), \overline{D_\zeta}] + L_{p_0}[|(g(w_n - w_l))_{\overline{\zeta}}| + |(g(w_n - w_l))_\zeta|, \overline{D_\zeta}] \\ \le S_n + S_{n-1} + \cdots + S_{l+1} \le (\eta^{n-1} + \eta^{n-2} + \cdots + \eta^l)S_1 \\ = \eta^l(1 + \eta + \cdots + \eta^{n-l-1})S_1 \le \eta^{N+1}\frac{1-\eta^{n-l}}{1-\eta}S_1 \le \frac{\eta^{N+1}}{1-\eta}S_1 \end{aligned} \tag{3.13}$$

for $n \ge l > N$, where N is a positive integer. This shows that $S(w_n - w_m) \to 0$ as $n, l \to \infty$. Following the completeness of the Banach space $B = C_\beta(\overline{D}) \cap W^1_{p_0}(D_m)$, there is a function $w_*(\zeta)g(\zeta) \in B$, such that when $n \to \infty$,

$$\begin{aligned} S(w - w_*) = C_\beta[g(w_n - w_*), \overline{D_\zeta}] + L_{p_0}[|(g(w_n - w_*))_{\overline{\zeta}}| \\ + |(g(w_n - w_*))_\zeta|, \overline{D_\zeta}] \to 0, \end{aligned}$$

from (3.6) it follows that $w_*(\zeta)$ is a solution of Problem M for (3.4), i.e. (3.1) for $t \in E$. It is easy to see that the positive constant δ is independent of t_0 $(0 \le t_0 < 1)$. Hence from Problem M for the complex equation (3.1) with $t = t_0 = 0$ is solvable, we can derive that when $t = \delta, 2\delta, \ldots, [1/\delta]\delta, 1$, Problem M for (3.1) are solvable, especially Problem M for (3.1) with $t = 1$ and $B(\zeta) = 0$, namely (2.4) has a unique bounded solution.

We summarize the above discussion as

Theorem 3.1. *Let the complex equation* (2.4) *satisfy the condition* (2.3) *and* (2.12). *By using the Newton Imbedding Method, we can derive the approximate solution $w_n(\zeta)$ of Problem M of complex equation* (2.4), *and Problem M has bounded solution $w(\zeta)$, such that $g(\zeta)w(\zeta) \in B = C_\beta(\overline{D_\zeta}) \cap W^1_{p_0}(D_\zeta)$, $\beta = 1 - 2/p_0$, $2 < p_0 \le p$, $g(\zeta) = \Pi^2_{j=1}(\zeta - \zeta_j)^\eta$, $0 < \eta < 1$.*

4. Error Estimates of the Approximate Solution

The function $w^t_n(\zeta) = w_n(\zeta)$ represents the approximate solution of Problem M of the complex equation (3.1), and satisfies the iteration equation (3.6).

Let $w = w(\zeta)$ be a solution of Problem M for the complex equation (2.4). From (2.4) and (3.6), it follows that

$$\begin{aligned}&(w-w^t_{n+1})_{\bar\zeta} = F(\zeta,w,w_\zeta) - t_0F(\zeta,w^t_{n+1},w_{n+1\zeta})\\&-(t-t_0)F(\zeta,w^t_n,w^t_{n\zeta}) - (1-t)F(\zeta,0,0) = (1-t)[F(\zeta,w,w_\zeta)\\&-F(\zeta,0,0)] + t_0[F(\zeta,w,w_\zeta) - F(\zeta,w^t_{n+1},w^t_{n+1\zeta})] + (t-t_0)\\&\times[F(\zeta,w,w_\zeta) - F(\zeta,w^t_n,w^t_{n\zeta})] = t_0[\tilde{Q}(w-w^t_{n+1})_\zeta\\&+\tilde{B}(w-w^t_{n+1})] + (1-t)[F(\zeta,w,w_\zeta)-F(\zeta,0,0)]\\&+(t-t_0)[F(\zeta,w,w_\zeta)-F(\zeta,w^t_n,w^t_{n\zeta})].\end{aligned}\tag{4.1}$$

It is obvious that $w - w^t_n$ satisfies the boundary condition

$$\mathrm{Re}[\overline{\lambda(\zeta)}(w-w^t_{n+1})] = 0,\ \zeta\in\Gamma^*.\tag{4.2}$$

On the basis of the estimate in [5], it can be obtained

$$\begin{aligned}&S(w-w^t_{n+1}) \le M[(1-t)(Q_0+K_0)S(w)\\&+|t-t_0|(Q_0+K_0)S(w-w^t_n)] = M_2^{n+1}|t-t_0|^{n+1}S(w-w^t_0)\\&+M_2(1-t)\frac{(1-M_2^{n+1}|t-t_0|^{n+1})S(w)}{1-M_2|t-t_0|},\end{aligned}\tag{4.3}$$

where $M = M(Q_0,K_0,p_0,D_m)$, $M_2 = M(Q_0+K_0)$, $w^t_0 = w(\zeta,t_0)$ is a solution of Problem M for (3.6) with $t = t_0$ and $B = (1-t_0)F(\zeta,0,0)$. Noting that the function $w(\zeta) - w^t_0(\zeta)$ is a solution of the following boundary value problem:

$$(w-w^t_0)_{\bar\zeta} = t_0[f(\zeta,w,w_\zeta) - f(\zeta,w^t_0,w^t_{0\zeta})] + (1-t_0)f(\zeta,w,w_\zeta),\tag{4.4}$$

$$\mathrm{Re}[\overline{\lambda(\zeta)}(w(\zeta)-w^t_{n+1}(\zeta))] = 0, \zeta\in\Gamma^*,\tag{4.5}$$

in which $f(\zeta,w,w_\zeta) = F(\zeta,w,w_\zeta) - F(\zeta,0,0)$, we can conclude

$$S(w) \le M_3 = M_3(Q_0,K_0,p_0,D_m),\tag{4.6}$$

$$\begin{aligned}S(w-w^t_0) &\le M(1-t_0)L_{p_0}[f(\zeta,w,w_\zeta),\overline{D_m}]\\&\le M(Q_0+K_0)(1-t_0)S(w) \le M_2M_3(1-t_0).\end{aligned}\tag{4.7}$$

From (4.3) (4.6) and (4.7), the estimate

$$\begin{aligned}&S(w-w^t_{n+1}) \le M_2^{n+2}|t-t_0|^{n+1}(1-t_0)M_3\\&+M_2(1-t)M_3\frac{[1-M_2^{n+1}|t-t_0|^{n+1}]}{1-M_2|t-t_0|}\\&=M_2M_3[M_2^{n+1}|t-t_0|^{n+1}|1-t_0|+(1-t)\frac{(1-M_2^{n+1}|t-t_0|^{n+1})}{(1-M_2|t-t_0|)}]\end{aligned}\tag{4.8}$$

is derived. Thus we have the following theorem.

Theorem 4.1. *Under the same conditions as in Theorem* 3.1, *if* $w(\zeta)$ *denotes the solution of Problem M for complex equation* (2.4), *and* $w_n^t(\zeta) = w_n(\zeta, t)$ *represents the approximate solution of Problem M for complex equation* (3.6), *then we have the following error estimates*

$$\begin{aligned} S(w-w_n^t) &= C_\beta[g(w-w_n^t), \overline{D_\zeta}] + L_{p_0}[|(g(w-w_n^t))_{\bar\zeta}| + |(g(w-w_n^t))_\zeta|, \overline{D_\zeta}] \\ &\le M_2 M_3 [M_2^n |t-t_0|^n (1-t_0) + \frac{1-M_2^n|t-t_0|^n}{1-M_2|t-t_0|}(1-t)], \end{aligned} \tag{4.9}$$

where M_2, M_3 *are real constants as started in* (4.3) *and* (4.6).

References

[1] V. N. Monakhov, *Boundary value problem with free boundaries for elliptic systems of equations,* Amer. Math. Soc., Providence, RI, 1983.

[2] V. N. Monakhov, Inverse mixed boundary value problem for several unknown arcs, Dokl. Akad. Nauk SSSR 141 (1961), 292-295.

[3] R. P. Gilbert and G. C. Wen, Free boundary problems occurring in plannar fluid dynamics, *Nonlinear Analysis* 13 (1989), 285-303.

[4] Z. L. Xu and G. Q. Zhang, An inverse boundary value problem for cavity flows in gas dynamics, *Nonlinear Analysis* 50 (2002), 717-725.

[5] G. C. Wen and H. Begehr, *Boundary value problems for elliptic equations and systems,* Longman Scientific and Technical, Harlow, 1990.

[6] G. C. Wen, *Approximate methods and numerical analysis for elliptic complex equations,* Gordon and Breach Science Publishers, Amsterdam, 1999.

AN EXPLICIT METHOD FOR SOLVING FUZZY PARTIAL DIFFERENTIAL EQUATION

A. FARAJZADEH

Department of Mathematics, University of Razi, Kermanshah, Iran

E-mail: faraj1348@yahoo.com

This paper deals with a numerical method for the solution of the fuzzy heat equation. First we express the necessary materials and definitions, then consider a difference scheme for the one dimensional heat equation. In forth section we express the necessary conditions for stability and check stability of our scheme. In final part we give an example for considering numerical results. In this example we obtain the Hausdorff distance between exact solution and approximate solution.

1. Introduction

The topics of numerical methods for solving fuzzy differential equations have been rapidly growing in recent years. The concept of fuzzy derivative was first introduced by Chang and Zadeh in [10]. It was following up by Dubois and Prade in [2], who defined and used the extension principle. Other methods have been discussed by Puri and Relescu in [4] and Goetschel and Voxman in [9]. The initial value problem for first order fuzzy differential equations have been studied by several authors [5–8,11] on the metric space (E^n, D) of normal fuzzy convex sets with the distance D given by the maximum of the Hausdorff distances between exact solution and approximate solution.

2. Materials and Definitions

We begin this section with defining the notation we will use in the paper. Let X be a collection of objects denoted generically by x. Then a fuzzy set $\widetilde{A}$ in X is a set of ordered pairs:

$$\widetilde{A} = \{(x, \mu_{\widetilde{A}}(x))|x \in X\},$$

where $\mu_{\widetilde{A}}$ is called the membership function or grade of membership of x in $\widetilde{A}$. The range of the membership function is a bounded subset of the nonnegative real numbers.

Definition 2.1. The set of elements belonging the fuzzy set $\widetilde{A}$ at least to the degree α is called the α-cut set:

$$A_\alpha = \{x \in X|\mu_{\widetilde{A}}(x) \geq \alpha\},$$

and $A'_\alpha = \{x \in X | \mu_{\widetilde{A}}(x) > \alpha\}$ is called strong α-cut.

Definition 2.2. The triangular fuzzy number $\widetilde{N}$ is defined by three numbers $\alpha < m < \beta$ as follows:

$$\widetilde{A} = (\alpha, m, \beta).$$

This representation is interpreted as membership function (Figure 1):

$$\mu_{\widetilde{A}}(x) = \begin{cases} \dfrac{x-\alpha}{m-\alpha}, & \alpha \leq x < m, \\ 1, & x = m, \\ \dfrac{x-\beta}{m-\beta}, & m < x \leq \beta, \\ 0, & \text{otherwise.} \end{cases}$$

If $\alpha > 0\,(\alpha \geq 0)$, then $\widetilde{A} > 0\,(\widetilde{A} \geq 0)$, and if $\beta < 0$ $(\beta \leq 0)$, then $\widetilde{A} < 0\,(\widetilde{A} \leq 0)$.

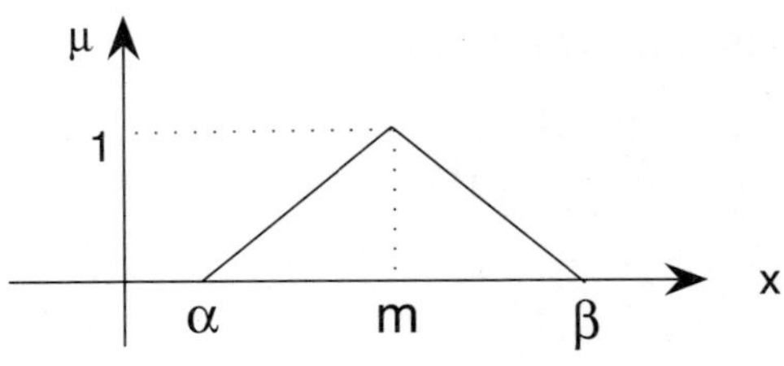

Definition 2.3. An arbitrary fuzzy number is showed by an ordered pair of functions $(\underline{a}(r), \overline{a}(r))$, $0 \leq r \leq 1$, which satisfies the following requirements:

1. $\underline{a}(r)$ is a bounded left semicontinuous non-decreasing function over $[0, 1]$,
2. $\overline{a}(r)$ is a bounded left semicontinuous non-increasing function over $[0, 1]$,
3. $\underline{a}(r) \leq \overline{a}(r)$, $0 \leq r \leq 1$.

In particular, if $\underline{a}$, $\overline{a}$ are linear functions, we have a triangular fuzzy number.

A crisp number a is simply represented by $\underline{a}(r) = \overline{a}(r) = a,\ 0 \le r \le 1$.

Definition 2.4. For arbitrary fuzzy numbers $u = (\underline{u}(r), \overline{u}(r))$ and $v = (\underline{v}(r), \overline{v}(r))$, we have algebraic operations as follows:

1. $ku = \begin{cases} (k\underline{u}, k\overline{u}), & k \ge 0, \\ (k\overline{u}, k\underline{u}), & k < 0, \end{cases}$
2. $u + v = (\underline{u}(r) + \underline{v}(r), \overline{u}(r) + \overline{v}(r))$,
3. $u - v = (\underline{u}(r) - \overline{v}(r), \overline{u}(r) - \underline{v}(r))$,
4. $u \cdot v = (\min s, \max s)$, in which

$$s = \{\underline{u}\ \underline{v}, \underline{u}\overline{v}, \overline{u}\underline{v}, \overline{u}\ \overline{v}\}.$$

Remark. Since the α-cut of fuzzy numbers is always a closed and bounded interval, so we can write $\widetilde{A}_\alpha = [\underline{a}(\alpha), \overline{a}(\alpha)]$ for all α.

Definition 2.5. Assume that $u = (\underline{u}(r), \overline{u}(r)),\ v = (\underline{v}(r), \overline{v}(r))$ are two fuzzy numbers. The Hausdorff metric D_H is defined by

$$D_H(u, v) = \sup_{r \in [0,1]} \max\{|\underline{u}(r) - \underline{v}(r)|, |\overline{u}(r) - \overline{v}(r)|\}. \tag{1}$$

This metric is a bound for error. By it we obtain the difference between exact solution and approximate solution.

3. Finite Difference Method

In this section we solve the fuzzy heat equation by an explicit method. Assume that $\widetilde{U}$ is a fuzzy function of the independent crisp variables x and t. We define

$$I = \{(x, t) | 0 \le x \le 1,\ 0 \le t \le T\}.$$

A α-cut of $\widetilde{U}(x, t)$ and its parametric form, will be

$$\widetilde{U}(x, t)[\alpha] = [\underline{U}(x, t; \alpha), \overline{U}(x, t; \alpha)].$$

We let that the $\underline{U}(x, t; \alpha)$, $\overline{U}(x, t; \alpha)$ have the continuous partial differential quotient, therefore $(D_t - a^2 D_x^2)\, \overline{U}(x, t; \alpha)$, and $(D_t - a^2 D_x^2)\, \underline{U}(x, t; \alpha)$ are continuous for all $(x, t) \in I$, all $\alpha \in [0, 1]$.

Now we consider the heat equation

$$(D_t - a^2 D_x^2)\widetilde{U} = \widetilde{0}, \tag{2}$$

with boundary conditions and initial condition:

$$\begin{cases} \widetilde{U}(0, t) = \widetilde{U}(1, t) = \widetilde{0}, \\ \widetilde{U}(x, 0) = \widetilde{f}(x). \end{cases} \tag{3}$$

We divide the domain $[0,1]\times[0,T]$ into $M\times N$ mesh with spatial step size $h=\frac{1}{N}$ in x-direction and $k=\frac{T}{M}$ in t-direction. The gride points are given by (Figure 2):

$$x_i = ih,\ \ i = 0,1,\cdots,N,$$

$$t_j = jk,\ j = 0,1,\cdots,M.$$

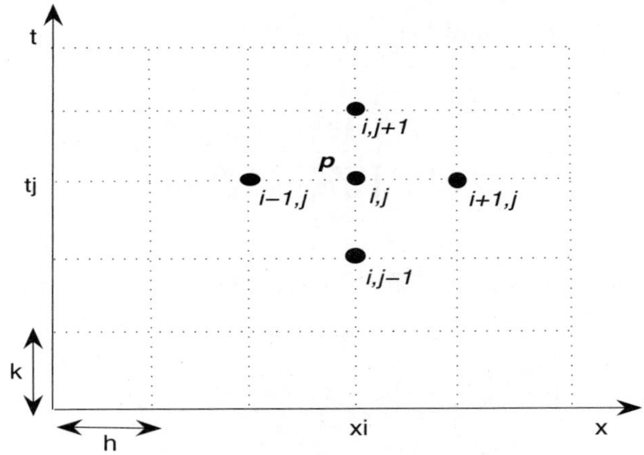

Denote the value of $\widetilde{U}$ at the representative mesh point $p(x_i,t_j)$ by

$$\widetilde{U}_p = \widetilde{U}(x_i,t_j) = \widetilde{U}_{i,j},$$

and also parametric form of fuzzy number $\widetilde{U}_{i,j}$ is

$$\widetilde{U}_{i,j} = (\underline{U}_{i,j},\overline{U}_{i,j}).$$

We have

$$\begin{cases} (D_t)\widetilde{U}_{i,j} &= (\underline{D_tU_{i,j}},\overline{D_tU_{i,j}}),\\ (D_x^2)\widetilde{U}_{i,j} &= (\underline{D_x^2U_{i,j}},\overline{D_x^2U_{i,j}}). \end{cases}$$

Then by Taylor's expansion, we obtain

$$\begin{cases} \underline{D_x^2\widetilde{U}_{i,j}} \simeq \dfrac{\underline{u}_{i-1,j+1} - 2\overline{u}_{i,j+1} + \underline{u}_{i+1,j+1}}{h^2},\\ \overline{D_x^2\widetilde{U}_{i,j}} \simeq \dfrac{\overline{u}_{i-1,j+1} - 2\underline{u}_{i,j+1} + \overline{u}_{i+1,j+1}}{h^2}. \end{cases} \tag{4}$$

And also for $(D_t)\widetilde{U}$ at p, we have

$$\begin{cases} \underline{D_t\widetilde{U}_{i,j}} \simeq \dfrac{\underline{u}_{i,j+1} - \overline{u}_{i,j}}{k}, \\ \overline{D_t\widetilde{U}_{i,j}} \simeq \dfrac{\overline{u}_{i,j+1} - \underline{u}_{i,j}}{k}. \end{cases} \tag{5}$$

Parametric form of heat equation will be

$$\begin{cases} \underline{(D_t)\widetilde{U}} - a^2\overline{(D_x^2)\widetilde{U}} = \widetilde{0}, \\ \overline{(D_t)\widetilde{U}} - a^2\underline{(D_x^2)\widetilde{U}} = \widetilde{0}. \end{cases} \tag{6}$$

By (4) and (5), the difference scheme for heat equation is

$$\begin{cases} \dfrac{\underline{u}_{i,j+1} - \overline{u}_{i,j}}{k} - a^2\dfrac{\overline{u}_{i-1,j+1} - 2\underline{u}_{i,j+1} + \overline{u}_{i+1,j+1}}{h^2} = 0, \\ \dfrac{\overline{u}_{i,j+1} - \underline{u}_{i,j}}{k} - a^2\dfrac{\underline{u}_{i-1,j+1} - 2\overline{u}_{i,j+1} + \underline{u}_{i+1,j+1}}{h^2} = 0. \end{cases} \tag{7}$$

By above equations we obtain

$$\begin{cases} -r\underline{u}_{i-1,j+1} + (1+2r)\overline{u}_{i,j+1} - r\underline{u}_{i+1,j+1} = \underline{u}_{i,j}, \\ -r\overline{u}_{i-1,j+1} + (1+2r)\underline{u}_{i,j+1} - r\overline{u}_{i+1,j+1} = \overline{u}_{i,j}, \end{cases} \tag{8}$$

where

$$r = \frac{ka^2}{h^2}, \tag{9}$$

$\widetilde{U} = (\underline{u}, \overline{u})$ is the exact solution of the approximating difference equations, and x_i $(i = 0, 1, \cdots, N)$ and t_j $(j = 0, 1, \cdots, M)$.

Since the boundary values are know at x_0 and x_N, we have $2(N-1)$ equations with $2(N-1)$ unknown. Therefore equations can be written in matrix as

$$\begin{pmatrix} 0 & -r & & & 2r+1 & & & \\ -r & 0 & -r & & & 2r+1 & & \\ & & \ddots & & & & \ddots & \\ & & -r & 0 & & & & 2r+1 \\ 2r+1 & & & & 0 & -r & & \\ & 2r+1 & & & -r & 0 & -r & \\ & & \ddots & & & \ddots & \ddots & \\ & & & 2r+1 & & & -r & 0 \end{pmatrix}$$

$$\times \begin{pmatrix} \underline{u}_{1,j+1} \\ \underline{u}_{2,j+1} \\ \vdots \\ \underline{u}_{N-1,j+1} \\ \overline{u}_{1,j+1} \\ \overline{u}_{2,j+1} \\ \vdots \\ \overline{u}_{N-1,j+1} \end{pmatrix} = \begin{pmatrix} \underline{u}_{1,j} \\ \underline{u}_{2,j} \\ \vdots \\ \underline{u}_{N-1,j} \\ \overline{u}_{1,j} \\ \overline{u}_{2,j} \\ \vdots \\ \overline{u}_{N-1,j} \end{pmatrix}.$$

Then we have

$$A\begin{pmatrix} \underline{U}_{j+1} \\ \overline{U}_{j+1} \end{pmatrix} = \begin{pmatrix} \underline{U}_{j} \\ \overline{U}_{j} \end{pmatrix} \Rightarrow \begin{pmatrix} \underline{U}_{j+1} \\ \overline{U}_{j+1} \end{pmatrix} = A^{-1}\begin{pmatrix} \underline{U}_{j} \\ \overline{U}_{j} \end{pmatrix}, \tag{10}$$

where

$$A=\begin{pmatrix} E & F \\ F & E \end{pmatrix}, E=\begin{pmatrix} 0 & -r & & & \\ -r & 0 & -r & & \\ & & \ddots & & \\ & & & -r & 0 \end{pmatrix}, F=\begin{pmatrix} 2r+1 & & \\ & \ddots & \\ & & 2r+1 \end{pmatrix}. \tag{11}$$

4. Stability of Fuzzy Heat Equation

Definition 4.1. The largest of the eigenvalues of matrix A, is showed by $\rho(A)$.

Remark 4.1. The necessary and sufficient condition for the difference equations to be stable is $\rho(A) \leq 1$ [3].

Remark 4.2. If A^{-1} is the inverse of matrix A, then $\rho(A^{-1}) = \frac{1}{\rho(A)}$ [1].

Remark 4.3. The eigenvalues of a $N \times N$ tridiagonal matrix

$$\begin{pmatrix} a & b & & & \\ c & a & b & & \\ & \ddots & \ddots & \ddots & \\ & & c & a & b \\ & & & c & a \end{pmatrix}$$

are

$$\lambda_s = a + 2\sqrt{bc}\cos\frac{s\pi}{N+1}, \quad s = 1, \cdots, N. \ [3]$$

Theorem 4.1. *Let matrix A have a spacial structure as follows*

$$A = \begin{pmatrix} E & F \\ F & E \end{pmatrix}.$$

Then the eigenvalues of A are union of eigenvalues of $E+F$ and eigenvalues of $E-F$. [12]

Now, by using Theorem 4.1, we want to prove stability of our difference scheme. For this, it is sufficient to show in (11) $\rho(A^{-1}) < 1$. Thus by theorem (4.1), we find eigenvalues of $E+F$ and $E-F$, namely

$$E + F = \begin{pmatrix} 1+2r & -r & & \\ -r & 1+2r & -r & \\ & \ddots & \ddots & \\ & & -r & 1+2r \end{pmatrix},$$

and

$$E - F = \begin{pmatrix} -1-2r & -r & & \\ -r & -1-2r & -r & \\ & \ddots & \ddots & \\ & & -r & -1-2r \end{pmatrix}.$$

Moreover we obtain

$$E + F = I + rS,$$
$$E - F = -I - rS',$$

where

$$S = \begin{pmatrix} 2 & -1 & & \\ -1 & 2 & -1 & \\ & \ddots & \ddots & \\ & & -1 & 2 \end{pmatrix}, \quad S' = \begin{pmatrix} 2 & 1 & & \\ 1 & 2 & 1 & \\ & \ddots & \ddots & \\ & & 1 & 2 \end{pmatrix}.$$

Thus

$$\lambda_s = \lambda_{s'} = 2 + 2\cos\frac{k\pi}{N} = 4\cos^2\frac{k\pi}{2N}, \quad k = 1, 2, \cdots, N-1.$$

Hence the eigenvalues of $E+F$ and $E-F$ are

$$\lambda_{E+F} = 1 + 4r\cos^2\frac{k\pi}{2N}, \quad k = 1, 2, \cdots, N-1,$$

$$\lambda_{E-F} = -1 - 4r\cos^2\frac{k\pi}{2N}, \quad k=1, 2, \cdots, N-1,$$

we know

$$\rho(E+F) = \max_k |1 + 4r\cos^2\frac{k\pi}{2N}|, \quad k = 1, 2, \cdots, N-1,$$

$$\rho(E-F) = \max_k |-1 - 4r\cos^2\frac{k\pi}{2N}|, \quad k=1, 2, \cdots, N-1.$$

Since $\rho(E+F) = \rho(E-F)$, thus

$$\rho(A) = \max_k |1+4r\cos^2\frac{k\pi}{2N}|, \rho(A^{-1}) = \frac{1}{\max|1+4r\cos^2\frac{k\pi}{2N}|} < 1, \forall r > 0.$$

Therefore our difference scheme is unconditionally stable.

5. Numerical Example

In this section we test the proposed difference method on an example, whose exact solution is known to us.

Consider the fuzzy heat equation

$$\frac{\partial\widetilde{U}}{\partial t}(x,t) = 4\frac{\partial^2\widetilde{U}}{\partial x^2}(x,t), \quad 0 < x < 1, \ t > 0.$$

Subject to the boundary conditions and the initial condition

$$\widetilde{U}(0,t) = \widetilde{U}(1,t) = 0, \ t > 0, \ \widetilde{U}(x,0) = \widetilde{f}(x) = \frac{2}{\pi}\widetilde{K}\sin\pi x,$$

and $\widetilde{K}[\alpha] = [\underline{k}(\alpha), \overline{k}(\alpha)] = [\alpha - 1, 1 - \alpha]$. Which is easily seen to have an exact solution for

$$\frac{\partial\underline{U}}{\partial t}(x,t;\alpha) = 4\frac{\partial^2\underline{U}}{\partial x^2}(x,t;\alpha), \ \frac{\partial\overline{U}}{\partial t}(x,t;\alpha) = 4\frac{\partial^2\overline{U}}{\partial x^2}(x,t;\alpha),$$

namely

$$\underline{U}(x,t;\alpha) = \frac{2}{\pi}\underline{k}(\alpha)e^{-4\pi^2 t}\sin\pi x, \ \overline{U}(x,t;\alpha) = \frac{2}{\pi}\overline{k}(\alpha)e^{-4\pi^2 t}\sin\pi x.$$

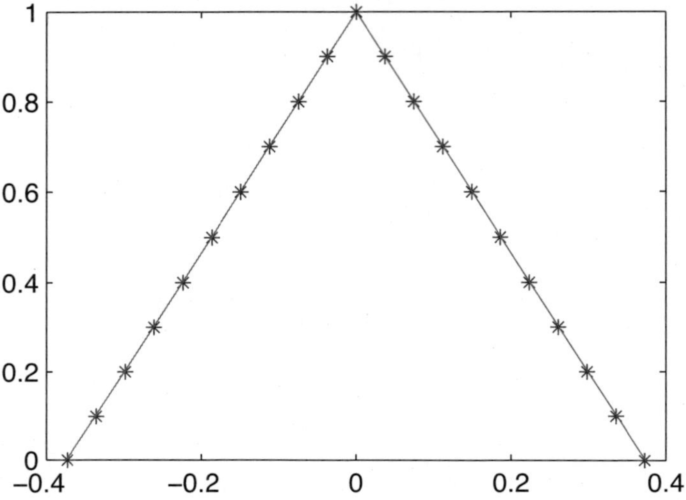
1
0.8
0.6
0.4
0.2
0
−0.4
−0.2
0
0.2
0.4

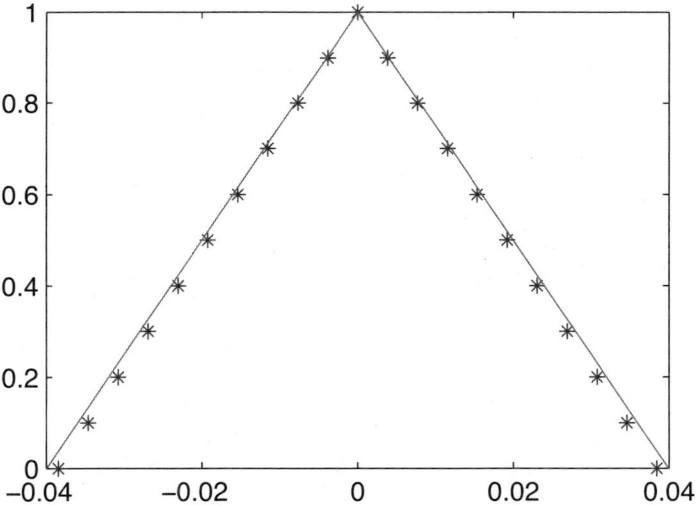
1
0.8
0.6
0.4
0.2
0
−0.04
−0.02
0
0.02
0.04

The exact and approximate solutions are shown in Figure 3 at the point $(0.2, 0.00099)$ with $h = 0.1$, $k = 0.00001$ and in Figure 4 at the point $(0.02, 0.0099)$ with $h = 0.01$, $k = 0.0001$. The Housdorff distance between solutions in first case is 0.0014 and in second case is 0.0015.

6. Conclusions

Our purpose in this article is solving fuzzy partial differential equation (FPDE). We presented an explicit method for solving this equation, and we considered necessary conditions for stability of this method. In last section we gave an example for considering numerical results. Also we compared the approximate solution and exact solution. Then we obtained the Hausdorf distance between them in two case.

References

[1] Biswa Nath Data, *Numerical linear algebra and applications,* Brooks/Cole Publishing Co., Pacific Grove, CA, 1995.

[2] D. Dubois and H. Prade, Towards fuzzy differential calculus, part 3, *differentiation, Fuzzy Sets and Systems,* 8 (1982), 225-233.

[3] G. D. Smith, *Numerical solution of partial differential equations, finite difference method,* The Clarendon Press, Oxford University Press, New York, 1985.

[4] M. L. Puri and D. A. Ralescu, Differentials of fuzzy functions, *J. Math. Anal. Appl.,* 91 (1983), 321-325.

[5] O. Kaleva, Fuzzy differential equations, *Fuzzy Sets and Systems,* 24 (1987), 301-307.

[6] O. Kaleva, The cauchy problem for fuzzy differential equations, *Fuzzy sets and systems,* 35 (1990), 389-396.

[7] P. Diamond and P. Kloeden, *Metric spaces of fuzzy sets,* World Scientific, Singapore, 1994.

[8] P. E. Kloeden, Remarks on Peano-like theorems for fuzzy differential equations, *Fuzzy Sets and Systems,* 44 (1991), 161-163.

[9] R. Goetschel and W. Voxman, Elementary fuzzy calculus, *Fuzzy Sets and Systems,* 18 (1988), 31-43.

[10] S. L. Chang and L.A. Zadeh, On fuzzy mapping and control, *IEEE Trans, Systems Mah Cybemet,* 2 (1972), 30-34.

[11] S. Seikkala, On the fuzzy initial value problem, *Fuzzy Sets and Systems,* 24 (1987), 319-330.

[12] T. Allahviranloo, N. Ahmadi, E. Ahmadi and Kh. Shams Alketabi, Block Jacobi two-stage method for fuzzy systems of linear equations, *Applied Mathematics and Computation,* 175 (2006), 1217-1228.

THE SOLUTION OF A PARTIAL DIFFERENTIAL EQUATION WITH NONLOCAL NONLINEAR BOUNDARY CONDITIONS

A. FARAJZADEH[a] and A. HOSSEIN POUR[b]
[a,b] *Islamic Azad university, Kermanshah Branch, Iran*
[a] *E-mail: faraj1348@yahoo.com*
[a] *E-mail: azadeh.hoseinpour@gmail.com*

This paper deals with a numerical method for the solution of the heat equation with non-linear nonlocal boundary conditions. Here non-linear terms are approximated by Richtmyer's linearization method. The integrals in the boundary equations are approximated by the composite Simpson rule. A difference scheme is considered for the one-dimensional heat equation. In final part the numerical results produced by this method is compared.

1. Introduction

This paper is concerned with the numerical solution of the heat equation

$$u_t - u_{xx} = f(x,t), \ x \in (0,1), \ t \in (0,T]. \tag{1}$$

Subject to the nonlocal boundary conditions

$$\begin{cases} u(0,t) = \displaystyle\int_0^1 k_0(x)u^p(x,t)dx + g_0(t), \\ u(1,t) = \displaystyle\int_0^1 k_1(x)u^p(x,t)dx + g_1(t), \end{cases} \tag{2}$$

and the initial condition

$$u(x,0) = g(x), \ x \in [0,1]. \tag{3}$$

Where f, k_0, k_1, g_0, g_1 and g are known functions. Over the last few years, many other physical phenomena were formulated into non-local mathematical models [1,2]. Hence the numerical solution of parabolic partial differential equations with non-local boundary specifications is currently an active area of research. The non-local problems are very important in the transport of reactive and passive contaminates in aquifers, an area of active interdisciplinary research of mathematicians, engineers, and life scientists. We refer the reader to [3,4] for the derivation of mathematical models and for the precise hypotheses and analysis. Parabolic problems with non-local boundary specifications also arise in quasi-static theory of thermoelasticity [5,6]. An interesting collection of non-local parabolic problems in one-space dimension is discussed in [8]. In the present paper, the difference schemes for the one-dimensional heat equation with constant coefficients and nonlocal boundary conditions is investigated. We selected the implicit difference scheme as the main object of study in this paper.

2. The Finite Difference Schemes

We divide the domain $[0,1]\times[0,T]$ into $M\times N$ mesh with spatial step size $h=\frac{1}{N}$ in x-direction and the time step size $k=\frac{T}{M}$ respectively. Where M is a positive integer and N is a positive even integer. The gride points are given by

$$x_n = nh, \qquad n = 0,1,\cdots,N,$$
$$t_m = mk, \quad m = 0,1,\cdots,M.$$

We define the following difference operators:

$$\begin{aligned} u_{i-\frac{1}{2}}^k &= \frac{1}{2}(u_i^k+u_{i-1}^k), \quad \delta_x u_{i-\frac{1}{2}}^k = \frac{1}{h}(u_i^k-u_{i-1}^k),\\ u_i^{k-\frac{1}{2}} &= \frac{1}{2}(u_i^k+u_i^{k-1}), \quad \delta_t u_i^{k-\frac{1}{2}} = \frac{1}{k}(u_i^k-u_i^{k-1}),\\ \delta_x^2 u_i^k &= \frac{1}{h^2}(u_{i+1}^k-2u_i^k+u_{i-1}^k). \end{aligned}$$

The inner product $<.,.>$ for $(N+1)$-dimensional vectors is defined by

$$< u, v > = \frac{h}{3}\sum_{i=0}^{N/2-1}(u_{2i}v_{2i}+4u_{2i+1}v_{2i+1}+u_{2i+2}v_{2i+2}),$$

and also

$$k_0^* = (k_0(x_0), k_0(x_1), \cdots, k_0(x_N)),$$
$$k_1^* = (k_1(x_0), k_1(x_1), \cdots, k_1(x_N)).$$

Our difference scheme for (1) is as follows [9]:

$$\begin{gathered} \frac{1}{12}(\delta_t u_{n-1}^{m-\frac{1}{2}} + 10\delta_t u_n^{m-\frac{1}{2}} + \delta_t u_{n+1}^{m-\frac{1}{2}}) - \delta_x^2 u_n^{m-\frac{1}{2}} = f_n^m,\\ 0\le m\le M,\ 0\le n\le N. \end{gathered} \tag{4}$$

And for boundary conditions and the initial condition, we define the following difference operators

$$\begin{cases} u_0^m = < k_0^*, u^m > + g_0(t_n),\\ u_N^m = < k_1^*, u^m > + g_1(t_n), \end{cases} \tag{5}$$

and

$$u_n^0 = g(x_n), \quad 0 \le n \le N.$$

Lemma 2.1([12]). *Let* N *be an even integer,* $h = \frac{1}{N}$, $x_n = nh$, $0 \le n \le N$. *If* $g(x) \in C^4[0,1]$. *Then*

$$\int_0^1 g(x)dx - \frac{h}{3}\sum_{i=0}^{N/2-1}[g(x_{2i}) + 4g(x_{2i+1}) + g(x_{2i+2})]$$
$$= -\frac{1}{180}h^4\frac{d^4g(x)}{dx^4}\,|_{x=\zeta}\ , \quad \zeta \in (0,1).$$

By Taylor's expansion about the point (n, m)

$$\begin{aligned}(u_n^{m+1})^p &= (u_n^m)^p + k\frac{\partial(u_n^m)^p}{\partial t} + \cdots \\ &= (u_n^m)^p + k\frac{\partial(u_n^m)^p}{\partial u_n^m}\frac{\partial u_n^m}{\partial t} + \cdots \\ &= (u_n^m)^p + p(u_n^m)^{p-1}(u_n^{m+1} - u_n^m) + \cdots,\end{aligned}$$

hence to terms of order k, we have

$$(u_n^{m+1})^p = p(u_n^m)^{p-1}(u_n^{m+1}) + (1-p)(u_n^m)^p. \tag{6}$$

Now, we approximate the integrals in the nonlocal boundary condition (2) by simpson's rule. Hence we have

$$\begin{aligned} u_0^{m+1} = \frac{h}{3}[k_0(x_0)(u_0^{m+1})^p + 4k_0(x_1)(u_1^{m+1})^p \\ +2k_0(x_2)(u_2^{m+1})^p + \cdots + 4k_0(x_{N-1})(u_{N-1}^{m+1})^p \\ + k_0(x_N)(u_N^{m+1})^p] + g_0^{m+1}. \end{aligned} \tag{7}$$

By (6), we obtain

$$\begin{aligned} a_0^m u_0^{m+1} + a_1^m u_1^{m+1} + a_2^m u_2^{m+1} + \cdots \\ + a_{N-1}^m u_{N-1}^{m+1} + a_N^m u_N^{m+1} = L_N^m, \end{aligned} \tag{8}$$

where

$$\begin{cases} a_0^m = phk_0(x_0)(u_0^m)^{p-1} - 3, \quad a_N^m = phk_0(x_N)(u_N^m)^{p-1}, \\ a_{2n-1}^m = 4phk_0(x_{2n+1})(u_{2n+1}^m)^{p-1}, \quad n = 0, 1, \cdots, \dfrac{N}{2} - 1, \\ a_{2n}^m = 2phk_0(x_{2n})(u_{2n}^m)^{p-1}, \quad n = 1, 2, \cdots, \dfrac{N}{2} - 1, \end{cases} \tag{9}$$

and

$$L_N^m = (p-1)hk_0(x_0)(u_0^m)^p + 4(p-1)hk_0(x_1)(u_1^m)^p$$
$$+2(p-1)hk_0(x_2)(u_2^m)^p + \cdots + 4(p-1)hk_0(x_{N-1})(u_{N-1}^m)^p$$
$$+(p-1)hk_0(x_N)(u_N^m)^p - 3g_0^{m+1},$$

and also

$$b_0^m u_0^{m+1} + b_1^m u_1^{m+1} + b_2^m u_2^{m+1} + \cdots + b_{N-1}^m u_{N-1}^{m+1} + b_N^m u_N^{m+1} = Q_N^m, \quad (10)$$

where

$$\begin{cases} b_0^m = phk_1(x_0)(u_0^m)^{p-1}, \quad b_N^m = phk_1(x_N)(u_N^m)^{p-1} - 3, \\ b_{2n-1}^m = 4phk_1(x_{2n+1})(u_{2n+1}^m)^{p-1}, \quad n = 0, 1, \cdots, \dfrac{N}{2} - 1, \\ b_{2n}^m = 2phk_1(x_{2n})(u_{2n}^m)^{p-1}, \quad n = 1, 2, \cdots, \dfrac{N}{2} - 1, \end{cases} \quad (11)$$

and

$$Q_N^m = (p-1)hk_1(x_0)(u_0^m)^p + 4(p-1)hk_1(x_1)(u_1^m)^p$$
$$+2(p-1)hk_1(x_2)(u_2^m)^p + \cdots + 4(p-1)hk_1(x_{N-1})(u_{N-1}^m)^p$$
$$+(p-1)hk_1(x_N)(u_N^m)^p - 3g_1^{m+1}.$$

By (4), we obtain

$$\begin{aligned} &(1-6r)u_{n-1}^{m+1} + (10+12r)u_n^{m+1} + (1-6r)u_{n+1}^{m+1} \\ &= (1+6r)u_{n-1}^m + (10-12r)u_n^m + (1+6r)u_{n+1}^m + 12kf_n^m, \end{aligned} \quad (12)$$

where

$$r = \frac{k}{h^2}.$$

We now consider the following difference scheme (Crank-Niklson)

$$\delta_t u_n^m + \frac{1}{2}(\delta_x^2 u_n^{m+1} + \delta_x^2 u_n^m) = \frac{1}{2}(f_n^{m+1} + f_n^m),$$

and obtain

$$u_{n+1}^{m+1} - \frac{2+2r}{r}u_n^{m+1} + u_{n-1}^{m+1} = M_n^m, \quad (13)$$

where

$$M_n^m = -u_{n+1}^m + \frac{2r-2}{r}u_n^m - u_{n-1}^m - \frac{2k}{r}f(x_m, \frac{1}{2}(t_n + t_{n+1})).$$

By the left-hand side of (12) and the right-hand side of (13) and (8), (10), we obtain the following matrix equation

$$\begin{pmatrix} a_0^m & a_1^m & a_2^m & \cdots & a_{N-1}^m & a_N^m \\ \alpha & \beta & \alpha & & & \\ & & \ddots & \ddots & & \\ & & & & \beta & \alpha \\ b_0^m & b_1^m & b_2^m & \cdots & b_{N-1}^m & b_N^m \end{pmatrix} \begin{pmatrix} u_0^{m+1} \\ u_1^{m+1} \\ \vdots \\ u_{N-1}^{m+1} \\ u_N^{m+1} \end{pmatrix} = \begin{pmatrix} L_N^m \\ \vdots \\ M_n^m \\ \vdots \\ Q_N^m \end{pmatrix}, \qquad (14)$$

where

$$\alpha = 1 - 6r, \quad \beta = 10 + 12r, \quad n = 1, \cdots, N-1,$$

and a_i, b_i $(i = 0, 1, \cdots, N)$ are given by (9) and (11).

3. Numerical Example

In this section, we test the proposed difference method on an example, whose exact solution is known to us. The right-hand side functions as well as the nonlocal boundary value conditions and initial value conditions are obtained from the exact solution. The systems of linear algebraic equations have been solved by using the Gaussian pivot method.

$$u_t - u_{xx} = \frac{-2(x^2 + t + 1)}{(t+1)^3}, \quad 0 < x \le 1, \ t \in (0, T]$$

subject to the nonlocal boundary conditions and the initial condition

$$u(0,t) = \int_0^1 k_0(x) u^2(x,t) dx - \frac{1}{6(t+1)^4},$$

$$u(1,t) = \int_0^1 k_1(x) u^2(x,t) dx + \frac{6t^2 + 12t + 5}{6(t+1)^4},$$

$$u(x,0) = x^2, \quad x \in (0,1].$$

It is easily seen to have the exact solution

$$u(x,t) = \left(\frac{x}{t+1}\right)^2.$$

The results with $h = 0.05,\ 0.005$ and $r = 0.4$ using the finite difference formulate discussed in Section 2 are shown in following table. In this table, we present the error for $x = 0.1$ and $t = 0.01,\ 0.02,\ 0.03, \cdots, 0.1$.

t	Exact	Error	
		h=0.05	h=0.005
0.0000	0.0100	0.0000	0.0000
0.0100	0.0098	0.0093	0.0098
0.0200	0.0096	0.0091	0.0096
0.0300	0.0094	0.0090	0.0094
...	...	...	
0.1000	0.0083	0.0079	0.0083

4. Summary and Concluding Remarks

In this paper, a new numerical method was applied to the one-dimensional diffusion equation with non-linear nonlocal boundary conditions replacing standard boundary conditions. These techniques applied well for one-dimensional diffusion with integral conditions. One example with closed form solution is studied carefully in order to illustrate the possible practical use of this method.

References

[1] W. Allegretto, Y. Lin and A. Zhou, A box scheme for coupled systems resulting from micro-sensor thermistor problems, *Dynam. Contin. Discr. Impuls. Syst.*, 5 (1999), 209-223.

[2] J. R. Cannon, The solution of the heat equation subject to the specification of energy, *Quart. Appl. Math.*, 21 (1963), 155-160.

[3] J. R. Cannon, *The one-dimensional heat equation,* Encyclopedia Math. Appl., Vol. 23, Addison-Welsey, Menlo Park, CA, 1984.

[4] J. R. Cannon and A. L. Matheson, A numerical procedure for diffusion subject to the specification of mass, *Internat. J. Engrg. Sci.*, 31 (1993), 347-355.

[5] J. R. Cannon and J. van der Hoek, Implicit finite difference scheme for the diffusion of mass in porous media, in *Numerical Solution of Partial Differential Equations,* Ed. J. Noye, North-Holland, Amsterdam, 1982, 527-539.

[6] J. R. Cannon and J. van der Hoek, Diffusion subject to specification of mass, *J. Math. Anal. Appl.,* 115 (1986), 517-529.

[7] J. R. Cannon and Y. Lin, A Galerkin procedure for diffusion equations with boundary integral conditions, *Internat. J. Engrg. Sci.*, 28(7) (1990), 579-587.

[8] J. R. Cannon and H. M. Yin, On a class of non-classical parabolic problems, *Differential Equations,* 79 (1989), 266-288.

[9] Z. Zhong Sun, A high-order difference scheme for a nonlocal boundary-value problem for the heat equation, *J. Computational Methods in Applied Mathematics,* 1 (2001), 398-414.

APPLYING FRACTIONAL DERIVATIVE EQUATIONS TO THE MODELING OF SUBDIFFUSION PROCESS

TADEUSZ KOSZTOŁOWICZ[a] and KATARZYNA D. LEWANDOWSKA[b]
[a] *Institute of Physics, Jan Kochanowski University, Kielce, POLAND*
E-mail: tadeusz.kosztolowicz@ujk.edu.pl
[b] *Department of Radiological Informatics and Statistics, Medical University of Gdańsk, Gdańsk, POLAND*
E-mail: kale@gumed.edu.pl

Subdiffusion is a fractional Brownian motion described by linear equation with fractional Riemann–Liouville time derivative. We present the procedure of solving the subdiffusion equation for the system with infinitely thin membrane. The procedure exploits the Green's function and the boundary conditions at the membrane, which are taken in the general form as a linear combination of particles concentration and flux.
Keywords: Fractional partial differential equations, fractional processes.
AMS No: 35R11, 60G22.

1. Formulation of the Problem

Subdiffusion is a diffusion process, where a random walker waits anomalously long time to make a finite jump. Subdiffusion is characterized by the relation

$$\left\langle (\Delta x)^2 \right\rangle = \frac{2D_\alpha}{\Gamma(1+\alpha)} t^\alpha ,$$

where D_α is the subdiffusion coefficient, α is the subdiffusion parameter $(0 < \alpha < 1)$ and $\left\langle (\Delta x)^2 \right\rangle$ is a mean square displacement of the particle after time t

$$\left\langle (\Delta x)^2 \right\rangle = \int_{-\infty}^{+\infty} x^2 G(x,t;x_0)dx,$$

$G(x,t;x_0)$ denotes the Green's function, which is the solution of subdiffusion equation with the initial condition $G(x,0;x_0) = \delta(x - x_0)$ and appropriate boundary conditions. The Greens function is normalized

$$\int_{-\infty}^{+\infty} G(x,t;x_0)dx = 1,$$

and it can be treated as a probability density of finding a diffusing particle at point x and time t, under condition that at the initial time $t = 0$ the particle was at point $x = x_0$.

Subdiffusion is described by the equation with fractional time derivative

$$\frac{\partial C(x,t)}{\partial t} = D_\alpha \frac{\partial^{1-\alpha}}{\partial t^{1-\alpha}} \frac{\partial^2 C(x,t)}{\partial x^2}, \tag{1.1}$$

with $0 < \alpha < 1$, where $C(x,t)$ is the particles concentration and $\partial^\alpha/\partial t^\alpha$ denotes the Riemann-Liouville fractional time derivative defined for $\alpha > 0$ by the following relation

$$\frac{\partial^\alpha C(x,t)}{\partial t^\alpha} = \frac{1}{\Gamma(n-\alpha)} \frac{\partial^n}{\partial t^n} \int_0^t dt' \frac{C(x,t')}{(t-t')^{1+\alpha-n}},$$

the integer number n fulfills the relation $n - 1 < \alpha \leq n$. Definitions and properties of the fractional derivatives are widely discussed in many books and papers, see for example [1–4].

We solve the equation (1.1) in a one-dimensional system with a infinitely thin membrane located at $x = 0$. In the following we use the notation

$$C(x,t) = \begin{cases} C_1(x,t), & x < 0, \\ C_2(x,t), & x > 0. \end{cases}$$

We need four boundary conditions to solve the subdiffusion equation (1.1). Two of them are fixed at the membrane. First bonduary condition demands the continuity of the flux at the membrane

$$J_1(0^-,t) = J_2(0^+,t) \, (\equiv J(0,t)), \tag{1.2}$$

where the subdiffusive flux $J(x,t)$ is given by the generalized Fick's law

$$J_i(x,t) = -D_\alpha \frac{\partial^{1-\alpha}}{\partial t^{1-\alpha}} \frac{\partial C_i(x,t)}{\partial x}, \; j = 1,2. \tag{1.3}$$

There is no obvious choice of second boundary condition. We assume that the second boundary condition is given by a linear combination of concentrations and flux at the membrane

$$b_1 C_1(0^-,t) + b_2 C_2(0^+,t) + b_3 J(0,t) = 0, \tag{1.4}$$

where

$$b_1 b_3 \leq 0, \;\; b_2 b_3 \geq 0. \tag{1.5}$$

Equations (1.4) and (1.5) provide the commonly used membrane boundary conditions listed below

kind of boundary condition	b_1	b_2	b_3
fully absorbing membrane	0	$\neq 0$	0
	$\neq 0$	0	0
partially absorbing membrane	0	$\neq 0$	$\neq 0$
	$\neq 0$	0	$\neq 0$
partially permeable membrane	$\neq 0$	$\neq 0$	0
	$\neq 0$	$\neq 0$	$\neq 0$
fully reflecting membrane	0	0	$\neq 0$

Two remaining boundary conditions should be consistent with the initial condition

$$C(x,0) = f(x).$$

2. Green's Function

Solution of the equation (1.1) can be obtained from the formula

$$C(x,t) = \int_{-\infty}^{\infty} f(x_0)G(x,t;x_0)dx_0, \tag{2.1}$$

where $G(x,t;x_0)$ is the solution of the equation (1.1) with the boundary conditions (1.2), (1.4) and

$$G(\pm\infty,t;x_0) = 0, \tag{2.2}$$

and the initial condition

$$G(x,0;x_0) = \delta(x-x_0), \tag{2.3}$$

$\delta(x)$ is Dirac's delta function.

We solve the equation (1.1) by means of the Laplace transform method. The Laplace transforms $L[f(t)] \equiv \hat{f}(s) \equiv \int_0^\infty f(t)e^{-st}dt$ of the equations (1.1), (1.3), (1.4) and (2.1) are

$$\hat{C}(x,s) - f(x) = D_\alpha s^\alpha \frac{\partial^2 \hat{C}(x,s)}{\partial x^2} - D_\alpha \frac{\partial^{\alpha-1}}{\partial t^{\alpha-1}} \frac{\partial^2 C(x,t)}{\partial x^2}\bigg|_{t=0}, \tag{2.4}$$

$$\hat{J} = -D_\alpha s^{1-\alpha} \frac{d\hat{C}(x,s)}{dx}, \tag{2.5}$$

$$b_1\hat{C}_1(0^-,s) + b_2\hat{C}_2(0^+,s) + b_3\hat{J}(0,s) = 0, \tag{2.6}$$

$$\hat{C}(x,s) = \int_{-\infty}^{\infty} f(x_0)\hat{G}(x,s;x_0)dx_0, \tag{2.7}$$

respectively. The last term of (2.4) can be omitted according to Theorem 2.1.

Theorem 2.1. *For a bounded function C, there is*

$$\left.\frac{d^{\alpha-1}C(x,t)}{dt^{\alpha-1}}\right|_{t=0} = 0,$$

when $0 < \alpha < 1$.

Proof. Let $|C(x,u)| \le A$ for $u \in (0,t]$. Then

$$\left|\frac{\partial^{\alpha-1}C(x,t)}{\partial x^{\alpha-1}}\right| \le \frac{A}{\Gamma(1-\alpha)}\int_0^t (t-u)^{-\alpha}du = \frac{A}{\Gamma(1-\alpha)}t^{1-\alpha} \longrightarrow_{t\to 0} 0.$$

From (2.2)–(2.6) we get (here $x_0 < 0$) [5–7]

$$\hat{G}_1(x,s;x_0) = A(s)\mathrm{e}^{x\sqrt{s^\alpha/D_\alpha}} + \frac{1}{2\sqrt{D_\alpha}s^{1-\alpha/2}}\mathrm{e}^{-|x-x_0|\sqrt{s^\alpha/D_\alpha}}, \tag{2.8}$$

$$\hat{G}_2(x,s;x_0) = B(s)\mathrm{e}^{-x\sqrt{s^\alpha/D_\alpha}} + \frac{1}{2\sqrt{D_\alpha}s^{1-\alpha/2}}\mathrm{e}^{-|x-x_0|\sqrt{s^\alpha/D_\alpha}}, \tag{2.9}$$

where

$$B(s) = -A(s) = \frac{b_1 + b_2 + b_3\sqrt{D_\alpha}s^{1-\alpha/2}}{2\sqrt{D_\alpha}s^{1-\alpha/2}(b_1 - b_2 - b_3\sqrt{D_\alpha}s^{1-\alpha/2})}\mathrm{e}^{x_0\sqrt{s^\alpha/D_\alpha}}. \tag{2.10}$$

3. Example of a Solution

Let us solve the equation (1.1) for the initial condition

$$f(x) = \begin{cases} C_0, & x < 0, \\ 0, & x > 0, \end{cases} \tag{3.1}$$

and the boundary conditions (1.2), (1.4) and

$$C_1(-\infty,t) = C_0, \quad C_2(\infty,t) = 0.$$

From (2.7)–(2.10) and (3.1), we get

$$\hat{C}_1(x,t) = \frac{C_0}{s}\left(1 - \frac{b_1}{b_1 - b_2 - b_3\sqrt{D_\alpha}s^{1-\alpha/2}}\mathrm{e}^{x\sqrt{s^\alpha/D_\alpha}}\right), \tag{3.2}$$

$$\hat{C}_2(x,t) = \frac{C_0}{s}\frac{b_1}{b_1 - b_2 - b_3\sqrt{D_\alpha}s^{1-\alpha/2}}\mathrm{e}^{-x\sqrt{s^\alpha/D_\alpha}}. \tag{3.3}$$

Let us calculate the inverse Laplace transform of (3.2) and (3.3).

Theorem 3.1. *Let $\alpha, \beta > 0$ and $\nu \in \mathcal{R}$. Then,*

$$L^{-1}\left[s^{\nu}e^{-as^{\beta}}\right] = \frac{1}{\beta a^{\frac{1+\nu}{\beta}}} H^{1\,0}_{1\,1}\left(\frac{a^{\frac{1}{\beta}}}{t}\,\middle|\,\begin{matrix} 1 & 1 \\ \frac{1+\nu}{\beta} & \frac{1}{\beta}\end{matrix}\right) (\equiv f_{\nu,\beta}(t;a)), \qquad (3.4)$$

where H is the Fox function and $f_{\nu,\beta}(t;a)$ can be expressed by the series

$$f_{\nu,\beta}(t;a) = \frac{1}{t^{1+\nu}}\sum_{k=0}^{\infty}\frac{1}{\Gamma(-k\beta-\nu)k!}\left(-\frac{a}{t^{\beta}}\right)^{k}.$$

For the proof we refer the reader to [7].

1. $b_3 = 0$

Using Theorem 3.1 and $L^{-1}\{1/s\} = 1$, from (3.2) and (3.3) we obtain

$$C_1(x,t) = C_0\left[1 - \frac{b_1}{b_1 - b_2} f_{-1,\alpha/2}\left(t; \frac{|x|}{\sqrt{D_\alpha}}\right)\right],$$

$$C_2(x,t) = C_0\frac{b_1}{b_1 - b_2} f_{-1,\alpha/2}\left(t; \frac{x}{\sqrt{D_\alpha}}\right).$$

2. $b_3 \neq 0$

Singularities of (3.2) and (3.3) are $s = 0$ and $s = |\gamma|^{\frac{1}{1-\alpha/2}} e^{i\pi\frac{2m+1}{1-\alpha/2}}$, where

$$\gamma = \frac{b_1 - b_2}{b_3\sqrt{D_\alpha}}.$$

C_i $(i = 1, 2)$ are analytic functions at the point $s = \infty$. The series expansion of $b_1/(b_1 - b_2 - b_3\sqrt{D_\alpha}s^{1-\alpha/2})$ in the neighbourhood of the infinite point combined with (3.2) and (3.3) give

$$\hat{C}_1(x,s) = \frac{C_0}{s} - \frac{C_0\eta}{\sqrt{D_\alpha}} e^{-x\sqrt{s^{\alpha}/D_\alpha}}\sum_{n=0}^{\infty}\gamma^n \hat{g}_n(s), \qquad (3.5)$$

$$\hat{C}_2(x,s) = \frac{C_0\eta}{\sqrt{D_\alpha}} e^{x\sqrt{s^{\alpha}/D_\alpha}}\sum_{n=0}^{\infty}\gamma^n \hat{g}_n(s), \qquad (3.6)$$

where $\hat{g}_n(s) = s^{(\alpha/2-1)n+\alpha/2-2}$ and $\eta = -b_1/b_3$. The inverse Laplace transform reads

$$C_i(x,t) = \frac{1}{2\pi i}\int_{c-i\infty}^{c+i\infty} e^{st}\hat{C}_i(x,s)ds,$$

where $c > |\gamma|^{1/(1-\alpha/2)}$. The series occuring in (3.5) and (3.6) are uniform convergent on the line $(c-i\infty, c+i\infty)$, therefore we take the inverse Laplace transform of the series (3.5) and (3.6) term by term with using the formula (3.4). Finally, we get

$$C_1(x,t) = C_0\left[1 - \frac{\eta}{\sqrt{D_\alpha}}\sum_{n=0}^{\infty}\gamma^n f_{(\alpha/2-1)n+\alpha/2-2,\alpha/2}\left(t;\frac{|x|}{\sqrt{D_\alpha}}\right)\right],$$

$$C_2(x,t) = C_0\frac{\eta}{\sqrt{D_\alpha}}\sum_{n=0}^{\infty}\gamma^n f_{(\alpha/2-1)n+\alpha/2-2,\alpha/2}\left(t;\frac{x}{\sqrt{D_\alpha}}\right).$$

We add that the procedure presented in our paper has been used to solve the equations describing the subdiffusion in various physical systems such as the system with a thick membrane [8], the systems with chemical reactions [9] and the electrochemical one [10].

References

[1] K. B. Oldham and J. Spanier, *The fractional calculus,* Academic Press, New York-London, 1974.

[2] A. A. Kilbas, H. M. Srivastava and J. J. Trujillo, *Theory and applications of fractional differential equations,* Elsevier, Amsterdam, 2006.

[3] I. Podlubny, *Fractional differential equations,* Academic Press, San Diego, 1999.

[4] K. S. Miller and B. Ross, *An indroduction to the fractional calculus and fractional differential equations,* Jhon Wiley & Sons, INC., New York, 1993.

[5] T. Kosztołowicz, K. Dworecki and S. Mróczyński, Measuring subdiffusion parameters, *Phys. Rev. E.,* 71 (2005), 041105-1–041105-11.

[6] T. Kosztołowicz, K. Dworecki and S. Mróczyński, How to measure a subdiffusion parameter, *Phys. Rev. Lett.,* 94 (2005), 170602-1–170602-4.

[7] T. Kosztołowicz, From the solutions of diffusion equation to the solutions of subdiffusive one, *J. Phys. A: Math. Gen.,* 37 (2004), 10779–10789.

[8] T. Kosztołowicz, Subdiffusion in a system with a thick membrane, *J. Membr. Sci.,* 320 (2008), 492-499.

[9] T. Kosztołowicz and K. D. Lewandowska, Time evolution of the reaction front in a subdiffusive system, *Phys. Rev. E.,* 78 (2008), 066103-1–066103-2.

[10] T. Kosztołowicz and K. D. Lewandowska, Hyperbolic subdiffusive impedance, *J. Phys. A: Math. Theor.,* 42 (2009), 055004-1–055004-14.

GROWTH OF MODIFIED RIESZ POTENTIAL IN HALF SPACE[1]

YAN-HUI ZHANG[a] and GUAN-TIE DENG[b]

[a] *Department of Applied Mathematics, Beijing Technology and Business University, Beijing 100048, China*
E-mail: zhangyhchengd@yahoo.com.cn

[b] *Sch. Math. Sci. & Lab. Math. Com. Sys., Beijing Normal University, Beijing 100875, China*
E-mail: denggt@bnu.edu.cn

Growth at infinity is given for modified Riesz potential in the half space.

Keywords: Modified Riesz-potential, growth, estimate.
AMS No: 31B05, 31B10.

1. Introduction

Let $\mathbb{R}^n (n \geq 3)$ denote the n–dimensional Euclidean space with points $x = (x_1, x_2, \cdots, x_n) = (x', x_n)$, where $x' \in \mathbb{R}^{n-1}$ and $x_n \in \mathbb{R}$. The boundary and closure of an open set Ω of $\mathbb{R}^n$ are denoted by $\partial\Omega$ and $\bar{\Omega}$, respectively. The half-space is the set $H = \{x = (x', x_n) \in \mathbb{R}^n; x_n > 0\}$, whose boundary is ∂H. We identify $\mathbb{R}^n$ with $\mathbb{R}^{n-1} \times \mathbb{R}$ and $\mathbb{R}^{n-1}$ with $\mathbb{R}^{n-1} \times \{0\}$.

Denote by $B(x, \rho)$ and $\partial B(x, \rho)$ the open ball and the sphere of radius ρ with the center x in $\mathbb{R}^n$ respectively. Recall the Laplace operator

$$\Delta = \partial_1^2 + \partial_2^2 + \cdots + \partial_n^2$$

has the property that[1]

$$-\Delta(f)(\xi) = 4\pi^2 |\xi|^2 \hat{f}(\xi), \tag{1}$$

where f is a Schwartz function,

$$\hat{f}(\xi) = \int_{\mathbb{R}^n} f(x) e^{-ix\xi} dx$$

for $f \in L^1(\mathbb{R}^n)$. Motivated by (1), $(-\Delta)^{-\frac{\alpha}{2}} (0 < \alpha \leq n)$ can be defined as the operator by

$$I_\alpha(f)(x) = f * g_\alpha(x) = \int_{\mathbb{R}^n} f(x, y) g_\alpha(y) dy,$$

[1] Project supported by the Academic Human Resources Development in Institutions of Higher Learning under the Jurisdiction of Beijing Municipality(PHR201008257) and Scientific Research Common Program of Beijing Municipal Commission of Education(KM 200810011005) and Innovation Project for the Development of Science and Technology(201098).

where $g_\alpha(x) = C_{(n,\alpha)}|x|^{-n+\alpha}$, $C(n,\alpha) = 2^{-\alpha}\pi^{-\frac{n}{2}}\frac{\Gamma(\frac{n-\alpha}{2})}{\Gamma(\frac{\alpha}{2})}$ and the integral is convergent, if f is a function in the Schwartz class. The operator $I_\alpha = (-\Delta)^{-\frac{\alpha}{2}}$ is called Riesz Potential of order α, so $u = I_2(f)$ is the solution of $(-\Delta)u = f$.

Let V_n be the volume of the unit ball $B(0,1)$ in $\mathbb{R}^n$ and ω_{n-1} be the surface of the unit sphere $S^{n-1} = \partial B(0,1)$, we have $V_n = \dfrac{\pi^{\frac{n}{2}}}{\Gamma(\frac{n}{2}+1)}$ and $\omega_{n-1} = nV_n = 2\dfrac{\pi^{\frac{n}{2}}}{\Gamma(\frac{n}{2})}$, $C_{(n,2)} = \dfrac{\Gamma(\frac{n}{2}-1)}{2^2\pi^{\frac{n}{2}}} = \dfrac{\frac{n}{2}\Gamma(\frac{n}{2}-1)}{2^2\pi^{\frac{n}{2}}\frac{n}{2}} = \dfrac{\Gamma(\frac{n}{2})}{2n\pi^{\frac{n}{2}}}$.

The function

$$E(x) = -C_{(n,2)}|x|^{2-n}$$

is the fundamental solution of $(-\Delta)u = \delta$, where $\delta(x)$ is Dirac delta function (Dirac distribution) and $\omega_n = \frac{2\pi^{\frac{n}{2}}}{\Gamma(\frac{n}{2})}$ is the surface area of the unit sphere in $\mathbb{R}^n$.

The Green function $G(x,y)$ for the upper half space H is given by [1], namely

$$G(x,y) = E(x-y) - E(x-\tilde{y}),\ \ x, y \in \overline{H},\ x \neq y,$$

where $\tilde{}$ denotes reflection in the boundary plane ∂H just as $\tilde{y} = (y_1, y_2, \cdots, y_{n-1}, -y_n)$, then we define the Poisson kernel $P(x,y')$, when $x \in H$ and $y' \in \partial H$ by

$$P(x,y') = -\frac{\partial G(x,y)}{\partial y_n}\bigg|_{y_n=0} = \frac{2x_n}{\omega_n|x-(y',0)|^n}.$$

The Riesz kernel g_α inspired us to define the modified Riesz kernel G_α for the half space H by

$$G_\alpha(x,y) = \frac{1}{|x-y|^{n-\alpha}} - \frac{1}{|\bar{x}-y|^{n-\alpha}},\ \ 0 < \alpha \leq n,\ n > 2,$$

g_α is called Green function[2] of order α in H.

There is a positive (possibly infinite) Borel measure ν satisfied

$$\int_H \frac{y_n d\nu(y)}{(1+|y|)^{n-\alpha+2}} < +\infty, \tag{2}$$

if and only if the associated Green Potential

$$(G\nu)(x) = \int_H G_\alpha(x,y)d\nu(y),\ \ x \in H. \tag{3}$$

is not identically equal to $+\infty$ (see, for example [3]) for the case $0 < \alpha \leq n$. We shall assume through that a potential is not identically equal to $+\infty$, and use the notation $G\nu$ only under this assumption.

2. Preliminaries

In this section we shall give some lemmas. Here and later on, we use the convertion that A denotes a positive constant depending only on indicated parameters, and that the value may from one expression to the next. In addition, if f and g are any functions, then we write $f \cong g$ provided $cf \leq g \leq f/c$ for some positive const $c > 0$.

The next Lemma 2.1 gives straightforward estimates involving the Green function.

Lemma 2.1[2]. (1) *If* $\alpha = n = 2$, *then*

$$G_\alpha(x,y) \leq \begin{cases} \log \dfrac{3x_n}{|x-y|}, & y \in B(x, \frac{x_n}{2}),\ x \in H, \\ \dfrac{2x_n y_n}{|x-y|^2}, & x, y \in H. \end{cases} \tag{4}$$

(2) *If* $0 < \alpha < n$, $(n > 2)$, *then*

$$G_\alpha(x,y) \cong \frac{1}{|x-y|^{n-\alpha}} \frac{x_n y_n}{|\bar{x}-y|^2}, \quad x, y \in H_n.$$

Lemma 2.2[3]. *Let* μ *be a positive Borel measure in* $\mathbb{R}^n$, $\gamma \geq 0$, $\mu(\mathbb{R}^n) < \infty$, *for any* $\lambda \geq 5^\gamma \mu(\mathbb{R}^n)$, *set*

$$E(\lambda) = \{x \in \mathbb{R}^n : |x| \geq 2,\ \ M(d\mu)(x) > \frac{\lambda}{|x|^\gamma}\}.$$

Then there exists $x_j \in E(\lambda)$, $\rho_j > 0\,(j = 1, 2, \cdots)$, *such that*

$$E(\lambda) \subset \bigcup_{j=1}^{\infty} B(x_j, \rho_j), \tag{5}$$

and

$$\sum_{j=1}^{\infty} \frac{\rho_j^\beta}{|x_j|^\gamma} \leq \frac{3\mu(\mathbb{R}^n)5^\gamma}{\lambda}. \tag{6}$$

Remark. Lemma 2.2 is the generalization of Lemma in [4], whose proof is similar to one of Vitali Lemma.

In order to describe the asymptotic behavior of subharmonic functions in half-spaces, we establish the following theorem.

Main Theorem. *For Green potential* (3) *satisfying* (2) *in* $H(0<\alpha\le n)$, *there exists* $\rho_j>0$, $x_j\in H$, *such that* $\sum_{j=1}^{\infty}\frac{\rho_j^{n-\alpha}}{|x_j^{n-\alpha}|}<\infty$, *and*

$$v(x)=o(|x|),\quad \text{as } |x|\to\infty \tag{7}$$

holds in $H\backslash G$, *where* $G=\cup_{j=1}^{\infty}B(x_j,\rho_j)$, $\alpha>0$.

3. Proof of Main Theorem

Let μ be a positive Borel measure in $\mathbb{R}^n$, $\gamma\ge 0$. The maximal function $M(d\mu)(x)$ of order γ is defined by

$$M(d\mu)(x)=\sup_{0<r<\infty}\frac{\mu(B(x,r))}{r^{\gamma}},$$

then the maximal function $M(d\mu)(x)\colon \mathbb{R}^n\to[0,\infty)$ is lower semicontinuous, hence it is measurable.

When $n>2$ and $\alpha\le n$, define the measure $dm(y)$ and the kernel $K(x,y)$ by

$$dm(y)=\frac{y_n d\nu(y)}{1+|y|^{n+2-\alpha}},$$

and

$$K(x,y)=\frac{1+|y|^{n+2-\alpha}}{y_n}\left(\frac{1}{|x-y|^{n-\alpha}}-\frac{1}{|\bar{x}-y|^{n-\alpha}}\right).$$

Then the Green potential now can be represented in the form

$$G(x)=\int_H K(x,y)dm(y).$$

For any $\varepsilon>0$, there exists $R_\varepsilon>2$ by (2), such that

$$\int_{|y|\ge R_\varepsilon} dm(y)\le\frac{\varepsilon}{5^{n-\beta}}.$$

For every Lebesgue measurable set $E\subset\mathbb{R}^n$, the measure $m^{(\varepsilon)}$ is defined by

$$m^{(\varepsilon)}(E)=m(E\cap\{x\in\mathbb{R}^n: |x|\ge R_\varepsilon\})$$

satisfying

$$m^{(\varepsilon)}(\mathbb{R}^n)\le\frac{\varepsilon}{5^{n-\beta}}.$$

By Lemma 2.1,

$$\frac{1}{|x-y|^{n-\alpha}}-\frac{1}{|\bar{x}-y|^{n-\alpha}}\le\frac{x_ny_n}{|\bar{x}-y|^{n-\alpha}}\frac{1}{|x-y|^{n-\alpha}}\le\frac{x_ny_n|x\cdot y|^2}{|x-y|^{n-\alpha+2}},$$

then we have

$$|K(x,y)| \leq \frac{x_n(1+|y|^{n+2-\alpha})}{|\bar{x}-y|^2|x-y|^{n-\alpha}}.$$

Denote

$$G_1(x)=\int_{\{y:\delta<|x-y)|\leq\frac{x_n}{2}\}} K(x,y)dm^{(\varepsilon)}(y),$$

$$G_2(x)=\int_{\{y:\frac{x_n}{2}\leq|x-y|\leq 3|x|\}} K(x,y)dm^{(\varepsilon)}(y),$$

$$G_3(x)=\int_{\{y:|x-y|\geq 3|x|\}} K(x,y)dm^{(\varepsilon)}(y),$$

$$G_4(x)=\int_{\{y:1<|y|<R_\varepsilon\}} K(x,y)dm(y),$$

$$G_5(x)=\int_{\{y:|y|\leq 1\}} K(x,y)dm(y),$$

it is easy to derive

$$|G(x)| \leq |G_1(x)|+|G_2(x)|+|G_3(x)|+|G_4(x)|+|G_5(x)|. \tag{8}$$

Let $E_1(\lambda)=\{x\in\mathbb{R}^n : |x|\geq 2,\ \exists t>0,\ m^{(\varepsilon)}(B(x,t)\cap\mathbb{R}^n)>\lambda(\frac{t}{|x|})^{n-\beta}\}$. Therefore, if $|x|\geq 2R_\varepsilon$ and $x\notin E_1(\lambda)$, then we get

$$\begin{aligned} |G_1(x)| &\leq \int_{\{y:\delta<|x-y|\leq\frac{x_n}{2}\}} \frac{x_n(1+|y|)^{n+2-\alpha}}{|x-y|^{n+2-\alpha}} dm^{(\varepsilon)}(y) \\ &\leq \int_{\{y:\delta<|x-y|\leq\frac{x_n}{2}\}} Ax_n|x|^{n+2-\alpha} \int_\delta^{\frac{x_n}{2}} \frac{dm_x^{(\varepsilon)}(t)}{|t|^{n-\alpha+2}} \\ &\leq M\varepsilon|x|, \end{aligned} \tag{9}$$

where

$$m_x^{(\varepsilon)}(t)=\int_{|x-y|\leq t} dm^{(\varepsilon)}(y), \tag{10}$$

herein we explain that A is a positive constant independent of x,y, which can be denoted different constants at different places. Moreover we have

$$|G_2(x)| \leq \int_{\{y:\frac{x_n}{2}<|x-y|\leq 3|x|\}} \frac{x_n(1+|y|)^{n+2-\alpha}}{|x-y|^{n+2-\alpha}} dm^{(\varepsilon)}(y)$$

$$
\begin{aligned}
&\le Ax_n|x|^{n+2-\alpha}\int_{\frac{x_n}{2}}^{3|x|}\frac{dm_x^{(\varepsilon)}(t)}{|t|^{n-\alpha+2}}\\
&\le Ax_n|x|^{n+2-\alpha}[\frac{m_x^{(\varepsilon)}(t)}{t^{n+2-\alpha}}\Big|_{\frac{x_n}{2}}^{3|x|}+\int_{\frac{x_n}{2}}^{3|x|}\frac{m_x^{(\varepsilon)}(t)}{|t|^{n-\alpha+3}}dt\\
&\le Ax_n|x|^{n+2-\alpha}[\frac{\varepsilon t^{n-\beta}}{t^{n+2-\alpha}|x|^{n-\beta}}\,|_{t=3|x|}\\
&\quad+\int_{\frac{x_n}{2}}^{3|x|}\frac{\varepsilon t^{n-\beta}}{t^{n+3-\alpha}|x|^{n-\beta}}dt]\\
&\le A|x|^{n+2-\alpha}x_n\varepsilon|3x|^{\alpha-n-2}\\
&\quad+A|x|^{n+2-\alpha}x_n\varepsilon\int_{\frac{x_n}{2}}^{3|x|}t^{n-\beta-3}dt\\
&\le A\varepsilon x,
\end{aligned}
\tag{11}
$$

where $m_x^{(\varepsilon)}(t)$ is defined by (10), and

$$
\begin{aligned}
|G_3(x)|&\le\int_{\{y:|x-y|\ge 3|x|\}}\frac{x_n(1+|y|)^{n+2-\alpha}}{|\bar{x}-y|^2|x-y|^{n-\alpha}}dm^{(\varepsilon)}(y)\\
&=Ax_n\int_{\{y:|x-y|\ge 3|x|\}}\frac{dm^{(\varepsilon)}(y)}{|x-y|^{n-\alpha+2}}\\
&\quad+x_n\int_{\{y:|x-y|\ge 3|x|\}}\frac{|y|^{n+2-\alpha}}{|x-y|^{n+2-\alpha}}dm^{(\varepsilon)}(y)\\
&\le Ax_n\int_{3|x|}^{+\infty}\frac{dm_x^{(\varepsilon)}(t)}{|t|^{n-\alpha+2}}\\
&\quad+Ax_n\int_{\{y:|x-y|\ge 3|x|\}}\frac{1}{3^{n+2-\alpha}}dm^{(\varepsilon)}(y)\\
&\le A\varepsilon x,
\end{aligned}
\tag{12}
$$

in which $m_x^{(\varepsilon)}(t)$ is defined by (10), and

$$
\begin{aligned}
|G_4(x)|&\le\int_{\{y:1<|y|<R_\varepsilon\}}\left|\frac{1}{|x-y|^{n-\alpha}}-\frac{1}{|\bar{x}-y|^{n-\alpha}}\right|d\nu(y)\\
&\le 4x_nR_\varepsilon\int_{\{y:1<|y|<R_\varepsilon\}}\frac{d\nu(y)}{(\frac{|x|}{2})^{n+2-\alpha}}\\
&=O(x),\quad |x|\to\infty,
\end{aligned}
$$

$$
\begin{aligned}
|G_5(x)| &\leq \int_{\{y:|y|\leq 1\}} \frac{4x_n y_n}{|x-y|^{n+2-\alpha}} d\nu(y) \\
&\leq 4x_n \int_{|y|\leq 1} \frac{d\nu(y)}{(\frac{|x|}{2})^{n+2-\alpha}} \\
&= O(x), \quad |x| \to \infty.
\end{aligned} \tag{13}
$$

Thus, by collecting (8), (9), (11), (12) and (13), there exists a positive constant A independent of ε, such that if $|x| \geq 2R_\varepsilon$ and $x \notin E_1(\varepsilon)$, we have

$$|G(x)| \leq A\varepsilon|x|.$$

Let μ_ε be a measure in $\mathbb{R}^n$ defined by $\mu_\varepsilon(E) = m^{(\varepsilon)}(E \cap \mathbb{R}^n)$ for every measurable set E in $\mathbb{R}^n$. Take $\varepsilon = \varepsilon_p = \frac{1}{2^{p+2}}$, $p = 1, 2, 3, \cdots$, then there exists a sequence $\{R_p\}$: $1 = R_0 < R_1 < R_2 < \cdots$, such that

$$\mu_{\varepsilon_p}(\mathbb{R}^n) = \int_{|y|\geq R_p} dm(y) < \frac{\varepsilon_p}{5^{n-\alpha}}.$$

Take $\lambda = 3 \cdot 5^{n-\alpha} \cdot 2^p \mu_{\varepsilon_p}(\mathbb{R}^n)$ in Lemma 2.2, then there exist $x_{j,p}$ and $\rho_{j,p}$, where $R_{p-1} \leq |x_{j,p}| < R_p$, such that

$$\sum_{j=1}^{\infty} \left(\frac{\rho_{j,p}}{|x_{j,p}|}\right)^{n-\alpha} \leq \frac{1}{2^p},$$

if $R_{p-1} \leq |x| < R_p$ and $x \notin G_p = \cup_{j=1}^{\infty} B(x_{j,p}, \rho_{j,p})$, we have

$$|G(x)| \leq A\varepsilon_p |x|.$$

Thereby

$$\sum_{p=1}^{\infty}\sum_{j=1}^{\infty} \left(\frac{\rho_{j,p}}{|x_{j,p}|}\right)^{n-\alpha} \leq \sum_{p=1}^{\infty} \frac{1}{2^p}.$$

Set $G = \cup_{p=1}^{\infty} G_p$, thus the main theorem holds.

References

[1] Loukas Grafakos, *Classical and modern Fourier analysis*, China Machine Press, Beijing, 2006.

[2] L. Hörmander, *Notions of convexity*, Birkäuser, Berlin, 1994.

[3] R. D. Bergman and W. S. Cohn, Littlewood theorems for limits and growth of potentials along level sets of hölder continuous functions, *Amer. J. Math.*, 114 (1991), 185–227.

[4] Y. H. Zhang and G. T. Deng, Growth properties for a class of subharmonic functions in half space, *Acta Mathematica Sinica,* 51(2) (2008), 319-326.

[5] B. Y. Levi, *Lectures on entire functions,* American Mathematical Society, Providence, RI, 1996.

[6] S. Axler, P. Bourdon and W. Ramey, *Harmonic function theory,* Springer-Verlag, Berlin, 1992.

[7] Y. H. Zhang and G. T. Deng, Integral representation and asymptotic property of analytic functions with order less than two in the half plane, *Complex Variables Theory and Application,* 50(4) (2005), 283–297.

[8] J. L. Doob, *Class potential theory and its probabilistic counterpart,* Springer-Verlag, Berlin, 1984.

[9] Y. Mizuta, On the boundary behavior of subharmonic functions in a half space, *J. Math. Soc.,* 34 (1982), 269–278.

THE PROBLEM OF SMOOTH SOLUTIONS FOR A PARABOLIC TYPE COMPLEX EQUATION OF SECOND ORDER[1]

GUANG-SHAN XU[a,b] and ZHONG-TAI MA[a]

[a]*School of Math., Shandong Institute of Business and Technology, Yantai 264005, China*
E-mail: zhongtm@126.com

[b]*Institute of Applied Math., Naval Aeronautical Engineering Academy, Yantai 264001, China*
E-mail: liyan19616@yahoo.com.cn

In this paper, the problem of smooth solutions for a general second order parabolic type complex equation

$$u_{xx} - u_{yy} + 2iu_{xy} - 2iz(u_x + iu_y)_t = f(x,y,t), \quad (x,y,t) \in \mathbb{R}^3$$

is discussed, and the result–there are C^∞-functions f for which the equation above has no C^2-solutions is proved by using a method of function construction, moreover, the dependent relations between the solution u and the diffusion term f is given.

Keywords: Parabolic type complex equation, holomorphic solution, Cauchy-Kowaleski theorem, Schwarz reflection principle.

AMS No: 35A07, 35B30, 35K10.

1. Introduction

The analysis of nonlinear parabolic type equations has experienced a dramatic growth in the last ten years or so. The key factor in this has been the transition from linear analysis. It is well known to us, in dealing with the existence of solutions of partial differential equations, it was customary during the past years and it still is today in many applications. For large classes of equations this extension of the range of equation and solution has been carried out, since the beginning of the last century, in particular much attention has been given to linear partial differential equations and system of such. If the coefficients of a linear partial equation are holomorphic, then by the Cauchy-Kowaleski theorem in [3], we can always conclude that there exist plenty of solutions to the given differential equation. However, as pointed out by Lewy [6] in 1957, if the coefficients are not holomorphic, there may be not exist any solutions at all. Since Lewy's paper, there has been an extensive investigation into the conditions under which a given linear differential equation admits solutions, this can be referred to Egorov [1], Hömander [5], Nirenberg [7] and Nirenberg-Tréves [8–9] for past and current developments in this area. Here our aims are more modest, and we simply wish to present an example of a general parabolic type complex differential equation having no solution (cf. Garabedian [2] and Grushin

[1]This research is supported by NSF (No.Y2008A31)

[4]). On the side of complex equations, we want to especially mention G. C. Wen, he has first established a new method to research the real equations using the complex analysis method, in which avoided the using of the complicated functional relations, obtained and generalized a host of results (for example to see [10–15]). The connection of the present example with the subject of this article is in one sense negative–that is, instead of a solution being as nice as possible, i.e., holomorphic, it is as bad as possible, i.e., it does not even exist–and in another sense positive.

2. The Main Results and Its Proof

In this section, we deal with the following problem of smooth solutions for a general parabolic type complex equation of second order

$$u_{xx} - u_{yy} + 2iu_{xy} - 2iz(u_x + iu_y)_t = f(x, y, t), \quad (x, y, t) \in \mathbb{R}^3, \tag{1}$$

where $z = x + iy \in \mathbb{C}, i^2 = -1$. If $f(x, y, t)$ is a holomorphic function, then from the classical Cauchy-Kowaleski theorem [3], the equation $u_{xx} - u_{yy} + 2iu_{xy} - 2iz(u_x + iu_y)_t = f(x, y, t)$ would admit holomorphic solutions. But if f is not a holomorphic function, what would be happen in this cases? For this problem, here and henceforth, we denote the right hand of the equation (1) by using $\mathcal{L}$, i.e., $\mathcal{L}u = u_{xx} - u_{yy} + 2iu_{xy} - 2iz(u_x + iu_y)_t$, then first, we begin by giving one theorem as follows.

Theorem 1. *Suppose that $\varphi(t)$ be a real function of C^∞. Then, if φ is not holomorphic at $t = 0$ we have, the equation $\mathcal{L}u = \varphi'(t)$ has no C^2-solution in any neighborhood of $0 \in \mathbb{R}^3$.*

Proof. Suppose that $u = u(x, y, t)$ is a solution of equation $\mathcal{L}u = \varphi'(t)$ in $|z| < R, |t| < T$, where $z = x + iy$. In the region $|t| < T, 0 \le r < R$, let

$$g(r, t) = \int_{|z|=r} (u_x + iu_y)dz = \int_{|z|=r} u_x dx - u_y dy + i \int_{|z|=r} u_y dx + u_x dy, \tag{2}$$

we have

$$\begin{aligned} g(r, t) &= -2 \int_{|z|\le r} u_{xy}\, dxdy + i \int_{|z|\le r} (u_{xx} - u_{yy})dxdy \\ &= -2 \int_0^r \int_0^{2\pi} u_{xy} \rho d\rho d\theta + i \int_0^r \int_0^{2\pi} (u_{xx} - u_{yy}) \rho d\rho d\theta, \end{aligned} \tag{3}$$

by Green's theorem. Set $y = r^2$, $G(y, t) = g(r^2, t)$. Then $G_y = g_r r_y =$

$g_r/(2r)$, so that using the equation $\mathcal{L}u = \varphi'(t)$, we have

$$\begin{aligned} G_y &= \int_{|z|=r^2} iu_{xy}\,dz/z + \frac{1}{2}\int_{|z|=r^2}(u_{xx}-u_{yy})\,dz/z \\ &= \frac{1}{2}\int_{|z|=r^2}(u_{xx}-u_{yy}+2iu_{xy})\,dz/z \\ &= i\int_{|z|=r^2}(u_x+iu_y)_t\,dz + \frac{1}{2}\int_{|z|=r^2}\varphi'(t)\,dz/z \\ &= iG_t + \frac{1}{2}\varphi'(t)\cdot 2\pi i \\ &= i[G_t+\pi\varphi'(t)], \end{aligned} \tag{4}$$

that is $G_y = i[G_t + \pi\varphi'(t)]$ or $G_t + \pi\varphi'(t) = -iG_y$. If we set $F(y,t) = G(y,t) + \pi\varphi(t)$, then $F_y = G_y, F_t = G_t + \pi\varphi'(t) = -iG_y = -iF_y$, i.e., $F_t + iF_y = 0$ or $\bar{\partial}F = (F_t + iF_y)/2 = 0$. Thus F satisfies the Cauchy-Riemann equations, and since F is a C^2-function, F is a holomorphic function of $t + iy$. From $F(0,t) = G(0,t) + \pi\varphi(t) = \pi\varphi(t)$, we see that F is a real function on $y = 0$. By the Schwarz reflection principle, F can be extended to a holomorphic function in the region $D := \{-R^2 < y < 0, |t| < T\}$, by defining $F(t+iy) = \overline{F(t-iy)}$ for $y \leq 0$. Thus $\pi\varphi(t) = F(0,t)$ is holomorphic at $t = 0$, so that this proves the equation $\mathcal{L}u = \varphi'(t)$ has no C^2-solution in any neighborhood of $0 \in \mathbb{R}^3$, if φ is not holomorphic at $t = 0$. Therefore the proof of the Theorem 1 is completed.

From Theorem 1 we immediately have

Theorem 2. *Suppose that $\varphi(t)$ be a real function of C^∞ and if $u = u(x,y,t)$ is a C^2-solution of the general parabolic type complex equation*

$$u_{xx} - u_{yy} + 2iu_{xy} - 2iz(u_x + iu_y)_t = \varphi'(t), \quad (x,y,t) \in \mathbb{R}^3 \tag{5}$$

in a neighborhood $U(0)$ of the point $0 \in \mathbb{R}^3$. Then $\varphi(t)$ is holomorphic at $t = 0$.

By a change of coordinates, we have the following results.

Theorem 3. *Let $u = u(x,y,t)$ be C^2 near (x_0,y_0,t_0). Then the equation $\mathcal{L}u = \varphi'(t-t_0-2y_0x+2x_0y)$ implies that φ is holomorphic in a neighborhood of $0 \in \mathbb{R}^3$.*

Proof. Let $\bar{x} = x - x_0, \bar{y} = y - y_0, \bar{t} = t - t_0 - 2y_0x + 2x_0y$. Then we can easily to obtain

$$\begin{aligned} \mathcal{L}u &= u_{xx} - u_{yy} + 2iu_{xy} - 2iz(u_x + iu_y)_t \\ &= u_{\bar{x}\bar{x}} - u_{\bar{y}\bar{y}} + 2iu_{\bar{x}\bar{y}} - 2i(\bar{x}+i\bar{y})(u_{\bar{x}}+iu_{\bar{y}})_{\bar{t}} \\ &= \varphi'(\bar{t}). \end{aligned} \tag{6}$$

Hence $\mathcal{L}u = \varphi'(t - t_0 - 2y_0x + 2x_0y)$ has a solution in a neighborhood of $(\bar{x}, \bar{y}, \bar{t})$ if and only if $\mathcal{L}u = \varphi'(\bar{t})$ has a solution in a neighborhood of $0 \in \mathbb{R}^3$.

From the proof of the Theorem 3, we have the corollary as follows:

Corollary 4. *Suppose that $\varphi(t)$ be a real function of C^∞ and if $u = u(x, y, t)$ is a C^2-solution of the general parabolic type complex equation* (5) *in a neighborhood $U(x_0, y_0, t_0)$ of the point $(x_0, y_0, t_0) \in \mathbb{R}^3$. Then $\varphi(t)$ is holomorphic at t_0.*

Using the same method as the proof of Lewy [6], we have the following result.

Theorem 5. *There exits C^∞-function $f(x, y, t)$ such that the general parabolic type complex equation $\mathcal{L}u = f(x, y, t)$ has no C^2-solution anywhere in $\mathbb{R}^3$.*

References

[1] V. Egorov, On the solvability of differential equation with simple characteristics, *Russian Math. Surveys*, 26 (1971), 113–130.

[2] P. Garabedian, An unsolvable problem, *Proc. Amer. Math. Soc.*, 25 (1970), 207–208.

[3] P. Garabedian, *Partial differential equations,* Wiley, New York, 1964.

[4] V. Grushin, A certain example of a differential equation without solutions, *Math. Notes,* 10 (1971), 499–501.

[5] L. Hörmander, *Linear partial differential operators*, Springer-Verlag, Berlin, 1963.

[6] H. Lewy, An example of a smooth linear partial differential equation, *Ann. Math.*, 66 (1957), 617–628.

[7] L. Nirenberg, *Lectures on linear Partial differential equations,* Amer. Math. Soc., Providence, RI, 1973.

[8] L. Nirenberg and F. Tréves, On local solvability of linear Partial differential equations, I, Necessary conditions, *Comm. Pure. Appl. Math.*, 23 (1970), 1–38.

[9] L. Nirenberg and F. Tréves, On local solvability of linear Partial differential equations, II, Sufficient conditions, *Comm. Pure. Appl. Math.*, 23 (1970), 459–510.

[10] G. C. Wen, Oblique derivative problems for linear equations of second order, *Science in China* (*Series A*), 41 (1998), 285–295.

[11] G. C. Wen, *Nonlinear Partial differential Complex Equations,* Science Press, Beijing, 1999 (in Chinese).

[12] G. C. Wen, Boundary value problems for degenerate hyperbolic system of first order equations, *Complex Variables,* 47 (2002), 839–851.

[13] G. C. Wen, *Initial-boundary value problems for nonlinear parabolic equations in higher dimensional domains,* Science Press, Beijing, 2002.

[14] G. C. Wen, *Linear and quasilinear complex equations of hyperbolic and mixed type,* Taylor & Francis Inc., Londan, 2002.

[15] G. C. Wen and B. T. Zou, Initial-irregular oblique derivative problems for nonlinear parabolic complex equations of second order with measurable coefficients, *Nonlinear Anal.,* 39 (2000), 937–953.

SOME RESULTS OF THE RANGE ABOUT THE EXPONENTIAL RADON TRANSFORM[1]

JING LI and JIN-PING WANG*

Department of Mathematics, Ningbo University, Ningbo 315211, China

* *Corresponding author. E-mail: wangjinping@nbu.edu.cn(Jin-Ping Wang)*

In this paper, a new theorem to the range conditions for the exponential Radon transform is established, their equivalent relationship theorem is proved with a new method.

Keywords: Exponential Radon transform, range conditions; Fourier transform, inversion formula.

AMS No: 42C15, 44A12.

1. Introduction

The set of all oriented lines in the plane can be identified with the spaces $\mathbb{R} \times S^1$ by associating the pair $(p, \omega) \in \mathbb{R} \times S^1$ with the line $L(p, \omega) = \{x \in \mathbb{R}^2 | x \cdot \omega = p\}$. A parameterization of this line is then given by the mapping $t \mapsto p\omega + t\omega^\perp$, $\omega^\perp$ is obtained by rotating ϕ counterclockwise through a right angle, where $\omega = (\omega_1, \omega_2) = (\cos\phi, \sin\phi) \in S^1 \subset \mathbb{R}^2$ is a unit vector, μ is a real (usually positive) number, which plays the role of a parameter, dot product $x \cdot \omega = x_1\omega_1 + x_2\omega_2$, and p is a real number.

Let f be a smooth, compactly supported function in the plane $\mathbb{R}^2$. The Radon transform of f is the function Rf on $\mathbb{R} \times S^1$ defined so that $Rf(p, \omega)$ is the integral of f along the line $L(p, \omega)$. If μ is a real number, the exponential Radon transform $R_\mu f$ is defined by the weighted integral

$$R_\mu f(p, \omega) = \int_{-\infty}^{+\infty} f(p\omega + t\omega^\perp) e^{\mu t} dt. \tag{1}$$

Note that the ordinary Radon transform is obtained as a special case of the exponential Radon transform when $\mu = 0$.

Both the Radon transform and the exponential Radon transform, as well as the still more general attenuated Radon transform, arise in applications to medical imaging, see [3, 6, 7]. It is then of interest to invert and characterize range of the transform. For the ordinary Radon transform was studied by J. Radon and A. Hertle [5] and a generalization to the exponential Radon transform was derived by O. Tretiak and C. Metz [2]; V. Aguilar, L. Ehrenpreis and P. Kuchment, see [1, 4]. A further generalization to the attenuated Radon transform was recently developed by R. Novikov and Xiaochuan Pan, see [9, 10].

[1]This research is supported by the NSFC (No. 60872095), and Ningbo Natural Science Foundation (2008A610018, 2009B21003, 2010A610100).

The aim of this paper is to find range conditions, and to prove their equivalent relationship theorem.

2. Main Results

Let $\hat{g}(\xi,\omega)$ be the Fourier transform of $R_\mu f(p,\omega)$ with respect to the first variable. Then

$$\hat{g}_l(\xi)(\xi+\mu)^l \text{ is an even function of } \xi \text{ for any integer } l, \tag{A}$$

where

$$\hat{g}(\xi,\omega(\phi)) = \sum_{-\infty}^{\infty} \hat{g}_l(\xi)e^{il\phi} \tag{2}$$

is the Fourier expansion of $\hat{g}(\xi,\omega(\phi))$ with respect to the angle ϕ (as before, $\omega(\phi) = (\cos\phi\ \sin\phi)$). The range conditions for R_μ are obtained:

Theorem 1. *Let $g(p,\omega)$ be a function on $\mathbb{R}\times S^1$, and μ be a positive real. Then $g = R_\mu f$ for some function $f \in C_0^\infty(\mathbb{R}^2)$ if and only if $g \in C_0^\infty(\mathbb{R}\times S^1)$, and the Fourier transform $\hat{g}(\xi,\phi)$ with respect to p variable of $g(p,\omega(\phi))$ satisfies the following condition for all real σ:*

$$\hat{g}\left(i\sigma, \phi - \arcsin\frac{\sigma}{\sqrt{\sigma^2+\mu^2}}\right) = \hat{g}\left(-i\sigma, \phi + \arcsin\frac{\sigma}{\sqrt{\sigma^2+\mu^2}}\right). \tag{3}$$

Here we employ the relation $\omega = (\cos\phi, \sin\phi)$ between ω and ϕ.

In [1], they used the projection-slice theorem, which is introduced in [3] to prove the necessary condition of the theorem and implied Paley-wiener theorem, constructed a new function to prove the sufficient condition of the theorem. In the following we only give the proof for its necessity.

Proof. Necessity. The inclusion $g \in C_0^\infty(\mathbb{R}\times S^1)$ for $f \in C_0^\infty(\mathbb{R}^2)$ is obvious. Let us assume that the attenuation coefficient μ is strictly positive (the case of a negative μ can be handled exactly the same way). The projection-slice theorem for R_μ reads

$$\widehat{R_\mu f}(\xi,\omega) = (2\pi)^{\frac{1}{2}}\tilde{f}(\xi\omega + i\mu\omega^\perp), \tag{4}$$

where hat denotes the one-dimensional Fourier transform with respect to the p variable, and tilde denotes the two-dimensional Fourier transform. If we put purely imaginary values $\xi = i\sigma$ into (4), and get

$$\widehat{R_\mu f}(i\sigma,\omega) = (2\pi)^{\frac{1}{2}}\tilde{f}(i(\sigma\omega + \mu\omega^\perp)).$$

The vectors $\nu = \sigma\omega + \mu\omega^\perp$ obviously cover the exterior of the disk of radius μ centered at the origin in $\mathbb{R}^2$, and a vector ν of the above form belongs to the line tangent to this disk at the point $\mu\omega^\perp$. Every point a outside of the disk can be reached along two different tangent lines, and hence it can be represented as ν in two different ways:

$$a = \sigma\omega_1 + \mu\omega_1^\perp = -\sigma\omega_2 + \mu\omega_2^\perp.$$

If the polar angle of $\omega_1^\perp$ is ϕ, then the polar angle of $\omega_2^\perp$ is

$$\phi + 2\arcsin\frac{\sigma}{\sqrt{\sigma^2+\mu^2}}.$$

Hence, if a function $r(\sigma,\phi)$ can be represented as

$$r(\sigma,\phi) = \text{const}\cdot\tilde{f}(i(\sigma\omega + \mu\omega^\perp)),$$

then it automatically satisfies the following condition:

$$r(\sigma,\phi) = r(-\sigma, \phi + 2\arcsin\frac{\sigma}{\sqrt{\sigma^2+\mu^2}}). \tag{5}$$

In order to make the relation more symmetric, we use the angle $\phi - \arcsin(\sigma/\sqrt{\sigma^2+\mu^2})$ instead of ϕ in (5)

$$r\left(\sigma, \phi - \arcsin\frac{\sigma}{\sqrt{\sigma^2+\mu^2}}\right) = r\left(-\sigma, \phi + \arcsin\frac{\sigma}{\sqrt{\sigma^2+\mu^2}}\right).$$

So we get

$$\hat{g}\left(i\sigma, \phi - \arcsin\frac{\sigma}{\sqrt{\sigma^2+\mu^2}}\right) = \hat{g}\left(-i\sigma, \phi + \arcsin\frac{\sigma}{\sqrt{\sigma^2+\mu^2}}\right).$$

This finishes the proof of the necessity of the theorem.

Due to the simple geometric meaning, the analog of (3) can be easily written in the case of an angle-depending attenuation $\mu(\omega)$. In this case, we get the following necessary range conditions:

Theorem 2. *Let $g(p,\omega)$ be a function on $\mathbb{R}\times S^1$, μ be a positive real, and $g = R_\mu f$ for some function $f \in C_0^\infty(\mathbb{R}^2)$. Then $g \in C_0^\infty(\mathbb{R}\times S^1)$, and the Fourier transform $\hat{g}(\xi,\omega)$ with respect to p variable of $g(p,\omega)$ satisfies the following condition for all real σ_1,σ_2:*

$$\hat{g}(i\sigma_1,\omega_1) = \hat{g}(i\sigma_2,\omega_2), \tag{6}$$

for any σ_j, ω_j such that

$$\sigma_1\omega_1 + \mu(\omega_1)\omega_1^\perp = \sigma_2\omega_2 + \mu(\omega_2)\omega_2^\perp. \tag{7}$$

Proof. The inclusion $g \in C_0^\infty(\mathbb{R} \times S^1)$ for $f \in C_0^\infty(\mathbb{R}^2)$ is obvious. In the above proof of Theorem 1, instead of the disk of radius μ, we get a curve with parametric representation

$$a = \sigma\omega + \mu(\omega)\omega^\perp,$$

where $\mu(\omega)$ is an angle-depending attenuation. At every point of this curve we draw the line which is perpendicular to the radius vector of the point. If two lines intersect at some point, then we have for any σ_j, ω_j such that

$$a = \sigma_1\omega_1 + \mu(\omega_1)\omega_1^\perp = \sigma_2\omega_2 + \mu(\omega_2)\omega_2^\perp.$$

If a function $r(\sigma, \omega)$ can be represented as

$$r(\sigma, \omega) = \text{const} \cdot \tilde{f}(i(\sigma\omega + \mu\omega^\perp)),$$

and we have

$$r(\sigma_1, \omega_1) = r(\sigma_2, \omega_2).$$

Since the projection-slice theorem (4) and let us put purely imaginary values $\xi = i\sigma$ into (4), and get

$$\widehat{R_\mu f}(i\sigma, \omega) = (2\pi)^{\frac{1}{2}} \tilde{f}(i(\sigma\omega + \mu\omega^\perp)).$$

Then we have

$$\hat{g}(i\sigma_1, \omega_1) = \hat{g}(i\sigma_2, \omega_2).$$

Remark. From the proofs of Theorems 1 and 2, we get condition (3) is equivalent to (6) and (7), and they have the similar geometric meaning.

Theorem 3. *Let $\hat{g}(\xi, \omega)$ be the Fourier transform of $R_\mu f(p, \omega)$ with respect to the first variable,*

$$\hat{g}_l(\xi)(\xi + \mu)^l \text{ is an even function of } \xi \text{ for any integer } l. \tag{A}$$

Then we have

$$\hat{g}\left(i\sigma, \phi + \arcsin\frac{\sigma}{\sqrt{\sigma^2 + \mu^2}}\right) = \hat{g}\left(-i\sigma, \phi - \arcsin\frac{\sigma}{\sqrt{\sigma^2 + \mu^2}}\right). \tag{8}$$

Here we also employ the relation $\omega = (\cos\phi, \sin\phi)$ between ω and ϕ.

Proof. Since (A): $\hat{g}_l(\xi)(\xi + \mu)^l$ is an even function of ξ for any integer l, so we get

$$\hat{g}_l(\xi)(\xi + \mu)^l = \hat{g}_l(-\xi)(-\xi + \mu)^l. \tag{9}$$

Since we assume the function f is compactly supported, we may plug any complex ξ into (9). Let us use for this purpose purely imaginary values $\xi = i\sigma$. Then we get in (9) the values

$$\hat{g}_l(i\sigma)(\mu + i\sigma)^l = \hat{g}_l(-i\sigma)(\mu - i\sigma)^l. \tag{10}$$

In order to prove our result, we use $\frac{-\sin\phi+i\cos\phi}{\cos\phi+i\sin\phi}$ instead of i in (10), then (10) can be written as

$$\hat{g}_l(i\sigma)\left(\mu+\frac{-\sin\phi+i\cos\phi}{\cos\phi+i\sin\phi}\sigma\right)^l=\hat{g}_l(-i\sigma)\left(\mu-\frac{-\sin\phi+i\cos\phi}{\cos\phi+i\sin\phi}\sigma\right)^l. \tag{11}$$

We have (11) multiplied by $(\cos\phi + i\sin\phi)^l$, and get

$$\begin{aligned}&\hat{g}_l(i\sigma)[\mu(\cos\phi + i\sin\phi) + \sigma(-\sin\phi + i\cos\phi)]^l\\ &=\hat{g}_l(-i\sigma)[\mu(\cos\phi+i\sin\phi)-\sigma(-\sin\phi+i\cos\phi)]^l.\end{aligned} \tag{12}$$

Divide (12) by $(\sqrt{\sigma^2+\mu^2})^l$, and then

$$\begin{aligned}&\hat{g}_l(i\sigma)\,[\frac{\mu}{}\sqrt{\sigma^2+\mu^2}\,(\cos\phi + i\sin\phi) + \frac{\sigma}{\sqrt{\sigma^2+\mu^2}}(-\sin\phi + i\cos\phi)]^l\\ &=\hat{g}_l(-i\sigma)[\frac{\mu}{\sqrt{\sigma^2+\mu^2}}(\cos\phi+i\sin\phi)-\frac{\sigma}{\sqrt{\sigma^2+\mu^2}}(-\sin\phi + i\cos\phi)]^l,\end{aligned} \tag{13}$$

$$\begin{aligned}&\hat{g}_l(i\sigma)\,[\,\cos\alpha\,(\cos\phi + i\sin\phi) + \sin\alpha\,(-\sin\phi + i\cos\phi)\,]^l\\ &=\hat{g}_l(-i\sigma)[\cos\alpha(\cos\phi+i\sin\phi)-\sin\alpha(-\sin\phi+i\cos\phi)]^l,\end{aligned} \tag{14}$$

where $\sin\alpha = \frac{\sigma}{\sqrt{\sigma^2+\mu^2}}$, $\cos\alpha = \frac{\mu}{\sqrt{\sigma^2+\mu^2}}$. It is easy to know that (14) can be written as

$$\begin{aligned}&\hat{g}_l(i\sigma)\,[\,(\cos\alpha\cos\phi - \sin\alpha\sin\phi) + i\,(\cos\alpha\sin\phi + \sin\alpha\cos\phi)\,]^l\\ &=\hat{g}_l(-i\sigma)[(\cos\alpha\cos\phi+\sin\alpha\sin\phi)+i(\cos\alpha\sin\phi-\sin\alpha\cos\phi)]^l,\end{aligned} \tag{15}$$

so

$$\hat{g}_l(i\sigma)[\cos(\phi+\alpha)+i\sin(\phi+\alpha)]^l=\hat{g}_l(-i\sigma)[\cos(\phi-\alpha)+i\sin(\phi-\alpha)]^l.$$

Obviously, we get

$$\hat{g}_l(i\sigma)e^{il(\phi+\alpha)} = \hat{g}_l(-i\sigma)e^{il(\phi-\alpha)}, \tag{16}$$

and by summing about l on both sides of (16), then we have

$$\sum_{-\infty}^{\infty}\hat{g}_l(i\sigma)e^{il(\phi+\alpha)} = \sum_{-\infty}^{\infty}\hat{g}_l(-i\sigma)e^{il(\phi-\alpha)}.$$

Due to the definition of $\hat{g}(\xi, \omega(\phi))$ (as (2)), we get

$$\hat{g}(i\sigma, \phi + \alpha) = \hat{g}(-i\sigma, \phi - \alpha). \tag{17}$$

Since

$$\sin\alpha = \frac{\sigma}{\sqrt{\sigma^2 + \mu^2}}, \quad \cos\alpha = \frac{\mu}{\sqrt{\sigma^2 + \mu^2}},$$

so we have

$$\alpha = \arcsin\frac{\sigma}{\sqrt{\sigma^2 + \mu^2}}. \tag{18}$$

Substituting (18) into (17), furthermore we obtain the result

$$\hat{g}\left(i\sigma, \phi + \arcsin\frac{\sigma}{\sqrt{\sigma^2 + \mu^2}}\right) = \hat{g}\left(-i\sigma, \phi - \arcsin\frac{\sigma}{\sqrt{\sigma^2 + \mu^2}}\right).$$

Theorem 4. *Let conditions be stated as Theorem* 1*. Then the condition* (8) *is equivalent to* (3).

Proof. In the proof of Theorem 1, if the polar angle of $\omega_1^\perp$ is ϕ, then the polar angle of $\omega_2^\perp$ also can be

$$\phi - 2\arcsin\frac{\sigma}{\sqrt{\sigma^2 + \mu^2}}.$$

We can get

$$r(\sigma, \phi) = r(-\sigma, \phi - 2\arcsin\frac{\sigma}{\sqrt{\sigma^2 + \mu^2}}). \tag{19}$$

We use the angle $\phi + \arcsin(\sigma/\sqrt{\sigma^2 + \mu^2})$ instead of ϕ in (19)

$$r\left(\sigma, \phi + \arcsin\frac{\sigma}{\sqrt{\sigma^2 + \mu^2}}\right) = r\left(-\sigma, \phi - \arcsin\frac{\sigma}{\sqrt{\sigma^2 + \mu^2}}\right).$$

So

$$\hat{g}\left(i\sigma, \phi + \arcsin\frac{\sigma}{\sqrt{\sigma^2 + \mu^2}}\right) = \hat{g}\left(-i\sigma, \phi - \arcsin\frac{\sigma}{\sqrt{\sigma^2 + \mu^2}}\right).$$

Hence the condition (8) is equivalent to (3).

References

[1] V. Aguilar, L. Ehrenpreis and P. Kuchment, Range conditions for the exponetial Radon transform, *J. Analyse Math.*, 68 (1996), 1–13.

[2] O. Tretiak and C. Metz, The exponential Radon transform, *SIAM J. Appl. Math.*, 39 (1980), 341–354.

[3] F. Natterer, *The mathematics of computerized tomography*, Wiley, New York, 1986.

[4] P. Kuchment and S. Lvin, Paley-Wiener theorem for exponential Radon transform, *Acta Appl. Math.*, 18 (1990), 251–260.

[5] A. Hertle, On the range of the Radon transform and its dual, *Math. Ann.*, 267 (1984), 91–99.

[6] Jinping Wang and Jinyuan Du, The harmonic decomposition reconstruction for the exponential Radon transform, *Wuhan University J. of Natural Sciences,* Wuhan Univ. J. Nat. Sci. 7 (2002), no. 1, 1–5.

[7] Jinping Wang, Some results on the exponential Radon transform, *Acta Math. Sci. Ser. A Chin. Ed.*, 26 (2006), 31–38.

[8] P. Kuchment and S. Lvin, The range of the exponential Radon transform, *Soviet Math. Dokl.*, 42 (1991), 183–184.

[9] Xiaochuan Pan, C. Kao and C. Metz, A family of π-scheme exponential Radon transforms and the uniqueness of their inverses, *Inverse Problems,* 18 (2002), 825–836.

[10] R. Novikov, On the range characterization for the two-dimensional attenuated X-ray transform, *Inverse Problems,* 18 (2002), 677–700.

WEAK MINIMA AND QUASIMINIMA OF VARIATIONAL INTEGRALS IN CARNOT-CARATHÉODORY SPACE

LAN-RU HOU

Beijing Science Technology and Management College, Beijing 102206, China

E-mail: hlr03@163.com

In *Carnot-Carathéodory* space, we prove the existence of an exponent r_1 $(1 < r_1 < p)$, such that every very weak solution $u \in W^{1,r}_{loc}(\Omega)$ $(r_1 \leq r \leq p)$ of the equation $\operatorname{div}(|\nabla u|^{p-2}\nabla u) = 0$ is a Q-quasiminima of functional $\int_\Omega |Xu|^r dx$, and Q is independent of r.

Keywords: Weak minima, quasiminima, Carnot-Carathéodory space.

AMS No: 45A05.

1. Introduction

Let $\Omega \subset \mathbb{R}^n$ be a bounded region and $X = (X_1, \ldots, X_n)$ be a family of smooth vector fields in $\mathbb{R}^n$ defined on a neighborhood of Ω with real, C^∞ coefficients. The family satisfies the Hörmander condition, if there exists an integer m such that a family of commutators of the vector fields up to the length m, i.e. the family of vector fields

$$X_1, \ldots, X_n, [X_{i_1}, X_{i_2}], \ldots, [X_{i_1}, [X_{i_2}, [\ldots, X_{i_m}]] \cdots], \; i_j = 1, 2, \ldots, n,$$

spans the tangent space $T_x\mathbb{R}^n$ at every point $x \in \mathbb{R}^n$.

For $u \in \text{Lip}(\mathbb{R}^n)$, we define $X_j u$ by

$$X_j u(x) = \langle X_j(x), \nabla u(x) \rangle, \quad j = 1, 2, \ldots, n,$$

and set $Xu = (X_1 u, \ldots, X_n u)$. Its length is given by

$$|Xu(x)| = \left(\sum_{j=1}^{n} |X_j u(x)|^2 \right)^{1/2},$$

$X^* = (X_1^*, \ldots, X_n^*)$ is a family of operators, where X_j^* is a formal adjoint to X_j in L_2, i.e.

$$\int_{\mathbb{R}^n} (X_j^* u) v \mathrm{d}x = - \int_{\mathbb{R}^n} u X_j v \mathrm{d}x \quad \text{for functions } u, v \in C_0^\infty(\mathbb{R}^n).$$

Given $\mathbb{R}^n$ with the family of vector fields, we define a distance function ϱ. We say that an absolutely continuous curve $\gamma : [a, b] \longrightarrow \mathbb{R}^n$ is admissible, if there exist functions $c_j : [a, b] \longrightarrow \mathbb{R}\,(j = 1, \ldots, n)$, such that

$$\dot{\gamma}(t) = \sum_{j=1}^{n} c_j(t) X_j(\gamma(t)) < \infty \quad \text{and} \quad \sum_{j=1}^{n} c_j(t)^2 \leq 1.$$

Functions c_j do not need to be unique, because vector fields X_j do not need to be linearly independent. The distance $\varrho(x, y)$ between points x and y is defined as the infimum of those $T > 0$ for which there exists an admissible curve $\gamma : [0, T] \longrightarrow \mathbb{R}^n$, such that $\gamma(0) = x$ and $\gamma(T) = y$. If such a curve does not exist, we set $\varrho(x, y) = \infty$. The function ϱ is called the Carnot-Carathéodory distance. In general it does not need to be a metric. When the family $X_1, \ldots, X_n$ satisfies the Hörmander condition, then ϱ is a metric and we say that $(\mathbb{R}^n, \varrho)$ is a Carnot-Carathéodory space.

Here and subsequently all the distances will be with respect to the metric ϱ. The metric ϱ is locally Hölder continuous with respect to the euclidean metric. Thus the space $(\mathbb{R}^n, \varrho)$ is homeomorphic with the euclidean space $\mathbb{R}^n$, and every set which is bounded in Euclidean metric is also bounded in the metric ϱ. The reverse implication is not true. However, if $X_1, \ldots, X_n$ have globally Lipschitz coefficients, then it is known that every bounded set with respect to ϱ is also bounded in euclidean metric.

In the following, all the balls B are balls with respect to the C. C. metric. If $\sigma > 0$ and $B = B(x, r)$, then σB will denote a ball centered in x of radius $\sigma \cdot r$. By diamΩ we will denote the diameter of the set Ω. We will consider the Lebesgue measure in the Carnot-Carathéodory space. As we change the metric, the measure of $B(x, r)$ is no longer equal to the familiar $\omega_n r^n$. However, the important fact is that the Lebesgue measure in the Carnot-Carathéodory space satisfies the so-called doubling condition.

Theorem 1.1. *Let Ω be an open, bounded subset of $\mathbb{R}^n$. There exists a constant $C_d \geq 1$ such that*

$$|B(x_0, 2r)| \leq C_d |B(x_0, r)|, \tag{1.1}$$

provided $x_0 \in \Omega$ and $r < 5\text{diam}\Omega$.

The best constant C_d is known as the doubling constant and we call a measure satisfying the above condition a doubling measure. Iterating (1.1) we obtain a lower bound on $\mu(B(x, r))$. Given a first-order differential operator $X = (X_1, \ldots, X_n)$, the Sobolev space $W_X^{1,p}(\Omega)$ is defined in the following way:

$$W_X^{1,p}(\Omega) = \{u \in L^p(\Omega) : X_j u \in L^p(\Omega),\ j = 1, 2, \ldots, n\},$$

where $X_j u$ is the distributional derivative. The $W_X^{1,p}$ norm is defined by

$$\|u\|_{1,p} = \|u\|_p + \|Xu\|_p.$$

Smooth functions are dense in $W_X^{1,p}(\Omega)$. The existence of smooth cut-off functions is known. We have Sobolev and Poincaré type inequalities.

Theorem 1.2. *Let Q be a homogeneous dimension relative to Ω. There exist constants $C_1, C_2 > 0$, such that for every metric ball $B = B(x,r)$, where $x \in \Omega$ and $r \leq \operatorname{diam}\Omega$, the following inequalities hold:*

$$\left(\fint_B |u-u_B|^{s^*}dx\right)^{1/s^*} \leq C_1 r\left(\fint_B |Xu|^s dx\right)^{1/s} \quad for\ 1\leq s<Q,$$

where $s^* = Qs/(Q-s)$ and

$$\fint_B |u-u_B|^s dx \leq C_2 r^s \fint_B |Xu|^s dx \quad for\ 1 \leq s < \infty.$$

Consider the integral functional

$$\int_\Omega F(x, Xu(x)dx, \tag{1.2}$$

in which Ω is an open subset of $\mathbb{R}^n$ $(n \geq 2)$, $u\colon \Omega \to \mathbb{R}^m$ $(m \geq 1)$ and $F\colon \Omega \times \mathbb{R}^{mn} \to \mathbb{R}$ is a Carathéodory function, such that

$$|F(x,\xi)| \leq c|\xi|^p + \alpha(x)$$

for $p > 1$ and $\alpha(x) \in L^1(\Omega)$. If F satisfies the following Lipschitz type condition

$$|F(x,\xi+\eta) - F(x,\xi)| \leq c|\eta|[|\xi|^{p-1} + |\eta|^{p-1} + g(x)],$$

then the notion of the weak minimizer makes sense.

Definition 1.1. A mapping $u \in W^{1,r}_{loc}(\Omega, \mathbb{R}^m)$, $\max\{1, p-1\} \leq r \leq p$, is called a weak minimizer of the integral (1.2), if

$$\int_\Omega [F(x, Xu + X\Phi) - F(x, Xu)]dx \geq 0$$

for all $\Phi \in W^{1,r/(r-p+1)}(\Omega, \mathbb{R}^m)$ with compact support.

If we assume that F is differentiable with respect to the variable $\xi \in \mathbb{R}^{nm}$, then it can be proved that the weak minimizer of (1.2) solves the equation

$$\int_\Omega A(x, Xu)X\Phi dx = 0 \tag{1.3}$$

for all $\Phi \in W_0^{1,r/(r-p+1)}$, where $A(x,\xi) = X_\xi F(x,\xi)$. Let us note that $r/(r-p+1) > p$ for $r < p$. Then we can say that u is a very weak solution of the Euler-Lagrange system.

In this paper we will prove that very weak solution of equation of type (1.3) are quasiminima of the functional

$$I(u,\Omega)=\int_\Omega |Xu|^r dx.$$

Thus we consider a mapping $A:\Omega\times\mathbb{R}^m\times\mathbb{R}^{mn}\to\mathbb{R}^{mn}$, such that for $(x,u,\xi)\in\Omega\times\mathbb{R}^m\times\mathbb{R}^{mn}$

$$A(x,u,\xi)\xi\geq a|\xi|^p,\ \ p\geq 2, \tag{1.4}$$

$$|A(x,u,\xi)-A(x,u,\eta)|\leq b|\xi-\eta|(|\xi|+|\eta|)^{p-2}, \tag{1.5}$$

$$|A(x,u,0)|\leq h(x)+d|u|^{p-1}, \tag{1.6}$$

where a,b,d are positive constants and h is a non negative function of class $L^{r/(p-1)}(\Omega)$, with some $r\leq p$.

Let $u\in W^{1,r}_{loc}(\Omega,\mathbb{R}^m)$ be a very weak solution of the A-harmonic equation

$$\mathrm{div}A(x,u,Xu)=0. \tag{1.7}$$

Our main result is

Theorem 1.3. *For each A-harmonic system* (1.7)*, there exist exponents*

$$p-1<r_1=r_1(m,n,p,a,b,d)<p<r_2=r_2(m,n,p,a,b,d),$$

and a constant $Q=Q(m,n,p,a,b,d)$ such that, if $r_1\leq r\leq r_2$, then for any test mapping $\varphi\in W^{1,r}_0(B,\mathbb{R}^m)$ and a ball $B\subset\subset\Omega$, we have

$$\begin{aligned}&\int_B(|Xu|^r+|u|^r+h(x)^{r/(p-1)}dx\\ \leq Q&\int_B(|X(u+\varphi)|^r+|u+\varphi|^r+h(x)^{r/(p-1)})dx.\end{aligned} \tag{1.8}$$

The inequality (1.8) reads that a solution of (1.7) is a quasiminimum of the functional

$$\mathcal{J}(u,\Omega)=\int_\Omega(|Xu|^r+|u|^r+h(x)^{r/(p-1)})dx$$

(see Definition 1.2 below).

Let us consider the functional

$$\mathcal{G}(u,\Omega)=\int_\Omega G(x,u,Xu)dx, \tag{1.9}$$

in which Ω is a bounded domain in $\mathbb{R}^n$, $u = u(u^1, \cdots, u^m)$ is a mapping into $\mathbb{R}^m$ $(m > 1)$ and $G(x, u, z)$ is a *Charathéodory* function. The latter means that G is measurable in x for every (u, z) and continuous in (u, z) for almost every $x \in \Omega$.

Suppose that G satisfies the following growth and coercivity conditions

$$|z|^a - b|u|^a - a(x) \leq G(x, u, z) \leq \mu|z|^a + b|u|^a + a(x), \; a > 1,$$

where $a(x)$ is a given non negative function and b, μ are non negative constants.

Definition 1.2. A function $u \in W^{1,a}_{loc}(\Omega, R^m)$ is a Q-quasiminimum ($Q \geq 1$), for the functional $\mathcal{G}$ if for any ball $B \subset\subset \Omega$ and for any $\varphi \in W^{1,a}_0(B)$, we have $\mathcal{G}(u, B) \leq Q\mathcal{G}(u + \varphi, B)$.

It turns out that under the above assumptions every quasiminimum $u \in W^{1,a}_{loc}(\Omega, \mathbb{R}^m)$ of the functional (1.8) is a quasiminimum (with different constant Q) of the simple functional

$$\int_\Omega (|Xu|^a + b|u|^a + a(x))dx. \tag{1.10}$$

Let's recall the following regularity result of quasiminima.

Theorem 1.4. *Let $u \in W^{1,a}_{loc}(\Omega, \mathbb{R}^m)$ be a Q-quasiminimum and a quasiminimum of the functional (1.10) with $a(x) \in L^s(\Omega), s > 1$. Then there exists an exponent $q > a$, such that $u \in W^{1,a}_{loc}(\Omega, R^m)$. Moreover, there exists $R_0 > 0$, such that for every ball $B_R \subset \Omega$ with $R < R_0$*

$$\int_{B_{R/2}} (|Xu|^a + b|u|^a)^{q/a} dx \leq c\{[\int_{B_R} (|Xu|^a + |u|^a)dx]^{q/a} + [\int_{B_R} a(x)^{q/a} dx]^q\},$$

in which $\int_{B_R}$ stands for the integral mean over the ball B_R.

Proposition 1.1. *Let u be a Q-quasiminimum of $\mathcal{G}(u, \Omega)$ of class $W^{1,a}_{loc}$ (Ω, R^m) $(1 < a < q)$, with a constant Q independent of a. Then there exists a constant $c = c(q, Q, n)$, such that*

$$\int_{B_{R/2}} (|Xu|^a + |u|^a)dx \leq c[\int_{B_R} (|Xu|^t + |u|^t)dx]^{a/t} + c\int_{B_R} a(x)dx$$

for every t with $\max\{1, na/(n + a)\} \leq t < a$, for every pair of concentric balls $B_{R/2}, B_R$ with $B_R \subset\subset \Omega$.

Proof. Let $B_R \subset\subset \Omega$ and $R/2 < \sigma < s < R$. If $\varphi = -\eta(u - u_s)$, where $u_s = \fint_{B_s} u dx$ and $\eta \in C^\infty_0(B_s)$, $0 \leq \eta \leq 1$, $\eta = 1$ on B_s, $|X\eta| \leq 2(s - \sigma)^{-1}$, from the definition 1.2 it follows that

$$\int_{B_s}(|Xu|^a+|u|^a)dx \le Q_1[\int_{B_s}(1-\eta)^a|Xu|^a dx$$

$$+\int_{B_s}|X\eta|^a|u-u_s|^a dx+\int_{B_s}|u-u_s|^a dx+|u_s|^a|B_s|+\int_{B_s}a(x)dx,$$

where $Q_1=Q_1(Q,q)$.

Noting that $\int_{B_s}|u-u_s|^a dx \le |B_R|^{a/n}\int_{B_s}|Xu|^a dx$, if $R<1$ is such that $Q_1|B_R|^{a/n}<1$, we can subtract the term $\int_{B_s}|u-u_s|^a dx$ from the left hand side, and get

$$\int_{B_\sigma}(|Xu|^a+|u|^a)dx \le Q_2[\int_{B_s-B_\sigma}|Xu|^a dx$$

$$+\frac{1}{(s-\sigma)^a}\int_{B_s}|u-u_s|^a dx+|u_s|^a|B_s|+\int_{B_s}a(x)dx].$$

Since

$$\int_{B_s}|u-u_s|^a dx \le c\int_{B_R}|u-u_R|^a dx,\ c=c(n),$$

$$|u_s| \le 2^n \fint_{B_R}|u|dx,$$

we have

$$\int_{B_\sigma}(|Xu|^a+|u|^a)dx \le Q_3[\int_{B_s-B_\sigma}|Xu|^a dx$$

$$+\frac{1}{(s-\sigma)^a}\int_{B_R}|u-u_R|^a dx+|B_R|(\fint_{B_R}|u|dx)^a+\int_{B_R}a(x)dx].$$

Adding the term

$$Q_3\int_{B_\sigma}(|Xu|^a+|u|^a)dx$$

to both sides and applying Lemma 6.1 of [3], we obtain

$$\int_{B_{R/2}}(|Xu|^a+|u|^a)dx \le c[R^{-a}\int_{B_R}|u-u_R|^a dx$$

$$+\int_{B_R}a(x)dx+|B_R|\int_{B_R}|u|dx)^a],$$

where $c=c(q,Q,n)$.

Finally, if $\{1, na/(n+a)\} \le t < a$, by Sobolev-Poincaré inequality, we have the result

$$\int_{B_{R/2}}(|Xu|^a+|u|^a)dx \le cR^{(1-a/t)n}[\int_{B_R}(|Xu|^t+|u|^t)dx]^{a/t}+c\int_{B_R}a(x)dx.$$

If f is a function of class $L^a(\Omega)$, we have

$$\|f\|_a = (\int_\Omega |f|^a dx)^{1/a}.$$

In order to prove theorem 1.3, we also need the following lemma.

Lemma 1.1. *Let $\Omega \subset \mathbb{R}^m$ be a regular domain and $w : \Omega \to \mathbb{R}^m$, with $w \in W_0^{1,a}(\Omega, \mathbb{R}^m)$, $a > 1$, and let $-1 < \varepsilon < a - 1$. Then there exist $\Phi \in W_0^{1,a/(1+\varepsilon)}(\Omega, \mathbb{R}^m)$, and $H \in L^{a/(1+\varepsilon)}(\Omega, \mathbb{R}^m)$ such that H is divergence-free and*

$$|Xw|^\varepsilon Xw = X\Phi + H.$$

Moreover

$$\begin{aligned} \|H\|_{L^{a/(1+\varepsilon)}} &\le c(a,n,m)|\varepsilon| \|Xw\|_{L^a}^{1+\varepsilon}, \\ \|X\Phi\|_{L^{a/(1+\varepsilon)}} &\le c(a,n,m)\|Xw\|_{L^a}^{1+\varepsilon}, \end{aligned} \tag{1.11}$$

where the constant c in (1.11) is independent of a, provided $a \in K$, for some compact $K \subset (1, +\infty)$.

2. Proof of the Main Theorem

Suppose that $A : \Omega \times \mathbb{R}^m \times \mathbb{R}^{mn} \to \mathbb{R}^{mn}$ satisfies (1.4), (1.5) and (1.6). Let us consider the equation (1.7).

Definition 2.1. A function $u \in W_{loc}^{1,r}(\Omega, \mathbb{R}^m)$ is a very weak solution of the equation (1.7), if

$$\int_\Omega A(x,u,Xu)X\Phi = 0$$

for any $\Phi \in W_0^{1,r/(r-p+1)}(\Omega, \mathbb{R}^m)$.

Proof of Theorem 1.3. Let $u \in W_{loc}^{1,r}(\Omega, \mathbb{R}^m)$ be a very weak solution of (1.7). Fix a ball $B = B(x_0, R)$, such that $B(x_0, R) \subset\subset \Omega$ and a function $\varphi \in W_0^{1,r}(B, \mathbb{R}^m)$, by lemma 1.1, there exist $\Phi \in W_0^{1,r/(r-p+1)}(B, \mathbb{R}^m)$, $H \in L^{r/(r-p+1)}(B, \mathbb{R}^m)$ such that

$$X\Phi = |X\Phi|^{r-p} X\varphi - H, \tag{2.1}$$

$$\|H\|_{r/(r-p+1)} \le c|r-p| \|X\varphi\|_r^{r-p+1}, \tag{2.2}$$

$$\|X\Phi\|_{r/(r-p+1)} \le c\|X\varphi\|_r^{r-p+1}, \tag{2.3}$$

where c is independent of x_0, R and r ($p - 1 \le r \le p$). By definition 2.1 we have

$$\int_B A(x,u,Xu)X\Phi = 0. \tag{2.4}$$

According to (1.4), (1.5), (1.6), we have

$$a\int_B |X\varphi|^r dx \le \int_B A(x,u,X\varphi)|X\varphi|^{r-p}X\varphi dx$$
$$= \int_B [A(x,u,X\varphi)-A(x,u,Xu)]|X\varphi|^{r-p}X\varphi dx + \int_B A(x,u,Xu)H dx$$
$$\le b\int_B |Xu - X\varphi|(|X\varphi|+|Xu|)^{p-2}|X\varphi|^{r-p+1}dx$$
$$+\int_B [A(x,u,Xu)-A(x,u,0)]H dx + \int_B A(x,u,0)H dx$$
$$\le b[\int_B |Xu - X\varphi|(|X\varphi|+|Xu|)^{r-1}dx + \int_B |Xu|^{p-1}|H|dx]$$
$$+\int_B h(x)|H|dx + d\int_B |u|^{p-1}|H|dx.$$

Let $g(x) = (Xu - D\varphi)(x)$. Then by (2.2) and (2.3), we get

$$a\|X\varphi\|_r^r \le b\|g\|_r\|X\varphi + (g+X\varphi)\|_r^{r-1}$$
$$+bc|r-p|\|X\varphi\|_r^{r-p+1}(\|g\|_r + \|X\varphi\|_r)^{p-1}$$
$$+c|r-p|\|h\|_{r/(r-p)}\|X\varphi\|_r^{r-p+1} + dc|r-p|\|u\|_r^{p-1}\|X\varphi\|_r^{r-p+1}.$$

Moreover by Young's inequality, we have

$$a\|X\varphi\|_r^r \le c_1(\varepsilon^{r/(r-1)}+|r-p|)\|X\varphi\|_r^r + c_2|r-p|(\|u\|_r^r$$
$$+\|h\|_{r/(p-1)}^{r/(p-1)}) + c_3[(\frac{1}{\varepsilon})^r + 1]\|g\|_r^r. \tag{2.5}$$

On the other hand, by Sobolev's inequality, we get

$$\|\varphi\|_r^r \le (\frac{r(n-1)}{n-r})^r|\Omega|^{r/n}\|X\varphi\|_r^r \le (\frac{p(n-1)}{n-p})^p(|\Omega|+1)^{p/n}\|X\varphi\|_r^r. \tag{2.6}$$

Combing (2.5), we have

$$a\|X\varphi\|_r^r \le c'(\varepsilon^{r/(r-1)} + |r-p|)\|X\varphi\|_r^r$$
$$+c''(\|h\|_{r/(p-1)}^{r/(p-1)} + \|u-\varphi\|_r^r + \|Xu - X\varphi\|_r^r).$$

with $c' = c'(b,d,p,m,n)$, $c'' = c''(b,d,p,m,n)$.

We can choose r, ε such that

$$c'(|r-p| + \varepsilon^{r/(r-1)}) \le a/2,$$

so we can find r_1 and r_2, which are in the statement of the theorem, such that, for $r_1 \le r \le r_2$ and for some constant c independent of r,

$$\int_B |X\varphi|^r dx \le c \int_B (|Xu - X\varphi|^r + |u - \varphi|^r + h(x)^{r/(p-1)})dx. \tag{2.7}$$

Remark that

$$\int_B (|Xu|^r + |u|^r)dx \le 2^r \int_B (|Xu - X\varphi|^r + |u - \varphi|^r)dx + 2^r \int_B |X\varphi|^r + |\varphi|^r)dx.$$

Then, by using (2.6), (2.7), we can get the result.

References

[1] B. Bojarski and T. Iwaniec, Analytical foundations of the theory of quasiconformal mappings in $\mathbb{R}^n$, *Ann. Acad. Sci. Fenn. Ser. AI Math.*, 8 (1983), 257-324.

[2] N. Fusco and C. Sbordone, Higher integrability of the gradient of minimizers of functionals with nonstandard growth conditions, *Comm. Pure Appl. Math.*, 43 (1990), 673-680.

[3] F. W. Gehring, The L^p-integrability of the partial derivatives of a quasi-conformal mapping, *Acta Math.*, 130 (1973), 265-277.

[4] M. Giaquinta and E. Giusti, Quasi-minima, Ann. Inst. H. Poincaré, *Analyse Nonlinéaire,* 1 (1984), 79-107.

[5] E. Giusti, *Metodi diretti nel calcolo delle variazioni,* Boll. Un. Mat. Ital., 1994.

[6] D. Giachetti, F. Leonetti and R. Schianchi, On the regularity of very weak minima, *Proc. Royal Soc. Edinburgh,* 126 A (1996), 287-296.

[7] T. Iwaniec, P-harmonic tensors and quasiregular mappings, *Ann. Math.*, 136 (1992), 589-624.

[8] T. Iwaniec and C. Sbordone, Weak minima of variational integrals, *J. Reine Angew. Math.*, 454 (1994), 143-161.

[9] J. Lewis, On very weak solutions of certain elliptic systems, *Comm. Partial differential Equations,* 18 (1993), 1515-1537.

[10] N. Meyers and A. Elcrat, Some results on regularity for solutions of nonlinear elliptic system and quasiregular functions, *Duke Math. J.*, 42 (1975), 121-136.

[11] G. Taleltt, Best constant in Sobolev inequality, *Ann. Mat. Pura Appl.*, 110 (1976), 353-372.

[12] G. I. Moscariello, Weak minima and quasiminima of variational integrals, *Bollettino U.M.I.*, (7) 11-B (1997), 355-364.

REMARK ABOUT LOCAL REGULARITY RESULTS FOR MINIMA OF FUNCTIONALS[1]

XIU-JUAN XU[a] and XIAO-LI LIU[b]
[a,b]*College of Science, Hebei Polytechnic University, Tangshan, Hebei 063009, China*
[a]*E-mail: xxjluck@126.com*
[b]*E-mail: liuxiaoli_ok@163.com*

Local regularity results for the minima of variational integrals with the integrand satisfies the growth conditions of Muckenhoupt weight has been obtained by choosing the suitable cut-off function and the reverse Hölder inequality.
Key words: Muckenhoupt weight, minima of variational integral, regularity.
AMS No: 35J50, 35J60.

1. Introduction

Let $\omega(x)$ be a local integrable nonnegative function in $\mathbb{R}^n$ $(n \geq 2)$. A Radon measure μ is canonically associated with the weigh $\omega(x)$:

$$\mu(E) = \int_E \omega(x)dx. \tag{1.1}$$

Thus

$$d\mu(x) = \omega(x)dx,$$

where dx is n-dimensional Lebesgue measure. In what follows, the weight $\omega(x)$ and the measure μ are identified via (1.1).

We say that w(or μ)is p-admissible, if the following four conditions are satisfied:

(1) $0 < w < \infty$ almost everywhere in $\mathbb{R}^n$ and the measure μ is doubling, i.e. there is a constants $C_1 > 0$ such that $\mu(2B) \leq C_1\mu(B)$, where B is a ball in $\mathbb{R}^n$.

(2) If D is an open set and $\phi_i \in C^\infty(D)$ is a sequence of functions such that

$$\int_D |\phi_i|^p d\mu \to 0 \ \text{ and } \int_D |\nabla\phi_i - v|^p d\mu \to 0 \text{ as } i \to \infty,$$

where v is a vector-valued measurable function in $L^p(D;\mu;\mathbb{R}^n)$, then $v = 0$.

(3) There are constants $\chi > 1$ and $C_2 > 0$, such that

$$\left(\frac{1}{\mu(B)}\int_B |\phi|^{\chi p} d\mu\right)^{1/\chi p} \leq C_2 r \left(\frac{1}{\mu(B)}\int_B |\phi|^p d\mu\right)^{1/p},$$

[1]The research is supported by Natural Science Foundation of Hebei Province (A2010000910) and Tangshan Science and Technology projects (09130206c).

whenever $B = B(x_0, r)$ is a ball in $\mathbb{R}^n$ and $\phi \in C_0^\infty(B)$.

(4) There is a constant $C_3 > 0$, such that

$$\int_B |\phi - \phi_B|^p d\mu \le C_3 r^p \int_B |\nabla \phi|^p d\mu,$$

whenever $B = B(x_0, r)$ is a ball in $\mathbb{R}^n$, and $\phi \in C_0^\infty(B)$ is bounded. Here

$$\phi_B = \frac{1}{\mu(B)} \int_B \phi d\mu.$$

For the details of p-admissible to see [3].

Let Ω be a bounded open subset of $\mathbb{R}^n$. Consider the functional

$$I(u; \Omega) = \int_\Omega f(x, u, Du) dx, u \in W^{1,p}_{loc}(\Omega), \tag{1.2}$$

where the integrand $f : \Omega \times \mathbb{R} \times \mathbb{R}^n$ is a Carathéodory function. The regularity property for minima of the functional I is an important research content in the Modern analysis.

Definition 1.1. By a local minima of the functional I we mean functions $u \in W^{1,p}_{loc}(\Omega)$, such that for every $\Psi \in W^{1,p}(\Omega)$with supp$\Psi \subset\subset \Omega$, it results in

$$I(u, \text{supp}\Psi) \le I(u + \Psi, \text{supp}\Psi). \tag{1.3}$$

The aim of the present paper is to prove the local regularity property for minima of the functional I. Recently, Giachetti and Porzio proved in [1] the regularity theory for local minima of (1.2) with the functional I satisfies the condition

$$a \mid \xi \mid^p \le f(x, s, \xi) \le b \mid \xi \mid^p + \varphi_0(x), \tag{1.4}$$

where $\varphi_0(x) \in L^r_{loc}(\Omega)$, $p > 1$, $r > 1$. This result was extended by Hong-ya Gao et al.[2] to the local regularity result of the minima of function I with the more general growth conditions than (1.4). Precisely, the authors considered the minima of functionals, where f is a Carathéodory function satisfying the growth conditions

$$\mid \xi \mid^p - b \mid s \mid^a - \varphi_0(x) \le f(x, s, \xi) \le a \mid \xi \mid^p + b \mid s \mid^a + \varphi_1(x), \tag{1.5}$$

where $p > 1$, $\varphi_0(x) \in L^{r_1}_{loc}(\Omega)$, $\varphi_1(x) \in L^{r_2}_{loc}(\Omega)$, $r_1, r_2 > 1$, $a \ge 1$ and b is positive constant, $p \le a \le \frac{np}{n-p}$.

This paper is aimed at bringing weight function into the minima of function, and proving the regularity result of the minima of function (1.2), which meet the weighting growth condition

$$\alpha \mid \xi \mid^p \le \frac{f(x, s, \xi)}{\omega(x)} \le \beta \mid \xi \mid^p + \gamma \mid s \mid^a. \tag{1.6}$$

2. Lemma and Preliminary Knowledge

Definition 2.1[3]. Given a nonnegative locally integrable function ω, we say that ω belongs to the A_p class of Muckenhoupt, $1 < p < \infty$, if

$$\sup_Q \left(\frac{1}{| Q |} \int_Q \omega dx \right) \left(\frac{1}{| Q |} \int_Q \omega^{\frac{1}{1-p}} dx \right)^{p-1} = A_p(\omega) < \infty, \tag{2.1}$$

where the supremum is taken over all cubes Q of $\mathbb{R}^n$.

When $p = 1$, replace the inequality (2.2) with

$$M\omega(x) \leq c\omega(x)$$

for some fixed constant c and a.e. $x \in \mathbb{R}^n$, where M is the Hardy-Littlewood maximal operator.

It is well known that $A_1 \subset A_p$, whenever $p > 1$ (see [3]). We say that a weight ω is doubling, if there is a constant $c > 0$ such that

$$\mu(2Q) \leq c\,\mu(Q),$$

whenever $Q \subset 2Q$ are concentric cubes in $\mathbb{R}^n$, where $2Q$ is the cube with the same center as Q and with sidelength twice that of Q. Given a measurable subset E of $\mathbb{R}^n$, we will denote by $L^p(E, \omega)$, $1 < p < \infty$, the Banach space of all measurable functions f defined on E for which

$$\|f\|_{L^p(E,\omega)} = \left(\int_E | f(x) |^p \omega(x) dx \right)^{\frac{1}{p}}. \tag{2.2}$$

The weighted Sobolev class $W^{1,p}(E, \omega)$ consists of all functions f for which f and its first generalized derivative belongs to $L^p(E, \omega), 1 < p < \infty$. The symbols $L^p_{loc}(E, \omega)$ and $W^{1,p}_{loc}(E, \omega)$ are self-explanatory.

We need the following Lemma in the proof of the main theorem.

Lemma 2.1[4]. *Let $Q = Q(R)$ be any cube with side-length R, $\tau > 1$ and $u \in C^1(\bar{Q})$. Then there exist constants c, $\delta^* > 0$, such that for all $1 \leq K \leq K^* = \frac{n}{n+1} + \delta^*$,*

$$\left(\frac{1}{\mu(Q)} \int_Q | u - u(Q) |^{K\tau} d\mu \right)^{\frac{1}{K\tau}} \leq CR \left(\frac{1}{\mu(Q)} \int_Q | \nabla u |^{\tau} d\mu \right)^{\frac{1}{\tau}}, \tag{2.3}$$

where

$$u_Q = \frac{1}{\mu(Q)} \int_Q u d\mu.$$

It is obvious that (2.3) can be extended to functions $u \in W^{1,K\tau}(\bar{Q})$ by an approximation argument.

Lemma 2.2[5]. *Let $f(t)$ be a nonnegative bounded function defined for $0 < \tau_0 \leq t \leq \tau_1$. Suppose that for $\tau_0 \leq t \leq s \leq \tau_1$, we have*

$$f(t) \leq A(s-t)^{-a} + B + \theta f(s),$$

where A, B, a, θ are nonnegative constants and $\theta < 1$. Then there exists a constant c_0, depending only on a, θ, such that for every $\rho, R, \tau_0 \leq \rho < R \leq R_1$, we have

$$f(\rho) \leq c_0[A(R-\rho) + B].$$

We need the following Lemma in the proof of the main theorem.

Lemma 2.3[6]. *Suppose that ω is a doubling weight and non-negative $f \in L^{\tau}_{loc}(\Omega, \omega)\,(1 < \tau < \infty)$ satisfying*

$$\left(\frac{1}{\mu(Q)}\int_Q f^{\tau} d\mu\right)^{\frac{1}{\tau}} \leq C_1 \frac{1}{\mu(2Q)}\int_{2Q} f d\mu$$

for each cube Q such that $2Q \subset \Omega$, where the constant $c_1\,(\geq 1)$ is independent of the cube Q. Then there exist $q > \tau$ so that

$$\left(\frac{1}{\mu(Q)}\int_Q f^{q} d\mu\right)^{\frac{1}{q}} \leq C_2 \left(\frac{1}{\mu(2Q)}\int_{2Q} f^{\tau} d\mu\right)^{\frac{1}{\tau}},$$

where the constants $c_2\,(\geq 1)$ is independent of the cube Q. In particular, $f \in W^{1,q}_{loc}(\Omega, \omega)$.

3. Theorem and Its Proof

Theorem 3.1. *Suppose that $\omega \in A_1$ be a doubling Muckenhoupt weight. Then there exists $p_1 > p$, such that for any local minima $u \in W^{1,p}_{loc}(\Omega, \omega)$ of the functional I, we have $u \in W^{1,p_1}_{loc}(\Omega, \omega)$.*

Proof. Let $\omega \in A_1$ be a doubling Muckenhoupt weight and $u \in W^{1,p}_{loc}(\Omega, \omega)$ be a minima of the functional I. For an arbitrary cube $2Q \subset\subset \Omega$, take a cut-off function $\eta \in C_0^{\infty}(2Q)$, such that

$$0 \leq \eta \leq 1,\ \eta \equiv 1 \text{ in } Q,\ |\ D\eta\ | \leq \frac{c(n)}{R},$$

where R is the side-length of Q. Choose

$$\Psi = -\eta(u-c) - c,$$

where c is a constant to be determined later on. By the minimality of u, we obtain

$$\begin{aligned}\textstyle\int_{2Q} f(x,u,Du)dx &\leq \int_{2Q} f(x,u+\Psi,Du+D\Psi)dx \\ &= \int_{2Q} f(x,(1-\eta)(u-c),(1-\eta)Du-D\eta(u-c))dx.\end{aligned} \tag{3.1}$$

Using (1.6) and (3.1), we have

$$\begin{aligned}a \textstyle\int_Q \mid Du \mid^p d\mu &\leq \int_Q f(x,u,Du)dx \\ &\leq \int_{2Q} f(x,u,Du)dx, \\ &\leq \int_{2Q} f(x,(1-\eta)(u-c),(1-\eta)Du-D\eta(u-c))dx \\ &= \beta \int_{2Q} \mid (1-\eta)Du - D\eta(u-c) \mid^p d\mu \\ &\quad +\gamma \int_{2Q} \mid (1-\eta)(u-c) \mid^a d\mu.\end{aligned} \tag{3.2}$$

By the basic inequality

$$(a+b)^p \leq 2^{p-1}(a^p+b^p),\ a,b \geq 0,\ p \geq 1,$$

we have

$$\begin{aligned}a \textstyle\int_Q |Du|^p d\mu &\leq 2^p\beta \int_{2Q} |(1-\eta)Du|^p d\mu + 2^p\beta \int_{2Q} |D\eta|^p |u-c|^p d\mu \\ &\quad +\gamma \int_{2Q} |1-\eta|^2 |u-c|^2 d\mu.\end{aligned} \tag{3.3}$$

Taking $c = u_{2Q}$ and by Lemma 2.1 with $K\tau = p, \tau > 1, 1 \leq K \leq \dfrac{n}{n-1} + \delta^*$ for any $1 < K < \min\left\{p, \dfrac{n}{n-1}\right\}$, we obtain that

$$\int_{2Q} |D\eta|^p |u-c|^p d\mu \leq CR\left(\frac{1}{\mu(2Q)}\right)^{K-1} \left(\int_Q |Du|^\tau d\mu\right)^K. \tag{3.4}$$

Thus

$$\int_{2Q} |1-\eta|^a |u-c|^a d\mu \leq CR\left(\frac{1}{\mu(2Q)}\right)^{K-1} \left(\int_Q |Du|^\tau d\mu\right)^K, \tag{3.5}$$

substituting (3.4), (3.5) into (3.3), and adding

$$2^p\beta\int_Q |Du|^p\,d\mu$$

on both sides of previous inequality, we obtain

$$\int_Q |Du|^p\,d\mu \le \frac{2^p\beta}{\alpha+2^p\beta}\int_{2Q}|Du|^p\,d\mu + \frac{(2^p\beta+\gamma)CR}{\alpha+2^p\beta}\left(\frac{1}{\mu(2Q)}\right)^{K-1}\left(\int_Q |Du|^\tau\,d\mu\right)^K.$$

By Lemma 2.2, there exists a constant c_0 such that

$$\int_Q |Du|^p\,d\mu \le c_0\frac{(2^p\beta+\gamma)CR}{\alpha+2^p\beta}\left(\frac{1}{\mu(2Q)}\right)^{K-1}\left(\int_Q |Du|^{\frac{p}{K}}\,d\mu\right)^K. \tag{3.6}$$

Dividing by $\mu(Q)$ in both sides of the above inequality yields

$$\frac{1}{\mu(Q)}\int_Q |Du|^p\,d\mu \le c'\left(\frac{1}{\mu(2Q)}\right)\left(\int_Q |Du|^{\frac{p}{K}}\,d\mu\right)^K, \tag{3.7}$$

we are now in a position of using Lemma 2.3 to improve the degree of integrability of $|Du|$. Accordingly, there exists $p_1 > p$ such that $|Du| \in L^{p_1}_{loc}(\Omega,\omega)$. By Sobolev imbedding theorem we have $u \in W^{1,p_1}_{loc}(\Omega,\omega)$. This completes the proof of the theorem.

References

[1] D. Giachetti and M. Porzio, local regularity results for minima of functional of the calculus of variation, *Nonlinear Analysis*, 39 (2000), 4:463-482.

[2] Hongya Gao, Yuming Chu et al, Local regularity results for minima of functionals and solutions of nonlinear elliptic equations, *Acta Mathematica Sinica, Chinese Series* 51 (2008), 411-416.

[3] J. Heinonen, T. Kilpelainen and O. Martio, *Nonlinear potential theory of degenerate elliptic equations,* Oxford Univ Press, New York, 1993.

[4] F. Chiarenza and M. Frasca, A note on weighted Sobolev inequality, *Proc. Amer. Math. Soc.*, 93(4) (1985), 31-46.

[5] M. Giaquinta and E. Giusti, On the regularity of the minima of variational integrals, *Acta. Math.*, 148 (1982), 31-46.

[6] J. Kinnunen, Higher integrability with weights, *Ann. Acad. Sci. Fenn. Ser A. I. Math.,* 19(2) (1994), 355-366.

[7] Hongya Gao, Jinjing Qiao et al, Local regularity results for minima of anisotropic functionals and solutions of anisotropic equations, *J. Inequal. Appl.*, 2008, Art. ID 835736, 11 pp.

BOUNDARY VALUE PROBLEMS ON LOCALLY CONVEX SPACES

CHUAN-GAN HU

School of Mathematical Sciences, Nankai University, Tianjin 300071, China

E-mail: cghu@live.com

In this paper, the solutions of the Riemann boundary value problems on infinite dimensional locally convex spaces into other infinite dimensional locally convex spaces are presented, via piecewise holomorphic mappings on infinite dimensional locally convex spaces, except for infinite dimensional locally convex manifolds as a boundary, where a key concept-the infinite dimensional index of a mapping is defined.

Keywords: Infinite dimensional index, locally convex m-algebra, Cauchy type integral, locally convex manifold, holomorphic mapping, (L)-analytic.

AMS No: 30E20, 30E25, 30G30, 46G12, 46G20, 46T05, 46T25.

1. Introduction

The theory of classical singular integral equations and boundary value problems of holomorphic functions on a complex plane $\mathbb{C}$ into another complex plane has been summarized in [6]. Solutions of boundary value problems (BVP for short) on $\mathbb{C}$ into Banach spaces or locally convex spaces has been studied as well (see [3,4]). But the research into the solutions of BVP of holomorphic mappings on an infinite dimensional space into another infinite dimensional space are much more difficult. In this paper, solutions of BVP on an infinite dimensional locally convex m-algebra, except for locally convex and bounded differential manifolds with infinite dimensions as a boundary into another infinite dimensional locally convex m-algebra are explained.

We shall state solutions of BVP on an infinite dimensional complex locally convex space in the following aspects: Section 2 gives locally convex spaces for a class, the concepts and properties of holomorphic and meromorphic mappings on locally convex spaces. In Section 3, the concepts and properties of singular integral, the Hölder condition, and the Plemelj formulas on infinite dimensional spaces are obtained. In Section 4, the solutions of Riemann boundary value problems of holomorphic mappings on infinite dimensional spaces into other infinite dimensional spaces are constructed.

2. Locally Convex Spaces for A Class

2.1. Product spaces

In this subsection, the properties of product spaces are showed. Because every locally convex space is topologically isomorphic to a linear subspace of a topological product of Banach spaces by conclusion 18.3.(7) of Section 18 in [5], the topological product of Banach spaces is discussed in this subsection and the following subsections, in order to establish the solutions of boundary value problems on a set in infinite dimensional spaces into other infinite dimensional spaces. We introduce product spaces as follows. Let $\mathcal{E}_\iota$ and $\mathcal{F}_\iota$ be commutative and complex Banach algebras with the unit element e_ι and with the unit element $\dot{e}_\iota$ for any ι respectively.

Let $\mathcal{E} = \prod\limits_\iota \mathcal{E}_\iota$ and $\mathcal{F} = \prod\limits_\iota \mathcal{F}_\iota$. Let $\mathsf{a} : (0,1) \to (0,1)$ be a bijective mapping. Similarly, define $\prod\limits_{\iota'} z_{\iota'} = \prod\limits_\iota z_\iota \in \mathcal{E}(\mathcal{F})$. Here $\iota' = \mathsf{a}\iota$. Define the multiplication in $\mathcal{E}$ by $zw = \prod\limits_\iota z_\iota w_\iota$. Here $z_\iota, w_\iota \in \mathcal{E}_\iota$. From the definition of the multiplication we can derive that the inverse z^{-1} of z is written as $\prod\limits_\iota z_\iota^{-1}$ if z^{-1} exists in $\mathcal{E}$. In $\overline{\mathcal{E}}_\iota(\overline{\mathcal{F}}_\iota)$, the point at infinity is defined by $\|z_\iota\| = \infty$, denoted by ∞_ι. Define $\widehat{\infty} = \prod\limits_\iota \infty_\iota$. Let

$$\Omega = \left\{ \Big(\prod_{0<\iota<1, \iota \neq \tau_j} \mathcal{E}_\iota \Big) \times \prod_{j=1}^{k} \{ z_{\tau_j} : \|z_{\tau_j}\| < r_{\tau_j} \} : \begin{matrix} \tau_j \in (0,1), k \in \mathsf{N}^+ \\ 0 < r_{\tau_j} < \infty \end{matrix} \right\}. \tag{2.1}$$

Then Ω in (2.1) is the base of neighborhoods of $\hat{0}$ in $\mathcal{E}$. Here $\hat{0} = \prod\limits_\iota 0$ and N^+ is a set consisting of all positive integers. Any neighborhood in Ω is one-to-one correspondent to a τ-norm (i.e. semi-norm) $\|z\|_{\tau(k)}$, where $\|z\|_{\tau(k)} = \Big(\sum\limits_{j=1}^{k} \|z_{\tau_j}\|^2 \Big)^{\frac{1}{2}}$ for any positive integer k. We can get that $\mathcal{E}$ is Hausdorff by using all $\mathcal{E}_\iota$ being Banach spaces. In fact, assume $z_1 \neq z_2 \in \mathcal{E}$, then there exists a ι_0 such that $z_{1\iota_0} \neq z_{2\iota_0}$ in $\mathcal{E}_{\iota_0}$. It follows that there are two neighborhoods satisfying $\nabla_{r_{1\iota_0}}(z_{1\iota_0}) \bigcap \nabla_{r_{2\iota_0}}(z_{2\iota_0}) = \emptyset$, because $\mathcal{E}_{\iota_0}$ is Hausdorff. Here $\nabla_{r_{j\iota_0}}(z_{j\iota_0}) = \{z_{\iota_0} : \|z_{\iota_0} - z_{j\iota_0}\| < r_{j\iota_0}\}$ for $j = 1, 2$. So there exist two neighborhoods $\nabla_{r_j}(z_j)$ $(j = 1, 2) \subset \mathcal{E}$ such that $\nabla_{r_1}(z_1) \bigcap \nabla_{r_2}(z_2) = \emptyset$, where $\nabla_{r_j}(z_j) = \prod\limits_{\iota \neq \iota_0} \mathcal{E}_\iota \times \nabla_{r_{j\iota_0}}(z_{j\iota_0})$. Thus $\mathcal{E}$ is Hausdorff. We can check that $\mathcal{E}$ is a Hausdorff complex locally m-convex and commutative algebra with the unit element $\hat{e} = \prod\limits_\iota e_\iota$.

Let $\mathcal{L}(\mathcal{E}_\iota, \mathcal{F}_\iota)$ be a set consisting of all bounded linear mappings from a locally convex space $\mathcal{E}_\iota$ into $\mathcal{F}_\iota$. Let $\mathcal{L}(\mathcal{E}, \mathcal{F}) = \prod\limits_\iota \mathcal{L}(\mathcal{E}_\iota, \mathcal{F}_\iota)$. Define the operation

$$\sum_{l=1}^{n} \prod_{j=1}^{k} T_{\iota_l \tau_j} = \prod_{j=1}^{k} \sum_{l=1}^{n} T_{\iota_l \tau_j} \tag{2.2}$$

for any $T_{\iota_l\tau_j} \in \mathcal{L}(\mathcal{E}_{\tau_j}, \mathcal{F}_{\tau_j})$. For convenience we also define $z^\alpha = \prod_\iota z_\iota^{\alpha_\iota}$. Here $\alpha \in \mathcal{R} = \prod_\iota \mathsf{R}_\iota$, $\alpha_\iota \in \mathsf{R}_\iota = (-\infty, \infty)$ and $z \in \mathcal{E}$.

Let $D_\iota \subset \mathcal{E}_\iota$ be an open set for any ι. It follows that $D = \prod_\iota D_\iota \subset \mathcal{E}$ is a finitely open set (cf. [2]). Let the set $H(D, \mathcal{F})$ consist of all holomorphic mappings on D into $\mathcal{F}$.

Let D be a finitely open and connected subset of $\mathcal{E}$. A mapping $f : D \to \mathcal{F}$ is said to be differentiable (derivable), if for each point $z_0 \in D$, there exists a mapping $T \in \mathcal{L}(\mathcal{E}, \mathcal{F})$ such that

$$\lim_{z \to z_0} \frac{\|f(z) - f(z_0) - T(z - z_0)\|_{\tau(k)}}{\|z - z_0\|_{\tau(k)}} = 0$$

for arbitrary τ-norms $\|\cdot\|_{\tau(k)}$ on $\mathcal{E}$ and $\mathcal{F}$. Here $Tz = \prod_\iota T_\iota z_\iota$ and T is denoted by $Df(z_0)$.

Let D be a finitely open and connected subset of $\mathcal{E}$. A mapping $f : D \to \mathcal{F}$ is said to have an (L)-derivative $T \in \mathcal{L}(\mathcal{E}, \mathcal{F})$ at $z = z_0$, if for each $\varepsilon \in \mathcal{R}^+$, a $\delta \in \mathcal{R}^+$ can be found such that

$$\|f(z) - f(z_0) - T(z - z_0)\|_{\tau(k)} < \varepsilon_{\tau(k)} \|z - z_0\|_{\tau(k)}$$

as $\|z - z_0\|_{\tau(k)} < \delta_{\tau(k)}$. Here $\mathcal{R}^+ = \prod_\iota (0, \infty_\iota)$. If $f(z)$ has a (L)-derivative at each point of D, then it is said to be (L)-analytic in D.

Theorem 2.1. *A mapping $f \in H(D, \mathcal{F})$ if and only if f can be expressed as $f(z) = \prod_\iota f_\iota(z_\iota)$, where $f_\iota \in H(D_\iota, \mathcal{F}_\iota)$ and $D \subset \mathcal{E}$ is open.*

Proof. Assume that $f \in H(D, \mathcal{F})$. Let $\Pi_{\tau(k)} : \mathcal{F} \to \mathcal{F}_{\tau(k)}$ be a projective operator to the product of all $\mathcal{F}_{\tau_j}, j = 1, \ldots, k$. Here $\mathcal{F}_{\tau(k)} = \prod_{j=1}^k \mathcal{F}_{\tau_j}$ and $\mathcal{F}_{\tau(k)}$ is a commutative and complex Banach algebra with the unit element $\hat{\dot{e}}_{\tau(k)} = \prod_{l=1}^k \dot{e}_{\tau_l}$. Lemma 2.7 in [1] yields that $\Pi_{\tau(k)} f(z)$ is a holomorphic mapping on D into $\mathcal{F}_{\tau(k)}$. It follows that $\Pi_{\tau(k)} f(z) \in H(\Pi_{\tau(k)} D, \mathcal{F}_{\tau(k)})$. It is clear that a mapping $f : D \to \mathcal{F}$ is differentiable if and only if it is (L)-analytic in D. So from Theorem 13.16 in [7] we derive that f is (L)-analytic in $\Pi_{\tau(k)} D$ if and only if it is holomorphic in $\Pi_{\tau(k)} D$. Because $\mathcal{F}_{\tau(k)}$ is a Banach algebra with the unit element, Theorem 26.4.1 in [2] yields

$$\begin{aligned}
\Pi_{\tau(k)} f(z) &= \sum_{j=0}^{\infty} T_{\tau^j(k)} (z_{\tau(k)} - z_{0\tau(k)})^j \\
&= \left(\sum_{j=0}^{\infty} T_{\tau_1^j} (z_{\tau_1} - z_{0\tau_1})^j, \ldots, \sum_{j=0}^{\infty} T_{\tau_k^j} (z_{\tau_k} - z_{0\tau_k})^j \right) \text{ (by (2.2))} \\
&= (f_{\tau_1}(z_{\tau_1}), \ldots, f_{\tau_k}(z_{\tau_k}))
\end{aligned}$$

in $\nabla_{r_{\tau(k)}}(z_{0\tau(k)}) = \prod_{l=1}^{k} \nabla_{r_{\tau_l}}(z_{0\tau_l})$, where

$$T_{\tau^j(k)} = (T_{\tau_1^j}, \ldots, T_{\tau_k^j}) \in \mathcal{L}(\mathcal{E}_{\tau(k)}, \mathcal{F}_{\tau(k)}),$$

$$z_{\tau(k)} = (z_{\tau_1}, \ldots, z_{\tau_k}) \in \mathcal{E}_{\tau(k)},\ T_{\tau_l^j} \in \mathcal{L}(\mathcal{E}_{\tau_l}, \mathcal{F}_{\tau_l}),$$

and $f_{\tau_l}(z_{\tau_l})$ is the analytic continuation of $\sum\limits_{j=0}^{\infty} T_{\tau_l^j}(z_{\tau_l} - z_{0\tau_l})^j$, and where $r_{\tau(k)} = \prod\limits_{l=1}^{k} r_{\tau_l}$. So $\Pi_{\tau(k)} f(z) = \prod\limits_{l=1}^{k} f_{\tau_l}(z_{\tau_l})$. Using the arbitrariness of operator $\Pi_{\tau(k)}$ and Lemma 2.7 in [1], we obtain that $f(z)$ can be rewritten as $\prod\limits_{\iota} f_\iota(z_\iota)$.

Inversely, a mapping f can be written as the product form of all $f_\iota(\in H(D_\iota, \mathcal{F}_\iota))$. From Definition 5.1 of a holomorphic mapping in [7], we derive that

$$f(z) = \prod_{\iota} \sum_{j=1}^{\infty} P_{f_\iota}^{[j]}(z_\iota - z_{0\iota}) = \sum_{j=1}^{\infty} \prod_{\iota} P_{f_\iota}^{[j]}(z_\iota - z_{0\iota}) = \sum_{j=1}^{\infty} P_f^{[j]}(z - z_0)$$

converges uniformly in $\nabla(z_0)$ for any τ-norm and any $z_0 \in D$, where $P_{f_\iota}^{[j]}(z_\iota - z_{0\iota})(P_f^{[j]}(z - z_0))$ is an j-homogeneous continuous polynomial of $z_\iota - z_{0\iota}(z - z_0)$ to be generated by $f_\iota(f)$ in suitable neighborhood $\nabla(z_{0\iota})(\nabla(z_0))$ of the point $z_{0\iota}(z_0)$. Hence f is holomorphic in D according to the definition of a holomorphic mapping. The proof is finished.

A meromorphic mapping on an open set D without appearance in [1] and [7] may be defined as follows: A mapping $f \colon D \subset \mathcal{E} \to \overline{\mathcal{F}}$ is called meromorphic, if there exists a holomorphic mapping φ (with $\varphi_\iota \not\equiv c_\iota$ for any ι): $D \to \mathcal{F}$ such that φf is holomorphic and $\varphi^{-1}(z)$ exist in D except for the zeros of φ, where c_ι is a constant for any $\iota \in (0,1)$. A set of all meromorphic mappings from D into $\overline{\mathcal{F}}$ is denoted by $M(D, \mathcal{F})$. From Theorem 2.1 we derive that $f \in M(D, \mathcal{F})$ can be represented by $\prod\limits_{\iota} f_\iota(z_\iota)$, where $f_\iota \in M(D_\iota, \mathcal{F}_\iota)$. In fact, because φ is holomorphic in D and φ^{-1} exist except for its zeros, $\varphi^{-1} = \prod\limits_{\iota}[\varphi_\iota(z_\iota)]^{-1}$. Sitting $\varphi f = g$ we have $f = \varphi^{-1} g = \prod\limits_{\iota} g_\iota(z_\iota)[\varphi_\iota(z_\iota)]^{-1}$ without common factors between holomorphic mappings φ and g in D. A holomorphic (meromorphic) mapping f on D into $\mathcal{F}(\overline{\mathcal{F}})$ is said to be actual, if its each component $f_\iota \not\equiv c_\iota$. The zeros of φ is called the poles of f.

A holomorphic mapping f is called to have a zero of order k at the point z_0, if its each component f_ι has a zero of order k at $z_{0\iota}$. This property of f is called to be of the homogeneous order at a zero z_0 (the homogeneous order for short). A pole z_0 of a meromorphic mapping f is called to be of order k, if the holomorphic mapping φ has a zero of order k at the point z_0.

A zero (pole) z_0 of f is called to be isolated, if there exists a finitely open subset $D_0 \subset D$ containing z_0, such that $f(z) \neq \hat{0}$ ($f(z) \neq \widehat{\infty}$) for $z \in D_0 \backslash \{z_0\}$. Then there are the following results.

Theorem 2.2. *Let $f \in H(D, \mathcal{F})$. A holomorphic mapping f of the homogeneous order at any zero of f is actual if and only if any zero of f is isolated, where $D(\neq \emptyset) \subset \mathcal{E}$ is an open set.*

Corollary 2.3. *If a mapping $f \in M(D, \mathcal{F})$ with poles possesses the homogeneous order, then f is actual if and only if any pole of f is isolate.*

The following conclusion is obviously.

Proposition 2.4. *If f at a pole z_0 is of order m with the property of the homogeneous order, then f can be written as $\sum_{j=-m}^{\infty} T_j(z-z_0)^j$ with $T_{-m} \neq 0$ in some punctured neighborhood $\nabla_r(z_0)\backslash\{z_0\}$.*

In particular $\mathcal{F} = \mathcal{E}$, we can define the exponential mapping $\exp z = \hat{e} + \sum_{j=1}^{\infty} \frac{z^j}{j!}$. Theorem 2.1 yields $\exp z$ is an entire mapping, because each component is entire by Theorem 3.19.1 in [2]. The inverse mapping of $\exp z$ is called the logarithmic mapping, denoted by $\log z$. It follows that we have $\exp(\log z) = z$. Because $\prod_\iota \exp z_\iota = \exp \prod_\iota z_\iota$ by the definitions of the multiplication and the exponential mapping on $\mathcal{E}$ for any τ-norm, $\log z = \prod_\iota \log z_\iota$. Hence $\log z$ is also holomorphic in its domain by Theorem 2.1 and the uniqueness of the analytic continuation.

2.2. Integrals

Assume that $\gamma : \mathsf{A} = \prod_\iota \mathsf{A}_\iota \to \mathcal{E}$ is a strongly continuous mapping in the sense of any τ-norm, where $\mathsf{A}_\iota \subset \mathsf{R}_\iota$ is a closed interval, that $\Gamma \subset \mathcal{E}$ is the graph of γ, and that $f : \Gamma \to \mathcal{E}$ is a mapping. If

$$\mathrm{Var}[\gamma]_{\tau(k)} = \sup_n \sum_{l=1}^{n} \|\gamma(t_l) - \gamma(t_{(l-1)})\|_{\tau(k)} < \infty$$

for all possible finite partition of A, then γ is called to be of the strongly bounded variation, and $\mathrm{Var}[\gamma]_{\tau(k)}$ is called the strongly total variation, where $\{t_l\}_{l=1}^n \subset \mathsf{A}$. If the limit of the partial sum $s_n(\pi) = \sum_{l=1}^{n} f(z_l)[z_l - z_{(l-1)}]$ exists for any partition π, then this limit is called the integral on Γ, denoted by $\int_\Gamma f dz$, where $z_l = \gamma(t_l)$.

Let $\mathcal{P}_{\eta_\iota} \subset \mathcal{E}_\iota$ be a hyperplane via the origin for $\eta_\iota \in \Theta_\iota$. Let the interior of the graph Γ_{η_ι} of any closed simple smooth curve γ_{η_ι} be in a simply connected subdomain of a relative domain $\mathcal{P}_{\eta_\iota} \bigcap D_\iota \subset \mathcal{E}_\iota$. Here $D = \prod_\iota D_\iota \subset \mathcal{E}$ is an open set, $\bigcup_{\eta_\iota \in \Theta_\iota} \mathcal{P}_{\eta_\iota} \bigcap D_\iota = D_\iota$ is satisfied. The positive direction of $\Gamma \subset \mathcal{P}_\eta$ can be defined by the positive direction of Γ_ι for each ι. Here $\mathcal{P}_\eta = \prod_\iota \mathcal{P}_{\eta_\iota}$ and $\eta = \prod_\iota \eta_\iota \in \Theta \left(= \prod_\iota \Theta_\iota\right)$. Let $D_{0\iota}^+$ be the interior of Γ_ι and $D_0^+ = \prod_\iota D_{0\iota}^+$. Let $\overline{D_0^+} = \prod_\iota \overline{D_{0\iota}^+}$.

Lemma 2.5. *If $f : \Gamma \to \mathcal{E}$ is holomorphic and γ is the product of the simple and rectifiable curves γ_{η_ι}, then*

$$\int_\Gamma f(z)dz = \prod_\iota \int_{\Gamma_{\eta_\iota}} f_\iota(z_\iota)dz_\iota,$$

where $f(z) = \prod_\iota f_\iota(z_\iota)$.

Next the most main theorems on $D \subset \mathcal{E}$ can be derived from the definition of integral and those corresponding results with classical conclusions in complex analysis respectively. For examples, Cauchy's integral theorem and formula hold on $D \subset \mathcal{E}$.

3. Singular Integral

3.1. Manifolds

In this subsection, the concept of a manifold is developed to the case on $\mathcal{E}$. Let $\mathcal{A}$ be a set. A chart on $\mathcal{A}$ is a bijection $\mathbf{z}$ from a subset $U \subset \mathcal{A}$ to an open subset of a locally convex space $\mathcal{E}$. We sometimes denote $\mathbf{z}$ by $(U, \mathbf{z})$, to indicate the domain U of $\mathbf{z}$. An C^k atlas on $\mathcal{A}$ is a family of charts $\mathcal{B} = \{(U_j, \mathbf{z}_j) : j \in \mathsf{N}^+\}$ such that

(i) $\mathcal{A} = \bigcup\{U_j : j \in \mathsf{N}^+\}$, and

(ii) any two charts in $\mathcal{B}$ are compatible in the sense that the overlap mappings between members of $\mathcal{B}$ are the C^k diffeomorphism: for two charts $(U_j, \mathbf{z}_j)$ and $(U_l, \mathbf{z}_l)$ with $U_j \bigcap U_l \neq \emptyset$, we form the overlap mapping: $\mathbf{z}_{lj} = \mathbf{z}_l \circ \mathbf{z}_j^{-1} | \mathbf{z}_j(U_j \bigcap U_l)$, where $\mathbf{z}_j^{-1} | \mathbf{z}_j(U_j \bigcap U_l)$ means the restriction of $\mathbf{z}_j^{-1}$ to the set $\mathbf{z}_j(U_j \bigcap U_l)$. We require that $\mathbf{z}_j(U_j \bigcap U_l)$ is open and that $\mathbf{z}_{lj}$ is an C^k diffeomorphism.

Two C^k atlases $\mathcal{B}_1$ and $\mathcal{B}_2$ are equivalent, if $\mathcal{B}_1 \bigcup \mathcal{B}_2$ is an C^k atlas. An C^k differentiable structure D on $\mathcal{A}$ is an equivalence class of atlases on $\mathcal{A}$. The union of the atlases in D, $\mathcal{B}_{\mathsf{D}} = \bigcup\{\mathcal{B} : \mathcal{B} \in \mathsf{D}\}$ is the maximal atlas of D. If $\mathcal{B}$ is an C^k atlas on $\mathcal{A}$, the union of all atlases equivalent to $\mathcal{B}$ is called the C^k differentiable structure generated by $\mathcal{B}$.

A differentiable manifold $M \subset \mathcal{E}$ is a pair $(\mathcal{A}, \mathsf{D})$, where $\mathcal{A}$ is a Hausdorff space and D is an C^k differentiable structure on $\mathcal{A}$. We shall often identify M with the set $\mathcal{A}$ for notational convenience. If a covering by charts takes their values in a locally convex space, then we say that M is an C^k locally convex manifold on the locally convex space.

A mapping $f : M \to \mathcal{F}$ is said to be holomorphic, if there is a family $\{U_j\}_{j\in\mathsf{N}^+} \supset M$ of open sets such that the composite mapping $f\circ (\mathbf{z}_j|U_j)^{-1} : \mathbf{z}_j(U_j) \to \mathcal{F}$ is holomorphic for each j. Similarly, we can define a meromorphic mapping on M into $\overline{\mathcal{F}}$.

3.2. Hölder condition

Assume that $\mathcal{E}$ and $\mathcal{F}$ are the complete Hausdorff locally m-convex algebras. If a mapping $f : M(\subset \mathcal{E}) \to \mathcal{F}$ satisfies for any semi-norm

$$\|f(t') - f(t'')\|_{\tau(k)} \le A\|t' - t''\|^{\alpha}_{\tau(k)} \quad \text{for} \quad 0 < \alpha \le 1$$

for any $t', t'' \in M$, where A and α are constants, then f is called satisfying the vector-valued Hölder condition of order α. The set $VH_\alpha(M)$ (or VH_α) consists of all mappings of satisfying the vector-valued Hölder condition of order α on M. A semi-norm $\|f\|_{\tau(k)_\alpha}$ on VH_α is defined for any $f \in VH_\alpha$ and for any semi-norm on $\mathcal{E}$ and $\mathcal{F}$ as follows

$$\|f\|_{\tau(k)_\alpha} = \sup\{\|f(t)\|_{\tau(k)} : t \in M\} + \sup\left\{\frac{\|f(t') - f(t'')\|_{\tau(k)}}{\|t' - t''\|^{\alpha}_{\tau(k)}} : t', t'' \in M\right\}.$$

Using similar to the classical methods (see [6]) and the properties of the complete Housdorff locally m-convex algebra (see [7]) we can obtain the following properties:

Proposition 3.1. *Assume $f, \varphi \in VH_\alpha(M)$, the following conclusions hold:*

(1) *if $f \in VH(M)(= \bigcup\limits_{\alpha} VH_\alpha)$, then f is continuous on M;*
(2) *if $0 < \beta \le \alpha$ and $f \in VH_\alpha(M)$, then $f \in VH_\beta(M)$;*
(3) *if $\mathcal{F}$ is a complete Hausdorff locally m-convex algebra and if $f, \varphi \in VH_\alpha(M)$, then $f \pm \varphi, f\varphi \in VH_\alpha(M)$.*

3.3. The principal value

Let M be a locally convex manifold in $\mathcal{E}$, D a domain in $\mathcal{E}$ and $\partial D\,(= M)$ a boundary of D. Assume that $B(\xi_\iota, \delta_\iota)$ is a ball in $\mathcal{E}_\iota$ with a center ξ_ι and a radius δ_ι, i.e., $\{x : \|x - \xi_\iota\| < \delta_\iota\} \subset \mathcal{E}_\iota$, that $B(\xi, \delta) = \prod\limits_{\iota} B(\xi_\iota, \delta_\iota)$, and that $\sigma(\xi, \delta) = M \cap B(\xi, \delta)$.

Assume for convenience, that $M \subset \mathcal{E}$ is a closed manifold and $L_\eta = \mathcal{P}_\eta \bigcap M$ is a section of M satisfying $M = \biguplus\limits_{\eta\in\Theta} L_\eta$. Here $\biguplus$ means that the set Θ satisfies $M = \bigcup_{\eta\in\Theta} L_\eta$ and there is the unique $\eta \in \Theta$ for any point $t \in M$ such that $t \in L_\eta$. Let $L_\eta = \prod_\iota L_{\eta_\iota}$, and $L_{\eta_\iota}(\subset \mathcal{P}_{\eta_\iota})$ be a simple smooth closed curve in the sense of isomorphism.

Let t and ξ be on $M(\subset \mathcal{E})$. Define the integral

$$\int_M f(t)dt = \bigoplus_\eta \int_{L_\eta} f_\eta(t_\eta)dt_\eta = \bigoplus_\eta \prod_\iota \int_{L_{\eta_\iota}} f_{\eta_\iota}(t_{\eta_\iota})dt_{\eta_\iota}$$

to be in the sense of any semi-norm. Here $f(t) = \bigoplus_\eta f_\eta(t_\eta) = \bigoplus_\eta \prod_\iota f_{\eta_\iota}(t_{\eta_\iota})\,(\in VH_\alpha) : M \to \mathcal{E}$ and $f_\eta(t_\eta)$ denotes that the domain L_η of mapping $f(t)$ is restricted to $\mathcal{P}_\eta$ and $\bigoplus$ denotes $f(t) = f_\eta(t_\eta)$ as $t = t_\eta$. If $(t-\xi)^{-1} \in \mathcal{E}$ exists and the integral $\lim_{\delta\to 0} \int_{M-\sigma(\xi,\delta)} f(t)(t-\xi)^{-1}dt$ exists for any semi-norm, then the limit is called the principal value of the Cauchy type integral, denoted by

$$\int_M f(t)(t-\xi)^{-1}dt. \tag{3.1}$$

Theorem 3.2. *If $f(t)(\in VH_\alpha) : M \to \mathcal{E}$ and $\xi \in M$ and if $(t-\xi)^{-1}$ exists for $t, \xi \in M$ except for $t = \xi$, then integral* (3.1) *exists in $\mathcal{E}$.*

3.4. Plemelj formulas

Let $\varphi(t)(\in VH_\alpha) : M \to \mathcal{E}$. It follows that

$$\int_{L_\eta} \varphi_\eta(t_\eta)(t_\eta - z_\eta)^{-1}dt_\eta$$

is also holomorphic function of z_η for $z_\eta \notin L_\eta$ in $\mathcal{P}_\eta$ and is denoted by $\Phi_\eta(z_\eta)$. Applying classical Plemelj formulas in [6], we deduce that

$$\Phi^{\pm}_{\eta_\iota}(\xi_{\eta_\iota}) = \pm\frac{1}{2}\varphi_{\eta_\iota}(\xi_{\eta_\iota}) + \frac{1}{2\pi i}\int_{L_{\eta_\iota}} \varphi_{\eta_\iota}(t_{\eta_\iota})(t_{\eta_\iota} - \xi_{\eta_\iota})^{-1}dt_{\eta_\iota} \tag{3.2}$$

for any $\xi_{\eta_\iota} \in L_{\eta_\iota}$ and any ι, where $\Phi^{\pm}_{\eta_\iota}(\xi_{\eta_\iota})$ (either $\in H_\alpha$ as $0 < \alpha < 1$ or $\in H_{1-\varepsilon}$ as $\alpha = 1$) are the boundary values of $\Phi_{\eta_\iota}(z_{\eta_\iota})$ as z_{η_ι} tends to L_{η_ι} from the interior and the exterior of L_{η_ι} for any ι respectively. Let

$$\Phi^{\pm}_\eta(\xi_\eta) = \prod_\iota \Phi^{\pm}_{\eta_\iota}(\xi_{\eta_\iota}),\ \Phi_\eta(z_\eta) = \prod_\iota \Phi_{\eta_\iota}(z_{\eta_\iota}), \text{and } \Phi(z) = \bigoplus_\eta \Phi_\eta(z_\eta). \tag{3.3}$$

Relations (3.2) and (3.3) imply the following conclusion.

Theorem 3.3. (Generalized Plemelj formulas) *If* $\varphi(\in VH_\alpha) : M \to \mathcal{E}$ *and*

$$\sup\{\|\varphi_{\eta_\iota}\|_{H_\alpha} : \iota \in (0,1)\} < \infty,$$

and if $(t-\xi)^{-1}$ *exists for* $t, \xi \in M$ *except for* $t = \xi$, *then*

(i) $\Phi^{\pm}(\xi) = \pm\frac{1}{2}\varphi(\xi) + \frac{1}{2\pi i}\int_M \varphi(t)(t-\xi)^{-1}dt$, *where* $\Phi^+(\xi)$ *and* $\Phi^-(\xi)$ *are boundary values in* $VH_\alpha(0<\alpha<1)$ *or* $VH_{1-\varepsilon}$ *of a mapping* $\Phi(z)$ *as* z *tends to* L *from the interior and the exterior of* L *respectively, to be defined by* z_{η_ι} *tending to* L_{η_ι} *from the interior and the exterior of* L_{η_ι} *in the sense of isomorphism respectively;*

(ii) *the piecewise* $\mathcal{E}$*-valued mapping* $\Phi(z)$ *is holomorphic as* $z \notin M$.

4. Riemann Boundary Value Problems

In this section, the holomorphic solutions $\Phi(z)$ with form $\bigoplus_\eta \Phi_\eta(z_\eta)$ of the Riemann boundary value problem

$$\Phi^+(t) = G(t)\Phi^-(t) + g(t),\ t \in M, \tag{4.1}$$

satisfying $\|\Phi(\widehat{\infty})\|_{\tau(k)} < \infty$ for any semi-norm are found, where $G(t) = \bigoplus_\eta G_\eta(t_\eta) = \bigoplus_\eta \prod_\iota G_{\eta_\iota}(t_{\eta_\iota})$ and $g(t) = \bigoplus_\eta g_\eta(t_\eta) = \bigoplus_\eta \prod_\iota g_{\eta_\iota}(t_{\eta_\iota})\,(\in VH_\alpha) : M \to \mathcal{E}$ are given.

Let $\mathrm{Ind}_{L_\eta} G(t_\eta) = \prod_\iota \mathrm{Ind}_{L_{\eta_\iota}} G_{\eta_\iota}(t_{\eta_\iota}) \in \mathcal{R}$ ($\kappa_\eta = \prod_\iota \kappa_{\eta_\iota}$ for short) be the index of G. Let

$$X_{\eta_\iota}(z_{\eta_\iota}) = \begin{cases} X^+_{\eta_\iota}(z_{\eta_\iota}) = e^{\Gamma_{\eta_\iota}(z_{\eta_\iota})}, & \text{if } z_{\eta_\iota} \in D^+_{\eta_\iota}, \\ X^-_{\eta_\iota}(z_{\eta_\iota}) = (z_{\eta_\iota} - z_{0\eta_\iota})^{-\kappa_{\eta_\iota}} e^{\Gamma_{\eta_\iota}(z_{\eta_\iota})}, & \text{if } z_{\eta_\iota} \in D^-_{\eta_\iota}. \end{cases}$$

Here

$$\Gamma_{\eta_\iota}(z_{\eta_\iota}) = \frac{1}{2\pi i}\int_{L_{\eta_\iota}} [\log(t_{\eta_\iota} - z_{0\eta_\iota})^{-\kappa_{\eta_\iota}} G_{\eta_\iota}(t_{\eta_\iota})](t_{\eta_\iota} - z_{\eta_\iota})^{-1} dt_{\eta_\iota}$$

for $z_{\eta_\iota} \notin L_{\eta_\iota}, z_{0\eta_\iota} \in D^+_{\eta_\iota}$, log denotes the principal value, the boundary ∂D_{η_ι} of D_{η_ι} is L_{η_ι}, $D^+_{\eta_\iota}$ and $D^-_{\eta_\iota}$ are the interior and the exterior of D_{η_ι}(a domain) $\subset \mathcal{P}_{\eta_\iota}$ respectively. Let $X(z) = \bigoplus_\eta \prod_\iota X_{\eta_\iota}(z_\iota)$, $P_\kappa = \bigoplus_\eta \prod_\iota P_{\kappa_{\eta_\iota}}(z_\iota)$, and $\kappa = \bigoplus_\eta \kappa_\eta$. Here $P_{\kappa_{\eta_\iota}}$ is a polynomial of degree κ_{η_ι}.

Classical corresponding methods in [6] yield the solutions of (4.1) as follows.

(i) If $\kappa \geq \hat{0}$, then

$$\begin{aligned}\Phi(z) &= \bigoplus_{\eta} \Phi_\eta(z_\eta) \\ &= \bigoplus_{\eta} \left\{ \frac{X_\eta(z_\eta)}{2\pi i} \int_{L_\eta} g_\eta(t_\eta)[(t_\eta - z_\eta)X_\eta^+(t_\eta)]^{-1} dt_\eta + P_{\kappa_\eta}(z_\eta)X_\eta(z_\eta) \right\} \\ &= \frac{X(z)}{2\pi i} \int_M g(t)[(t-z)X^+(t)]^{-1} dt + P_\kappa(z)X(z),\end{aligned}$$

where P_{κ_η} is any polynomial of degree κ_η and $P_\kappa = \bigoplus_{\eta} P_{\kappa_\eta}$ any polynomial of degree κ.

(ii) If $\kappa = -\hat{1}$, then (4.1) has a unique solution

$$\Phi(z) = \frac{X(z)}{2\pi i} \int_M g(t)[(t-z)X^+(t)]^{-1} dt, \tag{4.2}$$

where $\hat{1} = \prod_{\iota} 1$.

(iii) If $\kappa < -\hat{1}$ and $\int_L g(t)t^k[X^+(t)]^{-1}dt = \hat{0}$ for $k = \hat{0}, \hat{1}, \ldots, -\kappa - \hat{2}$, then (4.1) has a unique solution (4.2).

References

[1] S. Dineen, *Complex analysis in locally convex spaces,* North-Holland Math. Studies, Amsterdam, 1981.

[2] E. Hille and R. Phillip, *Functional analysis and semigroups*, AMS Colloquium Publications, Vol. 31, American Mathematical Society, Providence, RI, 1957.

[3] C. G. Hu, The disturbance of vector-valued doubly-periordic Riemann boundary value problems, *J. Math. Anal. Appl.,* 131 (1988), 373-391.

[4] C. G. Hu and C. C. Yang, *Vector-valued functions and their applications,* Kluwer Acad. Publ., Dordrecht, 1992.

[5] G. Köthe, *Topological vector spaces I*, Springer-Verlag, Berlin, 1983.

[6] J. K. Lu, *Boundary value problems of analytic functions,* World Scientific Publishing, Singapore, 1993.

[7] J. Mujica, *Complex analysis in Banach spaces,* North-Holland Math. Studied, Vol. 120, North-Holland Publishing Co., Amsterdam, 1986.